Causal Inference with Bayesian Networks

Build Bayesian Networks and Causal Inference Models with R and Python

Yousri El Fattah and Reza Bagheri

<packt>

Causal Inference with Bayesian Networks

Portfolio Director: Sunith Shetty
Relationship Lead: Nilesh Kowadkar
Project Manager**:** Yash Basil
Content Engineer**:** Gowri Rekha
Technical Editor: Seemanjay Ameriya
Indexer: Manju Arasan
Production Designer: Jyoti Kadam
Growth Lead: Shruthi Shetty

First published: May 2026

Production reference: 1270526

Published by Packt Publishing Ltd.
Grosvenor House
11 St Paul's Square
Birmingham
B3 1RB, UK.

ISBN 978-1-83508-498-4
`www.packtpub.com`

To my parents—your unwavering love and guidance have shaped every chapter of my life. This book is a humble tribute, a small reflection of the countless gifts and encouragement you have given me.

- Yousri El Fattah

I would like to dedicate this book to my family, who are my source of love and encouragement.

– Reza Bagheri

Contributors

About the authors

Dr. Yousri El Fattah is CEO of Causal Computing and an expert in machine intelligence, causal modeling, control systems engineering, and data science. He has taught artificial intelligence and control systems at multiple universities and has led research and development projects for aerospace and industrial automation companies. A former senior scientist at Rockwell and Teledyne Technologies, he is the author of a book on learning systems and numerous technical articles published in leading journals, including Machine Learning, Artificial Intelligence, IEEE Transactions, and ASME Transactions. He is a licensed professional engineer in control systems engineering in California and holds Ph.D. degrees in information and computer science and aeronautical engineering.

I would like to express my heartfelt appreciation to Gowri Rekha, the book editor, for her invaluable assistance throughout this process. Her insightful feedback and thoughtful suggestions were both constructive and deeply encouraging. I am also grateful to Lorenzo Toni, the technical reviewer, whose detailed input on several chapters helped elevate the quality of the content across the book. Their contributions have made a meaningful difference.

Reza Bagheri is a data scientist with over a decade of experience applying machine learning, statistical modeling, and advanced analytics to problems across mining, energy, and market research. He lives in Canada and currently works as a data scientist at Ipsos. Reza holds a Ph.D. in Chemical Engineering from the University of Alberta and is the author of several peer-reviewed publications. He has also written numerous blog posts on data science and machine learning and spoken at leading conferences. Reza is passionate about teaching statistics and machine learning.

About the reviewer

Lorenzo Toni is a Lead Data Scientist at Amazon, where he leads measurement and experimentation for AI-driven products serving over 200 million users worldwide. He is also the founder of Kodonics, an AI consulting firm based in New York, and a Lecturer in Applied Machine Learning at Columbia University, where he teaches graduate-level courses on AI and data science.

Table of Contents

Chapter 2: Probability and Bayes' Theorem 35

Chapter 3: Bayesian Networks 83

Chapter 6: Join Tree Clustering 251

Chapter 7: Probabilistic Inference with Join Tree Clustering 303

Chapter 8: Probabilistic Inference with Relational Database Models 355

Chapter 12: Causal Inference in Economic Research 551

Preface

Causal questions are at the heart of nearly all applied sciences. Does smoking cessation extend life? Does a minimum wage increase reduce employment? Does a treatment work better for some patients than for others? Traditional statistical methods that focus on correlations in observed data cannot address these questions. Drawing valid causal conclusions requires a separate framework that distinguishes correlation from causation. This book is about that framework and the computational tools that can bring it to life.

Concepts are reinforced with worked examples and code implementations in R and Python, using packages such as `pgmpy`, `CausalModels`, and `causallib`. As a result, the reader gains both a solid understanding of each topic and the code needed to implement it. The book is structured to move from foundations to applications. It first develops the mathematical framework of causal inference and then applies it to different case studies.

The first four chapters review probability and Bayes' theorem, introduce Bayesian networks and their structural properties, and develop the structural causal model framework. The middle chapters focus on relational database representations of probabilistic models, join tree clustering, belief propagation, and variable elimination. *Chapters 9* and *10* cover causal inference with structural models and observational data. They introduce the key tools in causal inference, including Pearl's do-calculus, the back-door and front-door criteria, the potential outcomes framework, and the counterfactual framework.

The final chapters shift from theory to applications, covering causal inference with machine learning (meta-learners) and three case studies drawn from economics (the Lalonde study of the National Supported Work Demonstration), epidemiology (smoking cessation and mortality in the National Health Epidemiologic Follow-up Study), and social science (the Card and Krueger minimum-wage study), all implemented in Python.

Who this book is for

This book is for scientists, researchers and students in different fields such as data science, social science, economics, health care and epidemiology who want to learn causal inference. You are expected to have basic knowledge of R and Python and their scientific libraries, along with knowledge of basic probability and statistics, to get the most out of this book.

What this book covers

Chapter 1, A Guided Tour of Book Topics, gives an introduction to Bayesian networks, structural causal models, and causal inference.

Chapter 2, Probability and Bayes Theorem, discusses probability theory, including its axioms and interpretations, and introduces Bayes' theorem as a way to update uncertainty about beliefs based on observed evidence.

Chapter 3, Bayesian Networks, introduces Bayesian networks and reviews their main structural properties. It explains d-separation and the concept of conditional independence between network variables and describes examples of important classes of Bayesian networks.

Chapter 4, Structural Causal Models, covers structural equation modeling (SEM) and structural causal modeling (SCM) as frameworks for representing relationships among observed and unobserved variables. It explains their key components, highlights SCM's causal interpretation using directed acyclic graphs and intervention concepts and includes practical examples with R and Python.

Chapter 5, Relational Database Models, introduces a relational database framework for representing and reasoning with probabilistic graphical models, extending traditional databases with weighted relations and expressing inference using relational algebra and SQL. It covers dependencies, efficient acyclic structures, representations of Bayesian networks and influence diagrams, and includes practical examples and programming exercises.

Chapter 6, Join Tree Clustering, explains join tree (junction tree) clustering as a method for transforming Bayesian networks into tree structures for efficient exact inference. It covers key graph transformations, core structural concepts, and practical implementation with R examples.

Chapter 7, Probabilistic Inference with Join Tree Clustering, discusses inference methods using belief propagation with join tree representations for Bayesian and Markov networks. It explains the algorithm for inference with join tree clustering, illustrates step-by-step inference with examples, and includes a practical Python implementation using pgmpy.

Chapter 8, Probabilistic Inference with Relational Database Models, presents an approach to probabilistic inference that represents Bayesian Networks as relational databases and extends relational algebra operators to weighted relations. It describes a variable elimination algorithm for compositional queries, applies it to posterior probability in Bayesian Networks and maximum expected utility in influence diagrams, and reinforces the concepts with practical examples implemented in R.

Chapter 9, Causal Inference with Structural Causal Models, introduces Pearl's do-calculus, a set of rules for rewriting interventional and observational distributions to determine whether causal effects are identifiable from observational data. It also covers back-door and front-door criteria and demonstrates symbolic estimation of causal effect queries with practical R examples.

Chapter 10, Causal Inference with Observational Data, presents the potential outcomes framework (Neyman-Rubin causal model), which uses counterfactual notation to define a causal effect as the difference in potential outcomes across counterfactual scenarios for a given population. It reviews the assumptions under which causal effects are identifiable from observable data, surveys methods for estimating causal estimands, and demonstrates them on a real-life dataset using the CausalModels package in R.

Chapter 11, Causal Inference with Machine Learning, covers four meta-learning algorithms for estimating the Conditional Average Treatment Effect (CATE) and analyzing heterogeneous treatment effects (HTEs) where treatment effects vary across individuals and subgroups. It includes Python implementations of meta-learners and their performance analyses.

Chapter 12, Causal Inference in Economic Research, analyzes the **National Supported Work** (NSW) Demonstration program and the Lalonde dataset to estimate the causal effect of subsidized transitional work experience on disadvantaged individuals using regression-based standardization, **inverse probability of treatment weighting** (IPTW), and propensity score matching. All methods are implemented in Python, with demonstrations of how the causallib package can be used to estimate the causal effect.

Chapter 13, Causal Inference in Epidemiology, applies causal survival analysis to the **NHEFS** (**National Health Epidemiologic Follow-up Study**) dataset to estimate the effect of smoking cessation on mortality. It demonstrates how the lifelines and `causallib` packages in Python can be used to perform the causal survival analysis.

Chapter 14, Causal Inference in Social Science Research, revisits the 1994 Card and Krueger minimum wage study, using fast-food employment data from New Jersey and Pennsylvania to estimate the causal effect of a minimum wage increase by treating Pennsylvania's trend as the counterfactual for New Jersey. The analysis employs linear regression and inverse probability of treatment weighting (**IPTW**), implemented in Python using the `causallib` package.

To get the most out of this book

This book assumes familiarity with R and Python and their scientific libraries, as well as basic knowledge of statistics and probability theory.

Download the example code files

The code bundle for the book is hosted on GitHub at `https://github.com/PacktPublishing/Causal-Inference-with-Bayesian-Networks`. We also have other code bundles from our rich catalog of books and videos available at `https://github.com/PacktPublishing`. Check them out!

Conventions used

There are a number of text conventions used throughout this book.

`CodeInText`: Indicates code words in text, database table names, folder names, filenames, file extensions, pathnames, dummy URLs, user input, and Twitter handles. For example: "We create the experimental Lalonde dataset and store it in a dataframe called `nsw`."

A block of code is set as follows:

```
def create_bn(nodes, edges):
    model = BayesianNetwork()
    model.add_nodes_from(nodes)
    model.add_edges_from(edges)
    return model
```

Any command-line input or output is written as follows:

```
pip install jupyterlab
```

Bold: Indicates a new term, an important word, or words that you see on the screen. For instance, words in menus or dialog boxes appear in the text like this. For example: "**Survival analysis** is a branch of statistics that deals with analyzing the time until an event occurs."

Warnings or important notes appear like this.

Tips and tricks appear like this.

Get in touch

Feedback from our readers is always welcome.

General feedback: If you have questions about any aspect of this book or have any general feedback, please email us at `customercare@packt.com` and mention the book's title in the subject of your message.

Errata: Although we have taken every care to ensure the accuracy of our content, mistakes do happen. If you have found a mistake in this book, we would be grateful if you reported this to us. Please visit `http://www.packt.com/submit-errata`, click **Submit Errata**, and fill in the form.

Piracy: If you come across any illegal copies of our works in any form on the internet, we would be grateful if you would provide us with the location address or website name. Please contact us at `copyright@packt.com` with a link to the material.

If you are interested in becoming an author: If there is a topic that you have expertise in and you are interested in either writing or contributing to a book, please visit `http://authors.packt.com/`.

Share your thoughts

Once you've read *Causal Inference with Bayesian Networks*, we'd love to hear your thoughts! Scan the QR code below to go straight to the Amazon review page for this book and share your feedback.

`https://packt.link/r/1835084982`

Your review is important to us and the tech community and will help us make sure we're delivering excellent quality content.

Free benefits with your book

This book includes free benefits designed to support your learning and help you apply what you learn effectively. Activate them now for instant access (see the *How to unlock* section for instructions).

Here's a quick overview of what you can instantly unlock with your purchase:

Complete Code Bundle

Download the full source code for this book's examples and projects.

DRM-Free PDF Version

Download DRM-free PDF and ePub copies of this book.

7-Day Packt Library Access

Get 7-day unlimited access to 8,000+ books and videos. No credit card required.

Available for first-time Packt+ trial users only.

Next-Gen Reader Access

Read this book on Packt Reader with progress sync, dark mode and note-taking.

How to unlock

Scan the QR code (or go to packtpub.com/unlock), search for this book by name, confirm the edition, and then follow the steps on the page.

UNLOCK NOW

Note: Have your invoice handy. Purchases made directly from the Packt website don't require an invoice.

1

A Guided Tour of Book Topics

Bayesian network (**BN**) models provide a framework to aid systems engineers and decision-makers by harnessing expert knowledge and leveraging empirical data. This allows them to make informed decisions and examine what-if scenarios.

We will give an overview of causal inference with BN and statistical methods via randomized controlled trials. We will discuss connections between machine learning and causal inference.

In this chapter, we're going to cover the following topics:

- Expert systems and Bayesian networks
- Why should we care about causality?
- Why causal graphical models?
- Overview of causal inference with BN and statistical methods
- Discussion of causal inference and machine learning

Expert systems and Bayesian networks

This section presents an overview of expert systems and BNs. We begin with a subsection on expert systems, where we give historical background, describe expert systems and related topics on knowledge acquisition, rule-based systems, advantages/ disadvantages of the rule-based paradigm, and finally, a list of application areas. We follow this with a subsection on BNs, where we describe a formulation of probabilistic expert systems using BNs, the advantages and disadvantages of the BN paradigm for expert systems, and a selective list of application areas of expert systems using BNs.

Expert systems

We begin with a historical background of expert systems. Next, we describe what an expert system is. Then, we briefly describe the knowledge acquisition task. Following this, we give an overview of the rule-based paradigm and how it works. Afterward, we list the advantages and disadvantages of the rule-based approach. Finally, we conclude with a list of application areas of expert systems.

Historical background

The 1970s witnessed a shift in AI toward what became known as **expert systems** or **knowledge-based systems**. The change involved 1) emphasizing the representation of knowledge and the creation of knowledge bases for specific, narrow domains of expertise and 2) having the reasoning component of the expert system (called an **inference engine**) decoupled from the knowledge base; an inference engine can be an off-the-shelf commercial software called an **expert system shell**.

Feigenbaum promoted what he called the knowledge principle (*knowledge is power*) when developing the first real-life, large-scale expert system called **DENDRAL** (Buchanan, 1978) in the late 1970s, which embodied task-specific knowledge about the chemistry domain. Other landmark expert systems developed in the 1970s and 1980s include **MYCIN** (Shortliffe, 1976) and **R1** (also called **XCON**, for **eXpert CONfigurer**) (McDermott, 1980).

The new shift brought up by the advent of expert systems exposed the need for improvement in representing knowledge and handling uncertainty.

The MYCIN expert system development introduced the notion of the certainty factor and adopted a particular calculus for propagating uncertainty. The undisciplined use of rules and heuristics in early knowledge-based systems made it difficult to establish guarantees about the system's performance metrics on accuracy, consistency, and efficiency. Knowledge acquisition also proved to be a limiting factor and even a bottleneck for developing expert systems, emphasizing the need for machine learning.

AI emphasized symbolic processing and qualitative reasoning until the end of the 1980s, when research began incorporating new approaches based on probability, graph theory, and statistics to knowledge representation and reasoning.

BNs have become a very active area of research in AI, with a growing, broad range of applications.

With the preceding brief overview, let us now define an expert system.

What is an expert system?

Expert systems (ESs) are knowledge-intensive applications with problem-solving expertise in specialized fields. They can process information and produce advice or recommendations for solving problems like human experts. An ES aims to give problem-solving assistance or decision-making aid in a complex area where human skills are scarce or judgments are error-prone. ESs are also called **knowledge-based systems** (KBSs), and we will use both terms alternately.

A KBS has two main components: a **knowledge base** (KB) and an **inference engine** (IE). The KB contains an explicit representation of knowledge about problem-solving or decision-making expertise in the task area of the specialized domain. The IE is the reasoning component of the KBS that uses explicitly represented knowledge to solve problems and make decisions. The main ideas behind KBS are the following:

- We can incrementally expand the behavior of the KBS by adding new pieces of knowledge and defining new tasks for the KB.
- We can debug faulty behavior by locating the erroneous element of the KB.
- We can explain the system's behavior by tracking the IE's inferencing steps:
    - The explicit representation of knowledge in the KB, independent of how the IE uses it, makes it possible to build the KBS and adjust its behavior correspondingly incrementally.
    - Having defined an ES, let us describe the knowledge acquisition task next.

Knowledge acquisition

Knowledge acquisition (**KA**) is a crucial step in ES development. KA is a bottleneck in the development of KBS. Traditional methods rely on a knowledge engineering process in which a knowledge engineer interviews human experts in the specialty field to extract domain knowledge that the experts use to solve problems and make decisions. The conventional knowledge engineering approach to KA can be slow, inefficient, and impractical. *Figure 1.1* showcases a sample ES:

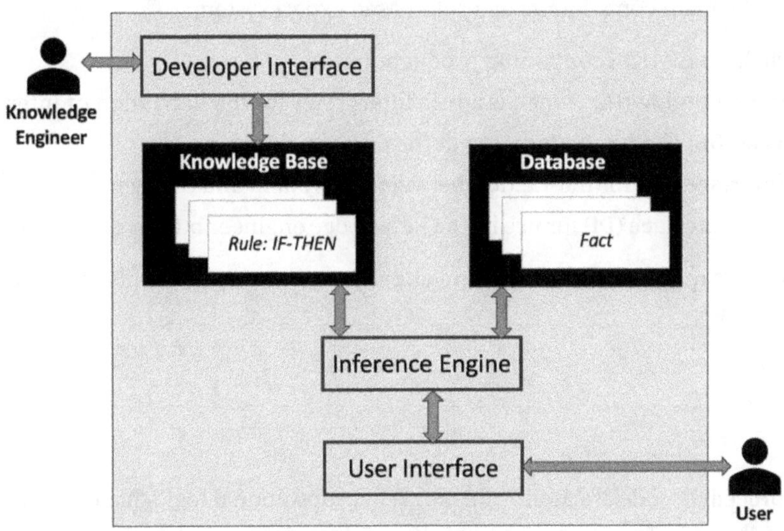

Figure 1.1: Expert system

The KA continues beyond the development phase when maintaining and upgrading the KBS. Supervised ML methods can extract decision trees and automatically induce rules from use cases that the expert classified. ML methods make the KB upgrading task easier because one can rerun the learning system as more examples accumulate. Data mining and techniques from knowledge discovery from databases can automate parts of the KA process, such as discovering association rules from historical use cases.

With the preceding definition of an ES and the description of the KA task, let us delve into the rule-based paradigm adopted by most implemented ES applications.

Rule-based systems

The rule-based approach is the most widely used paradigm for developing expert systems. The trend began in the 1960s and 70s with ESs, such as MYCIN, DENDRAL, and R1 (rule-based), as mentioned earlier. The success stories of those applications and the availability of commercial off-the-shelf expert system shells that are rule-based explain the popularity of the rule-based paradigm.

How production rule systems work

Many expert systems are rule-based systems or production rule systems. A production rule is an `IF-THEN` rule. The `IF` part contains a set of conditions, and the `THEN` part is the rule conclusion, which can add an assertion to the database or execute an action. A production rule is triggered when all its `IF` conditions are satisfied by matching the facts in the database. When the triggered rule is applied or *fired*, its conclusion gets executed.

The inference engine has metarules (i.e., rules about rules) to control the rule selection process and methods for conflict resolution at any step during the inference. Typically, the inferencing process goes in cycles, with each cycle containing three sequential steps: match, select, and execute. The cycle repeats until the problem is solved.

Figure 1.1 shows the main components of a rule-based expert system:

- A **knowledge base** (**KB**) containing production rules
- An **inference engine** (**IE**), which is a rule interpreter for the derivation of inferences
- A **database** (**DB**) containing facts about the domain object
- A **user interface** (**UI**) for interaction between the system and the user
- A **developer interface** (**DI**) for use by the knowledge engineer building and maintaining the KB

Rule-based systems can support deductive and abductive reasoning. A simple example of a deductive inference is as follows:

```
Fact: A=True
Rule: IF A THEN B
Inference: B=True
```

A and *B* in the preceding code are atomic formulas in propositional logic that evaluate to either `True` or `False`. Given that *A* is `True` and that *A* implies *B*, then we infer that *B* is *True*. Expert systems for prediction tasks are examples of deductive systems. Deductive reasoning is monotonic, because derived inferences are always valid and do not change their truth value with subsequent inferences.

The following is an example of abductive reasoning:

```
Fact: B=True
Rule: IF A THEN B
Rule: IF C THEN B
Inference: A=True and C=True are both plausible
```

Given that *B* is *True*, abductive reasoning will conclude that *A* is True and *C* is *True* are both plausible hypotheses. The hypotheses are not guaranteed to be valid, only plausible. Expert systems performing diagnosis tasks are examples of abductive reasoning systems in which reasoning tries to form plausible explanations for abnormal observations. Abductive reasoning is non-monotonic as hypotheses may be later retracted or validated on further evidence.

Expert knowledge in many real-world domains and specialty fields is typically imprecise, uncertain, or anecdotal. Expert systems have adopted various formalisms for handling uncertainty:

- Certainty factor
- Dempster-Shafer theory
- Fuzzy logic
- Truth maintenance systems
- Probability theory

In this book, we focus on probability theory and BNs as a framework for modeling and reasoning in knowledge-based systems. We will discuss the BN approach next. For now, let us conclude by mentioning the advantages and disadvantages of the rule-based approach.

Advantages and disadvantages

As pointed out earlier, many expert systems in the real world are rule-based. The rule-based approach is popular due to several advantages. However, it also has some shortcomings/limitations that are considered disadvantages. We will start with the advantages and then the disadvantages.

Here are the advantages:

- Rules are appealing because of their modularity. They can be added or deleted independently.
- Rules are natural to the way experts explain their problem-solving expertise.
- Rules duplicate the way experts use rules of thumb to do the following:
    - Factor a problem into subproblems
    - Prove some hypothesis
    - Deduce information from problem-specific data

Here are the disadvantages:

- The undisciplined use of rules can make it challenging to establish guarantees about the system's performance metrics on accuracy, consistency, and efficiency.
- Knowledge acquisition can be a limiting factor and even a bottleneck for developing expert systems. The following issues can hamper the conventional knowledge engineering approach of interviewing the expert:
    - The expert is busy or not always available
    - Expert judgment needs to be more consistent and precise
- Different experts can have different judgments.

- As the number of production rules increases, maintaining and organizing the rules becomes difficult, and the inference becomes very slow.

After reviewing the rule-based paradigm of expert systems, let's follow up in the next section with a mention of the principal categories of expert system application areas.

Application areas of expert systems

Expert systems have been applied in many domains to perform specialized tasks, including the following:

- **Classification**: Classify high-level objects from observed attributes
- **Diagnosis**: Infer causes of malfunction or disease from observable data
- **Monitoring**: Compare the behavior observed to the behavior expected
- **Process control**: Control of a physical process based on monitoring
- **Design**: Configure a system according to specifications
- **Scheduling**: Order and assign time to task execution
- **Planning**: Develop a course of action to achieve goals or desired effects

With this, we conclude this section on expert systems. Let's highlight the main takeaway points.

Expert systems presented a landmark in AI development that began in the 1970s and adopted the rule-based paradigm and heuristic-based search methods until the end of the 1980s, with the advent of Bayesian networks, when research began incorporating new paradigms based on graphical models incorporating probability, graph theory, and statistics to knowledge representation and reasoning. A rule-based ES consists of a knowledge base, a set of production rules, and an inference engine that contains procedures for chaining the rules and making inferences. A rule-based ES has advantages and disadvantages. The main advantage is that the rule-based paradigm mimics how human experts explain their problem-solving expertise. A primary disadvantage is the knowledge acquisition bottleneck. Application areas of ESs include classification, diagnosis, monitoring, process control, planning, and scheduling.

Next, we will delve into the crucial Bayesian network paradigm for expert systems.

Bayesian networks

We begin with a brief description of what a BN is. Next, we discuss how to use BNs to model probabilistic expert systems; just as we did for the rule-based, we follow with a list of advantages and disadvantages of BNs and mention some expert system application areas using BNs.

What is a Bayesian network?

The backbone of a BN is a **directed acyclic graph** (DAG), represented by a pair (G, P). G is the DAG, a crucial element that models the causal influence between variables, while P is a complete probability distribution on the random variable represented by the V nodes in G.

A DAG is a graphical depiction illustrating causal influence between variables, modeling causal effect relations in a given domain. A graph G is a pair (V, E), where V is a set of nodes, and E is a set of edges, where an edge is a link between a pair of nodes in V. In a **directed graph (digraph)**, the edges are directed; an edge $X \rightarrow Y$ in E goes from a parent node X (representing a cause) to a child node Y (representing an effect). A directed path in a digraph is an ordered sequence of nodes where directed edges connect every pair of consecutive nodes. In a DAG, we cannot visit a node twice along any directed path, meaning the DAG can have no directed cycles.

The DAG representing a BN can have both causal and probabilistic interpretations. The probabilistic interpretation of a BN is that the presence of directed edges between a node X and its set of parent nodes, pa(X), means that the probability of X depends on the value assigned to its parents. The absence of directed edges corresponds to conditional independencies in the probability model P of the BN.

In a later chapter on Bayesian networks, you will learn more formal definitions of a BN. It suffices to know that the DAG, G, representing a BN specifies a factorization scheme of the complete probability distribution P of the BN. A factorization of complete probability distribution means that the probability model of the BN is the product of the local probability distribution for every node in the BN, conditioned on its parents in the graph.

Bayesian networks allow us to answer queries about conditional independencies between sets of network variables. We can make probabilistic inferences and predictions and update beliefs based on observations of network variables.

With this informal definition and introduction to BNs, let us proceed to the next subsection, which describes how BNs represent a viable paradigm for building probabilistic expert systems.

Probabilistic expert systems

Earlier in the section on rule-based systems, we pointed out their limitation in knowledge representation of uncertainty and their resorting to ad hoc methods for reasoning under uncertainty. In many applications, the `IF-THEN` rules of the expert system need to represent uncertainty, and express *usually*, *it is likely that*, *often*, or *sometimes* relations between their conditions and the actions. Representing uncertainty is critical in building expert systems in domains such as diagnostic applications in medical expert systems that trace a patient's symptoms back to possible causes for these symptoms.

BNs provide a viable framework for probabilistic expert systems that are natural, as the probability is the lingua franca of the BNs; modular, as the probability model is factored in local modular conditional probabilities; and mathematically sound, since it is founded on the Bayes theorem and the laws of probability.

We give an example to illustrate how to express a BN model using `IF-THEN` rules and point out that the structured approach of BN imposes restrictions on the rules to ensure model consistency and coverage.

Bayesian networks provide a framework for knowledge representation and inference for probabilistic expert systems. The nodes of the DAG of a BN represent all system variables. The edges of the graph represent the probabilistic influence between all variables. A directed edge $X \rightarrow Y$ indicates that X has a direct influence on Y. The lack of directed edges between X and Y indicates conditional independencies.

The BN model encodes prior knowledge about a problem domain, including knowledge about uncertainties or degrees of belief. The rules of Bayesian inference can be applied to incrementally update degrees of belief as new information about network variables is collected. We will explain how this gets done when we review algorithms for probabilistic inference in *later chapters* of the book.

Critical data may need to be included in many diagnostic situations, which can pose a problem for rule-based expert systems. In contrast, a BN system will be able to track what information it has and attempt to provide the best possible explanation.

The expert knowledge encapsulated in BNs, represented as **conditional probability tables (CPTs)**, is not only equivalent but also more comprehensive than the `IF-THEN` rules in production rules systems. Each node and its associated CPT in a BN correspond to a set of `IF-THEN` rules. The CPT for the X variable assigns a probability to each possible world for its parents, $pa(X)$; by *world*, we mean a situation in which every parent variable has a value instantiation.

As an example, consider an alarm used for detecting burglaries. An earthquake may also trigger it. The BN has three variables: B (burglary), E (earthquake), and A (alarm). See *Figure 1.2*. The graph has two directed edges from B to A and E to A, since a burglary event can cause an alarm, and an earthquake can cause an alarm.

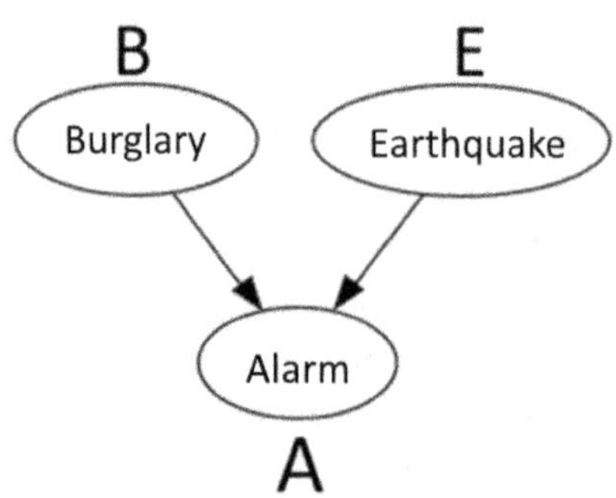

Figure 1.2: Alarm example BN

Assume that all variables are binary, taking one of two values: `True` (on) or `False` (off). Thus, we have four possible worlds enumerating all events that may cause an alarm event, that is, the alarm being on or off. For each possible world, w, the CPT assigns a probability of whether the alarm is triggered (i.e., its value is on or off):

$$P(A = \text{off} \mid B_w, E_w),$$

$$P(A = \text{on} \mid B_w, E_w),$$

Here, B_w, E_w are atomic propositions corresponding to the events of B, E in world w. Note the following:

$$P(A = \text{on} \mid B_w, E_w) = 1 - P(A = \text{off} \mid B_w, E_w)$$

Table 1.1 shows an example of the conditional probability of the event that the alarm is off:

w	B_w	E_w	$P(A = \text{off} \mid B_w, E_w)$
0	False	False	1
1	False	True	0.3
2	True	False	0.1
3	True	True	0

Table 1.1: The CPT for node A in the alarm example BN serves as a representation of possible worlds

Each row in *Table 1.1* presents a unique tuple that describes a possible world, where A's parents have specific instantiations from their value domains. Each row is equivalent to an `IF-THEN` rule. For example, the `IF-THEN` rule identical to the first row in the table is as follows:

If (B = False)and (E = False)Then A = off [Probability 1]

From the preceding example, a BN is a probabilistic model that assigns probabilities to sets of `IF-THEN` rules. The premise in the rule's `IF` part is a propositional sentence on a set of causes, and the rule's `THEN` part is an atomic proposition on an effect of those causes.

From this discussion, it follows that Bayesian networks guide rule-based systems to ensure sound representation of expert knowledge. The following two conditions on the propositional sentences of the rules' premises should be satisfied:

- [*Model consistency*]: The propositional sentences must be mutually exclusive to ensure that the rules will not conflict. Sentences α and β are mutually exclusive if they are never true in the same world.
- [*Model coverage*]: The propositional sentences must be exhaustive (i.e., each world satisfies at least one of the sentences) to ensure that the CPT implied by the rules is valid.

The following subsection will explore the advantages and disadvantages of BNs as an expert system paradigm.

Advantages and disadvantages of BNs

In this subsection, we mention some of the main advantages and disadvantages of BNs for KBSs, just as we did earlier for rule-based systems.

Here are the advantages:

- The graphical models of BNs enable a modular representation of knowledge. We can incrementally add/remove nodes and edges to their DAGs.
- Thanks to the BN graphical model's acyclic nature, we can always provide a definitive answer to a query, as there can be no circular inference paths. This certainty can be a significant advantage in KBSs.

- Based on the BN model's structural parameters, we can predict and control the computational complexity of inference. This predictability can instill confidence in the performance of BN-based expert systems.

Here are the disadvantages:

- Quantifying the model requires assigning condition probabilities of the CPT for each network variable; coming up with the numbers is challenging.
- The computational complexity for inferencing with BNs can suffer from the *curse of dimensionality* when there are many parents for nodes in the DAG of the BN model.

With this overview of the main advantages and disadvantages of BNs, let's proceed to the next subsection, which describes sample applications of probabilistic expert systems adopting the BN paradigm.

Applications of Bayesian networks

Probabilistic expert systems using directed graphical models of Bayesian networks have been successfully applied to various problems, including fault diagnosis, medical diagnosis, service robots, and fraud detection.

In the following, we provide an overview of those four application areas. We begin with the fault diagnosis application, where we describe how a model-based approach using structured graphical models, such as BNs, can efficiently diagnose and troubleshoot multiple faults in digital circuits.

Fault diagnosis

There is a growing demand for intelligent expert systems to assist consumers, service technicians, and field engineers in detecting, identifying, troubleshooting, and repairing engineering artifacts, digital devices, automated control systems in process industries, and automated robots in automated factory work cells.

Let us consider the combinatorial digital circuits domain of engineering devices. Reasoning about multiple faults in circuit diagnosis is challenging because the number of situation-action pairs that need to be covered in a rule-based expert system grows exponentially with the number of gates in the digital circuit.

Figure 1.3 shows a graphical depiction of the schematic from the netlist of a real-life example, circuit C432, from benchmark combinatorial circuits widely used in the fault diagnosis and testing community (Brglez and Fujiwara 1996). The circuit has 36 inputs, 7 outputs, and 160 components.

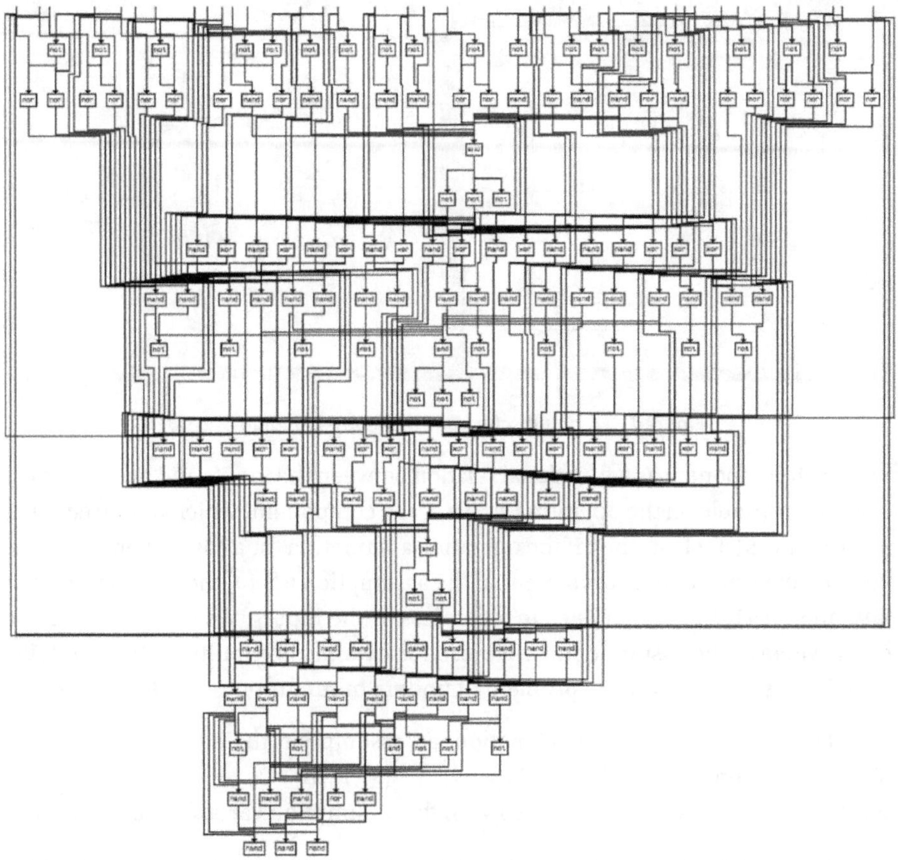

Figure 1.3: Schematic of circuit C432: 36 inputs, 7 outputs, and 160 components (Dechter and El Fattah 2001) (Zoom in on the image in the book's free PDF for better clarity)

The model-based approach, with models represented in Bayesian probabilistic networks or deterministic networks in the language of relational databases or propositional logic, offers a computationally more efficient and practical framework for expert system development in fault diagnosis applications. This practicality makes it a viable solution for the complex fault diagnosis task in engineering artifacts and digital devices.

El Fattah (EL Fattah 1993) presented a model-based approach to improving the troubleshooting of multiple faults. It provided a computational framework consisting of a new utility-based way to integrate diagnosis with repair, along with learning methods that combine associational reasoning with model-based reasoning. This approach learns associational rules for diagnosis and helps integrate them with the model-based diagnosis of multiple faults (El Fattah and O'Rorke, *Explanation-Based Learning for Diagnosis* 1993).

Let us illustrate how the structure-based approach using graphical models can be applied to the diagnosis task. Consider the tiny circuit in *Figure 1.4*, consisting of three components, two AND gates, and one OR gate; the circuit has four inputs (V1, V2, V4, and V5) and one output (V7).

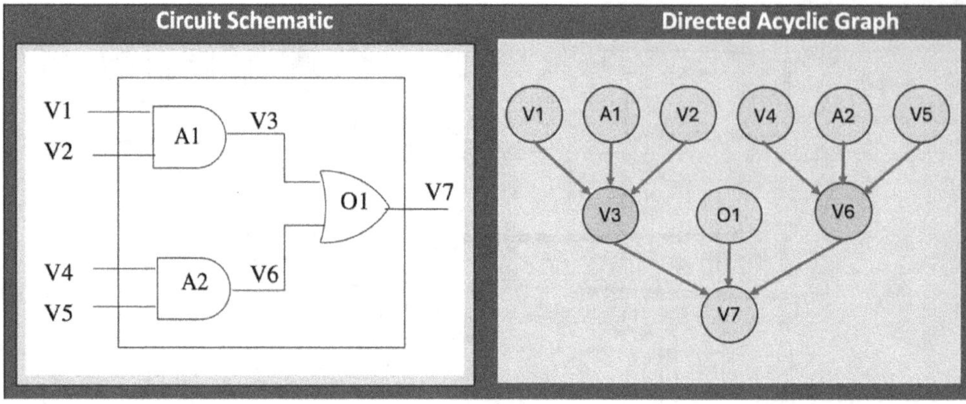

Figure 1.4: A circuit schematic and corresponding DAG

We can model each component as a functional relation between the inputs and output binary variables and an assumption variable on the operating status of the component, which has three values: ok, $sa0$ (stuck-at-zero), and $sa1$ (stuck-at-one). If the operating assumption for a component is ok, the output is the logic function expected of the component; if the assumption is $sa0$, the output is $False$ regardless of the inputs, and if $sa1$, the output is $True$ regardless of the inputs. We can assign an additive cost function on the value of the assumption variable: *0* for ok and *1* for faulty ($sa0$ or $sa1$). We pose the diagnosis problem as an optimization problem of finding minimum-cost explanations.

The diagnosis task is to construct an explanation, an assumption tuple $(A_1 = a_1, \cdots, A_m = a_m)$, that can be extended to a consistent solution $(V_1 = v_1, \cdots, V_n = v_n, A_1 = a_1, \cdots, A_m = a_m)$. The cost of an explanation is the sum of the costs associated with the assumption variables, as follows:

$$C(\{A_1 = a_1, \cdots, A_m = a_m\}) = \sum_{A_j \in A} c(A_j = a_j)$$

Finding a minimum-cost solution with the specified cost function amounts to finding a minimal cardinality diagnosis. For more details, see *(El Fattah and Dechter, Diagnosing tree-decomposable circuits* 1995).

Mapping observation to diagnosis

El Fattah (Y. El Fattah 1998) described a general elimination algorithm that takes observations as input and produces diagnoses as output. For the circuit example, the observations are the input-output variables ($V1, V2, V4, V5$, and $V7$), and the diagnosis is the value assignment to the assumption variables ($A1, A2$, and $O1$). The elimination algorithm compiles the circuit model to rules that associate observations with diagnoses. The following are the diagnostic rules compiled for the circuit example:

```
;;; Rule 1
(:if (:and (:or (= V1 F)(= V2 F))
           (:or (= V4 F) (= V5 F)) (= V7 T))
(:or (A1 = sa1) (A2 = sa1) (O1 = sa1)))
;;; Rule 2
(:if (:and (:or (= V1 F)(= V2 F))
```

```
(= V4 T)
(= V5 T)
(= V7 F))
(:or (A2 = sa0) (O1 = sa0)))
;;; Rule 3
(:if (:and (= V1 T)
(= V2 T)
(:or (= V4 F)(= V5 F))
(= V7 F))
(:or (A1 = sa0) (O1 = sa0)))
;;; Rule 4
(:if (:and (= V1 T)(= V2 T)(= V4 T)(= V5 T)(= V7 F))
(O1 = sa0))
```

Figure 1.5 shows an **ordered binary decision diagram (OBDD)** representing the compiled diagnostic rules.

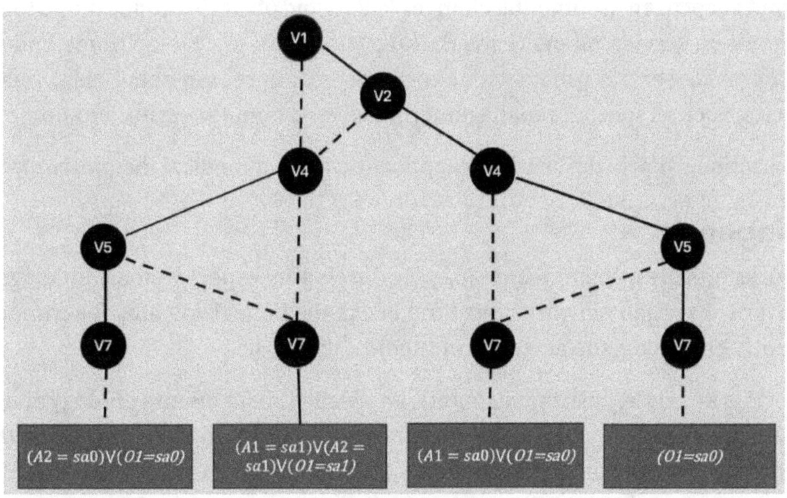

Figure 1.5: OBDD represents four diagnostic rules; nodes are observed binary variables, which are the circuit's input-output variables; solid lines represent the observed value 1 (True) and dashed lines binary 0 (False). The leaf nodes represent diagnostic conclusions.

The nodes represented by the circles represent the circuit's observed variables, and the edges represent the observed value, which is binary. The solid lines correspond to binary *1* (*T* or *True*) and the dashed lines to binary *0* (*F*or *False*). We begin at the root node and follow one of the lines from there until we reach a drain leaf node, which assigns the diagnostic explanation for the observed values. Assume we have the observation $(V1 = F, V2 = T, V4 = T, V5 = T, V7 = F)$. We begin at the root node, $V1$, follow the dotted branch to node $V4$, follow the solid line to $V5$, the solid line to $V7$, and the dashed line to the leaf node, concluding the diagnosis that either `A2` is stuck at zero, or `O1` is stuck at zero.

Having delved into the fault diagnosis application and described it in sufficient detail, let us briefly describe other significant application areas of Bayesian networks. Next, we will describe the application to service robots.

Service robots

The future is bright for service robots (Ceccarelli 2012), particularly in their ability to perform tasks with complete reliability in operation, safety, and full autonomy. These robots are designed to exhibit the social and interactive skills necessary for a variety of applications, such as providing support for elderly people.

Robinson and Nejat (Robinson and Nejat 2022) present an analysis of design recommendations for **socially assistive robotics** (SARs), and surveys use **dynamic Bayesian networks** (DBNs) for task and user state classification.

Bayesian networks provide technology for intelligent service robots to become fully autonomous, self-deciding robots and robots exhibiting social intelligence. The robot should reason about courses of action to achieve set goals and execute plans to perform various tasks. The robot must be autonomous, self-organize, and adapt to environmental changes (e.g., avoid obstacles and respond to environmental cues). Recent work on service robots (Park H. 2012) uses BNs to develop home and service robots. BN models can provide service robots with expert systems to reason about tasks related to various service categories, such as schedule management, delivery, home security, and home management.

In the next subsection, we will describe the application area of medical diagnosis.

Medical diagnosis

Bayesian networks provide a potent technology for developing expert systems for diagnosing various medical conditions. Bayesian networks form the backbone of decision aids for clinicians and physicians, significantly enhancing the accuracy of medical diagnosis.

Medical diagnosis is a complex task that demands knowledge about disease etiology and empirical and experiential data and relies on incomplete and uncertain information. Decisions in this healthcare field carry significant weight, impacting patients and practitioners alike. However, using Bayesian networks, with their unparalleled ability to significantly reduce the risk of errors, provides a sense of reassurance and security to clinicians and physicians in the diagnostic process.

Extensive research and development have been conducted using BN models to diagnose various medical conditions. A BN for medical diagnosis has the following components:

- Nodes representing unobserved variables that are the diseases, such as blood disorders
- Nodes representing the patient's observed findings, such as symptoms, signs, and test results
- Directed edges between various nodes that describe causal or probabilistic relationships

Collins and Fenton (Collins and Fenton 2020) describe a BN causal model for the early diagnosis and prediction of endometriosis. The BN model parameters were derived from multiple medical observational studies and trial data.

Lucas et al. (Lucas, Segaar and Janssens 1989) described a medical expert system named HEPAR, intended as a supportive tool for diagnosing liver and biliary tract disorders. The HEPAR system gathers and processes clinical data on patients with liver disorders. Onisko (A. Onisko, *Probabilistic Causal Models in Medicine: Application to Diagnosis of Liver Disorders* 2002) describes the development of the HEPAR system, HEPAR II, aimed at reducing the need for hepatic biopsy using Bayesian network models. The HEPAR II network was built based on the knowledge of human experts in the domain of hepatology, emphasizing the collaborative nature of Bayesian networks and human expertise. The numerical parameters are obtained from a clinical dataset and a human expert, underlining the significant role of human input in developing these systems (Onisko, Druzdzel and Wasyluk, *A Probabilistic Causal Model for Diagnosis of Liver Disorders* 1998).

Figure 1.6 shows a simple Bayesian network due to Onisko (Onisko, Druzdzel and Wasyluk, *A Probabilistic Causal Model for Diagnosis of Liver Disorders* 1998) representing a sample model fragment. The network models two diseases: steatosis and toxic hepatitis. Both diseases have a common cause (a history of alcohol abuse) and a common effect (elevated ALT). All variables are binary, with values indicating the presence or the absence of a given finding. Directed edges in the DAG indicate the presence of a causal relationship. For example, the presence of a history of alcohol abuse affects the presence of steatosis of the liver. This representation of **common cause** and **common effect** in the network model demonstrates how Bayesian networks can capture complex relationships in medical diagnosis. In this case, the common cause, a history of alcohol abuse, can lead to the development of both steatosis and toxic hepatitis, and the expected effect, elevated ALT, is a symptom that can be observed in patients with either of these diseases.

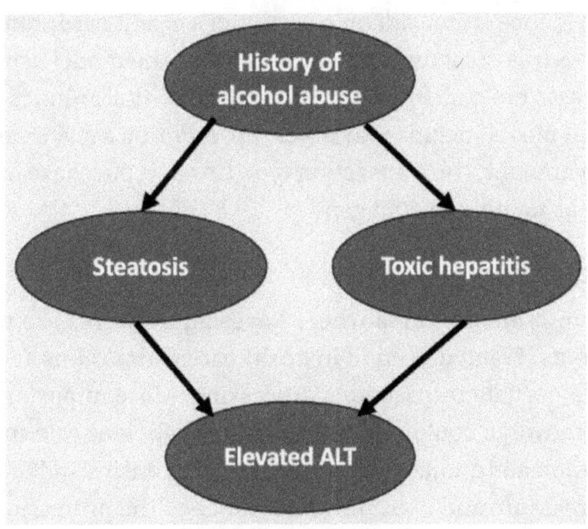

Figure 1.6: Example Bayesian network model fragment from liver disease domain; steatosis and toxic hepatitis diseases have a common cause and share a common effect (A. Onisko, Probabilistic Causal Models in Medicine: Application to Diagnosis of Liver Disorders 2002)

Onisko (A. Onisko 2008) gives a general review of Bayesian networks in medicine and states the problems of diagnosis, prognosis, and treatment as probabilistic inference tasks on Bayesian networks. Onisko formulates medical diagnosis as a problem of finding the **most probable explanation** (MPE) in Bayesian networks. Explanations involve computing the posterior probability of diseases based on the evidence. The MPE task is to find the disease, $\widehat{D}$, with the highest posterior probability:

$$\widehat{D} = \underset{D_i}{\operatorname{argmax}} P(D_i|E)$$

$P(D_i|E)$ is the probability of the disease, D_i, given the evidence, E, representing the set of observed findings that a patient presents with, such as symptoms, signs, and laboratory test results.

Our last application area is fraud detection, which we will describe next.

Fraud detection

This subsection cites two application areas for Bayesian network expert system development in fraud detection. Fraud detection aims to identify suspicious activities and transactions, which amounts to a binary classification problem. In this context, a binary classification problem is classifying a transaction as legitimate or fraudulent based on certain features or characteristics.

The first application is fraud detection in credit card transactions; the second is fraud detection for processing healthcare claims.

Credit card fraud detection

Credit card fraud detection involves the analysis of the spending behavior of customers in high-risk locations. Moso and Kenei (Moso and Kenei 2018) demonstrate the applicability of how a Bayesian network model can be used for making inferences about whether a transaction is likely to be legitimate or fraudulent based on evidence from data on a customer's credit card transactions. The demonstration adopts the Naive Bayes tree-structured model and trains the model using historical transactions based on a synthetic dataset created from a random sample that mimics hypothetical credit card transactions. The Bayesian model includes variables representing a customer's transaction attributes, including the transaction amount, the transaction type (foreign purchase or internet purchase), and customer attributes such as name, age, and gender.

Healthcare fraud detection

Healthcare fraud detection is another area where Bayesian networks are proposed to aid in the development of expert systems. Fraud detection in healthcare claims aims at detecting and identifying fraudulent activities in claims data transactions processing. More importantly, a strong argument is made on how Bayesian networks could play a crucial and efficient role in aiding downstream processes involving investigation and audits, providing a comprehensive and thorough approach to fraud detection, which should reassure and boost the confidence of the professionals involved.

Kumaraswamy et al. (Kumaraswamy, et al. 2024) demonstrate how, after validation with a real-life insurance claims dataset, the relational structure captured in the Bayesian network model provides benefits not available in methods based only on statistical machine learning methods such as logistic regression and random forest. They conclude that an expert system based on the **Bayesian belief network (BBN)** model provides supporting information that is highly interpretable to auditors and investigators, empowering them with the knowledge they need and boosting their confidence in their investigative capabilities.

To conclude this section on expert systems and Bayesian networks, let's reiterate the main takeaway points to reinforce your understanding of the key concepts learned in this section.

An expert system (ES), or knowledge-based system (KBS), consists of a knowledge base (KB) representing expertise in some domain and an inference engine (IE) capable of interpreting user queries, making inferences based on the knowledge stored in the KB, and answering the user queries in a form interpretable by the user.

One of the most crucial tasks in expert system development is knowledge acquisition (KA). This process involves extracting information from a variety of sources, including human domain experts, historical database records, documents and reports, and statistical datasets. The importance of this task cannot be overstated, as it forms the foundation of the entire system.

Many expert systems are built using the rule-based paradigm, where knowledge is represented in a KB consisting of sets of production rules: `IF <premise> THEN <action>`. The IE of a rule-based system is based on heuristics and control procedures for chaining the production rules: forward, backward, or mixed.

Rule-based systems offer two significant advantages. Firstly, human experts can often express their knowledge with `IF-THEN` rules, making the system more intuitive and user-friendly. Secondly, rules support modularity, allowing them to be added and deleted as the system evolves. However, rule-based systems also have their limitations, particularly as the number of rules grows: maintaining the KB becomes daunting, and the inference becomes very slow.

The Bayesian network paradigm for probabilistic expert systems uses graphical models representing the factorization of a probability model of a knowledge domain. BN models support modularity, where each element of the factored representation is a conditional probabilistic dependency of some variable on a subset of other variables. BN models are described by DAGs. The inference in BN systems is based on the probability laws and Bayes' theorem. Inference algorithms can leverage the structure of the DAG of the BN to ensure that computationally efficient inference and computational complexity guarantees can be provided in advance.

Building BN models is a complex task, particularly in challenging domains such as medical diagnosis. Here, the integration of multiple sources of information is necessary, and more stringent performance requirements must be met, making the process challenging. However, the rewards of overcoming these challenges are significant, as BN models can provide valuable insights and assist in decision-making.

In the next section, we will examine why we care about causality in the context of Bayesian networks, which is a central theme of the book. When interpreted causally, we will see that the directed edges in the DAG of the BN will lead to causal inferences based on the causal assumptions implied by those edges. Those causal inferences will serve to answer causal queries that statistical data alone cannot answer, and can assist in decision-making and predicting the effects of treatments and policies.

Why should we care about causality?

This section is organized as follows. We begin with a subsection on the role of causality in decision-making, where we highlight its role in decision-making and policy setting. We follow with a subsection on causal inference, where we discuss what it means that X causes Y, briefly introduce the two main problems of causal inference, and informally describe causal effect estimation tasks.

The following subsection briefly highlights the role of causality in decision-making.

The role of causality in decision-making

Causal inference plays a fundamental role in decision-making and policy settings. Decision-making requires knowing the effects of our interventions by posing causal queries to guide the selection of our actions. The following are a few examples of causal queries whose answers invoke causal inference based on the best available empirical evidence and plausible causal assumptions. Will a new drug lead to an increased lifetime span? Does using a particular medical device improve **quality of life** (QoL)? Would the implementation of a particular marketing campaign increase sales?

Decision-making requires a model that lays out the variables that we can control, which we can manipulate, and the variables that we can influence through our control. Such a model is called a **causal model.** While we can conjecture the presence of a causal relationship, confirmation of our assumptions must be tested by conducting experimental trials, surveys, or collecting observational data.

Policymakers, social scientists, epidemiologists, economists, financial econometricians, and researchers and data scientists in many fields are interested in making causal inferences from experimental and observational studies.

In epidemiology, causal inference plays a central role in "*clarifying the causes of diseases*" and "*evaluating procedures for health promotion and prevention of public health practice*" (Coughlin 2010).

Let us proceed to the next topic on causal inference in the following subsection.

Causal inference

We begin by defining probabilistic causation. Next, we define the two problems of causal inference. We follow with a section introducing causal effect estimation tasks.

Probabilistic causation

Intuitively, causality is a relationship between an X variable and a Y variable such that X controls Y. By control, we mean that changing or adjusting X's value can influence Y's value. Y is a variable that is dependent on X, meaning its value is determined or influenced by the value of X, and X is called the independent variable.

Let $P(Y|X)$ be the conditional probability of Y given X and let $P(Y)$ be the unconditional probability of Y.

Probabilistic causation interprets the assertion that X is a cause of Y. The interpretation is defined as follows:

X is a cause of Y if $P(Y|X) > P(Y)$

The preceding inequality reveals profound practical insights. Given the knowledge of $X = x$, the probability of Y is higher than the unconditional probability of Y. This simple yet powerful concept, when fully understood, can be a game-changer in various real-world scenarios. For instance, if smoking causes lung cancer, the probability of having lung cancer in the general population must be lower than in the subpopulation of cigarette smokers.

There are ongoing debates about the definition of causation and causal analysis, particularly in the context of conditional probability. For example, see the debate on climate change and the culprits for global warming (Ridgeway 2020).

Some argue that causality is inherent to the notion of intervention, where conditioning embodies conceptualizing intervention where we $do(X = x)$, meaning forcing the value of X and not simply conditioning on observing X. The concept of intervention conditioning, which we will delve into later in this chapter, is at the heart of this debate.

Next, let us introduce the two problems of causal inference.

The two problems of causal inference

Causal inference is all about finding answers to causal queries. A causal query seeks to estimate the effect of the intervention on a subset of X variables on the values of another set of variables, Y. The intervention on the X variables is formally expressed as $do(X = x)$, where x is a value of the X domain. We can express the causal inference problem by computing Y's probability distribution conditioned on the intervention on X, written as $P(Y|do(X = x))$.

Two significant problems underlie causal inference. The first, the **identification problem**, is of paramount importance. The second, which is of crucial practical importance, is the **estimation problem** of estimating the causal effect if it is determined to be identifiable. This problem involves us in the real-world applications of causal inference.

Let us describe each of the two problems of causal inference.

Causal effect identification

The causal effect identification problem is the process of determining whether the causal effect is uniquely identifiable from any positive probability distribution $P(V) > 0$ on the observed V variables for all causal models M with the same structure of the causal relationships. In simpler terms, it's about figuring out whether we can accurately isolate and measure the impact of a specific variable on the outcome.

Causal effect estimation

The causal effect estimation is the problem of estimating the causal effect of X on Y, $P(Y|do(X))$, from observational data on the observed variables, V, provided that the causal effect X on Y is identifiable.

In the parlor of literature on causal inference, X is called **treatment**, Y is called **outcome**, and the causal effect of X on Y is the causal effect of treatment, or simply the **treatment effect**.

The term *treatment effect* originates from the fields of medicine and epidemiology, which are concerned with the causal effects of binary variables representing treatment, such as a vaccine, an experimental drug, or a new surgical procedure.

The term *treatment* is used more generally in causal inference to refer to an experimental variable that can be changed or manipulated by an external agent.

The following are instances of the causal effect estimation problem:

- The causal effect of subsidized training programs on earnings. Given a dataset describing the labor market circumstances of trainees and non-trainees' comparison groups, estimate the **average treatment effect** (ATE).
- Estimating the causal effect of microcredit availability on the amount of money borrowed by household groups (young households, older households) is a problem of the **conditional average treatment effect** (CATE).

Let us conclude this section by summarizing the takeaway points. Causal inference is fundamental to decision-making and policy setting, and it requires answers to causal questions to know the effects of interventions. Causal inference is based on a causal model that lays out our causal assumptions and observational data on the observed variables in the causal model. The two problems in causal inference are the identification problem and the estimation problem. The identification problem is to determine whether the effect of the intervention on particular model variables designated as a treatment on other variables designated as the outcome is identifiable. The estimation problem is the problem of estimating the causal effect of treatment on outcome, provided that the causal effect is identifiable. Examples of estimation problem tasks are computing the ATE and the CATE.

Let us proceed to our next topic in the following section, addressing the question of causal graphical models.

Why causal graphical models?

In the previous section on causality, we mentioned that causality amounts to a relationship between X and Y in which X can control Y. Note that X and Y are random variables. We can only collect data on the relationship between X and Y from observational data or via experimental trials.

It is impossible to answer causality questions based only on the statistical data or what amounts to a probability distribution on X and Y, $P(X,Y)$.

A causal graphical model M is associated with a directed acyclic graph G consisting of nodes and directed edges. Each node in the DAG represents a random variable, and a directed edge from node X to node Y means X is a direct cause of Y.

> **Note**
>
> A causal graphical model M plays a catalyst role in causal inference:
>
> - M is a generative model of the probability distribution $\boldsymbol{P(X,Y)}$.
> - G makes assumptions on the causal influence between model variables explicitly stated.
> - G encodes all conditional independencies between model variables.

Table 1.2 shows causal graphs for common causal phenomena between X and Y:

- **Causation:** The assumption here is that X directly causes Y, and no other variables influence either X or Y.
- **Mediation:** The assumption here is that a third variable, Z, mediates the causal influence of X on Y, and the three variables form a chain, $X \rightarrow Z \rightarrow Y$.
- **Collider:** The assumption is that X and Y are unconditionally independent, but each causes a third variable, Z, called a collider. Conditioning on Z creates a spurious association between X and Y, called **collider bias** in observational studies.
- **Confounder:** Here, a third variable, Z, called a confounder, is a cause for both X and Y. Unmeasured confounders (i.e., not controlling for the confounder) result in so-called **confounder bias** in observational studies.
- **Instrumental variable** (IV): The IV affects Y solely through the effect on X. The random assignment in a trial is an example of an instrument. IV is commonly employed to estimate causal effects in the presence of unmeasured confounding (Angrist JD, 1996).

Phenomenon	Causal graph	Description
Causation	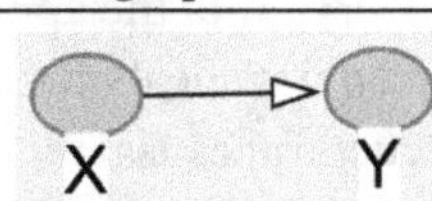	*X* causes *Y*
Mediation	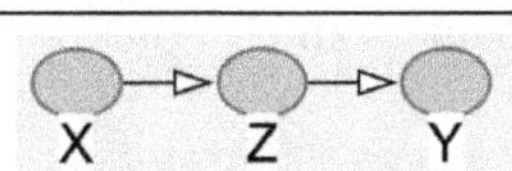	The effect of *X* on *Y* is mediated by *Z*
Collider	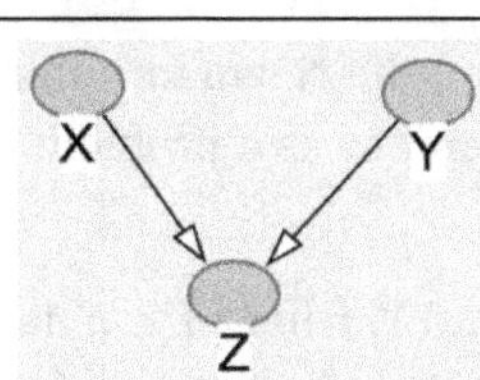	*Z* is a collider of *X* and *Y*
Confounder	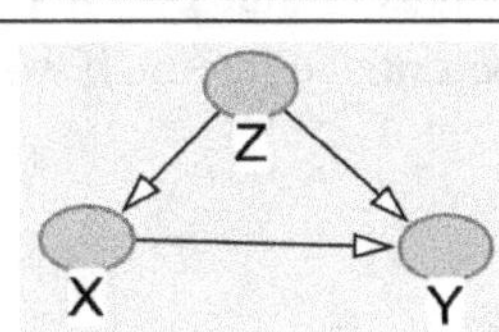	*Z* is a confounder of *X* and *Y*

Phenomenon	Causal graph	Description
Instrumental variable	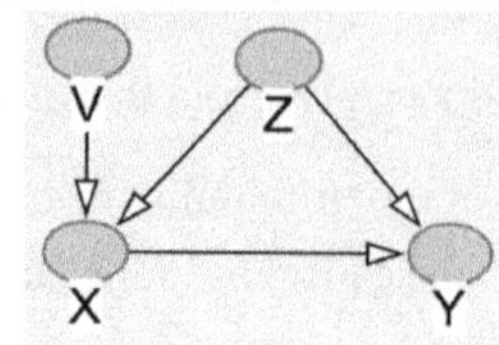	V is an instrument variable for X

Table 1.2: Causal graphs for common causal phenomena between X and Y

Let us conclude this section by summarizing the takeaway points. Causal inference requires causal assumptions to be made, as statistical inference with observational data alone cannot provide answers to causal queries. Causal graphical models enable handling causal queries by articulating all causal assumptions and making them explicitly represented using DAGs. The DAG articulates precisely the qualitative aspect of the cause-and-effect relationship between model variables. The nodes in the DAG represent model variables, and the directed edges point from cause to effect. Causal models play a critical role in qualitative causal analysis based on the DAG structure and in determining whether a specific causal effect is identifiable from observational data or the probability distribution of the observed model variables. DAGs describe several causal mechanisms, including causation, mediation, collider, confounder, and instrumental variable.

Moving on, let us proceed to our next topic in the following section, which will provide an overview of causal inference with Bayesian networks and statistical methods.

Overview of causal inference with BN and statistical methods

In this section, we will provide an overview of two main approaches to causal inference. The section is organized as follows. We begin with a general overview of the approach known as the Neyman-Rubin model or the potential outcomes, which is based on statistical methods. We follow this with an overview of the other approach, which is based on the structural causal model or the causal Bayesian networks.

Let us present the potential outcomes approach next.

Potential outcomes

The potential outcomes approach, also called the Neyman-Rubin model (Rubin 1974), is based on statistical methods. It formulates causal inference as a problem of statistical analysis with missing data or counterfactual outcomes.

Suppose we have two random variables, T and Y, where T is a dummy variable indicating the **exposure** or **treatment** status in a population. T takes a value of *1* if a unit U_i of the population receives treatment, and *0* if not. Define $Y_i(1)$ as the potential outcome if we give treatment for the unit U_i and $Y_i(0)$ if we do not. See *Figure 1.7*.

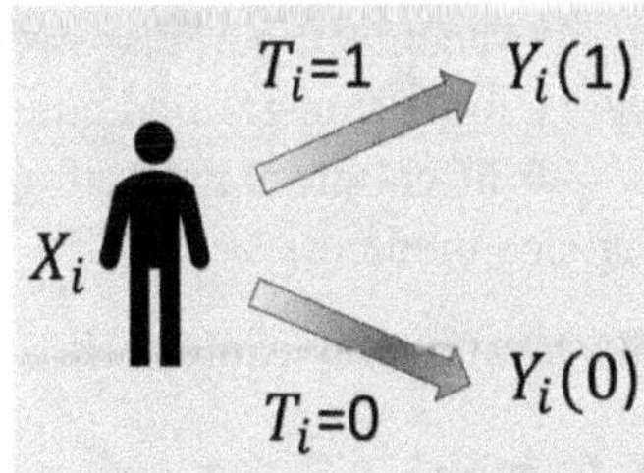

Figure 1.7: Potential outcomes of binary treatment

We can only observe one of the potential outcomes, not both, for any given unit, depending on whether it receives treatment. Hence, we have a *missing data problem,* with the data being the potential outcomes for the imaginary situation in which we reverse the actual treatment or non-treatment decision for all units. Because the missing data is for a condition that did not happen, we call the case the *counterfactual outcome.*

We can observe only one of the potential outcomes for any unit depending on the assignment variable, T, and we can define the observed outcome as the random variable:

$$Y = T\,Y(1) + (1 - T) \cdot Y(0).$$

Each unit U_i has features that can influence treatment decisions and outcomes. Let X be a k-dimensional vector of covariates with $k \geq 2$ representing a unit's features. X represents a set of characteristics such as sex, age, income, and so on.

The trial data denoted $\{(T_i, X_i, Y_i)\ i = 1 \ldots, n\}$ is a set of random variables of size n from the joint probability distribution, $P(T, X, Y)$.

Figure 1.8 shows a generative model for the probability distribution of the data. The graph shows that X has a causal effect on treatment and outcome variables. X is a **confounder.**

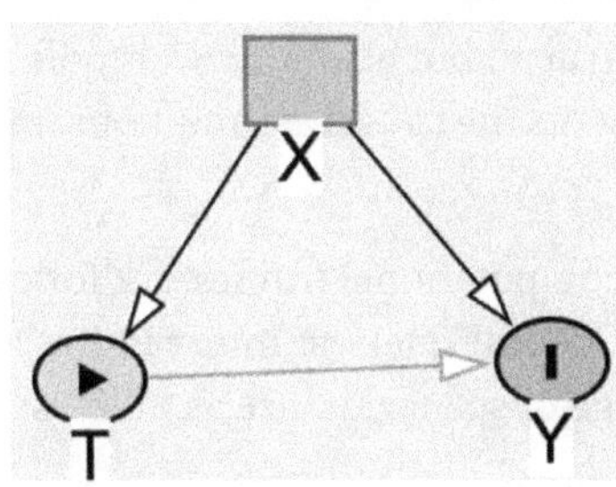

Figure 1.8: Generative model, where T is treatment, Y is outcome, and X is a covariate

Calculating the counterfactual outcome provides a means for the causal inference study. The study takes the trial data as input. It produces as output the causal effect of treatment by estimating the ATE or the CATE.

The ATE is the difference between the average outcome when everyone is exposed and the average outcome when nobody is. The CATE is the difference between the expected outcome under treatment and the expected outcome with no treatment, conditioned on the covariate vector that represents population segments.

For example, in a study on the COVID-19 vaccine, the ATE answers how protective the vaccine is on average. In contrast, the CATE answers a more fine-grained question on how protective the vaccine is for specific population subgroups, such as children. Estimating the CATE is helpful in many areas, such as direct marketing and personalized medicine

Next, we will overview the structural causal model framework.

Structural causal model (SCM)

We will begin by describing what an SCM is. We will then describe the types of causal inference queries that an SCM can help answer.

What is an SCM?

An SCM integrates structured equation models and graphical models to help analyze causal relationships. An SCM must satisfy the following conditions:

- The structure of the model must be a valid representation of reality
- The relationships between model variables must be acyclic
- The variables conditioned on their parents must be independent of their ancestors

The conditions make a graphical SCM map to an equivalent Bayesian network having the same graph and conditional probabilities derived from the structured equations of the SCM. Every instantiation of the exogenous variables in the structured equations of the SCM uniquely determines the values of all system variables. The source of uncertainty resides in the exogenous variables. Hence, a probability distribution on the exogenous variables defines a probability distribution on the system variables of the SCM. Thus, an SCM is a generative model of a complete probability distribution on its variables: the nodes of the directed acyclic graph or the causal graph.

What is an SCM good for?

SCMs are not just theoretical constructs, but also practical tools for reasoning about causality. The SCM's causal graph formally represents the causal assumptions, making them explicit and applicable to real-world scenarios.

SCMs provide a formal basis for a computational framework for causal inference. They are versatile, capable of answering two types of causal queries: intervention and counterfactual. These queries, which can be applied to a wide range of scenarios, are as follows:

- **Intervention questions**: What would happen if we gave this drug to this patient?
- **Counterfactual questions**: What would have happened if we had given this drug?

We can now proceed to our next topic, which is the discussion of causal inference and machine learning. In this section, we will discuss distinctions and synergies between the two areas.

Discussion of causal inference and machine learning

In this section, we discuss connections between causal inference and machine learning. The discussion covers two topics. The first is characterizing problems addressed by causal inference on the one hand and by **machine learning** (**ML**) on the other hand. The second is describing distinctions and synergies between ML and causal inference.

This section is organized as follows: the first focuses on characterizing problems of ML, and the second focuses on characterizing problems of causal inference. Subsequently, we delve into the differences between ML and causal inference before exploring the intriguing synergies between the two.

Let us begin with a discussion on characterizing problems of ML.

Characterizing problems of ML

ML begins with data and aims to find associations or correlations between data elements or identify clusters or patterns in the data. The following are two main ML tasks:

- **The predictive task**: The prediction task uses **supervised ML** methods, which involve training a computer algorithm using labeled datasets to learn models that use some variables to predict other unknown variables. For example, the training data can be examples of a patient's symptoms marked by the presence or absence of a particular disease, and the prediction model can then predict the likelihood of disease presence on new data.
- **The descriptive task**: The description task uses **unsupervised ML** methods, which involve a program looking for patterns in unlabeled data. Applications in recommendation systems and targeted marketing use unsupervised ML for customer segmentation, looking through online sales data, and identifying different types of clients making purchases.

Having characterized ML problems, let us characterize causal inference problems next.

Characterizing problems of causal inference

There are two main tasks of causal inference:

- **Identification**: This task is at the heart of the research design. Research design is your strategy to answer the research question. You have a research question, then think about the data you need to answer it, and the problems you could have establishing cause and effect.
- **Estimation**: You conduct an experiment where you intervene on units from the population, assigning them treatment or no treatment. You want to find the treatment effect by comparing the two potential outcomes. The challenge in causal inference is that we only observe one of the potential outcomes for each unit, not both. The non-observable or not-realized outcome is called the **counterfactual**, and the problem becomes one of answering a *what-if* question or predicting the counterfactual.

Let us proceed to the next theme in the discussion, which concerns distinctions between ML and causal inference.

Differences between ML and causal inference

Figure 1.9 compares traditional statistics, ML, and causal inference. Statistics deal with data, *D*, and their probability distributions, *P*, and can only deal with association questions. An example of an association question is *What is the likelihood that a client purchasing product A would also buy product B?* That association question is a task of predicting the probability of *B* if we see *A*. *Figure 1.9* depicts that association queries, *Q(P)*, deal with aspects of *P*, which is a model of the data, *D*.

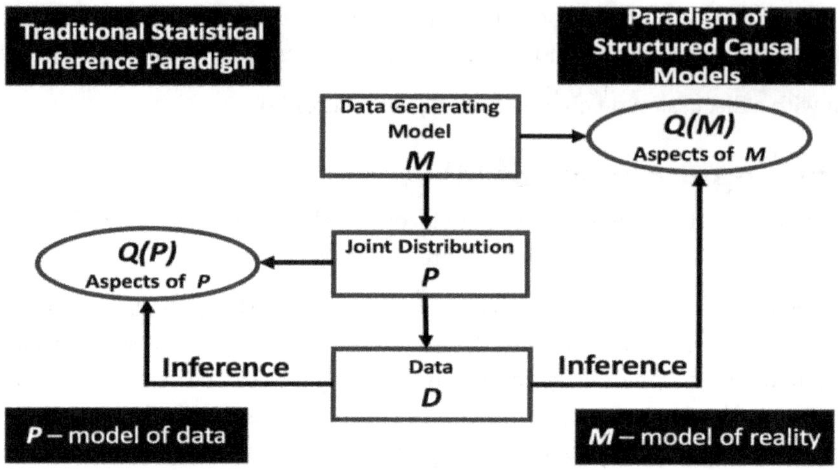

Figure 1.9: Comparison between causal inference and statistical inference

One of the biggest challenges of ML prediction models is their inability to generalize beyond the probability distribution of the training data. In other words, there is no guarantee that the prediction model could make accurate predictions when presented with new data whose probability distribution is significantly different from the training data.

On the other hand, causal inference uses a causal model, *M*, encoding causal assumptions of the data generative model. *M* is a generative model in that it defines a class of probability distributions of the data compatible with all the conditional independence conditions entailed by *M*. *Figure 1.9* explains that causal inferences are queries on aspects of *M* compared to statistical inferences on aspects of *P*.

Compared to statistical inferences, causal inference enables us to reason about change and deal with counterfactuals, which is impossible in the conventional statistical paradigm.

As an example of a *what-if* question, we can ask: *What is the likelihood that a client who bought product A would repurchase A if we were to double the price?* The example is a question about a changed probability distribution resulting from intervening by doubling the price of *A*. Note that doubling *A*'s cost is not the same as seeing the price of *A* double. Doubling *A*'s price amounts to intervention in model *M*, resulting in a different probability distribution for the data.

In the preceding paragraphs, we explained the difference between causal inference and ML. Now, it is time to discuss synergies between the two areas. We will discuss this topic next.

Causal inference and ML synergies

This section is organized as follows. First, we describe the integration of causal inference in ML, and then we discuss the integration of ML in causal inference with observational data.

Integration of causal inference in ML

The integration of causal inference and ML is a rapidly expanding area of research (Cloudera, 2020). There are multiple drives for that integration, as mentioned in OECD (2021):

- Causal inference integration with ML provides a better interpretation of learned patterns in input data. Correlation without causation identified in patterns by ML models may result in biased or ineffective decision-making.
- Incorporating causal reasoning in ML enables a better understanding of when to expect learned models to remain predictive over time and the conditions under which they may fail.
- Causal inference is essential to adapt ML models to new environments, settings, or populations.

State-of-the-art AI systems are required to provide accurate decision results and explain underlying reasons. An AI system making recommendations about products or services should be able to clarify on what grounds they are making the recommendations. The explanation lets a consumer understand why the lender rejected their loan application. The causes enable a decision-maker to know the actions they need to take to achieve the desired effects.

Causal inference has spurred research activity in the emerging area of explainable AI. An AI system that makes predictions based on ML models can use counterfactuals to determine the input data elements that need to change values and how much to make a different prediction. Then, the altered input will comprise the explanation. For example, an AI system predicting the rejection of a loan application can explain that if the credit score were higher than 700, then the rejection prediction would change to accepted.

Researchers in deep learning are increasingly recognizing the importance of causality models and attempting to use them in their neural network models for applications in robotics, natural language, and computer vision to enable adaptation to new tasks or environments (Rosemary Ke, 2022).

ML for causal inference

Earlier, in the section on potential outcomes, we mentioned the problem of estimating the ATE and CATE.

Many researchers have proposed several methods for using ML to estimate the CATE from observational data. We can group the proposed approaches into two categories: modified ML and meta-learner methods.

Modified ML methods customize existing ML algorithms to estimate the average causal effect.

A meta-learner is a model-agnostic algorithm that provides a recipe for estimating CATE using an arbitrary ML method (e.g., any supervised learning or regression estimator method). Example meta learners are the S-, T-, and X-learner.

Figure 1.10 shows a meta-learner function diagram. It takes as input the observational data, consisting of three-tuples for the treatment, *T*, the covariates, *X*, and the observed outcome, *Y*, and an arbitrary ML method. A meta-learner algorithm uses the base ML method to estimate the conditional average treatment effect as a function of the covariates' value. We will handle meta-learners in a later chapter in *Part 2* of this book.

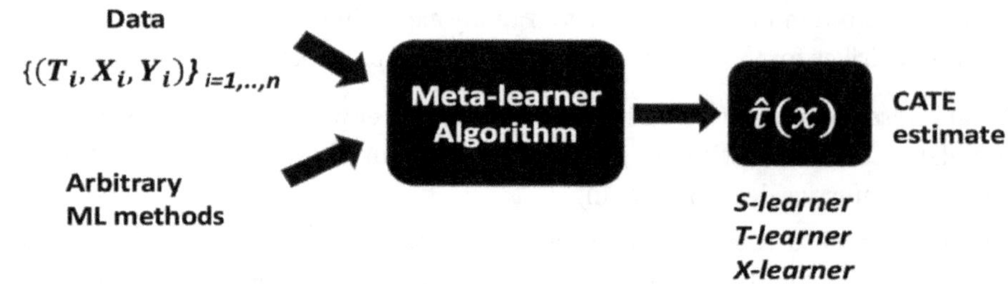

Figure 1.10: Meta-learner for estimating the CATE

With that, we conclude this discussion section. Let us highlight the main points covered. The main problems of ML are predictions and descriptions, while the main problems of causal inference are the identification and estimation of causal effects. Statistical ML deals with data, *D*, and its probability distributions, *P*, and can only deal with association questions.

Causal inference uses a causal model, *M*, encoding causal assumptions of the data generative model. Causal inference can reason about change and deal with counterfactuals and *what-if* questions.

Causal inference integrates into ML to provide an interpretation of the data as well as to enable generating explanations and adapting ML models to new environments, settings, or populations. Statistical ML integrates into causal inference to provide algorithms for estimating the CATE from observational data.

With this discussion section, we conclude this chapter, covering a guided tour of book topics. In the following section, we summarize the takeaway points covered in the chapter.

Summary

Expert systems, or knowledge-based systems, are knowledge-intensive applications with problem-solving expertise in specialized fields designed to process information and produce advice or recommendations for solving problems like human experts.

Expert systems allow incremental development, adjusting their behavior by incrementally updating the knowledge stored in the system. The conventional expert systems approach uses knowledge representation based on production rule systems. Bayesian networks are better suited for building probabilistic expert systems in complex domains requiring incremental uncertainty handling and reasoning with partial information.

Causal inference is fundamental to decision-making and policy setting, and it requires answers to causal questions to know the effects of interventions. Causal inference is based on a causal model that lays out our causal assumptions and observational data on the observed variables in the causal model. The two problems in causal inference are the identification problem and the estimation problem. The identification problem is to determine whether the effect of the intervention on particular model variables designated as a treatment on other variables designated as the outcome is identifiable uniquely from the probability distribution on observed variables. The estimation problem is the problem of estimating the causal effect of treatment on outcome, provided the causal effect is identifiable. Examples of estimation problem tasks are computing the ATE and the CATE.

Causal models are represented by causal graphs, which are directed acyclic graphs whose nodes are the domain variables and whose edges represent direct causal influence between pairs of variables.

Causal graphs explicate and explicitly represent all the causal assumptions. They allow qualitative causal analysis and depiction of underlying prototypical mechanisms, including causation, mediation, collider, confounder, and instrumental variable.

There are two main approaches to causal inference: one based on SCMs or causal BNs, and the other on potential outcomes, also known as the Neyman-Rubin model. The potential outcomes framework models causality as a manipulation (treatment, exposure, action, and intervention) applied to units from a population and uses statistical analysis to estimate measures of causal effect (e.g., the ATE and the CATE). The SCM/BN approach models intervention with the *do* operator and frames causal inference in probabilistic graphical models, incorporating the *do* operator semantics of the intervention conditioning and derived rules of the *do* calculus.

ML and causal inference use statistical data to perform core tasks. The main tasks are predictions and descriptions for ML and identification and estimation for causal inference.

ML deals with data, D, and its probability distributions, P, and can only deal with association questions. At the same time, causal inference uses a causal model, M, which is a generative model of P and can reason about change and handle counterfactuals and *what-if* questions.

There are synergetic connections between causal inference and ML. On the one hand, causal inference can enhance ML in two ways. First, it can provide ML with a better interpretation of learned patterns in data by differentiating causation from association patterns, thus avoiding biased or ineffective decision-making. Second, it can enable a better understanding of when to expect learned models to remain predictive over time and the conditions under which they may fail. On the other hand, ML methods contribute to causal inference by providing algorithms for estimating treatment effects such as the ATE and the CATE from observational data.

Code/packages

Table 1.3 gives a partial list of libraries/packages in R and Python that are useful for programming and coding related to modeling and reasoning with probabilistic graphical models in Bayesian networks, structured equation modeling, and structural causal models. We will use some of these libraries, as well as others introduced throughout the chapters, in hands-on practice.

Name	Purpose	Language	Link
py-bbn	Implementation of probabilistic and causal inference in Bayesian belief networks using exact inference algorithms	Python	https://github.com/vangj/py-bbn/blob/master/docs/source/index.rst
pgmpy	Structure learning, parameter estimation, approximate (sampling-based) and exact inference, and causal inference	Python	https://pgmpy.org
Causal graphical models	Describing and manipulating causal graphical models and SCMs	Python	https://github.com/ijmbarr/causalgraphicalmodels
gRain	Probability propagation in graphical independence networks, also known as Bayesian networks or probabilistic expert systems	R	https://cran.r-project.org/web/packages/gRain/gRain.pdf
dagitty	Graphical analysis of SCMs	R	https://github.com/jtextor/dagitty
ggdag	Visualizing and analyzing causal directed acyclic graphs	R	https://github.com/r-causal/ggdag
lmtest	Testing linear regression models	R	https://rdrr.io/cran/lmtest/
lavaan	Structural equation modeling, latent variable analysis	R	https://lavaan.ugent.be
causaleffect	Functions for identification and transportation of causal effects	R	https://github.com/santikka/causaleffect
dosearch	Causal effects from arbitrary observational and experimental probability distributions via do calculus	R	https://github.com/santikka/dosearch

Table 1.3: Libraries/packages for Bayesian networks and structural causal models

Table 1.4 lists the available libraries/packages for estimating causal effects using statistical and ML methods. Implementations of meta-learners and various ML and generic regression methods have been developed in R and Python. We will use some of these libraries in the code to estimate the average treatment effect, conditional treatment effect, and ML methods. Some of these libraries will be utilized for practice on application use cases and experimenting with open source datasets in *later chapters* of the book.

Name	Purpose	Language	Link
causallib	Inferring causal effects from observational data using statistical and ML methods	Python	https://github.com/BiomedSciAI/causallib
dowhy	Suite of tools for causal reasoning: modeling, identifying, estimating, and refuting	Python	https://github.com/py-why/dowhy
causalml	Causal inference methods using ML algorithms	Python	https://github.com/uber/causalml
EconML	Estimating heterogeneous treatment effects from observational data	Python	https://github.com/py-why/EconML#blogs-and-publications
causalToolbox	Estimation of heterogeneous treatment effects with ML	R	https://github.com/soerenkuenzel/causalToolbox

Table 1.4: Libraries/packages for causal inference and machine learning

References

- Angrist JD, Imbens GW, Rubin DB. 1996. *Identification of causal effects using instrumental variables.* J Am Stat Assoc. 444–55.
- Athey S, Wager S. 2017. *Efficient policy learning.* ArXiv preprint. `https://arxiv.org/abs/1702.02896`.
- Barr, A., and Feigenbaum, E. A. 1981. *The Handbook of Artificial Intelligence, vols. 1 and 2.* Los Altos: Kaufmann.
- Bollen, K.A. 2002. *Latent Variables in Psychology and the Social Sciences. Annual Review of Psychology.* Annual Review of Psychology 605–634.
- Brglez, F., and Fujiwara, H. 1996. *A neutral netlist of 10 combinatorial benchmark circuits and a target translator in Fortran.* Proc. IEEE International Symposium on Circuits and Systems.
- Brock, D.C. 2018. *Learning from Artificial Intelligence's Previous Awakenings: The History of Expert Systems.* AI Magazine 3–15.
- Buchanan, B., and Feigenbaum, E. 1978. *DENDRAL and Meta-DENDRAL: Their Applications dimension.* Artificial Intelligence 5–24.
- Ceccarelli, Marco . 2012. *Service Robots and Robotics: Design and Application.* Hershey, PA: Engineering Science Reference.
- Cloudera. 2020. *Causality for Machine Learning.* `https://ff13.fastforwardlabs.com/`.
- Collins, Rachel, and Fenton, Norman. 2020. *Bayesian network modelling for early diagnosis and prediction of Endometriosis.* medRxiv. 6 November. `https://www.medrxiv.org/content/10.1101/2020.11.04.20225946v1`.

- Coughlin, S.S. 2010. *Causal Inference and Scientific Paradigms in Epidemiology*. Bentham eBooks.
- Cowell, Robert G., A. Philip Dawid, Steffen L. Lauritzen, and David J. Spiegelhalter. 1999. *Probabilistic Networks and Expert Systems*. New York: Springer-Verlag.
- Curth, A., van der Schaar, M. 2021. *Nonparametric Estimation of Heterogeneous Treatment Effects: From Theory to Learning Algorithms*. Proceedings of the 24th International Conference on Artificial Intelligence and Statistics (AISTATS).
- Dechter, Rina, and Yousri El Fattah. 2001. *Topological parameters for time-space tradeoff*. Artificial Intelligence 93–118.
- Econ-ML. n.d. *Econ-ML user guide*. Accessed July 21, 2023. `https://econml.azurewebsites.net/spec/estimation/metalearners.html`.
- El Fattah, Y. 1998. *An Elimination Algorithm for Model-based Diagnosis*. Ninth International Workshop on Principles of Diagnosis (Dx98). Cape Cod, Massachusetts, USA.
- EL Fattah, Yousri. 1993. *Learning for diagnosis and repair*. Doctoral thesis. Irvine, California: University of California at Irvine.
- El Fattah, Yousri, and Paul O'Rorke. 1993. *Explanation-Based Learning for Diagnosis*. Machine Learning 35–70.
- El Fattah, Yousri, and Rina Dechter. 1995. *Diagnosing tree-decomposable circuits*. IJCAI'95: Proceedings of the 14th international joint conference on Artificial intelligence - Volume 2. San Francisco: Morgan Kaufmann. 1742–1748.
- Fayyad, U.M. 1996. *Advances in knowledge discovery and data mining*. Menlo Park, CA: AAAI Press.
- Felix, E. 2013. *Graphical Causal Models*. In *Handbook of Causal Analysis for Social Research*, by S. (ed.) Morgan, 245–273. New York: Sage Publications.
- Hahn PR, Murray JS, Carvalho CM. 2020. *Bayesian regression tree models for causal inference: Regularization, confounding, and heterogeneous effects*. Bayesian Analysis. DOI 10.1214/19-BA1195.
- Hansotia B, Rukstales B. 2002. *Incremental value modeling. Journal of Interactive Marketing*. Journal of Interactive Marketing.
- Hernán, M., Hsu, J., Healy B. 2019. *A Second Chance to Get Causal Inference Right: A Classification of Data Science Tasks*. Chance 42–49.
- Hitchcock, Christopher. 2021. *Probabilistic Causation*. The Stanford Encyclopedia of Philosophy. 21 March. `https://plato.stanford.edu/archives/spr2021/entries/causation-probabilistic/`.
- Jacob, D. 2021. *Jacob, D. CATE meets ML: Conditional average treatment effect and machine learning*. Accessed July 20, 2023. `https://www.econstor.eu/bitstream/10419/233509/1/1755344805.pdf`.
- Kennedy, E.H. 2020. *Optimal doubly robust estimation of heterogeneous causal effects*. ArXiv preprint. `https://arxiv.org/abs/2004.14497`.
- Kuang, K. et al. 2020. *Causal Inference*. Engineering 253–263.
- Kumaraswamy, Nishamathi, Tahir Ekin, Chanhyun Park, Mia K. Markey, Jamie C. Barner, and Karen Rascati. 2024. *Using a Bayesian Belief Network to detect healthcare fraud*. Expert Systems with Applications.

- Kunzel, S.R., Sekhona, I.S., Bickela , P.J., Yua, B. 2019. *Meta-learners for Estimating Heterogeneous Treatment Effects using Machine Learning*. Proc. of the national academy of sciences 4156–4165.
- Langarizadeh M, Moghbeli F. 2016. *Applying Naive Bayesian Networks to Disease Prediction: a Systematic Review*. Acta Inform Med. 364–369.
- Lee H, Aronson JK, Nunan D. 2019. *Catalogue Of Bias*. https://catalogofbias.org/biases/collider-bias/.
- Lucas, P. J. F., R. W. Segaar, and A. R. Janssens. 1989. *HEPAR: an expert system for the diagnosis of disorders of the liver and biliary tract*. Liver 266–275.
- Mano, M. M. 1988. *Computer Engineering - Hardware Design*. Prentice Hall.
- McDermott, J. 1980. *R1: An expert system in the computer systems domain*. Proceedings of AAAI-80.
- McLachlan S., Dube K., Hitman G., Fenton N., Kyrimi E. 2020. *Bayesian networks in healthcare: Distribution by medical condition*. Artificial Intelligence in Medicine.
- Moso, Juliet Chebet, and Jonah Kipcirchir Kenei. 2018. *Credit Card Fraud Detection using Bayes Theorem*. International Journal of Computer and Information Technology 184-189.
- Nie X, Wager S. 2017. *Quasi-oracle estimation of heterogeneous treatment effects*. ArXiv preprint. https://arxiv.org/abs/1712.04912.
- OECD. 2021. *Artificial Intelligence, Machine Learning and Big Data in Finance: Opportunities, Challenges, and Implications for Policy Makers*. https://www.oecd.org/finance/artificial-intelligence-machine-learningbig-data-in-finance.htm.
- Onisko, A. 2008. *Medical Diagnosis*. In *Bayesian Networks*, by Naim, P., Marcot, B., and Pourret, O., 15–32. Wiley.
- Onisko, Agnieszka . 2002. *Probabilistic Causal Models in Medicine: Application to Diagnosis of Liver Disorders*. 3 July.
- Onisko, Agnieszka , Marek J. Druzdzel, and Hanna Wasyluk. 1998. *A Probabilistic Causal Model for Diagnosis of Liver Disorders*. Intelligent Information Systems VII. Malbork, Poland.
- Park H., and Cho S., 2012. *A modular design of Bayesian networks using expert knowledge: Context-aware home service robot*. Expert Systems with Applications 39 2629–2642.
- Pearl, J., and Bareinboim, E. 2014. "External Validity: From Do-Calculus to Transportability Across Populations." Statistical Science 579–595.
- Pearl, J., Glymour, M., and Jewell, N. 2016. *Causal Inference in Statistics*. Wiley.
- Pearl, J. 2021. *Causal and Counterfactual Inference*. In *The Handbook of Rationality*, by Spohn, W. and Knauff, M., 427–438. The MIT Press.
- n.d. *CAUSAL INFERENCE IN STATISTICS A Gentle Introduction*. Accessed July 21, 2023. http://bayes.cs.ucla.edu/jsm-july2012-pdf.pdf.
- 1988. *Probabilistic Reasoning in Intelligent Systems: Networks of Plausible Inference*. San Francisco: Morgan Kaufmann.
- Pearl, J. 2009. *Causality*. Cambridge University Press.

- Powers, S., Qian, J., Jung, K., Schuler, A., Shah, N.H., Hastie, T., and Tibshirani, R. 2018. *Some methods for heterogeneous treatment effect estimation in high dimensions*. Statistics in Medicine 1767–1787.
- rRidgeway, John. 2020. *A Brief Primer on Causation – Climate Scepticism*. 14 March. `https://cliscep.com/2020/03/14/a-brief-primer-on-causation/`.
- Robinson, F., and G. Nejat . 2022. *An analysis of design recommendations for socially assistive robot helpers for effective human-robot interactions in senior care*. J Rehabil Assist Technol Eng.
- Rosemary Ke., N., and Bauer, K. 2022. *Causality and Deep Learning: Synergies, Challenges and Future*. ICML Tutorial.
- Rubin, D. B. 1974. *Estimating causal effects of treatments in randomized and nonrandomized studies*. Journal of Educational Psychology.
- Sackett, D.L. 1979. *Bias in analytic research*. Journal of Chronic Diseases 32: 51–63.
- Shortliffe, E. H. 1976. *Computer-Based Medical Consultation, MYCIN*. New York: Elsevier.
- Waterman, D. A. 1986. *A Guide to Expert Systems*. Addison-Wesley.

2

Probability and Bayes' Theorem

In this chapter, we give a formal definition of probability and state its main axioms. We discuss various interpretations of probability and emphasize its subjective interpretation as a measure of the degree of belief. You will learn about Bayes' theorem and understand how it serves as a mechanism for belief revision: updating uncertainty about beliefs based on observed evidence. You will see examples on the use of Bayes' theorem. The following are the main sections in this chapter:

- Defining probability and rules of counting
- Describing random variables
- Conditioning and conditional probability
- Characterizing independence and conditional independence
- Understanding Bayes' theorem
- Interpreting probability: frequentist vs. subjective

Technical requirements

To run the Python code, you need to: (1) install Python 3 following the installation and setup guide at `https://realpython.com/installing-python/`; (2) install an application called PyCharm, which is an **IDE**, or **Integrated Development Environment**. To get started, download and install the community version of PyCharm for your operating system from `https://www.jetbrains.com/pycharm/`.

To run the Python code for some examples, you must install the `SciPy` and `NumPy` libraries. The Python code for this chapter is in a Python project named `chapter2`, which can be found at this GitHub URL: `https://github.com/PacktPublishing/Causal-Inference-with-Bayesian-Networks/tree/main/Chapter_02/Python_code`.

To run the R code, you need to download and install R and RStudio IDE from this URL: `https://posit.co/download/rstudio-desktop/`. For hands-on programming with R and a guide on installing R and RStudio IDE, go to this URL: `https://rstudio-education.github.io/hopr/`.

The R code for this chapter can be found at this GitHub URL: `https://github.com/PacktPublishing/Causal-Inference-with-Bayesian-Networks/tree/main/Chapter_02/R_code`

Some practice examples for R in this chapter require you to install the `DescTools: Tools for Descriptive Statistics` library, which you can install from within RStudio.

Defining probability and rules of counting

This section defines probability and its basic properties based on Kolmogorov axioms. We then walk you through the combinatorics of counting in discrete sample spaces for different sampling categories. We demonstrate how to apply derived counting rules to calculate the probability on multiple examples. We also explain the accompanying code in Python and R to enforce the learning through hands-on practice and demonstrate how to automate solutions to similar problems.

Let us begin by defining probability. Probability is a measure of the uncertainty of events. An event is the outcome of an experiment, such as tossing a coin or flipping a die. By "outcome of an experiment," we mean that a random variable takes on a specific value, belongs to a set of discrete values, or lies within a range of continuous values. The values that a random variable can take form what we call a sample space. In the discrete variable case, as in the example of tossing a coin, the sample space has two values: heads or tails. In the real-valued variable case – variables such as temperature, pressure, height, weight, distance, and so on – we have the sample space as the space of real numbers. Formally, we define probability in terms of a probability space, as follows.

Definition 2.1: Probability space

A probability space (or probability model) is a triple $((\Omega, \mathcal{F}, P)$:

- Ω is the sample space, namely an arbitrary set representing the set of outcomes of an experiment.
- $\mathcal{F}$ is the event space, a σ-algebra on the set of subsets of Ω that includes the empty set, and is closed under complementation and countable unions. By closed under complementation, we mean that if $A_n \in \mathcal{F}$, then $A_n^c \in \mathcal{F}$, where A_n^c is the set complement of A_n for Ω. By closed under countable unions, we mean that if we have a set of countable events $A_n \in \mathcal{F}$ for all non-negative integers $n \in \mathbb{N}$, then we also have the unions of all the events A_n, also belonging to $\mathcal{F}$. That is,

$$\bigcup_{n \in \mathbb{N}} A_n \in \mathcal{F},$$

- P is a probability measure on the event space $\mathcal{F}$, meaning that P is a function assigning, for each event in the event space, a probability, a real number in the closed interval $[0,1]$.

Now that we have defined probability, let us present its axioms and properties.

Definition 2.2: Kolmogorov axioms

The Kolmogorov axioms consist of the following three **axioms of probability**:

- **Axiom 1:** For every event $A, P(A) \geq 0$.
- **Axiom 2:** $P(\Omega) = 1$.

- **Axiom 3:** If $A_1, A_2, \ldots$ is a sequence of pairwise disjoint events, then

$$P\left(\bigcup_{i=1}^{\infty} A_i\right) = \sum_{i=1}^{\infty} P(A_i)$$

The Kolmogorov axioms provide the mathematical conditions that all probability measures must respect. Axiom 1 says that the probability of all events must be a nonnegative number. Axiom 2 says that the probability of the event consisting of all outcomes in the event space is 1, meaning it is a sure thing that an outcome lying anywhere in the entire sample space is bound to occur. Axiom 3 allows us to compute the probability of the union of countable and pairwise disjoint events by simply summing the probabilities of all the events.

The following properties all follow from the probability axioms (Gravner 2011):

- The null event $\emptyset$ has probability zero,

$$P(\emptyset) = 0 \quad (2.1)$$

- For any event A, the probability of A is between zero and one

$$0 \leq P(A) \leq 1 \quad (2.2)$$

- For any event A, the probability of its complement event A^c is:

$$P(A^c) = 1 - P(A) \quad (2.3)$$

- For any two events A, B, if A is a subevent of B,

$$A \subset B \Rightarrow P(B) \geq P(A) \quad (2.4)$$

- For any two events A, B,

$$P(A \cup B) = P(A) + P(B) - P(A \cap B) \quad (2.5)$$

- For any three events A_1, A_2, A_3,

$$P(A_1 \cup A_2 \cup A_3) = P(A_1) + P(A_2) + P(A_3) - P(A_1 \cap A_2) - P(A_1 \cap A_3) - P(A_2 \cap A_3) + P(A_1 \cap A_2 \cap A_3) \quad (2.6)$$

- The **inclusion-exclusion formula** for the union of n events ($n > 3$),

$$P(A_1 \cup \cdots \cup A_n) = \sum_{i=1}^{n} P(A_i) - \sum_{1 \leq i < j \leq n} P(A_i \cap A_j) + \sum_{1 \leq i < j < k \leq n} P(A_i \cap A_j \cap A_k) + \cdots + (-1)^{n-1} P(A_1 \cap \cdots \cap A_n) \quad (2.7)$$

Having defined probability and the underlying Kolmogorov axioms and listed the main properties derived from them, it is essential to learn how to calculate probability by counting for discrete sample spaces. Finding probability by counting depends on how we draw samples from the sample spaces. The task of counting uses combinatorial mathematics to find mathematical formulas for finding the number of outcomes as a function of the number of samples and the size of the sample spaces.

Counting outcomes in sample spaces

Suppose we are conducting an experiment with a finite sample space having n possible outcomes. We assume that all n outcomes are equally likely.

The **equally likely model** is commonly adopted as a convenient assumption to analyze random experiments. It makes sense when there is no prior information about the likelihood of the various outcomes. See the section on the 'Interpretation of Probability' for more on this topic.

Finding the probability of an event A, in the equally likely model, translates to a problem of counting the number m of outcomes that A contains, which is the **cardinality** $|A|$ of the set A

$$m = |A|$$

together with the number n of events in the sample space, Ω,

$$n = |\Omega|$$

The probability of an event A consisting of m outcomes from a sample space of n outcomes is given by:

$$P(A) = \frac{m}{n} \quad (2.8)$$

To apply *Equation 2.8*, it is required that you be skilled at counting outcomes in events of all kinds.

Definition 2.3: sampling

Sampling from a set means choosing an element at random from that set. Each element of the set has an equal chance of being chosen.

- **Samples with replacement versus samples without replacement**: In the first case, repetition of the same element is allowed; in the second, repetition is not allowed. Intuitively, with replacement means that, after an element is drawn, the set remains unchanged, while without replacement, the set keeps shrinking by eliminating from the set the elements already drawn.
- **Ordered versus unordered samples**: In an ordered sample, the order of the elements drawn from the set matters, while in unordered samples, the order does not matter. For example, if the elements in the set are labeled with the letters "a", "b", ..., and so on, then a 2-element sample (a, b) will be different from the sample (b, a) in the case of ordered samples, while both samples are the same in the unordered case.

How to count

Definition 2.4: the multiplication principle

Let an experiment comprise k sequential steps and let $n_1, n_2, \ldots, n_k$ be the number of distinct ways a sample can be drawn from the set in each respective step. The count of the experiment's possible outcomes is given by

$$\#outcomes = n_1 \times n_2 \times \cdots \times n_k.$$

We now apply the multiplication principle to the cases of ordered and unordered samples with and without replacements. We derive formulas for counting the number of outcomes in each of the four cases. The formulas are based on combinatorics using permutations and combination functions.

Ordered samples without replacement

We model this as an experiment of k steps, where in each step, we draw an element from a set of n elements.

In the first step, the size of the set is n. The number of ways we can choose one element out of n elements is simply n. So the count for step 1, n_1= n.

In the second step, the size of the set is n-1, because of the non-replacement, and thus the count for step 2, n_2= n-1.

We can continue similar reasoning for all k steps. At step k, we would have n_k ways of selecting the kth element, where $n_k = n - k + 1$.

Hence, by the multiplication principle, the count of possible outcomes equals

$$\#outcomes = n \times (n-1) \times \ldots \times (n-k+1) = \frac{n!}{(n-k)!} = P(n,k)$$

The $P(n,k)$ function is the k-permutations of an n-element set.

Ordered samples with replacement

With replacement, the set of n elements remains unchanged, and each of the k steps amounts to drawing one element out of a set of n elements. Since there are n possible ways to select an element in the sample at each of the k steps, then by the multiplication principle, the total number of outcomes equals

$$\#outcomes = n^k$$

Unordered samples without replacement

The number of ways of ordering k elements is equal to factorial(k}. The number of ways of selecting an ordered sample of k elements out of a set of size n is $P(n,k)$. Hence, the number of ways of drawing an unordered sample of k elements out of the n-element set is given by

$$\#outcomes = \frac{P(n,k)}{k!} = \frac{n!}{k!\,(n-k)!} = \binom{n}{k} = C(n,k)$$

$C(n,k)$ is the k-combination of an n-element set. $C(n,k)$ is sometimes read as "n choose k," given by the binomial coefficient, $\binom{n}{k}$.

Unordered samples with replacement

The number of samples in this case requires some explanation. This case represents choosing k unordered samples with repetition from a set of n elements.

To calculate the number of samples, we need to recall the definition of **weak composition.** A weak composition of k into n parts is an n-tuple of nonnegative integers that sum to k. The tuples (2,0,0), (1,1,0), (1,0,1), (0,2,0), (0,1,1), and (0,0,2) are all weak compositions for 2 into 3 parts.

To depict a weak composition pattern: draw k X's and place the n – 1 bars in any order. There is no restriction on bars at the beginning, end, or consecutively, since parts=0 is OK. *Table 2.1* depicts all patterns of weak compositions of 2 into 3 parts:

(2,0,0)	(1,1,0)	(1,0,1)	(0,2,0)	(0,1,1)	(0,0,2)
XX||	X|X|	X||X	|XX|	|X|X	||XX

Table 2.1: Weak decomposition of k into n parts that sum to k where k=2 and n=3

There are $n + k - 1$ symbols, (the k X's plus the $n - 1$ bars). Choose k of them to be X's (or $n - 1$ of them to be bars):

$$\binom{n+k-1}{k} = \binom{n+k-1}{n-1}$$

Hence, the number of unordered samples with replacement from n objects is given by

$$\#outcomes = \binom{(n+k-1)}{k} = \frac{(n+k-1)!}{k!\,(n-1)!}$$

To illustrate the different sampling schemes, consider a list of 3 letters, ["a" "b" "c"], and that we sample 2 objects from the list. *Table 2.2* shows all combinations of such samples for the four cases: ordered/unordered, with/without replacement.

Unordered samples without replacement	**Unordered samples with replacement**	**Ordered samples without replacement**	**Ordered samples with replacement**
[1] "a" "b"	[1] "a" "a"	[1] "a" "b"	[1] "a" "a"
[2] "a" "c"	[2] "a" "b"	[2] "b" "a"	[2] "b" "a"
[3] "b" "c"	[3] "a" "c"	[3] "a" "c"	[3] "c" "a"
	[4] "b" "b"	[4] "c" "a"	[4] "a" "b"
	[5] "b" "c"	[5] "b" "c"	[5] "b" "b"
	[6] "c" "c"	[6] "c" "b"	[6] "c" "b"
			[7] "a" "c"
			[8] "b" "c"
			[9] "c" "c"

Table 2.2: Ordered and unordered sampling with and without replacement from a set of three letters

The table shows that the number of samples for the unordered case is strictly less than that for the ordered, with replacement (6 versus 9) and without replacement (3 versus 6). Also, the number of samples for the case with replacement is strictly greater than for without replacement for both the ordered (6 versus 3) and the unordered (9 versus 6).

Table 2.3 summarizes the results of sampling k elements from a set of size n for ordered and unordered samples and for cases with and without replacement.

	Without replacement	With replacement
Ordered sample	$\mathrm{P(n, k)} = \frac{n!}{(n-k)!}$	n^k
Unordered sample	$C(n,k) = \binom{n}{k} = \frac{\mathrm{P(n,k)}}{k!}$	$C(n+k-1,k) = \binom{n+k-1}{k} = \frac{(n+k-1)!}{k!\,(n-1)!}$

Table 2.3: The number of subsets of size k out of a list of n elements

Example 2.1: birthday problem

The birthday problem addresses the probability that among k randomly surveyed people, a pair would have the same birthday. The birthday problem is also named the birthday paradox (Buddies 2012) because analysis shows that for only 23 people, the probability is 50%, and for 57 or more people, the probability exceeds 99% (Birthday Problem n.d.)

We make some assumptions to derive the probability that a pair shares a birthday among k randomly surveyed people. We assume the year has 365 days (ignoring leap years) and consider all birthdays equally likely.

It is easier to compute the probability of no shared birthdays among k number of people. The probability equals the number of outcomes for sampling k dates out of n with no replacement (that is no duplicate dates) divided over the total number of outcomes of choosing k dates out of n with replacement (the same dates allowed). Using the formulas in *Table 2.1*,

$$\mathrm{P(no\ \ shared\ birthdays)} = \frac{\#\, ordered\ sample\ of\ size\ k\ with\ no\ replacement\ from\ n\ objects}{\#\, ordered\ sample\ of\ size\ k\ with\ replacement\ from\ n\ objects}$$

$$= \frac{P_n^k}{n^k}$$

The event of a shared birthday is simply the complement of the event of no shared birthdays.

$$\mathrm{P(a\ shared\ birthday)} = 1 - \mathrm{P(no\ \ shared\ birthdays)}$$

Practice in Python

The following is an implementation of the solution to the birthday problem that calculates the probability of a shared birthday among k people. You can practice with the code in the `birthday.py` file, part of the `chapter2` Python folder.

```
from math import perm
def no_shared_birthday_probability(k):
    a = perm(365, k)
    b: int = 365 ** k
    return a / b

def a_shared_birthday_probability(k):
    p = 1 - no_shared_birthday_probability(k)
    return p
```

The same Python file also includes a function for computing a list of the probability of a shared birthday for various values of the number of people, k. The code is as follows:

```
def shared_probabilities(list):
    probabilities = []
    for k in list:
        p = a_shared_birthday_probability(k)
        print(k, p)
        probabilities.append(p)
    return probabilities
```

We generate a list of probabilities for a range of values of k, generate a plot of the probability of a shared birthday versus k, and save the plot in the `ex1.png` file. The Python code is as follows:

```
import matplotlib.pyplot as plt
def make_plot():
    list = [10, 23, 41, 57, 70]
    probabilities = shared_probabilities(list)
    plt.plot(list, probabilities)
    plt.xlabel('number of people')
    plt.ylabel('a shared birthday probability')
    plt.title('Shared birthday probability')
    plt.savefig('ex1.png')
    plt.show()
```

Figure 2.1 depicts the saved plot from running the preceding code:

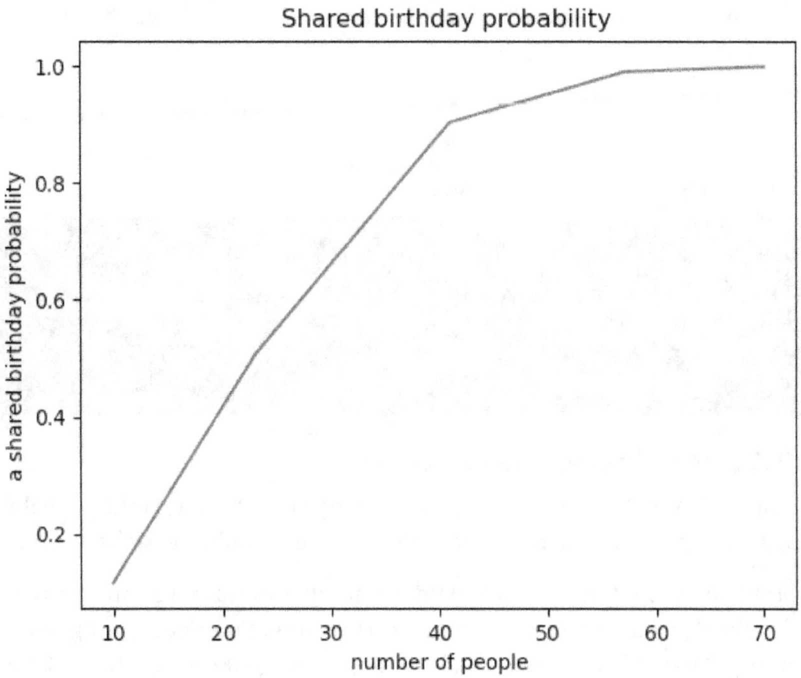

Figure 2.1: Probability of a shared birthday between k number of people

Practice in R

The following code implements the solution for the birthday problem using R. The code is available in the `birthday.R` file, which you can run in RStudio. The code computes the probability of a shared birthday among k people for various values of k equal 10, 23, 41, 57, 70. You will get the following output, and the code will produce a plot similar to the one in *Figure 2.1*. The code uses the `CombN` function from the `DescTools` library, which implements all the combinatorics functions in *Table 2.1*.

```
library(DescTools)
no_shared_birthday_probability <- function(k){
  n <- 365  # days in a year
  a <- CombN(n, k, repl=FALSE, ord=TRUE)
  b <- CombN(n, k, repl=TRUE, ord=TRUE)
  return (a/b)
}
a_shared_birthday_probability <- function(k){
  p <- 1 - no_shared_birthday_probability(k)
  print(p)
  return (p)
}
```

```
make_plot <- function(){
  nb_people <- c(10, 23, 41, 57, 70)
  probabilities <- Map(a_shared_birthday_probability, nb_people)
  plot(nb_people, probabilities)
}
make_plot()
```

The result from running the code is as follows:

```
[1] 0.1169482
[1] 0.5072972
[1] 0.9031516
[1] 0.9901225
[1] 0.9991596
```

Example 2.2: non-decreasing integers

If we are to sample with replacement k digits in the range 0 to 9, what is the probability that the sequence of k digits is in non-decreasing order? Calculate the probability for k=1 to 20.

To solve this problem, we first need to calculate the number of ways we can construct an integer of k digits such that the digits take values in the range 0 to 9, are in non-decreasing order, and repeated digits are allowed. Let us say k=2, then the k-digit integers can be 00, 01, 11, 12, 22, 23, and so on. Note that we disallow integers in decreasing order, such as 10, 21, 32, and so on.

The number of non-decreasing integers of k-digits is the same as the number of outcomes of unordered sampling with replacement k elements from a box containing 10 distinct objects labeled with digits from 0 to 9. An unordered sample ignores the order of the digits in the k-sequence, but we can have a one-to-one mapping of every unordered sample to an integer with k non-decreasing digits. *Table 2.1* shows the count of unordered samples with replacement of k out of n objects, where, in this case, n=10. Thus, the total count is given by

$$\binom{10+k-1}{k} = \frac{(9+k)!}{k!\,9!}$$

The total number of possible combinations of k digits is 10^k. Hence, the probability that randomly selecting k digits be in non-decreasing order is

$$\frac{(9+k)!}{k!\,9!\,10^k}$$

Practice In Python

The code in Python for this example is in the integers.py file. The code uses the comb function from the Python math library to define the function for the unordered sample with replacement. We then compute the probability in the nd_integers(k) function. The program creates a dictionary for the probability of k values from 2 to 10 and prints the result.

```
from math import comb

def count_samples_unordered_with_repl(n, k):
    return comb(n + k - 1, k)

def nd_integers(k):
    return count_samples_unordered_with_repl(10, k)

if __name__ == '__main__':
    probability = {}
    for k in range(2, 10) :
        probability[k] = nd_integers(k)/(10**k)
    print(probability)
```

The program output is as follows:

```
{2: 0.55, 3: 0.22, 4: 0.0715, 5: 0.02002, 6: 0.005005, 7: 0.001144, 8:
0.0002431, 9: 4.862e-05}
```

Practice in R

The code in R for this example is in the integers.R file. The code uses the DescTools library to compute nd_integers, the number of non-decreasing integers as a function of the number of digits, k. The code then iterates over a range of k from 1 to 20. It divides the number of non-decreasing integers by the total number of all possible integers having k digits to calculate the required probability. It stores the result in a dictionary that records the probability for every value of k. As you can see, the probability decreases exponentially from probability 0.55 for k=2 to probability 0.001 for k=7.

```
library(DescTools)
nd_integers <- function(k){
  # number of ways to draw non-decreasing digits of size k
  CombN(10, k, repl=TRUE, ord=FALSE)
}

for (k in 2:10) {
  p = nd_integers(k)/(10^k)
  cat("probability for k=", k, "is",  p,"\n")
}
```

The result from running the code is as follows:

```
probability for k= 2 is 0.55
probability for k= 3 is 0.22
probability for k= 4 is 0.0715
probability for k= 5 is 0.02002
probability for k= 6 is 0.005005
probability for k= 7 is 0.001144
probability for k= 8 is 0.0002431
probability for k= 9 is 4.862e-05
probability for k= 10 is 9.2378e-06
```

Let us conclude this section by highlighting the main takeaway points.

Probability is a measure of the uncertainty of events that are subsets in a sample space. The logical structure of a probability measure, which satisfies three Kolmogorov axioms, provides a systematic and reliable foundation for understanding, instilling confidence in your grasp of the subject.

Under the equally likely model, the probability of discrete and finite sample spaces can be computed by counting. This versatile method can be applied to various scenarios, from simple games to complex scientific experiments, making it a powerful tool in real-world situations. Events can be modeled as sampling from the discrete elements of the sample space. There are four categories of sampling: with and without replacements, ordered or unordered. There are combinatorics formulas for counting the outcomes of the discrete events in each of the four sampling categories.

In the next section, we will delve into the critical concept of random variables, which are key to understanding the probabilistic nature of real-world phenomena.

Describing Random Variables (RVs)

In the previous section, we set the mathematical definition of probability, a crucial foundation for understanding the concept of **Random Variables** (**RVs**), which is our focus in this section. RVs are functions of events in probability spaces and can either be discrete or continuous. Understanding these variables is key to mastering the field of probability and statistics.

The plan of this section is as follows. First, we present definitions for discrete and continuous RVs in a section on univariate probability, guiding you through each concept step by step. Univariate probability definitions involve a single RV. For discrete RVs, you will learn about the probability mass and cumulative distribution functions, demonstrate the definition with a worked example, and practice with code in R and Python. For continuous RVs, you will learn the definition of probability density functions and cumulative density functions in a similar structured manner. Next, in a section on multivariate probability, we generalize the distributions you learned about in the univariate probability section to apply to multiple variables and the definition of joint probability distributions. We follow this with a section on summary statistics of random variables, including expected value, variance, covariance, and Pearson correlation coefficient. We end with a final section on types of RVs, where you will learn about widely used distributions in practical applications, which lists the probability distributions for the Bernoulli, binomial, Poisson, geometric, and normal random variables. In that last section, you will also see examples of code for practice using R and Python.

Univariate probability

Univariate probability describes the probability distribution of a single RV, sometimes called the **marginal distribution** because the distribution concerns only the value domain of that variable. For a discrete RV, the probability distribution gives the probability that the RV takes on a specific value from its discrete domain. For continuous RVs, the probability that the RV takes a specific value is zero. We can assign a probability to only intervals of real numbers; the cumulative distribution function gives the probability that the RV has a value equal to or less than a specific value.

Discrete RVs

A discrete RV is one for which the sample space consists of a set of discrete values that are finite or countably infinite. Bernoulli and Poisson RVs are two instances of discrete RVs you will learn about later in this section. The Bernoulli RV can assume only one of two values. In contrast, the Poisson RV, corresponding to the number of arrivals in a given time interval, can take on one of the countably infinite values from the set of non-negative integers.

Let X be a discrete RV. The probability that X takes on the value x, $P(X = x)$, is defined as the sum of the probabilities of all sample points in Ω that are assigned the value$x$.

$$P(X = x) = \sum_{\omega \in \Omega : X = x} P(\omega) \quad (2.9)$$

This defines a function $f(x)$ called the **probability distribution** of X, also called the **Probability Mass Function** (**PMF**), for the discrete random variable X,

$$f(x) = P(X = x) \quad (2.10)$$

$f(x)$ can be represented by a formula or a table that assigns a probabilityfor every x in the domain of $X, dom(X)$. $f(x)$ must satisfy the following two conditions:

1. $f(x) \geq 0$ for each value $x \in dom(X)$
2. $\sum_x f(x) = 1$

The summation in (b) extends over all values within the domain of X.

Definition 2.5: Cumulative Distribution Function (CDF)

If X is a discrete RV, and $f(x)$ is the PMF of X, the CDF is given by

$$F(x) = P(X \leq x) = \sum_{t \leq x} f(t), \forall x \in dom(X) \quad (2.11)$$

Example 2.3: discrete Random Variable

Consider the sample space of tossing a fair coin three times. The RV X gives the number of heads recorded. Compute the probability mass function and the cumulative distribution function for X.

X is a discrete RV that takes values from the set {0, 1, 2, 3}. The sample space for the experiment is the set of all possible outcomes out of three tosses of the fair coin. Since the outcome of each toss is either heads (H) or tails (T), then by straightforward enumeration, the sample space is the set of 8 elements $\{TTT,\ TTH,\ THT,\ THH,\ HTT,\ HTH,\ HHT,\ HHH\}$ Assuming that all members of the sample space are equally likely, then we can calculate the probability that X=x using *Equation 2.9*. We also get the same result by simply counting the number of outcomes in the sample space in which we have x heads and dividing that count by 8, which is the cardinality of the sample space, as per *Equation 2.8*. *Table 2.4* shows the calculation and the resulting PMF.

$X = x$	$X = 0$	$X = 1$	$X = 2$	$X = 3$
$\{\varpi \in \Omega: \mathrm{X} = \mathrm{x}\}$	$\{TTT\}$	$\{TTH, THT, HTT\}$	$\{THH, HTH, HHT\}$	$\{HHH\}$
$f(x) = P(X = x)$	0.125	0.375	0.375	0.125
$F(x) = P(X \leq x)$	0.125	0.5	0.875	1

Table 2.4: The PMF and the CDF for the discrete RV X correspond to the number of heads in 3 fair coin tosses

Practice in Python

You can find the Python code for this example in the `rv_discrete.py` file.

The following `probability()` function automatically computes $P(X = x)$ as in *Table 2.4*. We use numeric values to designate the outcomes: 0 for tails and 1 for heads. We create 3 for loops, iterate over the set of outcomes for each toss, and sum the number of heads by adding all the outcomes that equal 1. We increment the respective members of the list count depending on whether the total number of heads in the three tosses equals 0, 1, 2, or 3. We divide the count list by the total sum to obtain the probability.

```
def probability():
    count = [0, 0, 0, 0]
    outcomes = [0, 1]
    for i in outcomes:
        for j in outcomes:
            for k in outcomes:
                lst = [i, j, k]
                s = sum(lst)
                if s == 0:
                    count[0] += 1
                elif s == 1:
                    count[1] += 1
                elif s == 2:
                    count[2] += 1
                elif s == 3:
                    count[3] += 1
    print(count)
```

```
    probability = np.divide(count, sum(count))
    print(probability)
    return probability
```

The following code defines the `make_plots` function, which takes three arguments:

- `Xk` is an array of integers
- `custm` is a custom discrete RV taking values in `xk`
- `filename` is the name of a file to which to save the plots

The `make_plots` function creates PMF and CDF plots for the custom RV.

```
def make_plots(xk, custm, filename):
    """
    :param xk: array of integers
    :param custm: custom discrete random variable
    :param filename: file name to save the figure tp
    :return: probability mass function (PMF) and cumulative distribution
function (CDF) plots
    """
    fig, ax = plt.subplots(2, 1)
    ax[0].plot(xk, custm.pmf(xk), 'ro', ms=12, mec='r')
    ax[0].vlines(xk, 0, custm.pmf(xk), colors='r', lw=4)
    ax[0].set_ylabel('PMF')
    ax[1].plot(xk, custm.cdf(xk), 'ro', ms=12, mec='r')
    ax[1].vlines(xk, 0, custm.cdf(xk), colors='r', lw=4)
    ax[1].set_ylabel('CDF')
    plt.show()
    fig.savefig(filename)
```

The following code first creates a custom discrete RV, `custm`, using the computed probability and the `scipy.stats.rv_discrete` library and then plots the PMF and CDF plots for `custm` using the `make_plots` function.

```
if __name__ == '__main__':
    pk = probability()
    xk = [0, 1, 2, 3]
    custm = stats.rv_discrete(name='custm', values=(xk, pk))
    make_plots(xk, custm, 'rv_discrete.png')
```

Figure 2.2 depicts the plots resulting from running the code.

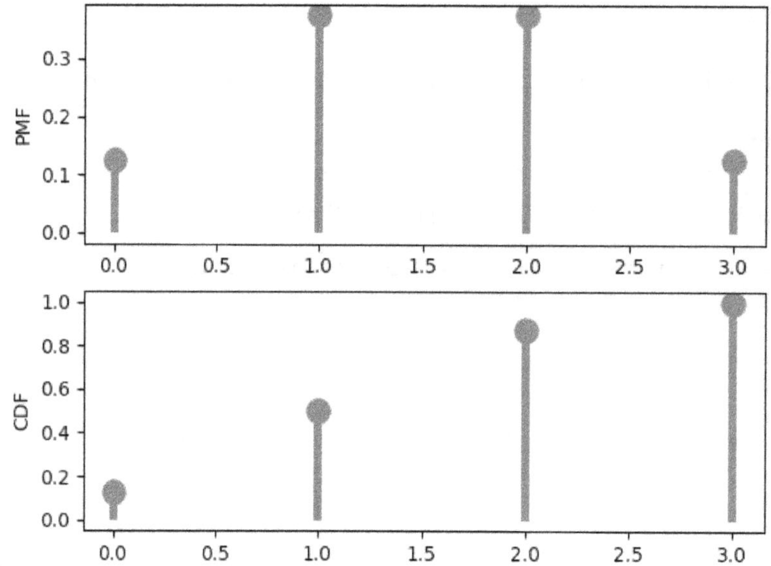

Figure 2.2: Number of heads in three fair coin tosses – PMF and CDF plots

Next, we create code using R to compute the PMF and the CDF for the number of heads in 3 fair coin tosses, as we did using Python.

Practice in R

The `rv_discrete.R` file has the entire code in R for *Example 2.3*. The code follows the same program logic as previously explained in the Python code. Here is a listing of the R program and the results from running the program, which is the PMF and the CDF for the number of heads in three fair coin tosses, identical to the distribution plotted in *Figure 2.2*.

```
probability <- function() {
  count <- c(0, 0, 0, 0)
  outcomes <- c(0, 1)
  for (i in outcomes) {
    for (j in outcomes) {
      for (k in outcomes) {
        lst <- c(i, j, k)
        s <- sum(lst)
        if (s == 0) count[1] <- count[1] + 1
        else if (s == 1) count[2] <- count[2] + 1
        else if (s == 2) count[3] <- count[3] + 1
        else if (s == 3) count[4] <- count[4] + 1
      }
    }
```

```
  }
  return(count / sum(count))
}

pmf = probability()
cdf <- c(pmf[1])
for (i in 2:length(pmf)) {
  cdf[i] <- cdf[i-1]+pmf[i]
}
```

The results from running the preceding R script are as follows:

```
> print(pmf)
[1] 0.125 0.375 0.375 0.125
> print(cdf)
[1] 0.125 0.500 0.875 1.000
```

Now that we have defined probability for discrete RVs, let's cover the definitions for continuous RVs next.

Continuous RVs

A continuous RV takes values in the set of real numbers $\mathbb{R}$. A continuous variable has contiguous values over an interval of the real line between $-\infty$ and $+\infty$. The probability of a continuous RV taking on a specific value x is zero for all x. This is unlike the discrete RV, for which $P(X = x)$ is well defined.

Definition 2.6: Probability Density Function (PDF)

The PDF of a continuous RV X is a function $f: \mathbb{R} \longrightarrow [0,1]$ with the following properties:

$$f(x) \geq 0, \forall x \in \mathbb{R}$$

$$\int_{-\infty}^{+\infty} f(x) = 1$$

Definition 2.7: Cumulative Density Function (CDF]

The CDF of a continuous RV is given by

$$F(x) = P(X \leq x) = \int_{-\infty}^{x} f(x)dx \tag{2.12}$$

Multivariate probability

In the general case, we have more than one RV, and we need to generalize the definitions for the univariate probability. The following are the definitions for the case of multivariate probability.

Discrete RVs

We'll now extend the PMF and CDF definition to the case of joint discrete RVs. A probability model generally consists of multiple discrete RVs that are probabilistically interdependent. We can derive the univariate distribution for a variable from its joint distribution by marginalizing all other variables in the joint distribution, which means summing the joint probability over all value domains of all other variables.

Definition 2.8: Joint Probability Mass Function

Let $X_1, \cdots, X_n$ be a set of n random variables having discrete-valued domains $dom(X_1), \cdots, dom(X_n)$. The joint PMF is defined as

$$f(x_1, \cdots, x_n) = P(X_1 = x_1, \cdots, X_n = x_n), \qquad x_i \in dom(X_i)\ \forall i \quad (2.13)$$

$f(x_1, \cdots, x_n)$ can be represented by a formula or a table that assigns a probability for every value tuple $(x_1, \cdots, x_n) \in dom(X_1) \times \cdots \times dom(X_n)$. $f(x_1, \cdots, x_n)$ must satisfy the following two conditions:

$$f(x_1, \cdots, x_n) \geq 0 \text{ for all } i, x_i \in dom(X_i)$$

$$\sum_{(x_1, \cdots, x_n) \in dom(X_1) \times \cdots \times dom(X_n)} f(x_1, \cdots, x_n) = 1$$

The summation in (b) extends over all joint value tuples of the n RVs from their respective domains.

Definition 2.9: Joint Cumulative Distribution Functions

Let $X_1, \cdots, X_n$ be a set of n RVs having discrete-valued domains $dom(X_1), \cdots, dom(X_n)$. The joint CDF is defined as

$$\boldsymbol{F(x_1, \cdots, x_n) = P(X_1 \leq x_1, \cdots, X_n \leq x_n)} \quad (2.14\text{-a})$$

Here,

$$\boldsymbol{P(X_1 \leq x_1, \cdots, X_n \leq x_n) = \sum_{\substack{(t_1, \cdots, t_n) \in dom(X_1) \times \cdots \times dom(X_n) \\ t_i \leq x_i \forall i}} f(t_1, \cdots, t_n)} \quad (2.14\text{-b})$$

Having defined multivariate probability for discrete variables, let us define the same for continuous variables next.

Continuous RVs

We similarly extend the PDF and CDF definition to the joint continuous variable case as we did for discrete RVs in the previous section. As before, a continuous RV takes values in the set of real numbers. The joint probability density function and cumulative density function describe a probability model for n continuous RVs on the value domain $\mathbb{R}^n$.

Definition 2.10: Joint Probability Density Function (PDF)

The joint PDF of a set of continuous RVs $X_1, \cdots, X_n$ is a function $f: \mathbb{R}^n \longrightarrow [0,1]$ with the following properties:

$$f(x_1, \cdots, x_n) \geq 0 \text{ for all } (x_1, \cdots, x_n) \in \mathbb{R}^n$$

$$\int_{-\infty}^{+\infty} \cdots \int_{-\infty}^{+\infty} f(x_1, \cdots, x_n) dx_1 \cdots dx_n = 1$$

Definition 2.11: Joint Cumulative Density Function (CDF)

The joint CDF of a continuous random variable is given by

$$F(x_1, \cdots, x_n) = P(\boldsymbol{X}_1 \leq \boldsymbol{x}_1, \cdots, \boldsymbol{X}_n \leq \boldsymbol{x}_n) = \int_{-\infty}^{t_1} \cdots \int_{-\infty}^{t_n} f(x_1, \cdots, x_n) dx_1 \cdots dx_n \quad (2.15)$$

In the next section, we will present basic summary statistics that define important statistical parameters in practice.

Summary statistics

Having defined the probability models for univariate and multivariate cases, you will learn how to calculate and interpret summary statistics of discrete and continuous RVs, including expected value, variance, covariance, and Pearson correlation coefficient, in this section.

Definition 2.12: expectation

The **expected value**, also called the **expectation**, **average**, or **mean**, of X is given by

$$\mu = E[X] = \begin{cases} \sum_i x_i P(X = x_i) = \sum_i x_i f(x_i) & \text{if } X \text{ is discrete} \\ \int_{-\infty}^{+\infty} X f(X)\, dX & \text{if } X \text{ is continuous} \end{cases} \quad (2.16)$$

The expected value for any real-valued function $g(X)$ of the discrete RV X, is given by

$$E[g(X)] = \begin{cases} \sum_i g(x_i)\, P(X = x_i) & \text{if } X \text{ is discrete} \\ \int_{-\infty}^{+\infty} g(X) f(X)\, dX. & \text{if } X \text{ is continuous} \end{cases}$$

Intuitively, the expected value of an RV is the asymptotic limit of the arithmetic mean of the RV when sampled a large number of times. From the law of large numbers, sampling the RV n times and computing the mean value of n samples, the absolute value of the error between the mean value of the n samples and the expected value of the RV goes to zero with increasing certainty as n goes toward infinity.

Definition 2.13: variance

Let X be an RV and let $\mu = E[X]$, then the *variance* of X is given by

$$Var(X) = E[(X - \mu)^2] \quad (2.17)$$

We can rewrite the preceding equation in another, easier form,

$$Var(X) = E[X^2] - 2\mu E[X] + \mu^2 = E[X^2] - \mu^2 \quad (2.18)$$

Definition 2.14: standard deviation

The *standard deviation* of X is the square root of its variance, and is given by

$$\sigma(X) = \sqrt{Var(X)} \quad (2.19)$$

X's variance and standard deviation are measures of the "spread-out" of its PMF or density function – the higher the variance, the higher the spread-out. The standard deviation measures the dispersion of values of the RV around its expected value. If an RV X has a normal (Gaussian) probability distribution, then on average, 95% of X's values are within a window of two standard deviations. The statistical realization of an RV will be scattered widely around its mean value if its standard deviation is high, and be more concentrated near the mean if the standard deviation is slight.

Definition 2.15: covariance

The covariance between two random variables X and Y is given by

$$Cov(X,Y) = E[(X-\mu_X)(Y-\mu_Y)] = \begin{cases} \sum_x \sum_y (X-\mu_X)(Y-\mu_Y) f(X,Y). \text{ if } X, Y \text{ are discrete} \\ \int_{-\infty}^{+\infty} \int_{-\infty}^{+\infty} (X-\mu_X)(Y-\mu_Y) f(X,Y)\, dx\, dy, \text{ if } X, Y \text{ are continuous} \end{cases} \quad (2.20)$$

The covariance between two RVs measures their interdependence. If two RVs are independent, then their covariance is zero. The covariance is a positive or negative number ranging from minus infinity to plus infinity. As the variance measures the variability of a single RV around its mean value, the covariance measures how the variability of two RVs around their mean values is linked. The covariance is positive if two RVs, X and Y, move around their mean values in the same direction; above in one is above in the other, and below in one is below in the other. The covariance is negative if two RVs, X and Y, move around their mean values in the inverse direction; above in one is below in the other, and below in one is above in the other. We also note that the covariance relation is symmetric,

$$Cov(X,Y) = Cov(Y,X) \quad (2.21)$$

Definition 2.16: correlation

The correlation is a statistical measure of the linear association between two RVs X and Y. It is defined by **Pearson's correlation coefficient:**

$$\rho(X,Y) = \frac{Cov(X,Y)}{\sqrt{Var(X)Var(Y)}} \quad (2.22)$$

Like covariance, the correlation measures the interdependency between two RVs. While the covariance takes values between minus infinity and plus infinity, the correlation has values between minus and plus one. The correlation is non-dimensional – as a metric, it does not depend on the units of measurement of the RVs. An RV measuring a distance can be in feet, meters, yards, or millimeters, resulting in order of magnitude differences when quantifying the covariance. Still, not the correlation, which is because the correlation is a quotient whose dividend and divisor have the same units of measurement. Like covariance, the correlation is also symmetric:

$$\rho(X,Y) = \rho(Y,X) \quad (2.23)$$

Having covered the basic definitions of probability and statistics for RVs, let us proceed to the next section to present some elemental distributions of discrete and continuous classes of RVs that are widely used in practice.

Types of RVs

In the following, you will learn about a few types of RVs commonly used in probabilistic modeling of practical applications. You will know their probability distribution and learn how to perform computer simulations with those variables using libraries in R and Python. All the examples for the Python practice in this section are in the `rv_stats.ipynp` Jupyter notebook included in the code folder for this chapter.

Bernoulli RV

A coin toss is an example of a Bernoulli RV, X, that can take only one of two values, 0 or 1, with probabilities $P\ (X\ =\ 1) =\ p$ and $P\ (X\ =\ 0) =\ 1\ -\ p$. The Bernoulli distribution has the following properties:

- $E[X] = p$
- $Var(X) = p(1 - p)$

Practice in Python

We use the `scipy.stats` library to simulate a Bernoulli(p) random variable, X. The following code generates an **Independent and Identically Distributed** (**IID**) sequence of Bernoulli(p) random variables. The series can simulate tossing a biased coin for which p=0.3. The following code generates a series of size 10.

```
from scipy.stats import bernoulli
r = bernoulli.rvs(p=0.3, size=10)
print(r)
```

The result is as follows:

```
[1 1 0 0 0 0 1 0 1 0]
```

The following code gets the mean and variance:

```
mean, var = bernoulli.stats(p=0.3, moments='mv')
print(f'mean = %.4f' % mean)
print(f'variance = %.4f' % var)
```

The result is as expected from the Bernoulli RV properties:

```
mean = 0.3000
variance = 0.2100
```

Binomial RV

A $Binomial(n, p)$ random variable X counts the number of successes in an experiment repeated n times, where the outcome each time is binary, either success or failure. If we flip a coin n times whose probability of turning heads is p, then the total number of heads in n tosses is a binomial(n, p). The binomial RV has the following properties:

- Its PMF is the function: $f(x) = P(X = x) = \binom{n}{x} p^x (1-p)^{n-x}, x = 0,1,\cdots,n$
- $E[X] = np$
- $Var(X) = np(1-p)$

Practice in Python

Let us recall *Example 3*: X is the number of heads in 3 fair coin tosses. We derived the PMF and the CDF from simple counting. Let us verify the previous results by simulating many repetitions of coin flips 3 times and taking the averages for the number of heads, being 0, 1, 2, and 3, respectively. We do this using the `Numpy.random.binomial` library as follows:

```
import numpy as np
n, p = 3, .5  # number of trials, probability of each trial
N = 200000
s = np.random.binomial(n, p, N)
# result of flipping a coin 3 times, tested 1000 times.
f0 =sum(np.random.binomial(n, p, N) == 0)/N
f1 =sum(np.random.binomial(n, p, N) == 1)/N
f2 =sum(np.random.binomial(n, p, N) == 2)/N
f3 =sum(np.random.binomial(n, p, N) == 3)/N
print(f0,f1,f2,f3)N
```

The result is a very close approximation to the exact probability mass function.

```
0.124325 0.374195 0.37572 0.12494
```

For the sake of practice, we also query for the mean and variance.

```
from scipy.stats import binom
mean, var = binom.stats(n, p , moments='mv')
print(f'mean = %.4f' % mean)
print(f'variance = %.4f' % var)
```

The result is as expected.

```
mean = 1.5000
variance = 0.7500
```

Poisson RV

A Poisson RV models the occurrences of events in phenomena where the happenings are all random and independent. Examples of Poisson RVs are arrivals at a shopping mall, phone calls to customer service, traffic on a bridge, machine breakdown, and so on.

A $Poisson(\lambda)$ random variable X with parameter $\lambda > 0$ represents the number of occurrences per time interval for which the average rate of occurrence is λ. The Poisson RV has the following properties:

$$P(X = x) = \frac{\lambda^x}{x!}e^{-\lambda}, \text{ for } x = 0,1,2,\cdots$$

$$E[X] = \lambda$$

$$Var(X) = \lambda$$

Practice in Python

The following code generates an IID sequence of 10 random variables having Poisson probability distribution with parameter $\lambda = 3$, meaning the average rate of arrivals or event occurrences is 3 per unit interval.

```
from scipy.stats import poisson
lam = 3
size = 10
r = poisson.rvs(mu=lam, size=10)
print(r)
```

The result is as follows:

```
[5 2 1 6 2 3 4 4 4 3]
```

Let us verify the Poisson RV mean and variance and compute the probability that more than three arrivals can occur per unit interval.

```
mean, var = poisson.stats(lam, moments='mv')
print(f'mean = {mean}, variance = {var}')
print(f'P(#arrovals>3)={(1- poisson(lam).cdf(3))}')
```

The results are as expected.

```
mean = 3.0, variance = 3.0
P(#arrovals>3)=0.35276811121776874
```

Notice that the probability that the Poisson RV has values higher than the mean value is less than 0.5 due to the skewness of the Poisson distribution. Skewness measures the asymmetry of a probability distribution around its mean. As lambda increases, the skewness decreases. For large lambda values, say greater than 1,000, the Poisson distribution becomes closely approximated by a normal distribution, which we will discuss next.

Geometric RV

Recall a Bernoulli(p) RV with two values: 1 for success and 0 for failure, where the probability of success is p. The Geometric(p) RV is the number of repeated experiments till a Bernoulli(p) RV has the first success value. For example, the number of trials required for the first coin flip to be heads is a Geometric(p) RV where p is the probability of a flip turning heads.

A $Geometric(p)$ random variable X has the following properties:

- $P(X = x) = p(1 - p)^{n-1}, n = 1,2, \cdots$
- $E[X] = \frac{1}{p}$
- $Var(X) = \frac{(1-p)}{p^2}$

Practice in Python

Suppose an oil drilling company estimates that the probability of finding oil in a particular area is 30% – let X be the number of drillings required until hitting the first successful oil wells. Calculate X's expected value, X's variance, and the probability that X will be more than 5.

```
from scipy.stats import geom
p= 0.3
X = geom(p)
print(f'E[X] = %.3f' % X.mean())
print(f'Var[X] = %.3f' % X.var())
print(f'Prob(X > 5)= %.3f' % (1-X.cdf(5)))
```

The results are as follows:

```
E[X] = 3.333
Var[X] = 7.778
Prob(X > 5)= 0.168
```

Normal RV

So far, we have dealt with probability distributions for discrete RVs. The normal probability distribution, called Gaussian, describes the value distribution of a continuous RV. The normal distribution has a bell shape, indicating that most values of the random variable c <Figure size 640x480 with 2 Axes>luster around the mean value. Earlier, we discussed the Poisson RV and mentioned the skewness of the Poisson probability mass function. By contrast, the normal probability density function is symmetric around the mean value. The normal distribution is the most significant probability distribution as it describes many natural phenomena. Also, the normal distribution represents the limiting behavior of the average of independent samples drawn from a probability distribution including non-normal distributions for a large number of samples.

A $Normal(\mu, \sigma)$ RV X takes value in $\mathbb{R}$ and has the mean value μ and the standard deviation σ. The normal RV is very useful as it is commonly used to model many real-world phenomenon. The normal RV has the following properties:

- $f(x) = \frac{1}{\sigma\sqrt{2\pi}} e^{-\frac{(x-\mu)^2}{2\sigma^2}}$
- $E[X] = \mu$
- $Var(X) = \sigma^2$

Practice in Python

The following code generates 10 random values from a normal distribution with mean = 5 and standard deviation = 2.5.

```
from scipy.stats import norm
mean = 5
std = 2.5
r = norm.rvs(mean, std, size=10)
print(r)
```

The result is the following list:

```
[4.58532401 4.33852441 3.6402468  3.03022195 4.67909111 3.11024266 8.18897661
5.57203407 4.09370954 6.37413483]
```

We now plot the probability density and cumulative distribution functions. The plots illustrate the properties of normal distributions mentioned previously: a bell shape that peaks at the mean value, symmetry of the density function, and most values cluster in a small window around the mean value. The following is the code to generate the plots.

```
import numpy as np
import matplotlib.pyplot as plt
from scipy.stats import norm
mean = 5
std = 2.5
#x-axis ranges from -3 and 3 with .001 steps
x = np.arange(mean-4*std, mean+4*std, 0.001*std)
y1 = norm.pdf(x, mean, std)
y2 = norm.cdf(x, mean, std)
fig, (ax1, ax2) = plt.subplots(1, 2)
fig.suptitle('Normal density and cumulative distribution functions')
ax1.plot(x, y1)
ax1.set_title('PDF')
ax2.plot(x, y2)
ax2.set_title('CDF')
```

```
plt.show()
fig.savefig('normal.png')
```

Figure 2.3 depicts the output plots that result from running the preceding code.

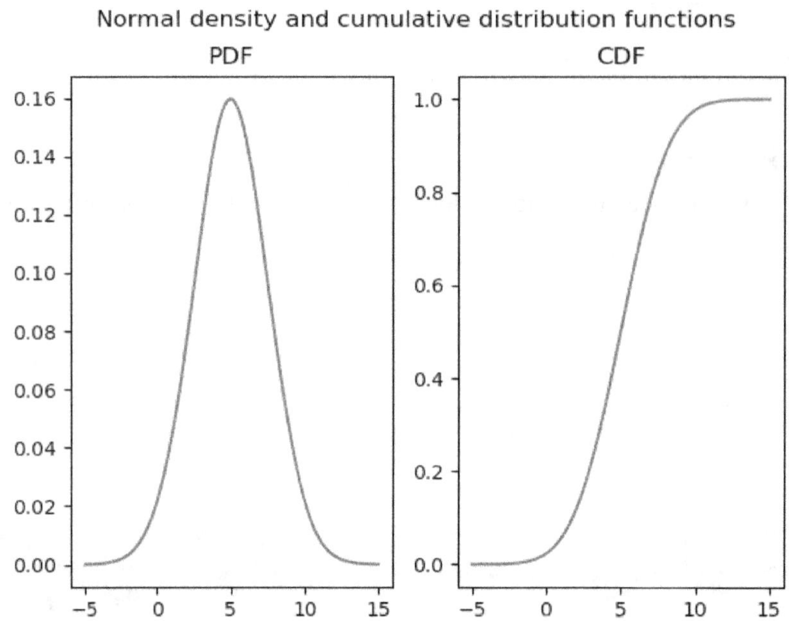

Figure 2.3: The Python code output PDF and CDF plots of normal RV Normal(5,2.5)

Let us now compute the quantile of the normal distribution; the quantile is the inverse of the cumulative distribution function. The following code determines the 0.025 and 0.975 quantiles, which returns the values for X for which the CDF equals 0.025 and 0.975, respectively. The quantiles give a window's lower and upper bounds containing 95% of X's data points.

```
upper = norm.ppf(0.975, loc=5, scale=2.5)
lower = norm.ppf(0.025, loc=5, scale=2.5)
print(f'95% of values are between {lower} and {upper}')
```

The result is as follows:

```
95% of values are between 0.10009003864986354 and 9.899909961350135
```

It is known that 95% of data points of normal RVs lie within the window $\mu \pm 1.96\sigma$, which is consistent with the computed result.

Let us highlight the main lessons learned in this section. RVs can be discrete or continuous. Univariate probability describes the probability distribution of a single RV, and multivariate probability represents the joint distribution of multiple RVs. A probability mass function defines the probability of discrete RV over its discrete value domain. A continuous RV has a probability density function over the real-valued domain. The sum of the PMF for discrete variables and the integral of the PDF for continuous variables gives the cumulative function that defines the probability that an RV does not exceed a specific value from its domain. We learned the formulas for RV summary statistics: the mean, the variance, the standard deviation, the covariance, and the correlation, and we discussed the meaning of those statistics. Finally, we reviewed several of the most practical probability distributions: Bernoulli, Binomial, Poisson, Geometric, and Normal. We acquired hands-on experience working with RVs using examples and executable code in R and Python.

In the following section, we will learn about conditioning and conditional probability, central to what comes next, when we learn about Bayes' theorem.

Conditioning and conditional probability

In this section, we will define what conditioning is – what it means that event A is conditioned on event B. You will learn about the mathematical basis and the interpretation of conditional probability. We will derive formulas for conditional probability conditioning on multiple events.

Conditioning

Conditioning event A on event B means that the occurrence of B carries information that influences the likelihood of A. Conditioning implies a probabilistic dependency between A and B and an association between A and B that `cov(A, B)>0`. An example of conditioning is when the earning report for a company is released; the information that the earnings are up or down influences the stock share price for the company. Another example is the association between symptoms and health outcomes; symptoms such as fatigue, pain, and shortness of breath are some conditions that influence the odds ratio of quality of life and the odds of functional disability (Walke LM 2007).

Conditioning on an event means restricting the sample space to include only outcomes that are consistent with the conditioning event, and to exclude all outcomes that are otherwise inconsistent. An outcome is consistent with an event if the outcome is an element in the set of outcomes for the event and is inconsistent otherwise.

Figure 2.4 is a visualization of conditioning an event A on an event B. The oval Ω is the sample space, and the ovals A and B represent the outcomes for the respective events. Conditioning on B means entering the universe where B has happened (sample space $\Omega' = B$). The subset of outcomes in A in the new sample space is the event E, the set intersection $A \cap B$. We write the event E as $A|B$.

Having interpreted the meaning of conditioning an event A on an event B, we proceed next to define the conditional probability of A given B formally.

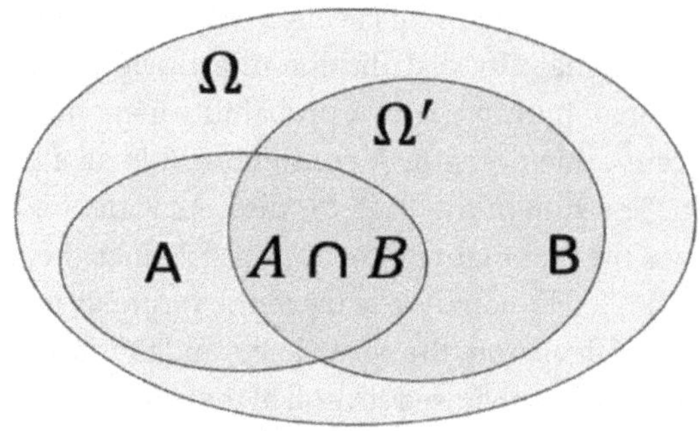

Figure 2.4: Event A conditioned on event B

Now that the concept of conditioning on events in sample spaces is clear, we can present the formal definition of conditional probability in the next section.

Conditional probability

As the previous section showed, conditioning on event B means two things. One is changing the original probability space from $(\Omega, \mathcal{F}, P)$ to $(\Omega', \mathcal{F}', Q)$, which we call the conditioning probability space where $\Omega' = B$, and $\mathcal{F}'$ is the event space over Ω'. The other is mapping the probability measure in the initial probability space to a new probability measure Q in the conditioning probability space. The mapping is such that Q respects all probability axioms, including that the probability of the set of outcomes spanning the entire conditioning sample space equals 1. The following equation defines the mapping from P to Q:

$$Q(E) = \frac{P(E)}{P(B)} \ \forall\, E \in \mathcal{F}' \quad (2.24)$$

The "scaling" of the original probability measure P is simply to ensure that $Q(\Omega') = 1$, by virtue of the probability axioms.

We use $P(A|B)$, a syntactic sugaring of the equivalent $Q(A \cap B)$, to denote the conditional probability of an event A given an event B. *Equation 2.24* leads us to conclude that for any two events, A and B, such that $P(B) > 0$, we can define the conditional probability of A given B with the following equation:

$$P(A|B) = \frac{P(A \cap B)}{P(B)} \quad (2.25)$$

Given a conditioning event B, the conditional probability $P(.|B)$ satisfies the three probability axioms, and all consequences of the axioms are valid for it. The conditional probability follows the same axioms of probability and all the laws derived from those axioms (e.g., *Equations 2.1* through *2.7*).

In cases where conditional probabilities are given, we can rewrite *Equation 2.25* to express the joint probability of two events in terms of their conditional probability,

$$P\,(A\ \cap\ B) = P\,(A|B)P(B) \quad (2.26)$$

A similar equation can be derived for the joint probability of three or more events. From *Equation 2.26*, we can write the joint probability of events $A, B. C$, as follows:

$$P(A \cap B \cap C) = P(A \cap B|C)P(C)$$

Applying *Equation 2.26* once more, we can rewrite the preceding equation as:

$$P(A \cap B \cap C) = P(A|B,C)P(B|C)P(C) \quad (2.27)$$

By induction, we can derive a general form of the preceding equation for any number of n events $A_1, \cdots, A_n$,:

$$P(A_1 \cap A_2 \cap \cdots \cap A_n) = P(A_1|A_2, \cdots, A_n)P(A_2|A_3, \cdots, A_n) \cdots P(A_{n-1}|A_n)P(A_n) \quad (2.28)$$

The pattern in the preceding equations applies to express the **joint probability** of any number of events in terms of respective conditional probabilities. The pattern is called **the chain rule.**

The chain rule is valid for any ordering of n variables. The order of the n variables will generally depend on the modeling of the probabilistic dependencies between the variables and the available conditional probabilities.

$P(A|B,C)$ denotes **conditioning on multiple events**, representing the probability of A occurring given that both B and C have occurred. We can apply the conditioning formula in *Equation (2.25)* on events that are themselves conditioning events! So, we can write $P(A|B,C)$ as the conditional probability of A conditioned on B|C, as follows:

$$P(A|B,C) = \frac{P(A \cap B|C)}{P(B|C)} \quad (2.29)$$

This equation states that the definition for the conditional probability of $P(A|B)$ still applies in the universe where C has occurred.

We have learned that conditional probability is mapping a probability space prior to conditioning to a conditioning probability space in which sample space contains only outcomes consistent with the conditioning event. Conditional probability follows all the laws of probability. We derived the mathematical formulas for conditional probability conditioning on multiple events. We learned about the chain rule, which expresses the joint probability using conditional probabilities. In later chapters, you will see many uses of the chain rule when dealing with Bayesian networks.

Conditioning and conditional probability are two core concepts central to Bayes' theorem, which we will explain later. Having understood the notion of conditioning on events, we are now ready to move on to formalize the concept of independence and conditional independence between multiple events, which we discuss next.

Characterizing independence and conditional independence

Firstly, you will learn about independence between multiple RVs. In the previous section, we learned about conditional probability, setting the stage for characterizing conditional independence between sets of RVs, which is the principal theme of this section. Conditional independence is a topic that represents a central thread in theory and computations with Bayesian networks, a theme that will recur throughout the following chapters.

Independent events

Let A and B be two events, $A, B \subseteq \Omega$, and $P(A \cap B)$ be the joint probability of A and B. Let $P(A)$ and $P(B)$ be the marginal probabilities of A and B, respectively. We can compute $P(A)$ by marginalizing $P(A \cap B)$ over B and similarly $P(B)$ by marginalizing $P(A \cap B)$ over A. We say that A and B are independent if:

$$P(A \cap B) = P(A)P(B) \quad (2.30)$$

To denote that two events A and B are independent, we write, $A \perp B$. Independence is a symmetric property so $A \perp B$ is equivalent to $B \perp A$,:

$$A \perp B \Leftrightarrow B \perp A \quad (2.31)$$

It follows from *Equations (2.25) and (2.30)* that if A, B are independent events such that $P(B) > 0$, then $P(A|B) = P(A)$,:

$$A \perp B. \implies P(A|B) = P(A) \quad (2.32)$$

Similarly, we can prove that if A and B are independent and $P(A) > 0$ then $P(B|A) = P(B)$,:

$$A \perp B. \implies P(B|A) = P(B) \quad (2.33)$$

Equations 2.32 and 2.33 interpret independence as meaning that if two events are independent, knowing that either event occurred does not add any information about the other event or change our prior beliefs about the other event.

Example 2.4: independent events

Let X and Y be discrete RVs with binary value domain $\{0, 1\}$. X and Y represent the indicator function that events A and B occur. Consider the joint probability of X and Y, as in *Table 2.5*. The following equations give the marginal probability of X and Y:

$$P(X = x) = \sum_{y \in Y} P(x, y) \quad (2.34)$$

$$P(Y = y) = \sum_{x \in X} P(x, y) \quad (2.35)$$

It follows that $P(X = 0) = P(X = 1) = 0.5$ is the marginal probability for X, and $P(Y = 0) = 0.6, P(Y = 1) = 0.4$ is the marginal probability for Y.

X and Y are not independent because $P(X, Y) \neq P(X)P(Y)$. For example, $P(X = 0, Y = 0) = 0.2$, while $P(X = 0)\,P(Y = 0) = 0.5 * 0.6 = 0.3$.

X = x	Y = y	P(x, y)
0	0	0.2
0	1	0.3
1	0	0.4
1	1	0.1

Table 2.5: Joint probability P(x,y)

We use *Equation 2.25* to compute the conditional probabilities $P(X|Y)$ and $P(Y|X)$. To calculate $P(X|Y)$, we divide $P(X.Y)$ by the marginal probability $P(Y)$, and to compute $P(Y|X)$, we divide $P(X, Y)$ by $P(X)$. *Table 2.6* gives the results of the computations.

X = x	Y = y	P(X = x | Y = y)	P(Y = y | X = x)
0	0	1/3	0.4
0	1	3/4	0.6
1	0	2/3	0.8
1	1	1/4	0.2

Table 2.6: Conditional probabilities P(X|Y) and P(Y|X)

X and Y are not independent because $P(X, Y) \neq P(X)P(Y)$. For example, $P(X = 0, Y = 0) = 0.2$, while $P(X = 0)P(Y = 0) = 0.5 * 0.6 = 0.3$.

To conclude, probabilistic independence between variables means that their joint probability equals the product of the marginal probabilities of all the variables. We also call this marginal independence. Observing one variable (knowing its value assignment) does not change the initial belief about the other variable if two random variables are marginally independent.

Moving on, we are going to learn about conditional independence next.

Conditional independence

We noted earlier that conditioning event A on event B is itself an event, which we denote with $A|B$. Also, the conditional probability is a probability we write as $P(A|B)$. Given this, all we learned in the previous section on probabilistic independence extends to conditional independence.

We say that two events, A and B, are conditionally independent given C if the two conditioning events, $A|C$ and $B|C$, are independent,

$$P(A \cap B|C) = P(A|C)P(B|C) \quad (2.36)$$

We have, from the chain rule,

$$P(A \cap B \cap C) = P(A \cap B|C)P(C) \quad (2.37)$$

Substituting $P(A \cap B|C)P$ from *Equation 2.36*, we get

$$P(A \cap B \cap C) = P(A|C)P(B|C)P(C) \quad (2.38)$$

We also have, from the chain rule,

$$P(A \cap B \cap C) = P(A|B, C)P(B|C)P(C) \quad (2.39)$$

Comparing *Equations (2.38) and (2.39)*, we conclude that A and B being conditionally independent given C implies that

$$P(A|B, C) = P(A|C) \quad (2.40)$$

Since A and B are interchangeable in the steps followed to derive *Equation 2.40*, we can also assert the following as a result of A and B being conditionally independent given C,

$$P(B|A, C) = P(B|C) \quad (2.41)$$

Equation 2.40 means that A is conditionally independent of B given C, and we write $A \perp B|C$. Hence,

$$P(A|B, C) = P(A|C) \Rightarrow A \perp B|C \quad (2.42)$$

Similarly, *Equation 2.41* means that B is conditionally independent of A given C, and we write

$$(B|A, C) = P(B|C) \Rightarrow B \perp A|C \quad (2.43)$$

Thus, conditional independence of A and B given C implies both *Equations 2.42 and 2.43*. Hence, conditional independence of A and B given C is symmetric. If A is conditionally independent of B given C, then also B is independent of A given C, and we write

$$A \perp B|C \Leftrightarrow B \perp A|C \quad (2.44)$$

Let us illustrate the concept of conditional independence with a practical example.

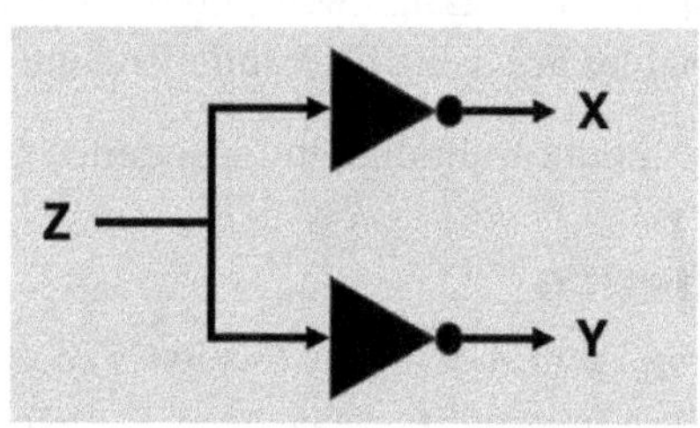

Figure 2.5: Two inverters with shared input

Example 2.5: conditional independence

Figure 2.5 shows a digital circuit of two inverters with shared input. An inverter is a NOT gate that takes a binary signal as input and produces the inverse of the logic state of the input. If the input is one, it outputs 0; if the input is 0, it outputs 2.5. In practice, logic gates can be unreliable and get stuck at 1 or 0 with some small probability. *Table 2.7* shows the conditional probabilities of X and Y given Z. X and Y are conditionally independent, given Z. Once we know the value of the input signal Z, we can ultimately determine the probability of X and Y independently of one another. For example, if Z=0, we know the probability of X=0 is 0.05 and X=1 is 0.95, and we also see the probability of Y=0 is 0.1 and Y=1 is 0.9; this implies that once we know the value assignment of Z, we can determine the probability of the joint event on X and Y. For example, knowing value assignment Z=0, we can calculate the probability of X=0 and Y=1 as the product of the probability of X=0, which equals 0.05, by the probability of Y=1, which equals 0.9, to obtain the result of 0.045.

Z	X	P(X|Z)	Y	P(Y|Z)
0	0	0.05	0	0.1
0	1	0.95	1	0.9
1	0	0.98	0	0.85
1	1	0.02	1	0.15

Table 2.7: Conditional probabilities of two inverters

Let us recap the lessons of this section. Marginal independence between two or more variables means the joint distribution of all variables equals the product of the marginal probabilities of all variables. Naturally, we can derive the marginal probabilities from the joint probability, but having the joint distribution restored from the marginals is possible only when the variables are independent. Marginal independence between two random variables, X and Y, implies that $P(X|Y) = P(X)$ and $P(Y|X) = P(Y)$, meaning knowing the value instantiation of one variable does not bring any added information about the probability distribution of the other variable. Conditional independence extends the concept of marginal independence to involve additional variables, the conditioning variables. Having three variables X, Y, and Z, we say that X and Y are conditionally independent given Z if the joint probability of X and Y conditioning on Z equals the product of the probabilities for X conditioning on Z and Y conditioning on Z. We learned that conditional independence, like marginal independence, is a symmetric relation. If X is independent of Y conditioning on Z, Y is independent of X conditioning on Z. We worked out two examples of marginal and conditional independence.

In the next section, we will learn about Bayes' theorem, building on what we have learned so far, particularly the topic of conditioning and conditional probability. Bayes' theorem is a cornerstone of the mathematical framework underlying theory and computation with Bayesian networks.

Understanding Bayes' theorem

First, we will introduce the law of total probability, which sets the stage for Bayes' theorem. Bayes' theorem builds on what we learned about conditional probability and consists of a formula that "reverses" the conditioning on multiple hypotheses. Following this, you will learn about the interpretation of Bayes' theorem as a mechanism for belief revision. You will see an implementation of Bayes' theorem in R and Python and use the implementation for hands-on practice with an example of belief revision.

Deriving the law of total probability

The law of total probability is a fundamental rule that enables us to express the probability of an event in terms of its conditional probabilities, where the conditioning is on a mutually exclusive set of events that spans the entire sample space.

To gain the central intuition, let A, B be two events and let B^c be the complement event of B. We can write

$$P(A) = P\big(A \cap (B \cup B^c)\big) = P(A \cap B) + P(A \cap B^c) \quad (2.45)$$

Using the chain rule, *Equation 2.26*, we can rewrite the preceding equation as

$$P(A) = P(A|B)P(B) + P(A|B^c)P(B^c) \quad (2.46)$$

The law of total probability generalizes the equation to the general case where we express the probability of an event in terms of its conditional probabilities of a set of pairwise disjoint events that partition the entire sample space.

Theorem: the law of total probability

Consider a set of pairwise disjoint events $B_1, \cdots, B_n$ spanning the entire sample space $B_1 \cup \cdots \cup B_n = \Omega$; precisely one of the events in the set always happens. We can express the probability of an event A as follows:

$$P(A) = P(A|B_1)P(B_1) + P(A|B_2)P(B_2) + \cdots + P(A|B_n)P(B_n) \quad (2.47)$$

Proof

From the chain rule, *Equation 2.26*, we can rewrite *Equation 2.47* as follows:

$$P(A) = P(A \cap B_1) + \cdots + P(A \cap B_n) \quad (2.48)$$

From the inclusion-exclusion formula, *Equation 2.7*, and the fact that $B_1, \cdots, B_n$ are pairwise disjoint, we rewrite *Equation 2.48* as follows:

$$P(A) = P\big((A \cap B_1) \cup \cdots \cup (A \cap B_n)\big) \quad (2.49)$$

From the algebra of sets on unions and intersections, we rewrite *Equation 2.49* as:

$$(A) = P\big(A \cap (B_1 \cup \cdots \cup B_n)\big) \quad (2.50)$$

Since $B_1 \cup \cdots \cup B_n$ equals the entire sample space Ω, we have:

$$P(A) = P(A \cap \Omega) \quad (2.51)$$

Hence, the law of total probability is valid, and the theorem is true.

Having derived the law of total probability, we will define Bayes' theorem next.

Defining Bayes' theorem

We can intuitively describe Bayes' theorem, which we will formally define next, as a mathematical formula for "reversed conditioning." The theorem considers pairwise disjoint events representing a complete set of explanatory hypotheses for an event labeled as evidence. The theorem provides a formula that takes as input a) the conditional probabilities of the evidence conditioning on each explanatory hypothesis, b) the prior probabilities of all explanatory hypotheses, and computes as output the posterior probabilities for all explanatory hypotheses.

Each posterior probability represents a "reversal of conditioning" as the posterior probability at the output conditions the hypothesis on the evidence. In contrast, the input conditional probabilities have the hypotheses conditioned on the evidence.

Theorem: Bayes' theorem

Consider a set of pairwise disjoint events $B_1, \cdots, B_n$ spanning the entire sample space $B_1 \cup \cdots \cup B_n = \Omega$; precisely one of the events in the set always happens. Given the **prior probabilities** $P(B_1), \ldots, P(B_n)$ and the conditional probabilities $P(A|B_1), \ldots, P(A|B_n)$ of an event A conditioning on each event $B_1, \cdots, B_n$ the **posterior probability** of each B_j given A is as follows:

$$\boldsymbol{P(B_j|A) = \frac{P(A|B_j)P(B_j)}{P(A)} = \frac{P(A|B_j)P(B_j)}{P(A|B_1)P(B_1) + \cdots + P(A|B_n)P(B_n)}} \quad (2.52)$$

Here, event B_j represents a **hypothesis**, $P(B_j)$ is its **prior probability**, and $P(B_j|A)$ is its **posterior probability**. $P(A)$ is the unconditional probability of event A.

Bayes' theorem provides a mechanism by which we can "revise" our probabilistic beliefs about hypotheses concerning the state of the world in light of the empirical evidence that accumulates.

The set of pairwise disjoint events $B_1, \cdots, B_n$ in Bayes' formula can be interpreted as all possible **hypotheses** about events causing A or explaining A. Given the prior probabilities of the alternative hypotheses and the evidence provided by event A, Bayes' theorem enables updating the probabilities of the various hypotheses. The update changes the probabilities of the hypotheses from the priors, $P(B_j)$ to the posteriors, $P(B_j|A)$.

We will elaborate next on how Bayes' theorem enables operationalizing belief revision.

Operationalizing belief revision

Let $P(H)$ be the prior probability of a hypothesis, H. For example, a patient has a particular illness. In this case, a doctor assigns a prior probability based on his background knowledge and assessment from examining the patient. The prior probability is subjective and thus can be assessed differently by different doctors.

Let the evidence be the random variable E, (e.g., the positive test result). Let the conditional probability $P(E|H)$ denote the conditional probability of E given H, (i.e., the test is positive given that illness is present. Let $P(E|H^c)$ be the probability that the test is positive, given that the illness is absent. Since H^c is the complement of H, the sum of their probabilities must equal one. Hence,

$$P(H^c) + P(H) = 1 \quad (2.53)$$

Using the law of total probability, *Equation 2.47*, the probability of the evidence, P(E), is determined by the equation:

$$P(E) = P(E|H)P(H) + P(E|H^c)P(H^c) \quad (2.54)$$

Substituting for $P(H^c)$ from *Equation 2.53*, we get

$$P(E) = P(E|H)P(H) + P(E|H^c)(1 - P(H) \quad (2.55)$$

Bayes' theorem updates the probability of H from the prior $P(H)$ to the posterior probability:

$$P(H|E) = \frac{P(E|H)P(H)}{P(E)} \quad (2.56)$$

$P(H|E)$ represents the updated belief that the patient has the illness, given that the test comes back positive.

The posterior probability $P(H|E)$ incorporates newly acquired knowledge from the observed events. This process of belief revision reflects a learning process, and many machine learning algorithms are based on Bayes' theorem, as we will touch on in a later section.

Example 2.6: belief revision

Assume a doctor believes there is a 30% chance the patient has a specific illness. The doctor orders a diagnostic test. The test returns a positive result 95% of the time for patients with that illness. The test yields a positive result 7% of the time for people who do not have that illness. Assuming the test is positive, what is the posterior probability that the patient has that illness?

For this example, we have two hypotheses:

H: the patient has a particular illness

H^c: the patient does not have that particular illness

Clearly, the two hypotheses represent two events spanning the entire sample space, as H^C is the complement of H concerning the sample space Ω. The prior probabilities from the data of *Example 2.6* are as follows:

$$P(H) = 0.30$$

$$P(H^c) = 1 - P(H) = 0.70$$

The test result E represents the evidence whose result was positive. From the data in the example, the conditional probability that the test result is positive given that the illness is present is 0.95, and the conditional probability that the test result is positive, conditioned on the illness being absent, is 0.07. That is:

$$P(E|H) = 0.95$$

$$P(E|H^c) = 0.07$$

We plug the numbers in *Equation 2.55* to compute the unconditional probability that the test is positive:

$$P(E) = P(E|H)P(H) + P(E|H^c)P(H^c) = 0.95 \times 0.3 + 0.07 \times 0.7 = 0.334 \quad (2.56)$$

We apply Bayes' formula, *Equation 2.56*, to compute the posterior probability that the patient has the particular disease given that the test was positive.

$$P(H|E) = \frac{P(E|H)P(H)}{P(E)} = \frac{0.95 \times 0.3}{0.334} = 0.853$$

Hence, the evidence from the test result bolsters the doctor's initial belief that the patient has that illness, increasing from a 30% chance at the outset to an 85% chance following the positive test result.

Practice in Python

The following code implements Bayes' formula from *Equation 2.52*. The code is part of the `bayes.py` Python file included in the Python project for this chapter on GitHub.

```
import numpy as np


def bayes_theorem(k, p_a_cond_b, p_b):
    """
    :param k: hypothesis index between 0 and n-1,
              where n is the number of hypotheses
    :param p_a_cond_b: probability of A conditioned
                       on the n hypotheses [P(A|B1),..,P(A|Bn)]
    :param p_b: probabilities of n hypotheses [P(B1),...,P(Bn)
    :return: the posterior probability of the kth hypothesis given A
    """
    x = np.array(p_a_cond_b)
    y = np.array(p_b)
```

```
    # calculate P(A)
    p_a = sum(x * y)
    return (p_a_cond_b[k] * p_b[k]) / p_a
```

To illustrate how to use the code to get the posterior probability that implements Bayes' formula, we input the data of *Example 2.6*, where we have an event A (the test) and two explanatory events, B_1 (patient has a particular illness) and its complement B_2 (patient does not have that particular illness). The prior probabilities $P(B_1) = 0.3$ and $P(B_2) = 0.7$. Thus, the input for the list of prior probabilities is `p_b = [0.3, 0.7]`. The conditional probabilities $P(A|B_1)$ and $P(A|B_2)$ correspond respectively to $P(E|H)$ and $P(E|H^c)$, vwhich are the conditional probabilities that the test is positive given that the patient has that particular illness, which equals 0.95, and 0.07 given that the patient does not have that particular illness. Thus, the input for the list of conditional probabilities is `p_a_cond_b = [0.95, 0.07]`. We call `bayes_theorem(0, p_a_cond_b, p_b))` to get the posterior probability for the first hypothesis $P(B_1|A)$, which corresponds to $P(H|E)$ and `bayes_theorem(1, p_a_cond_b, p_b))` to get the posterior probability for the second hypothesis $P(B_2|E)$, which corresponds to $P(H^c|E)$. The following is the implementation in Python of the steps given in the explanation.

```
p_b = [0.3, 0.7]
p_a_cond_b = [0.95, 0.07]
# calculate P(A|B[0])
print(f'P(A|B[0]) %.2f' % bayes_theorem(0, p_a_cond_b, p_b))
print(f'P(A|B[1]) %.2f' % bayes_theorem(1, p_a_cond_b, p_b))
```

The code output is identical to the expected results per the worked example:

```
P(A|B[0]) 0.853
P(A|B[1]) 0.147
```

Practice in R

You will find the following code in the R script of the `bayes.R` file included in this chapter's folder on GitHub. The code follows the same program logic previously explained for the Python code. Running the script, we get the same result for the posterior probabilities of the patient having the particular illness or not having the illness, given that the test result is positive.

```
bayes_theorem <- function(k, p_a_cond_b, p_b) {
  x <- as.numeric(p_a_cond_b)
  y <- as.numeric(p_b)

  p_a <- sum(x * y)
  return((p_a_cond_b[k] * p_b[k]) / p_a)
}

p_b <- c(0.3, 0.7)
p_a_cond_b <- c(0.95, 0.07)
```

```
cat(sprintf("P(A|B[1]) %.3f\n", bayes_theorem(1, p_a_cond_b, p_b)))
cat(sprintf("P(A|B[2]) %.3f\n", bayes_theorem(2, p_a_cond_b, p_b)))
```

When you run the script, the results are the same as expected.

```
P(A|B[1]) 0.853
P(A|B[2]) 0.147
```

In summary, the takeaway points to review what we learned in this section are as follows. Bayes' theorem is a formula that takes as input the prior probabilities on multiple hypotheses, and conditional probabilities of evidence given each of the hypotheses, and calculates as output the posterior probabilities of each of the hypotheses given the evidence. The multiple hypotheses must span the entire sample space, meaning one must happen and be mutually exclusive, saying only one can happen. Bayes' theorem provides a mechanism for updating beliefs by revising the initial beliefs in hypotheses represented by their prior probabilities to posterior beliefs incorporating the given evidence and represented by the posterior probabilities. We saw an implementation of Bayes' theorem in R and Python, and we practiced using the code on a belief revision example.

Turning to the final section, on interpreting probability, you will learn about two distinct interpretations: the frequentist interpretation, commonly used in statistics, and the subjective Bayesian interpretation.

Interpreting probability: frequentist vs. subjective

The Kolmogorov axioms formally define probability and set the calculus for deriving new probabilities when given specific probabilities as inputs. The axioms tell us nothing about how to develop the initial probability assignments or "where the numbers come from." Several schools of thought have endeavored to give various interpretations of probability to address the question.

We present here the highlights of two main interpretations: frequentist and subjective. Both interpretations are central to two distinct approaches to statistical inference, as we will discuss later in this section.

Frequentist interpretation

Long-term frequencies are central to probability's frequentist (or statistical) interpretation. In frequentist interpretation, probabilities are assigned based on experimentation or historical data. For example, to give a probability to an event A, we conduct an experiment, repeat it n times, and record the number of times n_A the event A happens. We then interpret the probability of A as the limit of the relative frequency of A as the number of experiments approaches an infinitely large number. Formally,

$$P(A) = \lim_{n \to \infty} \frac{n_A}{n} \quad (2.57)$$

In practice, we can repeat an experiment only a finite number of times, and the relative frequency provides only an estimate of the sought probability.

Frequentist interpretation thus links probability to repeatable experiments and defines probability only based on the outcomes of the experiments (e.g., flipping a coin and observing the result).

Subjective (Bayesian) interpretation

Subjective (or Bayesian) interpretation of probability centers on degrees of belief. In the Bayesian approach, we assign probability based on judgment or the degree of belief that we hold in the likelihood of the occurrence of an event.

As mentioned earlier, Bayes' theorem prescribes the belief revision mechanism in the subjective framework. As demonstrated in *Example 2.6*, belief revision signifies that, upon collecting evidence, we can update our prior beliefs in our hypotheses from the initial prior beliefs to updated posterior beliefs. The belief revision may boost or shrink the prior probabilities to become consistent with the evidence.

The subjective approach assigns a prior probability to a hypothesis based on personal assessment and background knowledge. The subjective approach assumes that the agent makes an assessment of its own degree of belief and is a rational decision maker. We'll discuss the meaning of a rational assessment of the degree of belief next.

Rational degree of belief

An agent's degree of belief is assumed to be rational according to this betting scenario due to (De Finetti 1974). If an agent's degree of belief assigns a probability P_A to an event A, then the agent must be indifferent between getting the cash amount of P_A dollars and a bet b_A that pays one dollar if A happens and zero dollars if it does not. That is, receiving the cash offer has the same utility for the agent as a bet b_A with the same expected dollar amount, P_A. For any event A, there should only be one price P_A for the bet b_A. A rational degree of belief must follow the laws of probability. If the agent is indifferent between P_A dollars and the bet b_A and between P_B dollars and the bet b_B then the agent must be indifferent also between $P_A + P_B$ and the bet $b_{A \cup B}$.

Having overviewed the frequentist and the subjective interpretations of probability, let us turn now to the implications of the two interpretations on approaches to statistical inference.

Statistical inference

We learned earlier about various RVs, and we saw their probability distributions depend on specific parameters that determine the RV statistics. For instance, *Bernoulli(p)*, *Binomial(p)*, and *Geometric(p)* RVs have their probability distributions depend on a parameter p, the $Poisson(\lambda)$ distribution depends on a parameter λ, and the normal distribution, $Normal(\mu, \sigma)$, depends on two parameters, μ and σ. We saw in the *Practice in Python* sections how we can generate an **Independent and Identically Distributed** (IID) sequence of RVs given the specification of their distribution parameters. Statistical inference is the problem of inferring the parameters of the generative probability distribution from an observed sequence of IID RVs generated from the distribution. To illustrate, a statistical inference problem for a *Bernoulli(p)* RV is as follows: given an IID sequence of Bernoulli RVs $\{X_1, X_2, \cdots, X_n\}$, we need to estimate the parameter p of the generative probability distribution.

The frequentist and the subjective interpretations of probability lead to two distinct approaches to statistical inference. Statistical inference consists of two main steps. The first step specifies a probabilistic model of observed data up to a parameter θ. The second step devises algorithms to estimate the value of θ from data.

Assume the observed data to be n IID samples, $X_1, \cdots, X_n$, drawn from a distribution function $f(X|\theta)$, representing the PMF if X is discrete or the PDF if continuous, conditioning on the parameter θ.

The frequentist framework

The frequentist approach devises an estimator for the parameter θ from the observational data. An example of an estimator for the parameter p from samples of a Bernoulli (p) distribution, where $\theta = p$, is given by the empirical mean:

$$\hat{\theta}_n = \frac{\sum_{i=1}^{n} X_i}{n} \quad (2.58)$$

Which can be expressed in the recursive form:

$$\hat{\theta}_n = \frac{n-1}{n}\hat{\theta}_{n-1} + \frac{1}{n}X_n \quad (2.59)$$

It is obvious that $E[\hat{\theta}_n] = p$.

Also, by the law of large numbers $\hat{\theta}_n$ converges in probability towards some fixed value p as n becomes very large. That is, for any small $\varepsilon > 0$, the probability that $|\hat{\theta}_n - p| \geq \varepsilon$ goes to zero as n goes to infinity.

Important note

In the frequentist approach, the estimator is a random variable, while we view the parameters as fixed, unknown quantities.

In the frequentist approach, when choosing an estimator, we are not restricted to a particular method; we can choose from various algorithms, though it could be unclear which one is best.

The primary goal of the frequentist approach is to establish probabilistic guarantees that the estimator will converge within a confidence interval around the unknown parameter "true" value.

The Bayesian framework

In the Bayesian framework, we consider the parameter θ to be an RV having a prior probability distribution $g(\theta)$, representing the initial belief about the value of θ. The observations of the RV, $X_1 = x_1, \cdots, X_n = x_n$, are the evidence about θ; the values $x_1, \cdots, x_n$ are given and fixed.

The Bayesian approach to statistical inference frames parameter estimation as a belief updating process following the laws of probability by applying Bayes' theorem to derive the posterior probability of θ from the conditional distribution $f(X|\theta)$ and the prior probability $g(\theta)$, as follows:

$$f(\theta|x_1, \cdots, x_n) = \frac{f(x_1, \cdots, x_n|\theta)g(\theta)}{\int f(x_1, \cdots, x_n|\theta')g(\theta')d\theta'} \quad (2.60)$$

The integral in the denominator does not depend on θ; it is effectively a constant, a normalization factor, not a function of θ. Thus, the posterior probability is proportional to the numerator. Also, the variables $X_1, \cdots, X_n$ are independent and identically distributed, so that we can write:

$$f(\theta|x_1, \cdots, x_n) \propto \prod_{i=1}^{n} f(x_i \mid \theta) g(\theta) \quad (2.61)$$

The Bayesian inference for the value of θ is the value that corresponds to the **maximum posterior probability** (MAP).

$$\vartheta_{MAP} = \arg\max_{\theta} \prod_{i=1}^{n} f(x_i \mid \theta) g(\theta) \quad (2.62)$$

Instead of the arg max for the product, it is more convenient to express the MAP problem as the sum of the log of the prior and the conditional distribution functions.

$$\vartheta_{MAP} = \arg\max_{\vartheta} \left(\sum_{i=1}^{n} \log\big(f(x_i|\theta)\big) + \log\left(g(\theta)\right)\right) \quad (2.63)$$

Important note

In the Bayesian approach, we view the parameters and their estimators as RVs.

The Bayesian approach frames the parameter estimation as an updating process of our beliefs about the parameters.

The prior probabilities represent the initial beliefs about the parameters, and the posterior probabilities represent the updated beliefs incorporating the evidence obtained from the observed variables.

Comparison

The following bullets summarize the comparison between the frequentist and the subjective (or Bayesian) interpretation of probability:

- The frequentist approach interprets probability as the limiting event frequency in an experiment repeated many times. The Bayesian approach interprets probability as a degree of belief based on a subjective state of knowledge.
- The frequentist approach assigns probability only to events we can observe in repeated experiments. The Bayesian approach gives probability to one-time events, including hypotheses and assumptions.
- The frequentist approach to statistical inference considers parameters as fixed, but unknown quantities, and the observations are random; the estimator can be any procedure with frequency guarantees.
- The Bayesian approach to statistical inference considers the parameters to be RVs, with prior distribution representing a subjective degree of belief; the estimator is a procedure for updating beliefs as a function of the observed data, which are fixed known quantities representing evidence on the parameters. The belief updating follows the laws of probability using Bayes' theorem.

Table 2.8 summarizes the distinct features of the frequentist versus Bayesian interpretations of probability.

	Frequentist	**Subjective (Bayesian)**
Probability	Limiting frequency in a repeated experiment	Degree of belief; subjective; follow the rules of probability
Inference method	Create procedures with frequency guarantees	Use Bayes' theorem to revise/update beliefs
Inference goal	Calculate confidence intervals	Calculate posterior probabilities
Parameter learning	Parameter ϑ fixed but unknown	Parameter ϑ RV with prior probability distribution

Table 2.8: Frequentist versus subjective interpretation of probability

In the following section, we present a selective review of two case studies that accentuate the comparison of the frequentist versus Bayesian approaches to statistical inference.

Case studies

The two approaches to statistical inference, frequentist and Bayesian, are used in many medical and clinical applications (Berger 2010). Fornacon-Wood et al. (Fornacon-Wood, et al. 2022) provide a practical explanation of the differences between frequentist and Bayesian approaches to statistical inference. They also describe the advantages or disadvantages of each approach and demonstrate how these two approaches may be implemented in a real-world example: a radiation therapy clinic. In this clinic, the two approaches could be used to analyze the effectiveness of different treatment regimens and the probability of adverse effects.

In the following, we review two case studies. One on multiple treatment comparisons and the other on medical device clinical trials.

Multiple treatment comparisons

A case study (Carlin, et al. 2013) compared Bayesian and frequentist statistical methods for multiple treatment comparisons. The study's context is pharmacological treatments for female **Urinary Incontinence (UI)**. The multiple treatment analysis aims to identify the more efficacious drugs and those with the highest odds of discontinuation due to adverse effects.

The multiple-treatment comparisons study employed a comprehensive approach, fitting fixed- and random-effect models of 8 pharmacological treatments to compute Bayesian and frequentist treatment ranks and odds ratios across all bivariate treatment comparisons.

The study led to the following conclusions. One, Bayesian methods are more flexible than frequentist methods. Two, Bayesian methods provide more valuable information for clinicians and health policymakers than frequentist methods. Bayesian methods offer direct probability statements about patient-centered outcome variables, such as the probability that one drug is the best among the top two drugs for an indication or the probability of experiencing a particular endpoint given the patient takes a particular drug. Combining these sets of results enables clinicians and policymakers to make informed decisions about which drugs should be used based on a joint assessment of their probabilities of being the most effective and safest.

Medical device clinical trials

According to a guidance document for Industry and FDA staff (U.S. Department of Health and Human Services 2010), frequentist statistical methods for medical device clinical trials have a limitation. While the trials may use prior information from previous studies at the design stage, information has no direct bearing, in the frequentist methods, on the inferences made at the data analysis stage. The Bayesian method, by contrast, has the prior information formally represented in Bayes' formula and thus directly affects the inferences made in the analysis stage. The use of prior information is beneficial in medical device clinical trials as the life cycle of device technology development is much shorter than that for drugs, making it necessary to shorten the time allocated to complete the evaluation of the safety and effectiveness of medical devices.

The FDA guidance document mentions that good prior information is often available for medical devices from various sources:

- Previous generations of the device, when modifications are minor
- Studies of the devices overseas
- Historical control data in a randomized controlled trial

A successful use of Bayesian methods in device trials is the TRANSCAN breast scanner device for diagnosing breast cancer (Pennello and Thompson 2008). In the TRANSCAN device trial, a reduced sample size for demonstrating effectiveness resulted from incorporating prior information from results from previous studies. According to Pennello (Pennello and Thompson 2008), the Bayesian analysis indicated that sensitivity and specificity in the small intended use study were decidedly greater for the TRANSCAN device used adjunctively with mammogram reading than for mammogram reading alone, and that using TRANSCAN would result in fewer biopsies. Pennello (Pennello and Thompson 2008) noted that these findings would only have been conclusive because of the prior information from the three previous studies.

Let us summarize the lessons learned in this section. The frequentist approach assigns probabilities to data, not hypotheses, whereas the Bayesian approach assigns probabilities to hypotheses (Fornacon-Wood, et al. 2022). The frequentist and Bayesian approaches to statistical inference lead to different methods when estimating parameters of the generative probability distribution of an observed IID sequence of RVs. The frequentist approach to statistical inference is based on null hypothesis significance testing. The frequentist method uses tools such as P-values and confidence intervals to test a null hypothesis against an alternative hypothesis. In contrast, the Bayesian approach is governed by Bayes' theorem, which incorporates prior information into the inferencing process, updating the posterior distribution for the parameter of interest as more data becomes available.

Cited case studies in multiple treatment comparisons and medical device clinical trials give empirical evidence that Bayesian statistical inference methods are superior to frequentist methods. The case studies find that Bayesian methods are more flexible than frequentist methods and that they provide substantially more valuable information for clinicians and health policymakers than frequentist methods. The Bayesian method can shorten the time for the trial by requiring fewer samples, which keeps pace with the technological flux in the medical device clinical trials domain.

The dominant theme of this book is the Bayesian approach, which we will expound on in the next chapter. In later chapters on causal inference, we will revisit the frequentist approach to statistical inference, which underlies many adopted machine learning algorithms.

With the conclusion of this final section, we wrap up this chapter, and it remains to summarize the overall topics and lessons learned next.

Summary

We covered several topics on probability and Bayes' theorem that set the foundation for the following chapters on Bayesian networks. Let us summarize the takeaway points of this chapter.

A probability is a measure of the uncertainty of events that are subsets in a sample space. The logical structure of a probability measure, founded on three Kolmogorov axioms, is not just a theoretical construct. It's a powerful tool that helps us understand and solve real-world problems. These axioms state that the probability of an event is a non-negative number, the event that spans the entire sample space has probability 1, and the probability of the union of n sequence of pairwise disjoint events is the sum of the probabilities of the n events. We asserted various formulas as properties that follow from the probability axioms. The assertions derive the probability of compound events represented by Boolean logic, such as negation, conjunction, and disjunction, corresponding respectively to set operations, set complement, set intersection, and set union. One particular helpful formula is the inclusion-exclusion formula, which computes the probability of the disjunction of n events (the union of n sets representing the events).

Under the equally likely model, probability on discrete and finite sample spaces can be computed by counting the outcomes of events produced by sampling. This model simplifies the computation process, making it easier to solve probability problems. There are four categories of sampling: with and without replacements, ordered or unordered. There are combinatorics formulas for counting the outcomes of the discrete events in each of the four sampling categories.

RVs are functions of events in probability spaces and can either be discrete or continuous. Probability distributions define RVs. Examples of probability distributions for discrete RVs include Bernoulli(p), Binomial(p), Geometric(p), and Poisson(λ)—the Normal(μ,σ) is an example of a continuous RV. RV summary statistics include the expected value, the variance, the standard deviation, the covariance, and Pearson's correlation coefficient.

Conditioning on an event means transforming the sample space to one where only outcomes consistent with the conditioning event are allowed. Conditional probability is fundamentally a probability and follows all the probability axioms and properties. The chain rule expresses the joint probability in terms of conditional probabilities.

Marginal independence between two or more variables means that the joint probability of the variables equals the product of their individual probabilities. Conditional independence between X and Y given Z means that the joint probability of X and Y given Z equals the product of X's probability given Z and Y's probability given Z. Conditional independence is symmetric – if X is conditionally independent of Y given Z, Y is conditionally independent of X given Z.

Based on the evidence, Bayes' theorem provides a formula for updating beliefs on pairwise disjoint hypotheses spanning an entire sample space. Bayes' formula computes the posterior probabilities of the hypotheses, conditioning on the evidence from the conditional probabilities of the evidence on each of the hypotheses and the prior probabilities of all hypotheses.

There are two distinct interpretations of probability. One, called the frequentist approach, assigns a probability to only events that can be observed in repeated experiments and considers probability to be the limiting relative frequency of the event in repeated experiments. The other, called the subjective or Bayesian interpretation, assigns a probability to one-time events, including assumptions or hypotheses. Probability in the Bayesian interpretation reflects the state of knowledge or degree of belief, which is subjective but must be rational, following all the laws of probability, as exemplified by the betting model.

The subjective and the frequentist interpretations led to two different approaches to statistical inference, which we discussed by comparing them when applied to the problem of estimating the generative distribution parameters θ from the observed IID sequence of the RV.

In the next chapter, on Bayesian networks, we will apply all that we learned in this chapter and see how Bayes' theorem sets the foundation for Bayesian networks. We will explore how Bayesian networks provide a computational framework for probabilistic reasoning that exploits graphical representations of conditional independencies between model variables. This will allow us to make more informed decisions based on probabilistic reasoning.

References

- Cao, Han, Chen Yao, and Ying Yuan, 2023. "*Bayesian approach for design and analysis of medical device trials in the era of modern clinical studies.*" *Med. Rev.* 408-424.
- Berger, Zackary, 2010. "*Bayesian and frequentist models: legitimate choices for different purposes of clinical research.*" *Eval Clin Pract.* 1045-7.
- n.d. Birthday Problem. Accessed 10 1, 2023: `https://pi.math.cornell.edu/~mec/2008-2009/TianyiZheng/Birthday.html`.
- Buddies, Science. 2012. *Scientific American.* March 29. Accessed October 1, 2023: `https://www.scientificamerican.com/article/bring-science-home-probability-birthday-paradox/`.
- Carlin, Bradley P., Hwanhee Hong, Tatyana A. Shamliyan, François Sainfort, and Robert L. Kane. 2013, *Case Study Comparing Bayesian and Frequentist Approaches for Multiple Treatment Comparisons.* March: `https://www.ncbi.nlm.nih.gov/books/NBK132729/`.
- Dawid, A.P. 1979. "*Conditional Independence in Statistical Theory.*" *Journal of the Royal Statistical Society* 1-31.

- De Finetti, B. 1974. *Theory of Probability*, Vols. 1 and 2. New York: Wiley.
- Fornacon-Wood, Isabella, Hitesh Mistry, Corinne Johnson-Hart, Corinne Faivre-Finn, James P.B. O'Connor, and Gareth J. Price. 2022. *"Understanding the Differences Between Bayesian and Frequentist Statistics." Int J Radiation Oncol Biol Phys* 1076–1082.
- Gravner, J. 2011. *Lecture Notes for Introductory Probability*. June 9. Accessed October 1, 2023: `https://www.math.ucdavis.edu/~gravner/MAT135A/resources/lecturenotes.pdf`
- Pennello , Gene , and Laura Thompson, 2008. *"Experience with reviewing Bayesian medical device trials." Journal of Biopharmaceutical Statistics* 81–115.
- U.S. Department of Health and Human Services. 2010. *Guidance for the Use of Bayesian Statistics in Medical Device Clinical Trials*. February 5: `https://www.fda.gov/regulatory-information/search-fda-guidance-documents/guidance-use-bayesian-statistics-medical-device-clinical-trials`.
- Verzani, J. 2002. *simpleR – Using R for Introductory Statistics*: `https://cran.r-project.org/doc/contrib/Verzani-SimpleR.pdf`.
- Wagenmakers, EJ., Lee, M., Lodewyckx, T., Iverson, G.J. 2008. *"Bayesian Versus Frequentist Inference."* In *Bayesian Evaluation of Informative Hypotheses.*, by H., Klugkist, I., Boelen, P.A. (eds) Hoijtink. New York: Springer.
- Walke LM, Byers AL, Gallo WT, Endrass J, Fried TR. 2007. *"The association of symptoms with health outcomes in chronically ill adults." J Pain Symptom Manage*. 58-66.

Join our community on Discord

Join our community's Discord space for discussions with the author and other readers: `https://packt.link/hYCWN`

3

Bayesian Networks

This chapter gives a formal definition of **Bayesian networks** (**BNs**) and reviews the main structural properties. We explain the notion of d-separation and the concept of conditional independence between network variables. We describe the **independence of causal influence** (**ICI**) condition in BNs and how it allows the factorization of a **conditional probability table** (**CPT**) into smaller pieces. We describe examples of important classes of BNs: namely, noisy-or and naïve Bayes.

In this chapter, we're going to cover the following topics:

- Defining Bayesian networks
- Graphical determination of independencies
- Graphoid axioms and Markov blankets
- Modeling and independence of causal influence

Technical requirements

To run the Python code, you must: (1) install Python 3, following the installation and setup guide at this URL: `https://realpython.com/installing-python/`; (2) install the PyCharm **Integrated Development Environment** (**IDE**). To get started, download and install the community version of PyCharm for your operating system from this URL: `https://www.jetbrains.com/pycharm/`.

The Python code in this chapter is in a Python project named `Chapter 3` that you can download from the GitHub URL: `https://github.com/PacktPublishing/Causal-Inference-with-Bayesian-Networks/tree/main/Chapter_03`

To run the code, you must also install the probabilistic graphical model package (pgmpy). The `pgmpy` package requires Python 3.7+, and you can find instructions on how to install it at `https://pgmpy.org/started/install.html`.

The code for this chapter also involves the installation of the graph layout package (`pygraphviz`), and you can find information on how to install it at `https://pygraphviz.github.io/documentation/stable/install.html`.

Defining Bayesian networks (BNs)

In *Chapter 2*, we learned about Bayes' theorem, conditional independence, and the various laws of probability. In this section, we will expand on those concepts and introduce BNs described as probabilistic graphical models. We'll begin with the background of probability and a short motivational introduction. Next, we'll formally define BNs and characterize their qualitative and quantitative components. Afterwards, we'll explain the local and global semantics of BNs, established by the local Markov property. Then, we'll present how a BN is a concise and compact description of a complete probability distribution. Following that, we'll describe an example of a BN representing a model for circuit diagnosis. We'll then give detailed instructions for hands-on practice using Python, the graphical modeling library `pgmpy`, and the graph layout library `pygraphviz` to implement the circuit diagnosis BN model and visualize its directed acyclic graph.

First, some background. Let us recall some results from *Chapter 2*– namely, the chain rule and the definition of conditional independence.

The chain rule

The joint probability for n random variables $X_1, X_2, \cdots, X_n$ is expressed in terms of conditional probabilities as follows:

$$P(X_1, X_2, \cdots, X_n) = P(X_1|X_2, \cdots, X_n)P(X_2|X_3, \cdots, X_n) \cdots P(X_{n-1}|X_n)P(X_n) \quad (3.1)$$

Conditional independence (CI)

We say that two random variables, X and Y, are conditionally independent given Z if:

$$P(X,Y|Z) = P(X|Z)P(Y|Z) \quad (3.2)$$

CI means that if we know Z, then knowing Y will not add any information about X, and knowing X will not add any information about Y:

$$P(X|Y,Z) = P(X|Z) \quad (3.3)$$

$$P(Y|X,Z) = P(Y|Z) \quad (3.4)$$

To express that two random variables, X and Y, are conditionally independent given Z, we write this as follows:

$$X \perp Y \mid Z \quad (3.5)$$

Motivation

As discussed in *Chapter 1*, BNs provide a structured approach for capturing expert knowledge in expert or knowledge-based systems. In those applications, there is a need to represent expert beliefs and judgments in the presence of many variables, often precluding exact specifications in the form of rules or relying on conventional statistical methods.

Inferences derived from BN models follow the laws of probability and Bayes' theorem. Therefore, they are guaranteed to be mathematically sound and consistent with the information and assumptions incorporated in the models.

In this chapter, we focus on knowledge representation and modeling with BNs. In *later chapters* of the book, we will deal with the topic of inferencing with BNs.

What is a Bayesian network?

Having recalled probability background and given this motivational introduction, let us formally define BNs, also called **Bayesian belief networks (BBNs)**. Before we do this, let us define directed acyclic graphs.

Definition 3.1: Directed Acyclic Graph (DAG)

A directed acyclic graph is a pair $G = (V, E)$, where $V = \{ V_1, V_2, \cdots, V_n\}$ is a set of nodes and $E = \{ (V_i, V_j) | 1 \leq i,j \leq n, \ i \neq j\}$ is a set of directed edges. We denote an edge directed from X_k to X_i as $X_k \rightarrow X_i$, and we say X_k is a **parent** of X_i or X_i is a **child** of X_k. If a node X_i has no parents, we say X_i is a **root** node, and if it has no children, it is a **leaf** node. We use $pa(X)$ and $ch(X)$ to denote, respectively, the set of all parents and all children of X in the DAG, G. A **directed path** comprises a sequence of directed edges, all oriented in the same direction and such that every edge points to the vertex that its successor edge points from. If there is a directed path from node X to node Y, we say Y is a **descendant** of X and X is an **ancestor** of Y. A node and all its parents constitute a **family**. A directed acyclic graph has no **directed cycles**, a directed path that begins and ends at the same node.

Figure 3.1 shows examples of directed graphs (digraphs) – one is acyclic, forming a DAG on the left, and the other, on the right, contains a directed cycle and thus is not a DAG.

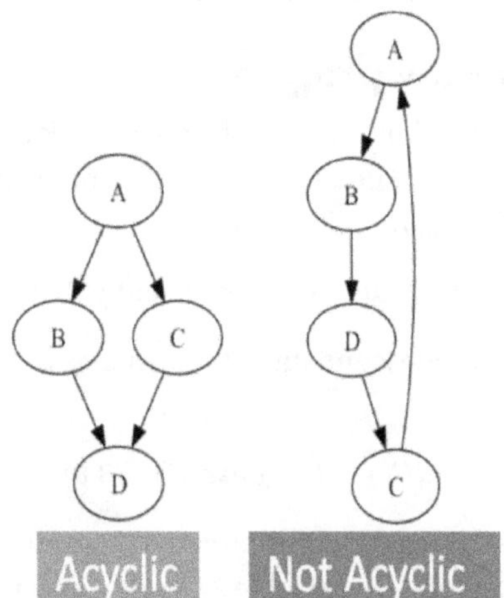

Figure 3.1 – Directed graphs – cyclic versus acyclic

Definition 3.2: Bayesian network (BN)

A BN is a pair (G, P) where G is a directed acyclic graph, P is a joint probability distribution, and G describes the factorization scheme of P. Formally, a BN consists of two components, one qualitative (G) and one quantitative (P):

- **Qualitative component:** G, a directed acyclic graph (DAG) whose nodes are the n random variables in the set $U = \{X_1, X_2, \cdots, X_n\}$, and the directed edges represent conditional dependencies or direct probabilistic influence between pairs of variables.
- **Quantitative component:** P, consists of the specification of the **conditional probability distributions (CPDs)** $\{P(X_i|pa(X_i)), i = 1, \cdots, n\}$, which are the probability of each X_i conditioned on the values of its parent variables. The following must be true by definition:

$$P(X_1, \dots, X_n) = \prod_{i=1}^{n} P(X_i|pa(X_i)) \quad (3.6)$$

Equation 3.6 links a BN's quantitative and qualitative aspects. The factored representation of the complete probability representation P in the form of the conditional probabilities in *Equation 3.6* is such that each conditional probability represents a family in the DAG, G. We will expound next on the local and global semantics underlying the factored representation of *Equation 3.6* of a complete probability distribution P. In essence, we will explain next the local Markov assumption that binds the local and global semantics of BNs.

Local and global semantics

Equation 3.6 specifies the global semantics, defining a complete joint probability distribution $P(X_1, \dots, X_n)$ as the product of the local conditional probabilities $P(X_i|pa(X_i))$. The directed acyclic graph G specifies the local semantics and the structure for the local conditional probabilities.

Definition 3.3: local Markov property

Given a BN (G, P), we say that the distribution P has the local Markov property for the directed acyclic graph G if each variable X representing a node in G is conditionally independent of all its non-descendants, given X's parents, pa(X). That is, for all X in U,

$$P(X|\, pa(X), nd(X)) = P(X|pa(X)) \text{ or equivalently } X \perp nd(X) \mid pa(X) \quad (3.7)$$

Here, $nd(X)$ denotes the set of all nondescendants of X, which is the complement of the descendants of X in the set U of all nodes in G,

$$nd(X) = U - desc(X) \quad (3.8)$$

To verify that *Equation 3.6* satisfies the local Markov property, consider the total ordering $d = (X_1, \dots, X_n)$ of the variables in U to be such that for each variable X_i, we have all its parents in the graph preceding X_i in d. At least one ordering d exists because the graph is acyclic; ordering d is a **topological or causal order** because every parent precedes its children.

We can assemble the joint probability from the local semantics as follows:

1. Choose a topological ordering $X_1, \ldots, X_n$ on the nodes of the graph G.
2. For $i = n$ to 1, set the conditional probability for X_i given all its predecessors in the ordering. By definition of topological ordering, the predecessors are composed of the parents of X_i and all its nondescendants.

$$P(X_i|X_1, \cdots, X_{i-1})) = P(X_i|pa(X_i), nd(X_i)) \quad (3.9)$$

By the chain rule, we can write the joint probability distribution on n variables as:

$$P(X_1, \ldots, X_n) = \prod_{i=1}^{n} P(X_i|X_1, \cdots, X_{i-1}) \quad (3.10)$$

Substituting from *Equation 3.9*, we get:

$$P(X_1, \ldots, X_n) = \prod_{i=1}^{n} P(X_i|pa(X_i), nd(X_i)) \quad (3.11)$$

Hence, *Equation 3.6* follows from *Equation 3.11* and the local Markov property, *Equation 3.7*.

As demonstrated, the local Markov property is the condition that enables the factored representation of a complete probability distribution as in *Equation 3.6*. The Markov condition articulates a set of conditional independencies in *Equation 3.7*. Those conditional independencies precisely enable *Equation 3.6* and enforce the identity between the BN's local and global semantics.

Next, we discuss how BNs concisely describe a complete (joint) probability distribution and how the factored representation of *Equation 3.6* is a compact representation of the joint distribution.

The compactness of the factored representation of the BN

To see the effect of the factorization of a joint probability distribution representation with BNs, let us consider the case where all n variables representing the network nodes are binary. The joint probability distribution of n variables requires a table of size 2^n. The CPT for a node X_i having k_i parents in the directed acyclic graph of the BN involves a table of 2^{k_i} rows for the probability of $X_i = True$ (the number for $X_i = False$ is one minus the probability of $X_i = True$). If a node X_i has no parents, we only need a table of one row. If each node in the graph has no more than k parents, then the complete specification of the network CPTs requires n tables of no more than 2^k rows. Thus, the factored representation of the BN will require space of the order $O(n\ 2^k)$, which grows linearly with n compared to the space requirement $O(2^n)$ for the joint distribution, which grows exponentially with n.

Hence, the compactness of the BN factored representation of its joint probability distribution P directly depends on the highest number of parents amongst the families in the DAG, G.

Suppose G is sparse, meaning the graph has few connections, and the number of parents for all families is small. In that case, the factored BN representation of P produces a significant reduction in space relative to the 2^n required for specifying P in a single table.

On the other hand, if G is dense, meaning the graph has many connections and many parents for at least one family, then the space saving from the BN factorization quickly becomes insignificant.

Quantifying the compactness of the BN representation

Given a $BN = (G,P)$ on n variables, there is an inverse relationship between the compactness of the BN representation of the complete probability distribution P and the number of parents k of the bulkiest family of the DAG G. The compactness is most significant if k is small and quickly becomes insignificant as k increases in order of magnitude relative to n.

Having defined BNs, discussed their foundation on the local Markov property, and quantified the BN representation's compactness, we will now consider examples to solidify the previous theoretical concepts. We will also practice implementing the examples using the Pgmpy library in Python.

Example 3.1: circuit diagnosis

Logic circuits for digital systems contain gates performing information-processing functions expressible in Boolean logic. Fault detection and diagnosis in combinational logic circuits is a challenging engineering application. *Figure 3.2* shows an example of a combinational circuit that outputs the exclusive-NOR (XNOR) of its inputs, given by the Boolean logic function (Mano, 1988):

$$F = X\,Y + \bar{X}\,\bar{Y} = \overline{X \oplus Y} \quad (3.11)$$

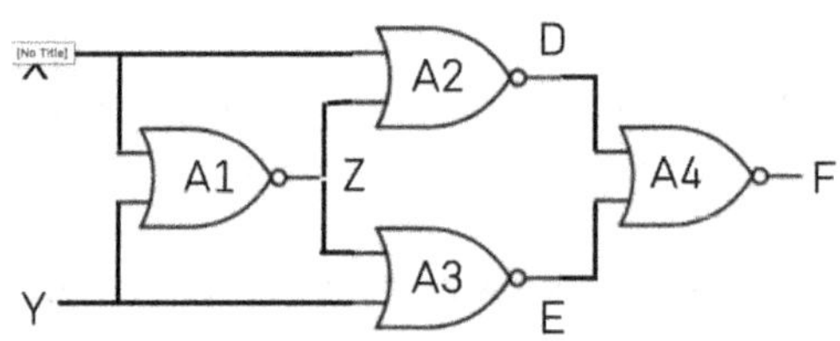

Figure 3.2 – Combinatorial digital circuit of Example 3.1

Table 3.1 shows the truth table for the exclusive NOR function of *Equation 3.11*.

X	0	0	1	1
Y	0	1	0	1
F	1	0	0	1

Table 3.1 – Exclusive-NOR (XNOR)

We demonstrate how to model the digital circuit in a Bayesian network, taking into account the normal and abnormal behaviors of the circuit to perform reasoning tasks for fault detection and diagnosis. To enable performing the reasoning tasks, we introduce an assumption variable for each gate that indicates each gate's operational state. As depicted in *Figure 3.2*, the assumption variables are A1 through A4 for the respective four gates.

Figure 3.3 shows the DAG for the BN model of the logic circuit. The DAG has 10 nodes and 12 edges. The DAG is sparse and has, at most, 3 parents per family.

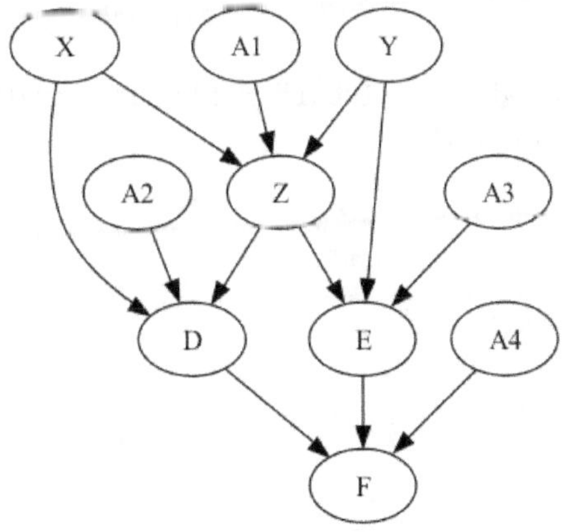

Figure 3.3 – Directed acyclic graph for the digital circuit BN model

Since all gates are identical, we only need to show the conditional probability for one gate, which will serve as the template for all four components.

Table 3.2 specifies the conditional probability for Z conditioning on A1, X, and Y. A1 denotes the operating condition of the gate; it takes the value 0 to indicate the gate is OK and 1 to indicate the gate is abnormal or faulty. X and Y are the inputs to the NOR gate and are Boolean, which takes the value 0 to indicate false and 1 to indicate true. The top three rows of *Table 3.2* enumerate all possible value assignments of the variables A1, X, and Y. The bottom two rows in *Table 3.2* give the conditional probability of the NOR gate's output for the given value tuple of the variables A1, X, and Y.

In the leftmost four columns of *Table 3.2,* we have A1=0 corresponding to the gate working correctly. In the rightmost four columns of *Table 3.2,* we have A1=1 corresponding to the gate being abnormal or faulty. The value of Z equals $(X \oplus Y)$ with a probability of 1 if A1=0. If A1=1, the value for Z is unpredictable, taking either the value 1 or 0 with an equal probability of 0.5.

A1	0	0	0	0	1	1	1	1
X	0	0	1	1	1	1	0	0
Y	0	1	0	1	0	1	0	1
Z(0)	0	1	1	1	0.5	0.5	0.5	0.5
Z(1)	1	0	0	0	0.5	0.5	0.5	0.5

Table 3.2 – Conditional probability table for Z given A1, X, and Y

Given our Bayesian model for the logic circuit, let us find the local independencies from the local Markov assumption that a node in the DAG is conditionally independent of all its nondescendants given all its parents. Let us state the local independencies for the D variable. The parents of D in the DAG are Z, X, and A2, and its non-descendants are A3, Y, A1, E, and A4. We can express this conditional independence as follows:

$$D \perp A3, Y, A1, E, A4 \mid Z, X, A2 \quad (3.12)$$

Practice in Python

The code in this section is included in the `circuit.py` file. First, we define the network structure by defining the nodes and edges for the DAG:

```
from pgmpy.models import BayesianNetwork
circuit: BayesianNetwork = BayesianNetwork()
circuit.add_nodes_from(["X", "Y", "Z","D","E","F","A1","A2","A3","A4"])
circuit.add_edge("X", "Z")
circuit.add_edge("Y", "Z")
circuit.add_edge("A1", "Z")
circuit.add_edge("X", "D")
circuit.add_edge("Z", "D")
circuit.add_edge("A2", "D")
circuit.add_edge("Z", "E")
circuit.add_edge("Y", "E")
circuit.add_edge("A3", "E")
circuit.add_edge("D", "F")
circuit.add_edge("A4", "F")
circuit.add_edge("E", "F")
```

We can create the DAG of the Bayesian network model as follows:

```
from pgmpy.base import DAG
G = DAG()
nodes = circuit.nodes
edges = circuit.edges
G.add_nodes_from(nodes)
G.add_edges_from(edges)
```

To find the local dependencies for a variable in the model, we use the `local_independencies` function in the `pgmpy` library. As an example, find the local dependencies for the variable `D`. Recall from the local Markov property that a variable is independent of all its non-descendants given all its parents – see *Equation 3.7*. Let us check the local independencies for node `D` entailed by the Markov condition.

```
print(G.local_independencies('D'))
```

The output is as expected from *Equation 3.12*.

```
(D ⊥ A3, E, A4, A1, Y | A2, X, Z)
```

Using the graph object G, we can pose various queries, as follows:

```
print(f'The nodes of the DAG are:\n {G.nodes()}')
print(f'The edges of the DAG are:\n {G.edges()}')
print(f'The root nodes of the DAG are:\n {G.get_roots()}')
```

```
print(f'The leaf nodes of the DAG are:\n {G.get_leaves()}')
print(f'The parents of the \'D\' node are:\n ', G.get_parents('D'))
print(f'the children of the \'D\' node are:\n ', G.get_children('D'))
print(f'number of nodes is: {G.number_of_nodes()}')
print(f'number of nodes is: {G.number_of_edges()}')
```

The results for the preceding queries are as follows:

```
The nodes of the DAG are:
 ['X', 'Y', 'Z', 'D', 'E', 'F', 'A1', 'A2', 'A3', 'A4']
The edges of the DAG are:
 [('X', 'Z'), ('X', 'D'), ('Y', 'Z'), ('Y', 'E'), ('Z', 'D'), ('Z', 'E'), ('D',
'F'), ('E', 'F'), ('A1', 'Z'), ('A2', 'D'), ('A3', 'E'), ('A4', 'F')]
The root nodes of the DAG are:
 ['X', 'Y', 'A1', 'A2', 'A3', 'A4']
The leaf nodes of the DAG are:
 ['F']
The parents of the 'D' node are:
  ['X', 'Z', 'A2']
the children of the 'D' node are:
  ['F']
number of nodes is: 10
number of nodes is: 12
```

Quantification of the circuit BN

As an example of how to define the prior probabilities for the root nodes, we show the code for the X node; the prior probability for Y, the other root node, is similar.

```
from pgmpy.factors.discrete import TabularCPD
cpd_X = TabularCPD(
    variable="X", variable_card=2, values=[[0.5], [0.5]]
```

As an example of how to define the conditional probabilities for the BN model, we list here the definition of the conditional probability for the Z node, as per *Table 3.2*. The code for the conditional probabilities of the nodes D, E, and F is similar to the Z node.

```
cpd_Z = TabularCPD(
    variable="Z",
    variable_card=2,
    values=[
            [0,1,1,1,0.5,0.5,0.5,0.5],
            [1,0,0,0,0.5,0.5,0.5,0.5],
    ],
```

```
        evidence=["A1", "X", "Y"],
        evidence_card=[2, 2, 2],
    )
```

The next step is to add all the prior probabilities and the conditional probabilities to the circuit model.

```
circuit.add_cpds(cpd_X,cpd_Y,cpd_Z,cpd_D,cpd_E,cpd_F)
```

The following code prints the conditional probability for the node F:

```
print(cpd_F)
```

Table 3.3 shows the printed output.

A4	A4(0)	A4(0)	A4(0)	A4(0)	A4(1)	A4(1)	A4(1)	A4(1)
D	D(0)	D(0)	D(1)	D(1)	D(0)	D(0)	D(1)	D(1)
E	E(0)	E(1)	E(0)	E(1)	E(0)	E(1)	E(0)	E(1)
F(0)	0.0	1.0	1.0	1.0	0.5	0.5	0.5	0.5
F(1)	1.0	0.0	0.0	0.0	0.5	0.5	0.5	0.5

Table 3.3 – Conditional probability table for the F node of the "circuit" BN model

Graph visualization

We use the `pygraphviz` library for laying out the DAG, drawing the DAG, and writing the DAG to a file in PNG format. In the following code, the `dagviz` function creates a digraph as a `pygraphviz AGraph` object, adds the nodes and edges to the digraph, and then writes the digraph to a file named filename in "dot" format, which is the graph specification format of `pygraphviz`. The last two lines of code of the `dagviz` function lay out the digraph, and next, write the visualization of the digraph to a file in the PNG graphic format named filename.

```
import pygraphviz as pgv
def dagviz(nodes, edges, filename):
    G = pgv.AGraph(strict=False, directed=True)
    G.add_nodes_from(nodes)
    for edge in edges:
        G.add_edge(edge)
    G.write(name + ".dot")
    # use dot
    G.layout(prog="dot")
    # write previously positioned graph to PNG file
    G.draw(filename + ".png")
```

The diagram in *Figure 3.3* is the output circuit.png file from executing the following code.

```
dagviz(nodes, edges, "circuit-dag")
```

The following is the output circuit.dot file, specifying the DAG in the dot specification language of pygraphviz.

```
digraph "" {
    X -> Z;
    X -> D;
    Y -> Z;
    Y -> E;
    Z -> D;
    Z -> E;
    D -> F;
    E -> F;
    A1 -> Z;
    A2 -> D;
    A3 -> E;
    A4 -> F;
}
```

Let us review the lessons learned in this section. Bayesian networks are a factored representation of a complete probability distribution. A BN is a pair (G, P) where G is a DAG whose nodes are the network variables, the edges indicate probabilistic influence between variables, and P is a joint probability distribution on the variables of the network. Each node and its parents in G define a "family" and a corresponding conditional probability distribution for the node given the values of all its parents. *Equation 3.6* represents the BN factorization of the complete probability distribution founded on the local Markov property. The local Markov property states that every node in the graph is conditionally independent of all its nondescendants, given the values of all its parents. The Markov property makes the BN a concise and compact description of a complete probability distribution. We worked out a BN model for circuit diagnosis to demonstrate how to build the qualitative and quantitative components of the BN model. We also implemented the circuit example in Python using the pgmpy library for probabilistic graphical models.

In the next section, we turn to an essential subject to further investigate the conditional independencies entailed in the graphical structure of the BN.

Graphical determination of independencies

In the previous section, *Defining Bayesian networks (BNs)*, we learned that the factored representation of *Equation 3.6* is valid only under the local Markov property that a variable is independent of its nondescendants given the values of all its parents in the DAG of the BN. The Markov property entails conditional independencies between network variables; we saw an example in *Equation 3.12*.

In this section, we will learn about a graphical method for determining conditional independencies from the topology of the DAG. We will learn about the condition of d-separation, which we define next. We'll follow this by explaining the connection between d-separation and conditional independence. Afterwards, we will describe a graph-based algorithm for implementing the condition of d-separation. We will demonstrate with examples how to apply the condition of d-separation for determining conditional independence between sets of BN variables. Finally, we will describe, with programming practice using Python, how to implement a working code solution to conditional independence queries on BN variables.

Defining d-separation

The d-separation is a graphical criterion for checking conditional independence between two sets of nodes, X and Y, dependent on knowing the values of a third set of variables, Z.

To gain basic intuition behind the d-separation, we'll consider three cases depicted in *Figure 3.4*. In the first case in the left-side box, we have Z on a path between X and Y where the directed edges diverge at Z, which means Z is a common cause for X and Y. In the second case in the box on the left, the arrows at Z are going in the same direction, which means Z is a mediator between X and Y. In both cases, Z is a non-collider where the edges on the path at Z are not colliding (going in opposite directions). In both cases framed in the left box in *Figure 3.4,* knowing Z will decouple X and Y; in other words, Z cuts or blocks the path between X and Y. The third case, framed in the box on the right in *Figure 3.4,* differs from the first two cases. In this case, knowing Z will couple X and Y. This is the case where Z is a collider, where the path between X and Y has directed edges colliding at Z. In the collider case, Z and all its descendants being unknown blocks the path between X and Y and keeps them marginally independent.

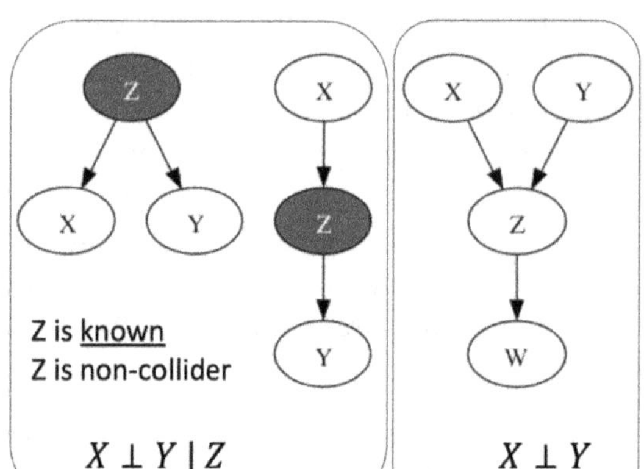

Figure 3.4 – Knowing Z makes X and Y independent on the two leftmost graphs but dependent on the rightmost

Definition 3.4: d-separation

Let X, Y, and Z be three disjoint sets containing variables that are nodes in the DAG G. A path between two nodes in X and Y is any sequence of edges in G, regardless of the edge direction that connects the node in X and the node in Y. We say that Z **d-separates** X and Y in G and write $X \perp_G Y \mid Z$ if Z blocks all paths between X and Y. Here is an explanation of the blocked path:

- There exists a node W in Z, and W is a non-collider; that is, W has two incident edges pointing in the same direction at W or pointing away from W (diverging at W).
- There is a collider node W (the incident edges of W along the path point toward W), and the union of W and all its descendants is not a subset of Z.
- We say that Z **d-connects** X and Y if Z does not d-separate X and$Y$.

Figure 3.5 depicts the condition of d-separation between two sets of variables, X and Y, by a set Z. The figure shows Z blocking all paths between X and Y. The top two paths in the figure are blocked because the incident arrows at a node in Z are not colliding; the bottom path has a collider node, and the collider node and all its descendants are not in Z.

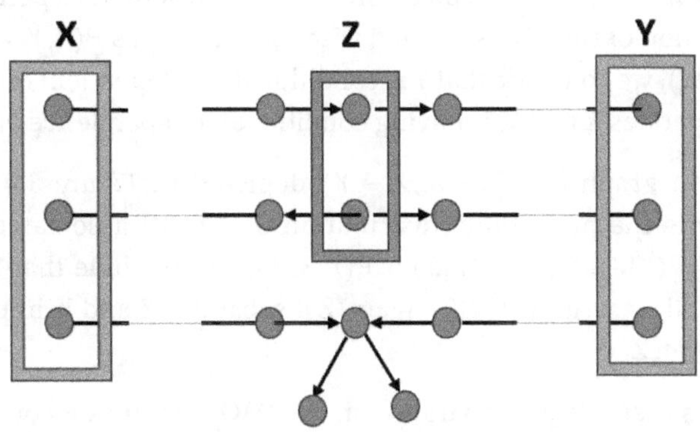

Figure 3.5 – All paths from X to Y are blocked; Z d-separates X and Y

It is apparent that if Z blocks all paths between X and Y, then it follows that any superset of Z that is disjointed with X and Y will also block all paths between X and Y. We conclude then that there must be a minimal d-separator set of X and Y, that is a set such that removing any node from the set makes X and Y not d-separated and, hence, d-connected instead. We formally define a minimal d-separator as follows.

Definition 3.5: minimal d-separator

Given a Bayesian network (G, P), we say that for any disjoint subsets of variables X, Y, Z, that Z is a minimal d-separator of X and Y if there exists no subset W of nodes in G such that W d-separates X and Y, and W is a proper subset of Z. That is,

$$X \perp_G Y \mid W \Rightarrow W \supseteq Z \quad (3.13)$$

In summary, we defined the condition of d-separation, a graphical criterion for a directed acyclic graph G, and three disjoint sets of graph nodes X, Y, and Z if Z d-separates X and Y. The condition of d-separation relates only to the DAG G of the BN; that is, the d-separation is a condition based only on the graph topology of G.

The following section will show connections between a BN's probability model P and the topology of the BN's DAG G.

Relating graph topology to the probability model

In this section, you will gain insight into the connection between the topology of a BN's DAG G and its probability model P. A relationship exists between the d-separation in graph G and the conditional independencies in P. The topology of G restricts the probability models of P that respect the independencies entailed by the d-separation in G. We will explain those relationships next.

Let us consider the graphs of the two d-separation cases depicted in *Figure 3.4* to illustrate the connection between d-separation and conditional independence.

Consider the graph $G = \{Z \to X, Z \to Y\}$ depicted in *Figure 3.4*, where Z is a parent of X and Y. We write the probability distribution of the BN associated with the graph G as $P(X,Y,Z) = P(Z)P(X|Z)P(Y|Z)$. Since $P(Y|Z) = P(Y|Z,X)$, we conclude that Y is conditionally independent of X given Z, and we write $Y \perp X|Z$. Hence, Z d-separates X and Y, implying conditional independence of X and Y given Z.

Similarly, for the chain graph $G = \{X \to Z, Z \to Y\}$ depicted in *Figure 3.4*, where Z is a mediator of X and Y, we write the probability distribution of the BN associated with the graph G as $P(X,Y,Z) = P(X)P(Z|X)P(Y|Z)$. Since $P(Y|Z) = P(Y|Z,X)$, we conclude that Y is conditionally independent of X given Z, and we write $Y \perp X|Z$. Again, Z d-separates X and Y, implying conditional independence of X and Y given Z.

The preceding indicates that every d-separation in the DAG G implies a corresponding conditional independence in probability model P of the Bayesian network (G, P).

Definition 3.6: global Markov property

Given a BN (G,P), we say that the distribution P has the global Markov property for the DAG G if for all disjoint subsets of network variables X,Y,Z represented by the nodes of G such that Z d-separates X and Y, we have

$$P(X,Y|Z) = P(X|Z)\,P(Y|Z) \qquad \text{or equivalently} \qquad X \perp Y|Z \quad (3.14)$$

Using the condition of d-separation between nodes of a DAG G, we can obtain the independencies that all joint distributions P that factorize according to the DAG G must satisfy.

Notably, a distribution that factorizes according to graph G could have more conditional independencies than those specified by d-separation in graph G. An example is the case of a distribution P on completely independent variables, which can factorize according to any DAG G with nodes being the variables in P and with any set of edges.

Practice in Python

You will find the complete code for this practice in the `circuit.py` file.

To query for all the d-separation assertions on pairs of nodes in the DAG of the "circuit" Bayesian model, we use the pgmpy `get_independencies()` method of the graph object. The number of all assertions between all pairs of variables in the graph is very large.

```
print(G.get_independencies())
```

We'll show only the first 10 results that you will get from running the query.

```
(A3 ⊥ Y, Z, A4, A2, X, A1, D)
(A3 ⊥ D, A4, A2, X, A1, Y | Z)
(A3 ⊥ A2, A4 | E)
(A3 ⊥ D, Z, A4, X, A1, Y | A2)
(A3 ⊥ D, Z, A4, A2, A1, Y | X)
(A3 ⊥ D, Z, A2, X, A4, Y | A1)
(A3 ⊥ Z, A4, A2, X, A1, Y | D)
(A3 ⊥ Z, A4, A2, X, A1, D | Y)
(A3 ⊥ Y, Z, A2, X, A1, D | A4)
(A3 ⊥ A2, A4 | E, Z)
```

To get all the conditional independencies in the probability model, we use the pgmpy `get_independencies` function. Since the probability model factors according to the graph G, d-separation in the graph implies conditional independence in the model. We will not show the result of the query here, as you will get the same results as the ones before.

```
print(circuit.get_independencies())
```

Recall from earlier that a minimum d-separator set is one such that removing any element from the set makes X and Y d-connected. To get the minimal d-separator, we use the pgmpy `minimal_dseparator` method of the DAG object. Let us run an example.

```
G.minimal_dseparator('E','X')
```

The answer is as follows:

```
{'Z', 'Y'}
```

It is easy to verify from *Figure 3.3* that `{'Z', 'Y'}` blocks all paths between `'E'` and `'X'` and removing either `'E'` or `'X'` will d-connect `'Z'` and `'Y'`.

To demonstrate that this is a minimal set, let us remove an element from the set and see if `'E'` and `'X'` become d-connected. We'll now check this as follows:

```
print(G.is_dconnected('E', 'X', ['Z']))
```

This is the answer:

```
True
```

The question now is how to operationalize the definition of d-separation. In the following, we describe a graph-based algorithm that takes the DAG for the BN model and the subsets of variables X, Y, and Z and outputs `True` or `False` depending on whether X and Y are d-separated by Z.

Algorithm 3.1: d-separations

Description: A graph-based algorithm determines whether Z blocks all paths between X and Y in a directed acyclic graph (DAG), G, consisting of three disjoint subsets, X, Y, and Z, returning either true or false.

Input: DAG G, and a query whether true or false: $X \perp_G Y|Z$ where X, Y, Z are disjoint subsets of nodes in graph G.

Output: True if Z d-separates X and Y; False otherwise.

1. Get the **ancestral graph** for the nodes X, Y, Z in G. The ancestral graph G^A is computed by removing all nodes in G that are not ancestors of nodes in X, Y, and Z. A node v is an ancestor of a node v' in the directed graph G if there is a directed path from v to v'.
2. Get the **moral graph** G^M by moralizing the ancestral graph G^A. The moralizing step means "marrying" the parents of every node in the DAG G^A. Marrying the parents means joining all pairs of parents for a node by adding an undirected edge between the pair. Convert all directed edges to be undirected.
3. Remove all nodes in Z from the moral graph.
4. Return `True` if no node in X is connected to a node in Y. Return *False* otherwise.
5. To illustrate the d-separation algorithm, we will consider the Bayesian network (BN) causal graph of the circuit shown in *Figure 3.3*. We will also pose the query whether the set $\{Y, Z\}$ d-separates E from X.
6. The following code in the `util.py` file gives the implementation of the *d-separation algorithm*. The `d_sep_g` function takes four parameters: `G` – a DAG object of class `pgmpy.base.DAG` and three disjoint subsets of nodes, `X`, `Y`, and `Z`. The function returns `True` if `Z` d-separates `X` and `Y` in `G` and returns `False` otherwise.
7. Here's an explanation of the code:
    1. We create the ancestral graph for the nodes `X`, `Y`, and `Z` in `G` and call the `dagviz` function, which we saw earlier in the section on graph visualization, to write the graph picture into a file.
    2. We moralize the ancestral graph, visualize the moral graph, and write the picture into a file.
    3. We remove the nodes in the `Z` list from the graph and call the `separated` function to determine whether the nodes in `X` separate from the nodes in `Y` in the mutated graph `G`, which we will explain next.

```
def d_sep_g(G, X, Y, Z):
    # ancestor graph
    anc_dag = G.get_ancestral_graph(nodes=X + Y + Z)
    dagviz(anc_dag.nodes(), anc_dag.edges(), "anc_g")
    # moralize
    gmoral = anc_dag.moralize()
    gviz(gmoral.nodes(), gmoral.edges(), "mor_g")
    # remove conditioning nodes
    gmoral.remove_nodes_from(Z)
    gviz(gmoral.nodes(), gmoral.edges(), "result_g")
    # check if nodes in X and Y are separated
    return separated(G, X, Y)
```

The `separated` function is straightforward. It takes three parameters: `G` – a DAG object of class `pgmpy.base.DAG`, `X` – a list of nodes, and `Y` – a list of nodes disjoint with `X`. The function loops through nodes in `X` and loops through nodes in `Y` and checks if there is an edge between a node in `X` and a node in `Y`. If it finds a connection, it returns `False`, and if it finds none, it returns `True`.

```
def separated(G, X, Y):
    res = True
    edges = G.edges()
    for x in X:
        for y in Y:
            if (x, y) in edges:
                print(" %s -- %s " % (x, y))
                res = False
                break
    return res
```

Example 3.2: d-separation

To demonstrate the *d-separation algorithm*, consider the circuit BN's DAG in *Figure 3.3*.

Figure 3.6 shows the application of the d-separation algorithm to determine if {Y, Z} d-separates E from X in the circuit DAG shown in *Figure 3.3*. The nodes in the set {Y, Z} have the color blue, E has the color yellow, and X has the color red.

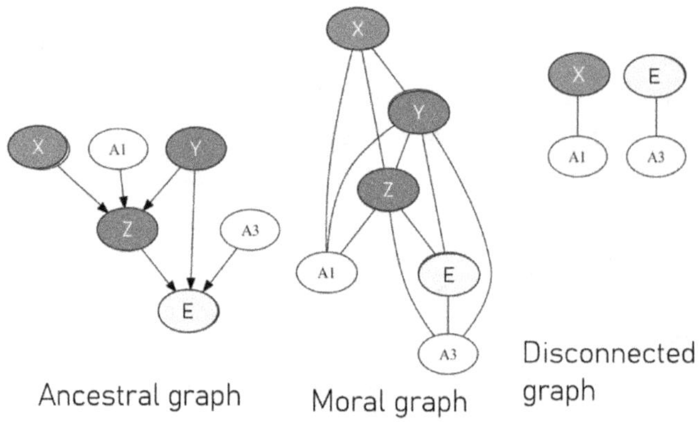

Figure 3.6 – The ancestral and moral graphs derived from the d-separation algorithm demonstrate that the set {Y,Z} d-separates E from X.

Viewing *Figure 3.6* from left to right, first, we see the ancestral graph for all the colored nodes. The ancestral graph deletes all nodes in the circuit's DAG, not ancestors of the colored nodes. The following step of the *d-separation algorithm* computes the moral graph, which is the undirected graph resulting from connecting (marrying) all parents in the ancestral graph. The final step in the *d-separation algorithm* involves removing the nodes in the set {Y, Z}. The action results in a disconnected graph, causing the nodes E and X to be in two separate subgraphs. Consequently, the output then verifies that {Y, Z} effectively d-separates E from X

To produce the graphs shown in *Figure 3.6* and the result from executing the d-separation algorithm in *Example 3.2*, we have the following practice in Python.

Practice in Python

We use the `d_sep_g` function, which we explained earlier. All the pertinent code is in the `circuit.py` file. We execute the following two lines of code to run the *d-separation algorithm* on *Example 3.2* in Python. The first obtains the DAG for the circuit BN model of *Example 3.1*. The second calls `d_sep_g` with the `G` parameter for the DAG; `['E'], ['X'],` for the start and end sets; and `['Y','Z']` for the separator set.

```
G = circuit_dag()
res = d_sep_g(G, ['E'], ['X'], ['Y','Z'])
print(res)
```

The results you get are the graphs shown in *Figure 3.6*, and the following output:

```
True
```

The program output verifies that the set `['Y','Z']` d-separates the elements `['E'], ['X']` in the circuit DAG illustrated in *Figure 3.3*.

From what we have learned to this point, it is crucial to emphasize the connection between the topology of the DAG G and the probability distribution P representing a BN. The local and global Markov properties are the BN properties that govern that connection.

Relating probability to graph topology

The probability distribution P that factors according to DAG G has the following properties:

- **Local Markov property** that each node in G is independent of all its nondescendants conditioning on all its parents.
- **Global Markov property** that each d-separation in G implies conditional independencies in P.

Figure 3.7 depicts the connection between topology and probability distributions in a BN. Given a DAG G, the graph's topology imposes restrictions on probability distributions P to encode in a BN (G, P). The more independencies in the graph, the more restrictions on the probability distributions P to encode in the BN.

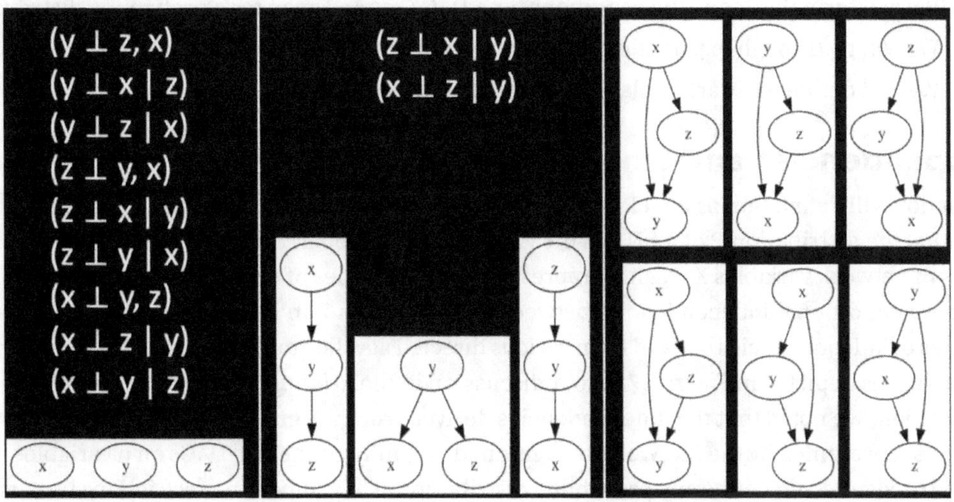

Figure 3.7 – Connection between graph topology and probability models

Figure 3.7 shows, at the bottom, three families of graph structures and, at the top, conditional independencies that probability distributions must obey. In the leftmost group, we have a DAG G with three nodes, X, Y, and Z, and no edges; the only distributions compatible with the graph's structure are the ones that guarantee X, Y, and Z are marginally independent. Moving to the right, the figure shows graphs with more connections. In the rightmost group, G has n nodes and is fully connected, which means G has the number of edges equal to n (n-1)/2. In that case, one can encode any joint probability distribution in the BN. *Figure 3.7* thus explains how graph topology impacts the class of probability distributions that factor according to the graph—the more connected the graph, the bigger the class of probability distributions represented in the BN.

To conclude, let us review the main lessons of this *Graphical determination of independencies* section. First, we learned the definition of d-separation in DAGs and a graph-based algorithm for determining, given three disjoint subsets of nodes in the graph, X, Y, Z, whether Z d-separates X and Y. Second, we learned that a probability distribution P has the global Markov property for the DAG G if for all subsets X, Y, Z of graph nodes such that Z d-separates X and Y, we have conditional independency between X and Y given Z in P. Third, we learned about the connection between the topology of the DAG of a BN and the complete probability distributions that the network represents, and that the local and global Markov properties govern the connection.

In the next section, we will learn about deriving more independencies based on graphoid axioms and the concept of the Markov blanket.

Graphoid axioms and the Markov blanket

In the previous section, *Graphical determination of independence*, we learned that all disjoint subsets X, Y, and Z of a BN (G, P) variables, such that Z d-separates X and Y in the DAG G, implies conditional independence between X and Y given Z in the probability distribution P. In this section, we will learn how to derive global independencies based on the graph structure:

- We will formally define the conditions for a DAG G as an I-map for a probability distribution P.
- We will go through explanations of the graphoid axioms for Bayesian networks.
- We will define the Markov blanket and explain its significance.

Independencies and I-maps

This section will define I-maps for a BN (G, P). I-maps are related to the global Markov property of BNs. Recall that the distribution P of a BN has the global Markov property for the DAG G if, for all disjoint subsets of network variables X, Y, and Z represented by the nodes of G, such that Z d-separates X and Y, we have conditional independencies between X and Y given Z in P as per *Equation 3.14*. We gave examples of independencies derived from various directed acyclic graph structures for a network with three variables depicted in *Figure 3.7*. In our discussion in the *Relating graph topology to the probability model* section, we noted that the independencies derived from the graphs restrict the probability distributions represented by the BN. We also mentioned that fully connected DAGs on n variables having $n(n-1)/2$ edges could define any probability distribution. They impose no restrictions because they yield a null set of independencies. However, we want to have the DAG of the BN representing the entire set of independencies that hold in the probability model P. With this informal introduction, let us define the I-map for the BN.

Definition 3.7: I-map

Let (G, P) be a BN. Let $I_G(X, Z, Y)$ be the relation on disjoint subsets X, Z, and Y of network variables such that Z d-separates X and Y in G.

$$I_G(X, Z, Y) = \{\text{all triples } \langle X, Z, Y\rangle \text{ such that Z d-separates } X \text{ and } Y \text{ in G}\} \quad (3.15)$$

Let $I_P(X, Z, Y)$ be the relation on disjoint subsets X, Z, and Y of network variables such that X and Y are conditionally independent given Z in P.

$$I_P(X, Z, Y) = \{\text{all triples } \langle X, Z, Y\rangle \text{ such that } X \perp Y \mid Z \text{ in } P\} \quad (3.16)$$

We say a graph G is an I-map of P if the relation I_G is a subset of I_P, that is,

$$\text{G is an I-map of P if } I_G(X, Z, Y) \subseteq I_P(X, Z, Y) \quad (3.17)$$

We note from the preceding definition that for a BN (G, P), we distinguish between the I-map for the DAG, G, and the I-map for the joint probability distribution P. The I-map for G depends only on the graph topology; the I-map for P depends on the conditional independencies entailed by the probability model for P. Recall that a BN is a graphical model representing a joint probability distribution of the network variables. If the probability model P factors according to G, then for every d-separation in G, we have corresponding conditional independence between the d-separated sets in P, which means that the I_G relation is a subset of the I_P relation as per *Equation 3.17*.

Example 3.3: I-map

We consider two BNs, BN1 and BN2, on three binary variables, X, Y, and Z. The DAG, G, for BN1, has three nodes and no edges, $G = \{X, Y, Z\}$, meaning that the three binary variables are marginally independent. The joint probability distribution

$$P(X = x, Y = y, Z = z) = P_X(x)\, P_Y(y)\, P_Z(z).$$

The I-map for BN1 consists of nine independencies:

$$(Y \perp Z, X) \quad (Y \perp X \mid Z) \quad (Y \perp Z \mid X) \quad (Z \perp Y, X) \quad (Z \perp X \mid Y)$$

$$(Z \perp Y \mid X) \quad (X \perp Y, Z) \quad (X \perp Z \mid Y) \quad (X \perp Y \mid Z)$$

The DAG for BN2 is $G = \{Y \to X, Y \to Z\}$. The joint probability distribution

$$P(X = x, Y = y, Z = z) = P_{X|Y}(x|y)\, P_{Z|Y}(z|y)\, P_Y(y).$$

The I-map for BN2 consists of two independencies:

$$(Z \perp X \mid Y) \quad (X \perp Z \mid Y)$$

We will verify that the preceding is the I-maps for the joint probability distribution of the two BNs next, with numeric examples using Python.

Practice in Python

You will find the entire code for the example in `imap_demo1()` and `imap_demo2()` in the `imap.py` file. We'll explain the code in detail for the BN1 model and show snippets from the code for BN2 because both demos share the same program logic. Recall that BN1 has three nodes and no edges. The `bn1(px, py, pz)` function creates the network – we do not show the code for that function as it is similar to what we showed earlier, in *Example 3.1*. We compute the eight-element array, `jpd`, by multiplying the marginals to get the probability of (X, Y, Z) for all binary value assignments (x, y, z) ordered from (0,0,0) to (1,1,1). We use the `numpy` library to do the math. Next, we construct the joint probability distribution object of the `pgmpy` library and check that the BN1 model, `m1`, is an I-map for the joint distribution. We print the I-map and the verification result.

```
px = np.array([[0.4], [0.6]])
py = np.array([[0.7], [0.3]])
pz = np.array([[0.1], [0.9]])
m1 = bn1(px, py, pz)
```

```
jpd = (
    (np.multiply(pz, (np.multiply(px, py.transpose())).flatten()))
    .transpose()
    .flatten()
)
JPD = JointProbabilityDistribution(["X", "Y", "Z"], [2, 2, 2], jpd)

check_imap = m1.is_imap(JPD)
print(f"Bayesian network m1 has \n nodes: {m1.nodes} edges: {m1.edges} ")
print(f"BN1 joint distribution P(x,y,z) is:\n{jpd}")

imap = m1.get_independencies()
print(f"BN1 I-map is:\n{imap}:")
print(f"Is BN1 an I-map of the joint distribution?\nAnswer: {check_imap}")
```

The output is as follows:

```
Bayesian network BN1 has
 nodes: ['X', 'Y', 'Z'] edges: []
BN1 joint distribution P(x,y,z) is:
[0.028 0.252 0.012 0.108 0.042 0.378 0.018 0.162]
BN1 I-map is:
(Y ⊥ Z, X)
(Y ⊥ X | Z)
(Y ⊥ Z | X)
(Z ⊥ Y, X)
(Z ⊥ X | Y)
(Z ⊥ Y | X)
(X ⊥ Y, Z)
(X ⊥ Z | Y)
(X ⊥ Y | Z):
Is BN1 an I-map of the joint distribution?
Answer: True
```

The following code fragment shows the creation of the BN2 model using the specification of the probabilities $P(Y), P(X|Y), P(Z|Y)$. Next, it computes the joint probability and checks that the BN2 is an I-map of the joint probability.

```
m2 = bn2(P_Y, P_X_Y, P_Z_Y)
jpd = []
for Y in [0, 1]:
    for X in [0, 1]:
        for Z in [0, 1]:
```

```
            p_yxz = np.multiply(np.multiply(P_Y[Y], P_X_Y[X, Y]), P_Z_Y[Z, Y])
            jpd.append(p_yxz[0])
prob = JointProbabilityDistribution(["Y", "X", "Z"], [2, 2, 2], jpd)
check_imap = m2.is_imap(prob)
print(f"Bayesian network BN2 has \n nodes: {m2.nodes} edges: {m2.edges} ")
print(f"BN2 joint distribution P(x,y,z) is:\n{jpd}")
print(f"Conditional independencies are:")
print(m2.get_independencies())
print(f"Is m2 an I-map of the joint distribution?\nAnswer: {check_imap}")
```

The output of the program is as follows:

```
Bayesian network BN2 has
 nodes: ['X', 'Y', 'Z'] edges: [('Y', 'X'), ('Y', 'Z')]
BN2 joint distribution P(x,y,z) is:
[0.14, 0.14, 0.21, 0.21, 0.024, 0.036, 0.096, 0.144]
Conditional independencies are:
(Z ⊥ X | Y)
(X ⊥ Z | Y)
Is m2 an I-map of the joint distribution?
Answer: True
```

In general, some independence assertions in P may not hold in G. If every assertion in P also holds in G, then graph G is a **perfect I-map** of P.

Definition 3.8: perfect I-map

Let (G, P) be a Bayesian network having $I\_G\ (X, Z, Y)$, the relation of all d-separations in G, and $I\_P\ (X, Z, Y)$, the relation of all conditional independencies in P. We say G is a perfect I-map for P if the two relations are identical, that is,

$$\text{G is a perfect I-map for P if } I_G(X, Z, Y) = I_P(X, Z, Y) \quad (3.18)$$

It is worth noting that it is not always possible to find a DAG G for a given joint probability distribution P, such that every conditional independency in P corresponds to a d-separation in G. That is, given a probability distribution P, we cannot always find a DAG, G, to represent a perfect map for P.

Another important observation is that there can be multiple DAGs on the same set of variables that share the same d-separation relation on their node subsets. *Figure 3.7* shows examples of various graphs having the same I-map.

Since there can be numerous graphical representations with identical I-maps, we can define equivalence classes of graph structures representing the same independencies, as stated in the following definition.

Definition 3.9: I-equivalence

Let G and H be two DAGs. Let U_G, I_G denote the set of nodes and the d-separation relation for G, and let U_H, I_H represent the set of nodes and the d-separation relation for H. We say that G and H are I-equivalent if the node sets for G and H are the same and all d-separations held in G are also valid in H and vice versa. That is, G and H are I-equivalent if

$$U_G = U_H$$

$$I_G = I_H$$

Example 3.4: I-equivalence

The following are three DAGs on the same set of variables but having different topologies:

$$G_1 = \{x \to y, y \to z\}$$

$$G_2 = \{y \to x, y \to z\}$$

$$G_3 = \{z \to y, y \to x\}$$

Figure 3.8 depicts all three graphs. In all three graphs, node y d-separates x and z, and therefore each graph has the same I-map given by:

$\{(z \perp x|y), (x \perp z|y)\}$

We then conclude that G_1, G_2, G_3, are all I-equivalent.

Practice in Python

You will find the entire code for the example in `equiv_demo()` in the `imap.py` file. The code for creating the BN from the nodes and edges of the DAG is as follows:

```
def create_bn(nodes, edges):
    model = BayesianNetwork()
    model.add_nodes_from(nodes)
    model.add_edges_from(edges)
    return model
```

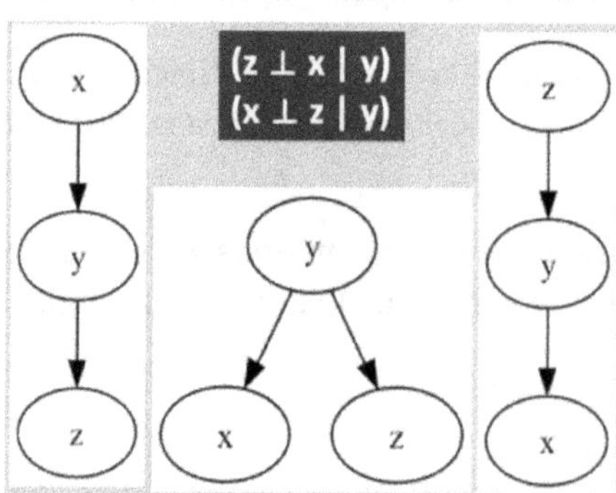

Figure 3.8 – I-map multiple graphical representation

Here, we list the code for `equiv-demo()`. The code first creates three models, `m1`, `m2`, and `m3`, then establishes the independency maps for each model represented by the sets: `set1`, `set2`, and `set3`. Finally, it checks for each model pair to see if their I-maps are identical and prints the results.

```
def equiv_demo():
    # first model
    m1 = create_bn(["x", "y", "z"], [("x", "y"), ("y", "z")])
    set1 = m1.get_independencies()
    # second model
    m2 = create_bn(["x", "y", "z"], [("y", "x"), ("y", "z")])
    set2 = m2.get_independencies()
    # first model
    m3 = create_bn(["x", "y", "z"], [("z", "y"), ("y", "x")])
    set3 = m3.get_independencies()
    if set1 == set2:
        print(f"m1 and m2 are I-equivalent:\n [{set1}] == [{set2}]")
    if set1 == set3:
        print(f"m1 and m3 are I-equivalent:\n [{set1}] == [{set3}]")
    if set2 == set3:
        print(f"m2 and m3 are I-equivalent:\n [{set2}] == [{set3}]")
```

The result from running the code is as follows:

```
m1 and m2 are I-equivalent:
 [(x ⊥ z | y) (z ⊥ x | y)] == [(x ⊥ z | y) (z ⊥x | y)]
m1 and m3 are I-equivalent:
 [(x ⊥ z | y) (z ⊥ x | y)] == [(x ⊥ z | y) (z ⊥x | y)]
m2 and m3 are I-equivalent:
 [(x ⊥ z | y) (z ⊥ x | y)] == [(x ⊥z | y) (z ⊥x | y)
```

Quiz

Are any of the graphs in *Figure 3.9* I-equivalent? Explain why and verify your answer by checking the I-maps of the three digraphs using `pgmpy`.

As we have seen, an I-map contains the global independencies from all the d-separations determined from the DAG. A BN is a representation of a joint probability distribution. For a BN to be an I-map for a joint distribution P, all independencies included in the I-map must be valid in P.

In the next section, we will present the five graphoid axioms that enable the derivation of global independencies in P based on the graph structure.

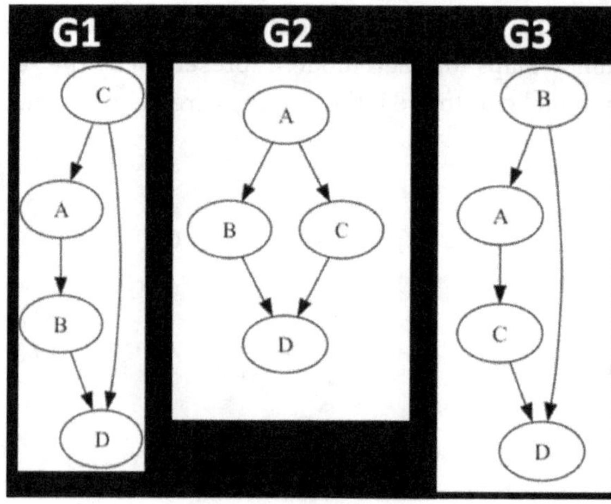

Figure 3.9 – Three digraphs

Graphoid axioms

In this section, we will learn about the graphoid axioms of symmetry, weak union, decomposition, contraction, and intersection. Those five axioms enable the derivation of new independencies from known ones. The axioms allow extending the independencies known from the local and global Markov properties.

The local Markov property (*Equation 3.7*) tells us that node X in graph G is independent of its non-descendants $nd(X)$ conditioning on its parents $pa(X)$. That is, for all nodes X in G, we have,

$$I(X, pa(X), nd(X)) \quad (3.19)$$

The global Markov property tells us that for a BN (G, P), if the set Z d-separates, the sets X and Y in G, X, and Y are conditionally independent in P. That is,

$$I(X, Z, Y) \text{ if Z d-separates } X \text{ and } Y \quad (3.20)$$

The graphoid axioms enable us to derive global independencies in the joint probability P based on the topology of graph G. We'll explain next each of the graphoid axioms.

Symmetry

The symmetry axiom is straightforward. It tells us that given Z, finding that X is conditionally independent of Y, that is, $I(X, Z, Y)$, must lead to concluding that Y is conditionally independent of X, that is, $I(Y, Z, X)$. In other words, if given Z, knowing Y does not influence belief in X, then knowing X also does not affect belief in Y. That is,

$$I(X, Z, Y) \Leftrightarrow I(Y, Z, X) \quad (3.21)$$

For instance, consider the graphs $G_1 = \{X \rightarrow Z, Z \rightarrow Y\}$, $G_2 = \{Y \rightarrow Z, Z \rightarrow X\}$, and $G_3 = \{Z \rightarrow X, Z \rightarrow Y\}$. Knowing that $I(X, Z, Y)$ holds allows us to infer that $I(Y, Z, X)$.

Decomposition

The decomposition axiom tells us that given Z finding that the union of two sets of variables Y and W is conditionally independent of X, that is, I (X,Z,Y ∪ W) must lead to concluding that given Z, each of Y and W are also conditionally independent of X, that is, I(X,Z,Y) and I(X,Z,W). In other words, if conditioning on Z two pieces of information, Y ∪ W, together not influencing belief in X, then conditioning on Z each piece of information individually is not going to affect belief in X. That is,

$$I\ (X, Z, Y\ \cup\ W) \Rightarrow I\ (X, Z, Y)\ \ and\ I\ (X, Z, W)\quad (3.22)$$

Applying this axiom to the local Markov property $I(X, pa(X), nd(X))$, we conclude that $I(X, pa(X), W)$ holds for all subsets W of $nd(X)$. Using the decomposition axiom, we can make assertions derived from the local Markov property to reason about subsets W of nondescendants of X,

$$I(X, pa(X), W)\ \ \forall W \subseteq nd(X)\quad (3.23)$$

Let us revisit *Example 3.1* of circuit diagnosis, having the DAG depicted in *Figure 3.3*. For the circuit's DAG, we have *Equation 3.12* from the local Markov property, which we rewrite as

$$I(\{D\}, \{Z, X, A2\}, \{A3, Y, A1, E, A4\})\quad (3.24)$$

The equation says that knowing $\{Z, X, A2\}$ makes $\{D\}$ conditionally independent of $\{A3, Y, A1, E, A4\}$. Since we can decompose the set $\{A3, Y, A1, E, A4\}$ in $\binom{5}{2}$ ways, *Equation 3.24* allows us to apply the decomposition axioms 10 times and derive 20 independencies. One instance of the derivations is the following:

$$I(\{D\}, \{Z, X, A2\}, \{A3, Y, A1, E, A4\}) \Rightarrow$$

$$I(\{D\}, \{Z, X, A2\}, \{A3\}) \text{ and } I(\{D\}, \{Z.X.A2\}, \{Y, A1, E, A4\})\quad (3.25)$$

Equation 3.25 derives two new independencies: $I(\{D\}, \{Z, X, A2\}, \{A3\})$and $I(\{D\}, \{Z.X.A2\}, \{Y, A1, E, A4\})$.

Weak union

The weak union axiom says that given Z, finding the union of Y and W does not influence belief in X must lead to the conclusion that given the union of Z and W will not make W affect belief in X. Alternatively stated, if two pieces of information Y and W are irrelevant to our belief in X, then learning one piece of information Y will not make W relevant for our belief in X. That is,

$$I\ (X, Z, Y\ \cup\ W) \Rightarrow I\ (X, Z\ \cup\ Y, W)\quad (3.26)$$

Let us apply this weak union axiom to the local Markov property $I(X, pa(X), nd(X))$. Let W be a proper subset of X's nondescendants $nd(X)$ and write $nd(X)$ as the union of W and its complement in $nd(X)$, $nd(X) - W$. Thus, given a subset $W \subset nd(X)$, we rewrite the local Markov property as

$$I(X, pa(X), (nd(X) - W) \cup W)\quad (3.27)$$

Substituting $(nd(X) - W)$ for Y in *Equation 3.26*, we conclude, based on the weak union axiom, that for all $W \subset nd(X)$,

$$I(X, pa(X), (nd(X) - W) \cup W) \Rightarrow I(X, pa(X) \cup (nd(X) - W), W)\quad (3.28)$$

Let us apply the weak union to *Example 3.1* of circuit diagnosis, having the DAG depicted in *Figure 3.3* and *Equation 3.24* of the local Markov property. We can partition the nondescendants $\{A3, Y, A1, E, A4\}$ as the union of two subsets $\{A3, A1, A4\}$ and $\{Y, E\}$. Upon applying the weak union, we get the following new independence relation.

$$I(\{D\}, \{Z, X, A2\}, \{A3, Y, A1, E, A4\}) \Rightarrow I(\{D\}, \{Z, X, A2, A1, A3, A4\}, \{Y, E\}) \quad (3.29)$$

Contraction

The contraction axiom tells us that if knowing Z makes X irrelevant to Y and knowing both Y and Z makes X irrelevant to W, then knowing Z makes X irrelevant to Y and W. In other words, given we know Z, if Y does not influence belief in X and after learning Y, the information W turns out not to affect the belief about X, then the two pieces of information Y and W must have been irrelevant to X from the start. That is,

$$I\,(X, Z, Y)\ and\ I\,(X, Z \cup Y, W) \Rightarrow I\,(X, Z, Y \cup W) \quad (3.30)$$

Example 3.5: contraction axiom

Figure 3.10 shows a DAG for a BN. We use the DAG to illustrate the contraction axiom. Consider the node subsets:

$$X = \{B\}, \qquad Y = \{T\}, \qquad Z = \{S\}, \qquad W = \{A, L, E, X\}$$

You can verify that X and Y are d-separated given Z, implying $I(\{B\}, \{S\}, \{T\})$, also that X and W are d-separated given $Z \cup Y$, implying $I(\{B\}, \{S, T\}, \{A, L, E, X\})$. According to the contraction axiom, we must conclude that X and $Y \cup W$ are independent given Z,' that is the assertion $I(\{B\}, \{S\}, \{T, A, L, E, X\})$. We will verify this using the *d-separation algorithm* implementation in Python next.

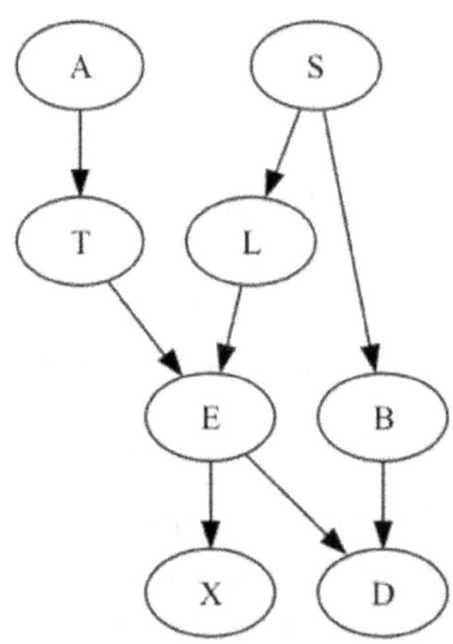

Figure 3.10 – DAG of a Bayesian network

Practice in Python

You will find the entire code for this practice in the `imap.py` Python file. Next, we'll display code snippets and explain their logic. First, we specify the nodes and edges and call the `create_bn` function to create the network.

```
nodes = ["A", "S", "T", "L", "E", "B", "X", "D"]
edges = [
    ("A", "T"),
    ("S", "L"),
    ("S", "B"),
    ("T", "E"),
    ("L", "E"),
    ("E", "X"),
    ("E", "D"),
    ("B", "D"),
]
m = create_bn(nodes, edges)
```

Next, we specify the sets X, Y, Z, W for the graphoid contraction example and make three consecutive calls for the d-separation algorithm explained earlier using the `d_sep` function, and finish by printing the results, verifying the contraction rule.

```
X = ["B"]
Y = ["T"]
Z = ["S"]
W = ["A", "L", "E", "X"]

res1 = d_sep(m, ["B"], ["S"], ["T"])
print(f"I({X}, {Z}, {Y}) is {res1}")

res2 = d_sep(m, ["B"], ["S", "T"], ["A", "L", "E", "X"])
print(f"I({X}, {Z} U {Y}, {W}) is {res2}")

res3 = d_sep(m, ["B"], ["S"], ["T", "A", "L", "E", "X"])
print(f"I({X}, {Z}, {W} U {Y}) is {res3}")
```

The results are as follows:

```
I(['B'], ['S'], ['T']) is True
I(['B'], ['S'] U ['T'], ['A', 'L', 'E', 'X']) is True
I(['B'], ['S'], ['A', 'L', 'E', 'X'] U ['T']) is True
```

Intersection

The intersection axiom says that given Z, if knowing W makes Y irrelevant to X and knowing Y makes W irrelevant to X, then the set union of Y and W must be irrelevant to X given Z. In other words, given Z, if knowing the information W makes Y not influence the belief about X and learning the information Y makes W not influence the belief about X, then Y and W's combined information must also not affect the belief about X. Formally,

$$I\ (X, Z\ \cup\ W, Y)\ \text{and}\ I\ (X, Z\ \cup\ Y, W) \Rightarrow\ I\ (X, Z, Y\ \cup\ W) \quad (3.31)$$

It is worth noting that the intersection axiom holds only for strictly positive probability distributions, for which all discrete value assignments have non-zero probability. The probability model for *Example 3.1* on circuit diagnosis does not have a strictly positive probability since the logic gates under non-faulty behavior have an output that is deterministic and given as a Boolean function of its inputs. The conditional probabilities have zero assignments to the logic gates' incorrect output.

Example 3.6: intersection axiom

Let us again consider the BN shown in *Figure 3.10* and use its DAG to illustrate the intersection axiom of *Equation 3.30* using the node subsets:

$X = \{E\}$ $\quad Y = \{A\}$ $\quad W = \{S\}$ $\quad Z = \{T, D, B, X, L\}$

You can verify that X and Y are d-separated given $Z \cup W$, meaning $I(\{E\}, \{T, D, B, X, L, S\}, \{A\})$, also that X and W are d-separated given $Z \cup Y$, meaning $I(\{E\}, \{T, D, B, X, L, A\}, \{S\})$. According to the intersection axiom, we must conclude that X and $Y \cup W$ are independent given Z, implying the assertion $I(\{E\}, \{T, D, B, X, L\}, \{A, S\})$. We will verify this using the *d-separation algorithm* implementation in Python next.

Practice in Python

You will find the entire code for this practice in the `imap.py` Python file. After creating the BN model as done in *Example 3.5*, we specify the sets X, Y, Z, and W for the graphoid intersection in *Example 3.6*, make three consecutive calls for the d-separation algorithm using the d_sep function, and finish by printing the results, verifying *Equation 3.30*.

```
res1 = d_sep(m, ["E"], ["A"], ["T", "D", "B", "X", "L", "S"])
print(f"I({X}, {Z} U {W}, {Y}) is {res1}")

res2 = d_sep(m, ["E"], ["S"], ["T", "D", "B", "X", "L", "A"])
print(f"I({X}, {Z} U {Y}, {W}) is {res2}")

res3 = d_sep(m, ["E"], ["S", "A"], ["T", "D", "B", "X", "L"])
print(f"I({X}, {Z}, {W} U {Y}) is {res3}")
```

The results are as follows:

```
I(['E'], ['T', 'D', 'B', 'X', 'L'] U ['S'], ['A']) is True
I(['E'], ['T', 'D', 'B', 'X', 'L'] U ['A'], ['S']) is True
I(['E'], ['T', 'D', 'B', 'X', 'L'], ['S'] U ['A']) is True
verified intersection axiom
```

As we conclude this section on graphoid axioms, the takeaway point is that through the five axioms of symmetry, decomposition, weak union, contraction, and intersection, we can derive new independence statements from known ones obtained from the Markov properties of Bayesian networks based on the topology of the DAG.

To complete the section topics, we will proceed to the last topic, the Markov blanket, which characterizes a particular independency relation providing a kind of "shield" between a node in a DAG and all the remaining nodes.

Markov blanket

In this section, we will learn about one further independence property in BNs, derived from the concept of the Markov blanket. Earlier, we learned about the local Markov property that tells us that, given its parents, a variable is conditionally independent of all its nondescendants in the DAG. The Markov blanket forms a conditioning such that, given the value assignments of all variables in the blanket, a variable is independent of all remaining variables in the network. That is, for any variable X, knowing the Markov blanket for X makes all variables outside the blanket irrelevant to X.

We begin this section by defining the Markov blanket and follow with a theorem that specifies the structure of the Markov blanket for nodes in a DAG. Next, we'll give an example using the BN in *Figure 3.10* and follow with practice in Python.

Defining the Markov blanket

The idea of the Markov blanket is to find a minimal subset of variables so that, when conditioning on those variables, a variable X becomes independent of all other variables in the BN.

A Markov blanket enables us to make inferences about a variable X based on the values of the variables within the blanket and ignore the irrelevant variables from outside the blanket.

Compare the Markov blanket with the local Markov property. The blanket implies that knowing the values of the variables in the blanket for a variable X makes the variables outside the blanket irrelevant to X. The local Markov property entails that knowing the values of the parents of X makes the nondescendants of X irrelevant to X. The first observation is that the blanket must be a superset of the parents of X. The second observation is that the independence condition from the Markov blanket for X includes nodes in a subset of the descendants of X. In contrast, the local Markov independence condition has only the nondescendants and precludes the descendants of X.

Definition 3.10: Markov blanket

Let G be the DAG of a BN having joint probability P on a set of variables U, which are the nodes of the DAG, G. We define the Markov blanket of a node X, denoted MB(X), as the minimal set of nodes such that knowing MB(X) makes all other nodes irrelevant to X. In other words, given MB(X), X is conditionally independent of every other node not in MB(X). Formally, MB(X) satisfies two conditions:

The Markov blanket independence condition:

$$I(X, MB(X), U - \{X\} - MB(X)) \quad (3.32)$$

No proper subset S of $MB(X)$ satisfies the independence condition:

$$I(X, S, U - \{X\} - S) \Rightarrow S \supseteq MB(X) \quad (3.33)$$

Definition 3.10 of the Markov blanket specifies the independence condition that the blanket must satisfy, but it does not articulate the set elements constituting the Markov blanket.

Next, we will learn about a remarkable theorem that characterizes, for any given DAG G, the minimal set $MB(X)$ for any graph node X. We'll follow with an example and practice in Python.

Theorem 3.1: Markov blanket

The Markov blanket, denoted as $MB(X)$, of a node X in the directed acyclic graph (DAG) G, consists of the node's parents, $pa(X)$; its children, $ch(X)$; and its spouses, which are the co-parents of its children, represented as $sp(X)$. We define the Markov blanket of X as

$$MB(X) = pa(X) \cup ch(X) \cup sp(X) \quad (3.34)$$

where:

$$sp(X) = \cup_{Y \in ch(X)} pa(Y) - \{X\} \quad (3.35)$$

We'll next give a worked example to apply *Theorem 3.1* on the Markov blanket and demonstrate the conditional independencies that result from conditioning on the Markov blanket.

Example 3.7: Markov blanket

Figure 3.11 shows the example Bayesian network used in the previous examples. Consider the variable L, highlighted in the figure by a thick boundary. By applying *Theorem 3.1*, we find that the Markov blanket of L is $\{S, T, E\}$. Recall that the Markov blanket of a node consists of the union of its parents, its children, and its spouses (that is, the other parents of its children). In this case, the parent of L is S, its child is E, and its spouse is T. Therefore, given its Markov blanket, L is conditionally independent of all nodes outside the blanket—namely, the uncolored nodes in *Figure 3.11*: $\{A, B, X, D\}$. Hence, we obtain the following independence relation:

$$I(\{L\}, \{S, T, E\}, \{A, B, X, D\})$$

We will practice this example using Python in the following section.

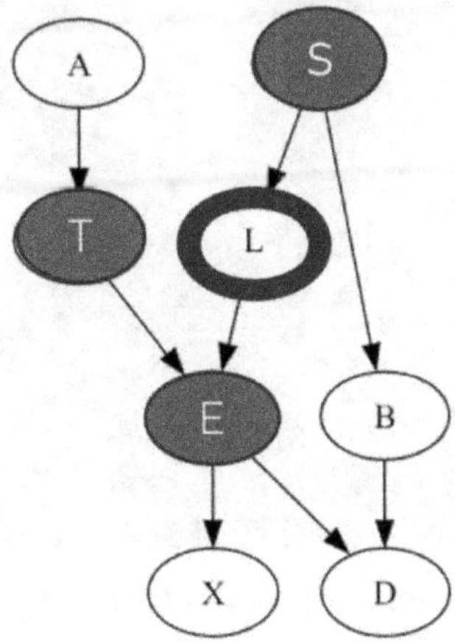

Figure 3.11 – The Markov blanket for L consists of S, T, and E

Practice in Python

You will find the entire code for this example in the `imap.py` file. Next, we'll show a code snippet for implementing *Example 3.7*. Let us walk through the code – we first create the BN model using the `example_bn` function. We use the `pgmpy` `get_markov_blanket` method to get the blanket for node L. The rest of the code verifies the independence relation implied by the blanket. The code computes the set of all nodes outside the blanket as the difference between the set of all nodes other than the node L and the set constituting the blanket. Next, we call the `d_sep` function based on *Algorithm 3.1* of d-separation to verify the conditional independence of L, and the outside set given the blanket, and print the result we expect to be true.

```
name = "mb_demo"
m = example_bn(name)
node = "L"
blanket = m.get_markov_blanket(node)
print(f"the Markov blanket for node {node} is {blanket}")
# find the set of nodes outside the blanket
all = set(m.nodes)
all.remove(node)
outside = list(all.difference(set(blanket)))
print(f"nodes outside the blanket: {outside}")
# verify the blanket independence condition
res = d_sep(m, [node], outside, blanket)
print(f"I([{node}], {blanket}, {outside}) is {res}")
print(f"verified Markov blanket")
```

Here is the program output.

```
created Bayesian network model mb_demo
Nodes:
['A', 'S', 'T', 'L', 'E', 'B', 'X', 'D']
Edges
[('A', 'T'), ('S', 'L'), ('S', 'B'), ('T', 'E'), ('L', 'E'), ('E', 'X'), ('E',
'D'), ('B', 'D')]
the Markov blanket for node L is ['T', 'S', 'E']
['X', 'B', 'A', 'D']
I([L], ['T', 'S', 'E'], ['X', 'B', 'A', 'D']) is True
verified Markov blanket
```

We have come to the end of this *Graphoid axioms and the Markov blanket* section, so let us review the highlights of what we covered.

The concepts of I-map and perfect I-map are of utmost importance in the study of probability models. We say a DAG G is an I-map of probability model P, or G is compatible with P if every d-separation in G implies a conditional independency in P. Similarly, G is a perfect I-map of P if every conditional independency in P implies a d-separation in G. These concepts are not just theoretical, but they have practical implications in the field of probability theory and statistics.

The conditional independence of a probability model P is underpinned by axioms that are of paramount importance in the field. These axioms, constituting a graphoid or semi-graphoid, include four semi-graphoid axioms: symmetry, decomposition, weak union, and contraction. The graphoid axioms, which are equally crucial, contain an additional axiom named intersection, which is satisfied if the probability model disallows zero probabilities.

In a BN, the concept of a Markov blanket plays a pivotal role. It defines the minimal set of nodes in the DAG G that render a node conditionally independent of all nodes outside the blanket. This practical concept is the key to understanding the blanket for a node X, which is the union of the parents of X, the children of X, and the spouses of X, the co-parents of X's children.

We'll proceed next to our last section on the modeling and independence of causal influence, where we will learn about several special classes of BNs having wide practical applications in modeling and machine learning.

Modeling and independence of causal influence

In the previous sections, we presented the fundamentals of the theory of probabilistic graphical models of BNs. We learned that specifying a BN comprises two components, one qualitative and the other quantitative. The qualitative component consists of a DAG, G, and the quantitative component is the factored representation of a joint probability distribution P over G.

To apply the theory to practice, we need to address the following questions:

1. Where do the qualitative and quantitative specifications come from?
2. How do we ensure maximum compactness of the factored representation of the BN?
3. How can we minimize the number of parameters required to quantify the conditional probability distributions or the CPDs?

In this section, we will address the preceding questions. Firstly, we will address the first question of where the network's specifications come from. We follow that by addressing the second and third questions, which are part and parcel of the modeling assumption of **independence of causal influence (ICI)**, or causal independence for short. You will learn about two BN models: noisy or and naive Bayes. Both models have wide use and significant applications in machine learning and diagnosis, and classification applications

Modeling

In this section, we address where the BN's specifications come from. There are two parts to the specification: qualitative and quantitative. The qualitative aspect has to do with specifying the structure of the DAG. The DAG encodes two pieces of information:

- The causal relationships between model variables
- The modular structure of the causal relationships

Suppose we have a model with three variables: A, B, and C. The modules are the subsets of the variables (e.g., {A, B}, {A, C}, {B. C}, or {A, B, C}). The causal relationships are the directed edges between the three nodes (e.g., $A \longrightarrow B,\ A \longrightarrow C,\ B \longrightarrow C$), or any other collection of directed edges respecting a total ordering on the variables to guarantee the directed graph is acyclic.

The qualitative specification of the model is not arbitrary, and there are sources of information that guide us in developing the proper structure for the BN. Those sources are as follows:

- **Prior knowledge of causal relationships**: When modeling engineering applications, there are well-defined causal paths from the system inputs to the system outputs and from control variables to the controlled variables. Hence, in those cases, we know the causes are the system inputs and the control variables from the actuators, and we know the effects from the system outputs, the measurements from the sensors, and so on.
- **Prior knowledge of modular relationships**: Again, when we model engineering systems components, we can identify the modules by the components' inputs and outputs and by the functional relationships between them. Those relationships specify the modular structure for the causal model. For example, to model an AND-gate in an electronic circuit, we represent the output as a node in a BN and have all the inputs as parents for the node in the DAG.
- **Assessments from experts**: In domains where the system is too complex to model at granular fidelity or ill-structured, such as in social studies or areas such as epidemiology and economy, we must rely on expert knowledge to distill information on the qualitative specification.

- **Learning from data:** Learning the qualitative specification of a BN purely from statistical data involves searching in the space of all possible structures. If we have n variables, a complete graph has `n(n-1)/2` edges, and each edge can have one of two directions. Hence, the hypothesis space is super-exponential. $2^{O(n^2)}$, and the problem is clearly very hard. There exist numerous algorithms for learning the structure of a BN from data, primarily relying on forms of heuristics such as the score-based methods and the K2 algorithm (Cooper, et al., 1992), (Ruiz, 2005).

Let us now address the second and third questions we posed earlier, which will be the topic of the next section, on independence of causal influence. There, you will learn about models that aim to maximize the compactness of the factored representation of the BN and minimize the number of parameters required to quantify the conditional probability distributions or the CPDs.

Independence of causal influence (ICI)

In this section, you will learn about the independence of causal influence assumption that establishes the basis for creating BN models structured as trees. This results in a) a compact representation of the probability model and b) fewer parameters for specifying the **conditional probability distribution (CPDs)**.

In the following, we will first describe the BN model following the causal independence assumption. Afterward, we will describe two instances of the ICI model, which have wide practical applications. The first instance is **noisy-or**, and the second is **naive Bayes**.

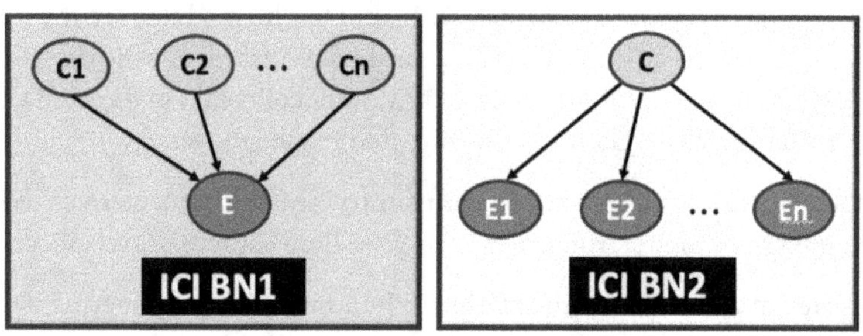

Figure 3.12 – Two ICI Bayesian networks

ICI models of Bayesian networks

The assumption of causal independence aims to reduce the complexity of the BN model. In its basic form, ICI-based models have a tree structure where we have independent causal influences. *Figure 3.12* shows two generic forms of ICI models.

- [**ICI BN1**] In this model, the tree has multiple causes and one effect. The causal influences from the different causes on the effect are independent. They are combined through some deterministic function, resulting in a CPD for the effect requiring parameters linear in the number of its parents, a massive reduction in the number generally required for the CPD, which is exponential in the number of parents. We will see an instance of this model in the noisy-or model.

- [ICI BN2] In this model, the tree has one cause, constituting the root of the tree, and multiple effects, which are the leaves of the tree. The causal influences from the different causes on the effect are independent. In this case, the CPDs are compact since each effect node has only one cause as its parent. We will see an instance of this in the naive Bayes model.

Having described the general idea of the independence of causal influence models, in the following sections, we will learn about the two most representative instances of such models. In the next section, we will explain the noisy-or model, which represents the ICI BN1 category of models.

Noisy-or model

Figure 3.13 shows the concept of a noisy-or Bayesian network. The noisy-or model has practical applications for modeling classifiers where Y represents the classifier's output, and the X variables represent the features of the object to be classified.

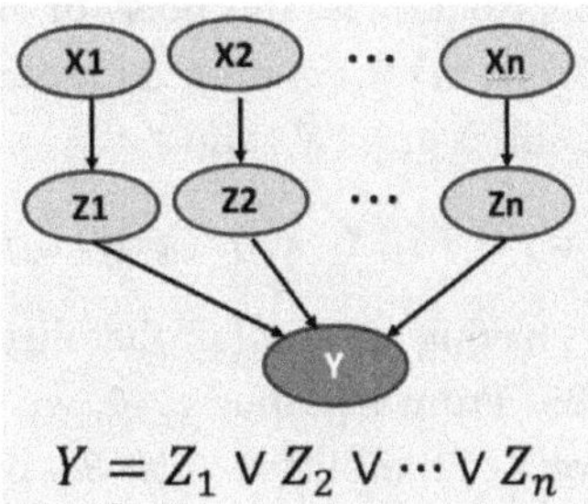

Figure 3.13 – The noisy-or model

We consider all the noisy-or model variables binary: on (1), off (0). The variables $X_1, \cdots, X_n$ are mutually independent causes. Each cause X_i independently influences the effect Y through a mediating variable Z_i. The influence links from every X_i to the mediator Z_i are probabilistic; each X_i causes Z_i unless an inhibitor prevents it, and the probability of that inhibition is q_i,

$$P(Z_i = 0 \mid X_i = 1) = q_i, \forall i \in \{1, \cdots, n\} \quad (3.36)$$

The effect Y is the logical-or of all the mediating variables $Z, \cdots, Z_n$ given by

$$Y = Z_1 \vee \cdots \vee Z_n \quad (3.37)$$

Hence, we can compute the conditional probability of Y, given the values of $X_1, \cdots, X_n$, as follows.

The probability of Y is off for the configuration of causes where a subset of causes is on is the product of inhibition probability for that subset by virtue of the causal independence assumption. Hence,

$$P(Y = 0 \mid X_1 = x_1, \ldots, X_i = x_i, \ldots, X_n = x_n) = \prod_{i:x_i=1} q_i \quad (3.38)$$

$$P(Y = 1 \mid X_1 = x_1, \ldots, X_i = x_i, \ldots, X_n = x_n) = 1 - \prod_{i:x_i=1} q_i \quad (3.39)$$

As we can see from *Equations 3.38 and 3.39*, the CPT for the noisy-or requires the specification of only n parameters instead of the necessary 2^n, where n is the number of the parent variables. Therefore, the independence of the causal influence assumption leads to a considerable reduction in the number of parameters required for quantifying the Bayesian network.

In the following, we give an example to work on the details of quantifying the CPD of a noisy-or model.

Example 3.8: noisy-or model

Consider a noisy-or model for a classifier node Y with inputs $X_1, \cdots X_n$ and respective inhibit probabilities:

$$q_1 = 0.6, q_2 = 0.2, q_3 = 0.1.$$

Plugging in the inhibit probabilities in *Equations 3.38* and *3.39*, we get the CPD for the classifier node Y, as shown in *Table 3.4*. Notice the highlighted cells in the table showing the three cases where only one input is active, and thus, the inhibit probability for that input directly gives the CPD for $Y(0)$. In all other cases where multiple inputs are on, we calculate the conditional probability from the inhibit probabilities according to *Equations 3.38* and *3.39*.

Let us compute the joint probability distribution for this noisy-or model. We use the following equation, which determines the joint probability from the conditional probability of Y given X_1, X_2, X_3, provided by the CPD, multiplied by the probabilities of X_1, X_2, and X_3:

$$P(X_1, X_2, X_3, Y) = P(Y \mid X_1, X_2, X_3)P(X_1)P(X_2)P(X_3) \quad (3.40)$$

Consider the $X1, X2$, and $X3$ variables having an equal probability of 0.5 for being on or off. Thus, we have $P(X_1)P(X_2)P(X_3) = 0.5^3 = 0.125$. From *Equation 3.40*, we can calculate the joint probability by simply multiplying the last two rows of the CPD in *Table 3.4* by 0.125, and the resulting two rows constitute a 16-element array quantifying the joint distribution.

In the following, we will practice implementing the noisy-or BN model of *Example 3.8* in Python. The purpose of the practice is twofold. First, calculate the joint distribution and check the results in *Table 3.4*. Second, check programmatically that the noisy-or BN is an I-map for the joint distribution.

X1		0	0	0	0	1	1	1	1
X2		0	0	1	1	0	0	1	1
X3		0	1	0	1	0	1	0	1
Y(0)		1	0.1	0.2	0.02	0.6	0.06	0.12	0.012
Y(1)		0	0.9	0.8	0.98	0.4	0.94	0.88	0.988

Table 3.4 – The Noisy-or CPD

Practice in Python

You can find the entire code in the `noisyor.py` file. First, we specify the structure of the noisy-or Bayesian network.

```
from pgmpy.models import BayesianNetwork
noisyor: BayesianNetwork = BayesianNetwork()
noisyor.add_nodes_from(["x1", "x2", "x3", "y"])
noisyor.add_edge("x1", "y")
noisyor.add_edge("x2", "y")
noisyor.add_edge("x3", "y")
```

Next, we define the CPDs for the four variables.

```
from pgmpy.factors.discrete import TabularCPD
cpd_1 = TabularCPD(
    variable="x1", variable_card=2, values=[[0.5], [0.5]])
cpd_2 = TabularCPD(
    variable="x2", variable_card=2, values=[[0.5], [0.5]])
cpd_3 = TabularCPD(
    variable="x3", variable_card=2, values=[[0.5], [0.5]])
cpd_y = TabularCPD(
    variable="y", variable_card=2,
    values=[[1,0.1,0.2,0.02,0.6,0.06,0.12,0.012],
            [0,0.9,0.8,0.98,0.4,0.94,0.88,0.988],],
    evidence=["x1", "x2", "x3"], evidence_card=[2, 2, 2],
)
```

We add the CPD specification to the model and check that the model is correct.

```
noisyor.add_cpds(cpd_1,cpd_2,cpd_3, cpd_y)
noisyor.check_model()
```

Now, we compute the joint probability distribution for the network. We then check whether the model is an I-map for the joint distribution.

```
import numpy as np
from pgmpy.factors.discrete import JointProbabilityDistribution

p = 0.5 * 0.5 * 0.5
val = [1, 0.1, 0.2, 0.02, 0.6, 0.06, 0.12, 0.012,
       0, 0.9, 0.8, 0.98, 0.4, 0.94, 0.88, 0.988]
jp = p * np.array(val)

print(f'joint probability distribution (JPD) P(Y,X1,X2,X3) is:\n{jp}')
print(f'sum of the JPD array equals \n{jp.sum()}')

JPD = JointProbabilityDistribution(['y','x1','x2','x3'],[2,2,2,2], jp)
noisyor.is_imap(JPD)
```

The program output is the following.

```
joint probability distribution (JPD) P(Y,X1,X2,X3) is:
[0.125  0.0125 0.025  0.0025 0.075  0.0075 0.015  0.0015 0.     0.1125
 0.1    0.1225 0.05   0.1175 0.11   0.1235]
sum of the JPD array equals
```

```
1.0

True
```

To summarize, we have described a noisy-or Bayesian network, worked on an example, and implemented the example using Python.

Another important instance of BN models based on the ICI assumption is the naïve Bayes model, which we will consider next.

Naïve Bayes

Naïve Bayes models are used in supervised machine learning to learn the classification of categorical data where the nodes have discrete values. Learning the network parameters takes place by feeding a collection of examples where each example consists of a value tuple for the set of features and their correct classification. After learning the network conditional probabilities, the naïve Bayes model predicts the category of new cases not previously included in the training dataset.

Naive Bayes is a particular case of the Bayesian model where the only edges are from the root node to the leaf nodes. The root node represents a classification variable, and the leaves represent features or attributes that depend on the classification variable. The only parameters that need to be specified are the conditional probabilities of each feature variable given the classification variable. The naïve Bayes model thus requires only a few linear parameters in the number of nodes.

Let us work an example to illustrate using the naive Bayes model in a classification problem.

Example 3.9: the weather problem

The weather problem is a fictitious dataset often used to illustrate machine learning methods (Witten, et al., 2011). The dataset concerns the conditions suitable for playing some unspecified game. This weather dataset has four attributes (the independent variables): outlook, temperature, humidity, and windy. The outcome, the classifier's output (the dependent variable), is whether to play. In the weather dataset, shown in *Table 3.5*, all four attributes have values that are symbolic categories.

The outlook can be sunny, overcast, or rainy; the temperature can be hot, mild, or cool; the humidity can be high or normal; and windy can be true or false.

outlook	**temperature**	**humidity**	**windy**	**play**
sunny	hot	high	FALSE	no
sunny	hot	high	TRUE	no
overcast	hot	high	FALSE	yes
rainy	mild	high	FALSE	yes
rainy	cool	normal	FALSE	yes
rainy	cool	normal	TRUE	no
overcast	cool	normal	TRUE	yes

outlook	temperature	humidity	windy	play
sunny	mild	high	FALSE	no
sunny	cool	normal	FALSE	yes
rainy	mild	normal	FALSE	yes
sunny	mild	normal	TRUE	yes
overcast	mild	high	TRUE	yes
overcast	hot	normal	FALSE	yes
rainy	mild	high	TRUE	no

Table 3.5 – The weather data

Having described the structure of the naive Bayes classifier for this weather problem, let us learn about the conditional probabilities from a training dataset and then apply the classifier to new data. We will do this in the following practice using Python.

Practice in Python: Naïve Bayes

The code for this practice is found in full in the `naivebayes.py` file included in the Python folder for this chapter. The dataset is included in the `play.csv` file.

First, we import the pandas and `NaïveBayes` libraries.

```
import pandas as pd
from pgmpy.models import NaiveBayes
```

Next, we create a `NaiveBayes` object named `weather` and read the dataset from the CSV file named `play.csv` into a pandas **DataFrame** called `values`, and print the data for viewing.

```
weather = NaiveBayes()
values = pd.read_csv("play.csv")
print(values)
```

To train the naive Bayes model named `weather`, we call the `fit` method of the `NaiveBayes` object named `play`. The `fit` method runs a supervised learning algorithm on the `values` DataFrame to estimate the CPDs of the Bayesian network model and print them for viewing.

```
weather.fit(values, "play")
for cpd in weather.get_cpds():
    print(cpd)
```

Having learned the CPDs from the data, we create the classifier BN by making the network structure and adding the learned CPDs. The following function creates the classifier BN and visualizes the graph using the `dagviz` function.

```
def create_classifier():
    classifier: BayesianNetwork = BayesianNetwork()
    classifier.add_nodes_from([
```

```
        "outlook", "temperature",
        "humidity", "windy", "play"
    ])
classifier.add_edges_from(
        [
            ("play", "outlook"),
            ("play", "temperature"),
            ("play", "humidity"),
            ("play", "windy"),
        ]
)
dagviz(classifier.nodes(), classifier.edges(), "weather")
```

Figure 3.14 shows the classifier BN with the learned CPDs. Having created the naive Bayes classifier BN, we will demonstrate next how to use the classifier BN to decide the classification of new data. The following code uses pgmpy belief propagation to perform the inference on the data. We will address inference with Bayesian networks in a later chapter.

Before we can use the classifier for inference, we need to create a dictionary for the state names of each variable so that we can query the classifier using the state names as the value assignments to variables instead of the state's index, so that, for example, we can write `outlook=sunny` instead of `outlook=0`. We must do that because pgmpy references the state names for network variables only by their index.

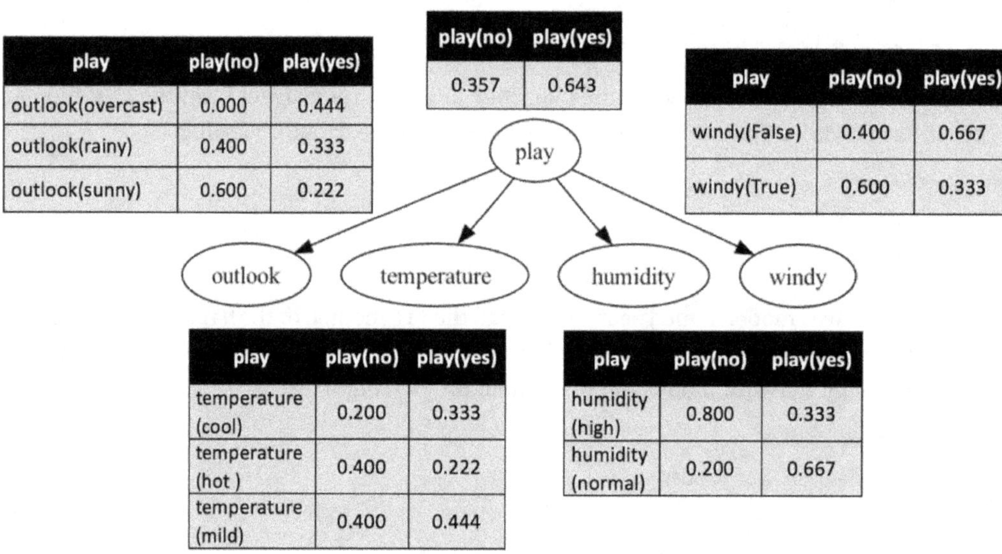

play(no)	play(yes)
0.357	0.643

play	play(no)	play(yes)
outlook(overcast)	0.000	0.444
outlook(rainy)	0.400	0.333
outlook(sunny)	0.600	0.222

play	play(no)	play(yes)
windy(False)	0.400	0.667
windy(True)	0.600	0.333

play	play(no)	play(yes)
temperature (cool)	0.200	0.333
temperature (hot)	0.400	0.222
temperature (mild)	0.400	0.444

play	play(no)	play(yes)
humidity (high)	0.800	0.333
humidity (normal)	0.200	0.667

Figure 3.14 – The naïve Bayes classifier BN and the CPDs learned from data

Before we can use the classifier for inference, we need to create a function that takes a list of variable value pairs where the value is a label – for example, `outlook = sunny` – and maps that list to another where all values are the indexes corresponding to the label in the variable domain – for example, `outlook = 0`. We must do that because `pgmpy` references the state names for network variables only by their index. We define the function named `map_evidence`, and you can see the relevant code in the Python file. Here is an example of `map_evidence` usage.

```
print(map_evidence({"outlook": "rainy", "windy": "False"}))
{'outlook': 1, 'windy': 0}
```

We are now ready to write the query function for the classifier. The query takes as input a value assignment for a subset of features and produces as output the probability of whether to play. The function uses the `BeliefPropagation` library for inference.

```
from pgmpy.inference import BeliefPropagation
def classify(evidence):
    belief_propagation = BeliefPropagation(classifier)
    evidence2 = map_evidence(evidence)
    res = belief_propagation.query(
        variables=["play"],
        evidence=evidence2,
    )
    return res
```

Finally, let us test the classifier on some examples. In the first example, let us determine whether to play for (*outlook = sunny and humidity = high*).

```
res1 = classify(classifier, {"outlook": "sunny", "humidity": "high"})
print(f"If outlook = sunny and humidity = high then\n{res1}")
# Output:
If outlook = sunny and humidity = high then
+---------+-------------+
| play    |   phi(play) |
+=========+=============+
| play(0) |      0.7881 |
+---------+-------------+
| play(1) |      0.2119 |
+---------+-------------+
```

In the second example, we ask whether to play for *(outlook = overcast and temperature = mild and windy = False)*

```
res2 = classify(
    classifier, {"outlook": "overcast", "temperature": "mild", "windy":
"False"}
```

```
)
print(
    f"If outlook = overcast and temperature = mild and windy = False then\
n{res2}"
)
# Output:
If outlook = overcast and temperature = mild and windy = False then
+---------+-------------+
| play    |   phi(play) |
+=========+=============+
| play(0) |      0.0000 |
+---------+-------------+
| play(1) |      1.0000 |
+---------+-------------+
```

We have come to the end of this section on modeling and independence of causal influence, so let us sum up the lessons learned.

Understanding the quantification of a Bayesian network is a crucial part of the modeling process. It involves populating the conditional probability tables with numeric values, which raises the question of where these numbers come from. We've identified three key sources of information for modeling with Bayesian networks: prior knowledge, expert assessments, and data learning. Each of these plays a significant role in the quantification process.

The complexity of the quantification task in the modeling process depends on the structure of the DAG describing the Bayesian network. One class of model structures that significantly facilitates the quantification process is founded on a modeling assumption called **independence of causal influence (ICI)**. The independence of causal influence is a modeling assumption to achieve two desirable modeling goals: maximizing the compactness of the CPDs required for quantifying the Bayesian network and minimizing the parameters needed to specify the CPDs.

We delved into two subclasses of ICI models, noisy-or and naive Bayes, which have many practical applications. Noisy-or is an ICI model with CPDs specified by minimal parameters. Naive Bayes is an ICI model with the most compact CPDs. We have explored a classifier application using noisy-or and a machine learning application using naive Bayes.

With the end of this section, we come to the end of this chapter on Bayesian networks and will proceed to summarize what we have learned in this entire chapter.

Summary

A Bayesian network is a factored representation of a complete probability distribution P according to a directed acyclic graph G. A BN satisfies the local Markov property that every node in graph G is conditionally independent of all its nondescendants, given all its parents in the distribution P.

The graphical condition of d-separation between disjoint subsets of nodes X, Y, and Z means that nodes in Z block all paths between nodes in X and nodes in Y. A path is blocked if it travels through a node in Z whose incident edges are non-colliding, or it passes through a collider node, but that node and all its descendants are outside Z. By edges colliding, we mean that the direction of the arrows is in opposing directions.

We use the notation $I_G(X, Z, Y)$ to denote all triples $< X, Z, Y >$ such that Z d-separates X and Y in G , $I_P(X, Z, Y)$ to denote all triples $< X, Z, Y >$ such that X is conditionally independent of Y given Z in P.

A Bayesian network (G, P) satisfies the global Markov property if I_G is a subset of I_P and we say G is an I-map for P or that G and P are compatible. G is a perfect map for P if every conditional independence in P implies a matching d-separation in G, $I_G \equiv I_P$.

A Bayesian network represented by a probability model P may have multiple DAGs compatible with P. The concept of I-equivalence is crucial here. Two DAGs are said to be I-equivalent if they have identical I-maps, a fact that should not escape your attention.

In probability theory, conditional independence relations typically satisfy a core set of axioms known as the semi-graphoid axioms. If the underlying probability distribution is strictly positive, these relations also satisfy an additional axiom, making them graphoid axioms. A Bayesian network I-map must adhere to the four semi-graphoid axioms: symmetry, decomposition, weak union, and contraction. Additionally, the graphoid axioms include an extra axiom known as the intersection axiom, which applies when the probability model disallows zero probabilities.

In a Bayesian network, the Markov blanket of node X is the minimal set of nodes, denoted as $MB(X)$, such that X is conditionally independent of every other node not in $MB(X)$. This concept is significant as it helps in understanding the conditional independence of nodes. It is a theorem that $MB(X)$ for Bayesian networks with a DAG G is the union of the parents of X in G, the children of X in G, and the spouses of X in G, the co-parents of X's children.

The independence of causal influence is a modeling assumption to achieve two desirable aims: 1) the compactness of the conditional probability distributions, the CPDs, and 2) the reduction in the number of parameters to specify the CPDs. We delved into two classes of ICI model, noisy-or and naive Bayes.

Learning the lessons in this chapter provides a solid foundation for the lessons of the next chapter on structural causal models. To perform causal inference, we need conditional dependencies and causal structures to determine the system's behavior under external interventions. In the next chapter, we extend the graphical probabilistic models in this chapter by combining graphical probabilistic models and structured equation models to provide a formal language for causal reasoning.

References

- Cooper G. F. and Herskovits E. *A Bayesian Method for the Induction of Probabilistic Networks from Data*. [Journal]. Machine Learning. 1992. pp. 309-347.
- Darwiche A. *Modeling and Reasoning with Bayesian Networks*. [Book]. New York, NY : Cambridge University Press, 2009.

- Langarizadeh M and Moghbeli F. *Applying Naive Bayesian Networks to Disease Prediction: A Systematic Review*. [Journal]. Acta Inform Med. 2016. pp. 364-369.
- Lauritzen S L and Spiegelhalter D J. *Local computations with probabilities on graphical structures and their application to expert systems*. [Journal]. J. Royal Statistics Society. 1988. pp. 157–194.
- Mano M. M. *Computer Engineering: Hardware Design* [Book]. [s.l.]: Prentice Hall, 1988.
- Ruiz C. *Illustration of the K2 Algorithm for Learning Bayes Net Structures* [Online]. 2005. `https://web.cs.wpi.edu/~cs539/s05/Projects/k2_algorithm.pdf`.
- Witten I H, Frank E and Hall M.A. *Data Mining* [Book]. Amsterdam: Elsevier, 2011.

4

Structural Causal Models

This chapter provides a practical introduction to two structural frameworks designed for representing and reasoning about complex relationships among observed and unobserved variables. We start with an overview of **Structural Equation Modeling (SEM)**, detailing its components and methods for analyzing fit statistics to either support or refute a hypothesized model. Following this, we introduce **Structural Causal Modeling (SCM)**, which utilizes structural equations similar to SEM but necessitates that the dependent variables (endogenous) be defined through functional equations involving both exogenous and endogenous variables. In a structural causal model, exogenous variables are described by a probability distribution and are independent of the endogenous variables. Unlike SEM, SCM inherently incorporates causality, where **Directed Acyclic Graphs (DAGs)** are used to show how dependent variables are related to each other. The representation of these graphs and their structural analysis are essential components of the SCM toolkit. Moreover, SCM provides formal causal semantics for Bayesian networks through the significant concept of intervention conditioning. Throughout the chapter, we include numerous illustrations, examples, and hands-on practice, featuring implemented code in R and Python.

In this chapter, we're going to cover the following topics:

- Structural Equation Modeling (SEM)
- Use case of SEM
- Representing with Directed Acyclic Graphs (DAGs)
- Structural Causal Modeling (SCM)
- Interpreting causal Bayesian networks
- Comparing SCM versus SEM

Technical requirements

The Python project for this chapter includes code for an SEM demo and tutorial notebooks, intended to support the hands-on practice outlined in this chapter. This project utilizes the `semopy` library (`https://semopy.com`). For setup and code execution, please refer to the instructions provided in the `README.md` file located in the Python project folder.

We will utilize the **DAGitty** tool and its R library throughout this chapter for examples and hands-on practice with DAGs in the context of SCM. You can access the DAGitty software and related resources at `https://www.dagitty.net`. It is essential to install the R software from the **Comprehensive R Archive Network** (**CRAN**), available at `https://cran.r-project.org`. Additionally, we recommend installing the **RStudio IDE** to run the **RMarkdown** documents containing the examples in this chapter. The free version of RStudio can be downloaded from `https://posit.co/products/open-source/rstudio/`. To execute the examples presented in this chapter, you will need to install various R libraries, which can be accomplished within the RStudio IDE. As we progress through the examples and R practice in this chapter, we will identify the specific libraries required.

Structural Equation Modeling (SEM)

In this section, we provide an overview of the fundamentals of SEM (Bollen, 1989). We start with basic definitions of SEM. Next, we describe the main components of SEM and present illustrative examples. We then include a section on **Confirmatory Factor Analysis** (**CFA**). Finally, we provide a tutorial on conducting CFA using Python with the `semopy` SEM library.

What is SEM?

In this section, we provide a gentle introduction to SEM. We begin with a short review of the statistical and structural elements of SEM, where you will learn about the tasks and statistical methods relevant to it. Next, you will become familiar with path models and path diagrams, which are key tools for the statistical and graphical representation and analysis of tasks in SEM.

Reviewing statistical and structural elements of SEM

SEM combines elements of **factor analysis** and **multivariate multiple regression** within a statistical framework designed to analyze and validate hypothesized theoretical models that describe relationships between conceptual constructs, named **latent variables**, not directly measured, and observed variables, named manifest variables or **indicators**, that are directly measured. Statistical testing will confirm the hypothesized influences between the model variables, which stem from prior research and reality assumptions.

The statistical multivariate framework of SEM allows researchers to assess how well the predictions and explanations of theoretical models align with evidence from observational datasets related to the observable model variables.

Multivariate multiple regression in SEM employs simultaneous linear equations to model relationships and predict multiple dependent variables, known as **endogenous variables**, based on a set of predictor variables, referred to as **exogenous variables**.

Multivariate multiple regression in SEM extends standard **multiple regression**, which generally predicts a single dependent variable based on a set of predictor variables. This method facilitates the testing of theoretical models that account for the covariance among multiple outcome variables. In contrast, standard multiple regression treats each dependent variable as separate, ignoring any correlations that may exist between them.

The goal of SEM is to construct models that explain the influence between variables in some domain validated by observational data. For example, SEM can help predict the influence of digital advertising on online sales in conducting a marketing campaign.

The two main tasks in SEM, once a model has been specified, consist of 1) determining whether the model parameters are identifiable from the observed variables and 2) estimating the parameters to quantify the strength of the influence between the variables.

Having overviewed the structural and statistical elements of SEM, we follow that, in the next section, by describing the elements of path models and path diagrams.

Representing SEM with path models and diagrams

Sewell Wright (1918) introduced the technique of path models that use correlation coefficients and regression analysis to model influence relationships among observed variables. SEM uses covariances between observed variables and statistical testing to confirm or disconfirm the model assumptions.

Path models in SEM are visualized with path diagrams, which use shapes and arrows to represent the relationships between variables. Path diagrams are directed graphs consisting of nodes and directed edges that follow the conventions and interpretations depicted in *Figure 4.1*.

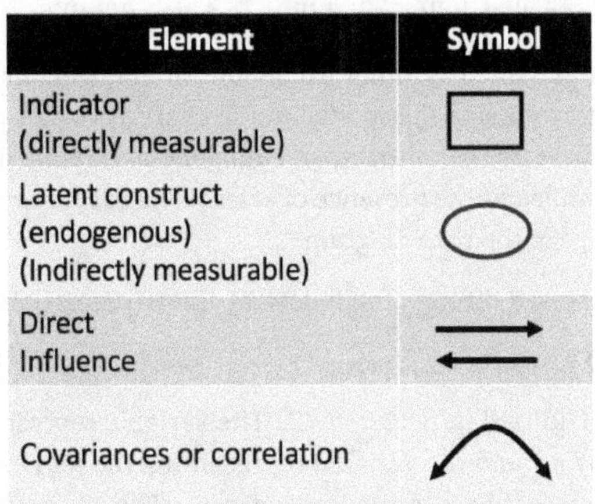

Figure 4.1 – Graphical representation of elements of a structured equation model

In a path diagram, the nodes stand for two different types of variables: observed variables, which are shown by boxes or rectangles, and unobserved (latent) variables, which are shown by circles or ellipses. One-way arrows represent the direct effect of one variable on another variable and correspond to a path coefficient representing a regression parameter for linear models.

The path diagram outlines hypotheses regarding direct and indirect influences among model variables. A directed edge from variable X to variable Y indicates that X directly influences Y. If X influences an intermediate variable Z, which in turn affects Y, then X indirectly influences Y. The proposed relationships between the model variables are based on previous research and assumptions about reality and are subject to confirmation through statistical testing.

Correlation between two distinct variables in the path diagram is represented by a curved, double-headed line between the node pair. Variance of a model variable is represented in the path diagram by a bidirectional curved line from and to the same node representing the variable.

Having provided a broad definition of SEM, we will now describe the components of SEM in the following section.

Describing components of SEM

A structural equation model is composed of two primary components: the **measurement model** and the **structural model**. We will begin by explaining the structural model, and then we will discuss the measurement model.

Structural model

The structural model is a fundamental aspect of a structural equation model, as it establishes the causal relationships between latent variables and the effect of independent (exogenous) variables on dependent (endogenous) variables. It assesses both direct and indirect effects within the model, including mediation and moderation analyses. Additionally, the model facilitates the calculation of explained and unexplained variance among dependent constructs. A high level of unexplained variance can compromise the model's validity, reliability, and overall utility.

In SEM, latent variables are classified as either exogenous or endogenous. An exogenous variable is influenced by external factors outside the model and is characterized by the absence of incoming directed edges in the path diagram. In contrast, an endogenous variable depends on other variables within the model, being identified by the presence of at least one parent variable, which is represented by an incoming directed edge in the path diagram.

The structural model is expressed through the following linear regression equation:

$$\eta_i = \sum_{j=1}^{n} \beta_{ij}\, \eta_j + \sum_{k=1}^{m} \gamma_{ik} \xi_k + \varepsilon_i \quad (i = 1, \cdots, n) \qquad (4.1)$$

The following notations are utilized in *Equation 4.1*: The variable representing the i-th endogenous latent variable is denoted by η_i, and the variable representing the k-th exogenous latent variable is denoted by ξ_k, The regression coefficients are denoted as β_{ij}. The γ_{ik} coefficients represent the effect of the k-th exogenous latent variable ξ_k on the i-th latent endogenous variable η_i. Furthermore, the model error associated with the i-th variable is denoted by ε_i.

Example: Structural model

Figure 4.2 presents the path diagram of a structural model based on a case study in the banking and commerce sector (Amani, 2020). This structural model seeks to examine the relationship between customer satisfaction and customer loyalty. It consists of four latent variables (unobserved constructs), with one exogenous variable, "customer satisfaction," serving as the independent variable. The remaining three variables are endogenous, or dependent. The latent variables "brand image" and "trust" each serve as mediators in the relationship between "customer satisfaction" and "customer loyalty."

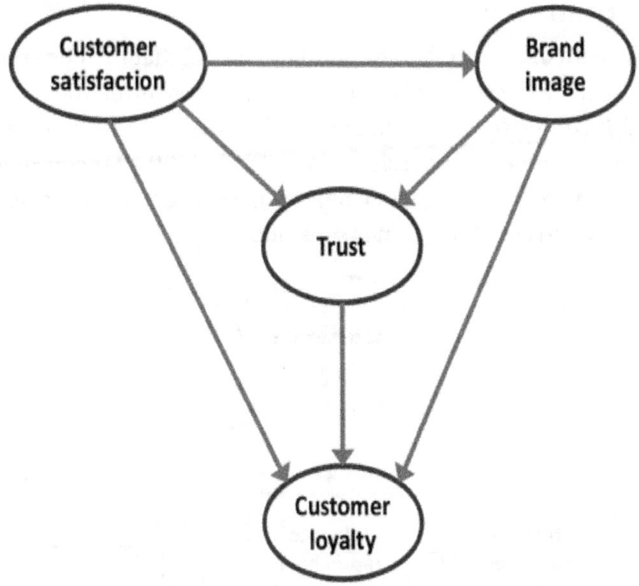

Figure 4.2 – Example structural mode from (Amani, 2020)

Having outlined the structural model, we will now describe the measurement model.

Measurement model

The measurement model consists of a system of linear regression equations that describe the relationship between latent (unobserved) variables and observed (measured) variables. Each observed variable is regressed onto the latent variable it aims to measure, along with a unique error term.

In an SEM measurement model, latent variables are unobserved constructs, indicators are the observed variables that are influenced by latent variables, and **factor loadings** are the regression coefficients that represent the strength of the relationship between the latent variable and its indicators. A high factor loading indicates a strong relationship, meaning the indicator reliably measures the latent construct.

The measurement model is expressed through linear equations that define the relationships between latent variables and their corresponding indicators, which are the observed or manifest variables. The measurement model for the exogenous latent variables is as follows:

$$x_i = \sum_{j=1}^{m} \lambda_{ij}\ \xi_j + \delta_i, \quad (i = 1, \cdots, v_x) \qquad (4.2)$$

The measurement model for the endogenous latent variables is this:

$$y_i = \sum_{j=1}^{n} \lambda_{ij}\ \eta_j + \delta_i, \quad (i = 1, \cdots, v_t) \qquad (4.3)$$

The following notation is used in *Equations 4.2* and *4.3*: The i-th indicator of the exogenous latent variables is denoted by x_i, while the indicator for the endogenous latent variable is denoted by y_i. The coefficients λ_{ij} represent the factor loading. Additionally, δ_i denotes the modeling error associated with the i-th indicator. The factor loading quantifies the change in an indicator that results from a one-unit change in the corresponding latent variable.

Example: measurement model

Figure 4.3 illustrates the path diagram of a measurement model that includes one latent variable and three indicators. This model assesses the latent variable, representing the unobserved construct of "customer satisfaction." It comprises three observed indicators obtained from a survey: "overall contentment with the product," "intent to repurchase," and "probability of recommending to others." The aim of the measurement model is to evaluate whether these three indicators effectively and reliably measure the underlying construct of "customer satisfaction."

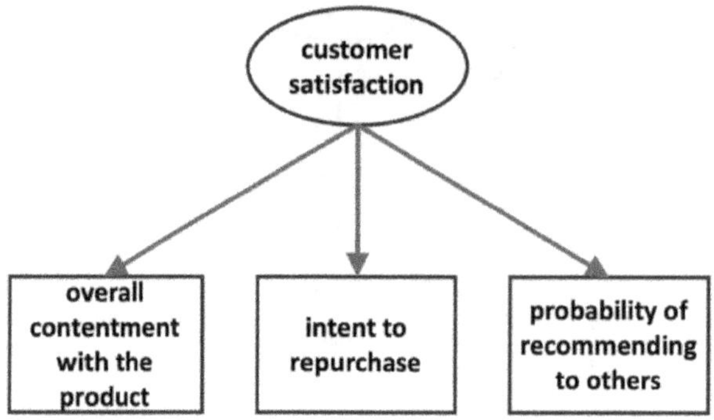

Figure 4.3 – Example measurement model

Following our discussion on the components of SEM, we will introduce confirmatory factor analysis in the next section.

Confirmatory Factor Analysis (CFA)

The measurement model utilizes **Confirmatory Factor Analysis** (CFA) to clarify the relationship between observed variables and their corresponding unobserved constructs, referred to as latent variables or factors. It specifies the extent to which each observed variable—also known as an indicator or manifest variable—correlates with its associated latent variable. Additionally, it accounts for measurement errors in the observed variables, which allows for more accurate estimations of correlations. CFA is employed to ensure that the indicators reliably represent their intended constructs before examining the relationships among the constructs.

In this section, we provide a brief theoretical background on CFA. CFA is a statistical procedure that investigates the proposed relationships between observed variables, known as indicators, and their underlying, unobserved constructs, referred to as latent variables or factors. In a one-factor CFA, each indicator is associated with a single factor. The factor loadings indicate the strength of the relationship between each indicator and its corresponding factor.

A one-factor CFA can be expressed through a simple linear equation for each indicator:

$$y_i = \lambda_i \eta + \varepsilon_i \qquad (4.4)$$

In this context, the symbols represent the following:

- y_i the i-th observed indicator
- λ_i the factor loading for the i-th indicator
- F the latent variable or factor
- ε_i the measurement error or unique variance associated with the i-th indicator

To identify the CFA model, it is essential to establish a scale for the factor. Common approaches to achieve this include (a) fixing the factor variance to 1 or (b) fixing one loading to 1.

In the following section, we will demonstrate how to use *semopy*, a SEM library in Python. We will outline the steps to specify a minimal CFA model, fit it with synthetic data, and extract the estimated model parameters along with the corresponding fit indices.

Practice using Python—CFA tutorial

This section provides an overview of the CFA tutorial in Python using the SEM library `semopy` (Igolkina & Meshcheryakov, semopy: A Python Package for Structural Equation Modeling, 2020). The code for this tutorial is available in the Jupyter file named `semopy_demo_tutorial.ipynb`, which is a component of the Python project for this chapter.

We intend this exercise as a pedagogical example to aid users in understanding the workflow, not as a comprehensive empirical study. The focus is on a basic CFA model estimated using the `semopy` library. The tutorial explains how to simulate data, specify and fit the model, inspect the results, compute fit indices, and generate a path diagram.

The tutorial illustrates the following key points:

- The tutorial demonstrates a lightweight simulation of a one-factor CFA model, in which a single latent variable (`eta`) explains three observed indicators (`y1`, `y2`, `y3`)
- The tutorial also covers the process of model fitting, which includes the extraction of parameter estimates and fit indices
- We will visualize the path diagram using either `Graphviz` or `Matplotlib`

In the next section, we will discuss the code utilized for simulating data in a one-factor CFA model.

Data simulation

The code in the notebook defines a helper function, `build_sample_data(n=120, seed=7)`, that draws a standard normal latent factor, eta ~ N(0, 1), and constructs indicators with specified loadings and residual noise. We set the model as follows: y1 = 0.8 eta + e1 (with e1 ~ N(0, 0.6^2)); y2 = 0.9 eta + e2 (with e2 ~ N(0, 0.5^2)); y3 = 0.7 eta + e3 (with e3 ~ N(0, 0.7^2)). The function returns a pandas DataFrame with columns `y1`, `y2`, and `y3`.

```
import pandas as pd
import numpy as np


def build_sample_data(n: int = 120, seed: int = 7) -> pd.DataFrame:
```

```
    rng = np.random.default_rng(seed)
    eta = rng.normal(size=n)  # Latent factor with variance ~ 1
    y1 = 0.8 * eta + rng.normal(scale=0.6, size=n)
    y2 = 0.9 * eta + rng.normal(scale=0.5, size=n)
    y3 = 0.7 * eta + rng.normal(scale=0.7, size=n)
    return pd.DataFrame({'y1': y1, 'y2': y2, 'y3': y3})

# Model specification (Lavaan-like syntax)
MODEL_SPEC = '''
# Measurement model
eta =~ y1 + y2 + y3

# Fix latent variance to 1 for identifiability
eta ~~ 1*eta
'''

data = build_sample_data()
data.head()
```

The results show the first five rows of the generated data, as follows:

y1	y2	y3
0.0546	-0.229	0.3533
-0.1156	0.6405	1.5187
-0.2905	-0.288	0.2225
-1.9111	-0.761	0.5843
-1.0426	-0.5546	1.4986

Next, we provide the code utilized to fit the one-factor CFA model, adhering to the specifications detailed in the previous step.

Fitting the model

The following code shows two steps for fitting the model. First, instantiate `semopy.Model` with the specification string. The code uses `MODEL_SPEC` to parse the specification and prepare an SEM object, `model`. Next, the `fit` method is applied to the SEM object, `model`, to estimate the model parameters by minimizing the SEM objective function, which is typically the maximum likelihood under multivariate normality.

```
from semopy import Model
model = Model(MODEL_SPEC)
res = model.fit(data)
```

Next, we will outline the code used to view the estimated model parameters obtained from the fitting step, as well as how to output the fit indices statistics.

Inspecting parameter estimates and fit indices

The following code retrieves a table summarizing the estimated model parameters in a way that is robust to semopy version differences.

```
# Estimates table
try:
    estimates = model.inspect()
except Exception:
    estimates = semopy.inspect(model) if hasattr(semopy, 'inspect') else None
print(estimates)
```

The code will output the first five rows of the data as follows:

```
   lval  op  rval  Estimate Std. Err z-value p-value
0  y1    ~   eta    1.0000        -       -       -
1  y2    ~   eta    0.6960   0.0711  9.7933  0.0000
2  y3    ~   eta    0.7717   0.0791  9.7621  0.0000
3  eta   ~~  eta    1.0000        -       -       -
4  y1    ~~  y1    0.1558   0.0727  2.1420  0.0322
5  y2    ~~  y2    0.3500   0.0579  6.0425  0.0000
6  y3    ~~  y3    0.4354   0.0717  6.0689  0.0000
```

The following code will display fit indices, which are statistical measures assessing how well a proposed model reproduces the relationships observed in a dataset. Common indices you may encounter include the **Comparative Fit Index (CFI)**, **Tucker-Lewis Index (TLI)**, **Root Mean Square Error of Approximation (RMSEA)**, **Goodness of Fit Index (GFI)**, **Adjusted Goodness of Fit Index (AGFI)**, and **Degrees of Freedom (DoF)**. No single index is sufficient; researchers should evaluate multiple indices to draw a comprehensive conclusion. The exact availability of these indices depends on the semopy version and any installed optional dependencies.

```
# Fit statistics
stats = None
try:
    stats = model.calc_stats() if hasattr(model, 'calc_stats') else semopy.
calc_stats(model)
except TypeError:
    try:
        stats = semopy.calc_stats(model, data)
    except Exception:
        stats = None
print(stats)
```

The output stats are as follows:

```
      DoF  DoF Baseline    chi2  chi2 p-value  chi2 Baseline     CFI     GFI  \
Value    1   3  8.2918  0.0040  127.6570  0.9415  0.9350

        AGFI     NFI     TLI  RMSEA     AIC      BIC  LogLik
Value 0.8051 0.9350 0.8245 0.2475 9.8618 23.7993  0.0691
```

The notebook includes detailed discussions and explanations regarding this practice, as well as accompanying Markdown documentation found in the Python project folder for this chapter.

Next, we move on to the final task in this demonstration: visualizing the path diagram for the model.

Visualizing the path diagram

The notebook parses the `MODEL_SPEC` to extract nodes and edges, then it first attempts to render the diagram using `Graphviz`. If `Graphviz` is not available—meaning neither the Python `Graphviz` package nor the dot executable can be found—the notebook defaults to using `Matplotlib`. This dual approach ensures that the diagram remains readable across different environments while keeping dependencies minimal. The code for this process is not included here due to space constraints. Running the code will produce the path diagram displayed in *Figure 4.4*.

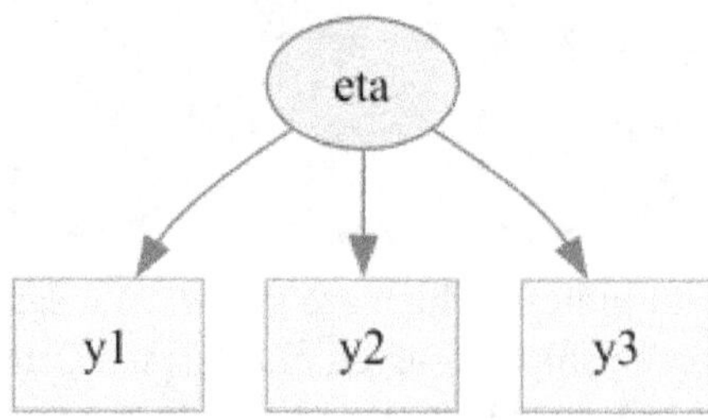

Figure 4.4 – Path diagram for semopy demo

To conclude this section, we'll summarize the key points. SEM is a multivariate statistical framework that enables researchers to analyze hypothesized relationships between variables, including both observed and latent (unobserved) constructs. SEM can illustrate relationships between latent and observed variables and is primarily used as a confirmatory method to test theoretical models. SEM consists of two components: a structural model and a measurement model. By incorporating latent variables, SEM allows researchers to account for measurement errors and utilize multiple indicators to more accurately capture the underlying concepts.

In the following section, you will explore a detailed use case of SEM applied to a real-life scenario.

Use case of SEM

In this section, we explore the application of SEM using a real-world example: **Bollen's political democracy model**. The primary goal of this model is to identify the causal factors that influence political democracy. Bollen's model posits that industrialization has a significant effect on the level of democracy.

We will begin by formally describing Bollen's structural equation model. Following that, we will provide a practical demonstration of how to fit this model programmatically using Bollen's dataset, which includes data from 75 countries. We will estimate the model parameters and examine the fit statistics.

Describing Bollen's political democracy model

Figure 4.5 illustrates the path diagram of Bollen's structural equation model, which encompasses both observed (measured) variables and unobserved (latent) variables.

In this path diagram, the latent variables—`dem60`, `dem65`, and `ind60`—are represented as nodes in oval shapes, while the measured variables—`y1`, `y2`, `y3`, `y4`, `y5`, `y6`, `y7`, and `y8`—appear as nodes within rectangular boxes. Additionally, this model incorporates a temporal dimension, reflecting levels of democracy at two distinct time points: 1960 and 1965. The path diagram demonstrates that `ind60` has a direct causal effect on `dem60`. Moreover, it indicates that `ind60` exerts both a direct effect and a longitudinal mediated effect on `dem65`, with `dem60` acting as the mediator. The term "longitudinal mediated effect" refers to the process in which a mediator variable explains how an initial exposure or independent variable influences an outcome over time, utilizing data collected at multiple time points.

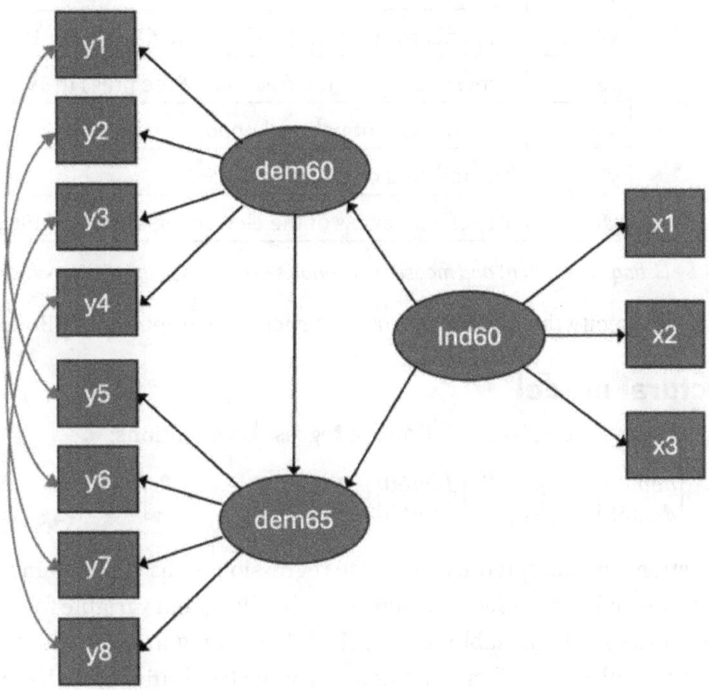

Figure 4.5 – Path diagram of Bollen's structural equation model

Table 4.1 provides a comprehensive list of all latent and observed variables, along with brief descriptions of each variable's representation.

Category	Name	Description
Exogenous latent variables	ind60	Industrialization (1960)
Exogenous latent variables	dem60	Political Democracy (1960)
	dem65	Political Democracy (1965):
Observed variables (measured)	x1	The gross national product (GNP) per capita in 1960
	x2	The inanimate energy consumption per capita in 1960
	x3	The percentage of the labor force in industry in 1960
	y1	Expert ratings of the freedom of the press in 1960
	y2	The freedom of political opposition in 1960
	y3	The fairness of elections in 1960
	y4	The effectiveness of the elected legislature in 1960
	y5	Expert ratings of the freedom of the press in 1965
	y6	The freedom of political opposition in 1965
	y7	The fairness of elections in 1965
	y8	The effectiveness of the elected legislature in 1965

Table 4.1 – Listing of the latent and measured variables in Bollen's political democracy model

In the next section, we specify the equations for the structural component of Bollen's model.

Bollen's structural model

The structural model is described by the following regression equations:

$$\begin{bmatrix} dem60 \\ dem65 \end{bmatrix} = \begin{bmatrix} 0 & 0 \\ \beta_{21} & 0 \end{bmatrix} \begin{bmatrix} dem60 \\ dem65 \end{bmatrix} + \begin{bmatrix} \gamma_{11} \\ \gamma_{21} \end{bmatrix} [ind60] + \begin{bmatrix} \varepsilon_1 \\ \varepsilon_2 \end{bmatrix} \quad (4.5)$$

The regression equations in *(4.5)* quantify the three regression paths depicted in the structural model in *Figure 4.5*. The directed edge $ind60 \rightarrow dem60$ says that the latent variable for industrialization in 1960 is a predictor of the latent variable for political democracy in 1960. The two directed edges $ind60 \rightarrow dem65$ and $dem60 \rightarrow dem65$ say that both the industrialization and the level of political democracy in 1960 are predictors for the level of democracy in 1965. The prediction error for `dem60` is the random variable ε_1, and for `dem65` is the random variable ε_2. The parameters $\beta_{21}, \gamma_{11}, and\ \gamma_{21}$ in *Equation 4.5* are the regression coefficients to be estimated from the observational data.

In the next section, we specify the equation for the measurement model.

The measurement model

Equation 4.6 gives the measurement model for the exogenous variable $ind60$.

$$\begin{bmatrix} x_1 \\ x_2 \\ x_3 \end{bmatrix} = \begin{bmatrix} \lambda^1_{x_1} \\ \lambda^1_{x_2} \\ \lambda^1_{x_3} \end{bmatrix} [ind60] + \begin{bmatrix} \delta_{x_1} \\ \delta_{x_2} \\ \delta_{x_3} \end{bmatrix} \quad (4.6)$$

The $\lambda^1_{x_i}$ coefficients are the factor loading for the i-th indicator x_i, $i = 1,2,3$. The δ_{x_I} are random variables representing the modeling error term for the indicator x_i, $i = 1,2,3$.

Similarly, *Equation 4.7* gives the measurement model for the endogenous variables $dem60$ and $dem65$.

$$\begin{bmatrix} y_1 \\ y_2 \\ y_3 \\ y_4 \\ y_5 \\ y_6 \\ y_7 \\ y_8 \end{bmatrix} = \begin{bmatrix} \lambda^1_{y_1} & 0 \\ \lambda^1_{y_2} & 0 \\ \lambda^1_{y_3} & 0 \\ \lambda^1_{y_4} & 0 \\ 0 & \lambda^2_{y_5} \\ 0 & \lambda^2_{y_6} \\ 0 & \lambda^2_{y_7} \\ 0 & \lambda^2_{y_8} \end{bmatrix} \begin{bmatrix} dem60 \\ dem65 \end{bmatrix} + \begin{bmatrix} \delta_{y1} \\ \delta_{y2} \\ \delta_{y3} \\ \delta_{y4} \\ \delta_{y5} \\ \delta_{y6} \\ \delta_{y7} \\ \delta_{y8} \end{bmatrix} \quad (4.7)$$

In the following section, we describe in a practice session with Python how to analyze the model fitted with observational data.

Practice using Python (political democracy use case)

The entire code for this practice is found in the *bollen_semopy.ipynb* notebook, included in the Python project for this chapter. This notebook demonstrates how to model the classic Bollen (1989) political democracy example using the *semopy* Python library.

1. Importing libraries and loading the dataset

   The code in the notebook begins by importing the necessary libraries that include `numpy`, `pandas`, and `semopy`. The Python version used is 3.13.0, and the versions of the libraries are as follows: `numpy`: 2.3.3; `pandas`: 2.3.2; and `semopy`: 2.3.11.

   After importing the necessary libraries, the first task is loading the Bollen (1989) Political Democracy dataset from `semopy's` bundled examples. The dataset contains various measures of political democracy and industrialization in 75 developing countries.

   ```
   import semopy as _semopy_ref

   data = _semopy_ref.examples.political_democracy.get_data()
   ```

2. Specifying the structural equation model

The next task is specifying the structural equation model (measurement + structural + residual covariances). The specification closely maps the path diagram in *Figure 4.5*. The code snippet for this task uses the `lavaan`-like syntax supported by `semopy`, as listed.

```
MODEL_SPEC = '''
# Measurement model
ind60 =~ x1 + x2 + x3
dem60 =~ y1 + y2 + y3 + y4
dem65 =~ y5 + y6 + y7 + y8
# Structural regressions
dem60 ~ ind60
dem65 ~ ind60 + dem60

# Residual correlations (same indicators across time)
y1 ~~ y5
y2 ~~ y6
y3 ~~ y7
y4 ~~ y8
'''
```

3. Fitting the model with the political democracy dataset

The following step is to fit the model with the dataset to estimate the regression coefficients and the error terms statistics. The code to perform this step is listed here.

```
# Fit the model
from semopy import Model
model = Model(MODEL_SPEC)
res = model.fit(data)
print('Fit completed. Optimizer result type:', type(res).__name__)
try:
    # If available, show optimizer success flag
    success = getattr(res, 'success', None)
    if success is not None:
        print('Converged:', bool(success))
except Exception:
    pass
```

The output from this step is the following.

```
Fit completed. Optimizer result type: SolverResult
Converged: True
```

4. Inspecting model parameters and common fit indices

Following the model fitting with the data, in this step, we are going to extract parameter estimates and common fit indices. The code is not listed here for lack of space. The output of this step will depend on the semopy version used when running the code. The line of code for extracting the entries of the parameter tables in new semopy versions is as follows:

```
estimates = model.inspect() # in new semopy versions
```

The output gives the estimates for the loadings in the measurement model, the regression coefficients in the regression equations of the structural model, the variance for all model variables (both the latent and the observed), and the covariances between the same y indicators across time. The output also gives the standard error, the p-value, and the z-value statistics associated with each estimate. The entire output is also saved in an HTML file named `bollen_use_case_report.html`.

In the following section, we will discuss and interpret the results of the report. For lack of space, we will not list the full code and the output table for this step, and list only the table for the estimated loadings and regression coefficients. The first three rows in the table in *Figure 4.6* show the three regression coefficients of the structural model, and the remaining rows show the estimates for the loadings of the measurement model.

	lval	op	rval	Estimate
0	dem60	~	ind60	1.4345
1	dem65	~	ind60	0.5069
2	dem65	~	dem60	0.8158
3	x1	~	ind60	1.0000
4	x2	~	ind60	2.1804
5	x3	~	ind60	1.8188
6	y1	~	dem60	1.0000
7	y2	~	dem60	1.3876
8	y3	~	dem60	1.0529
9	y4	~	dem60	1.3677
10	y5	~	dem65	1.0000
11	y6	~	dem65	1.3170
12	y7	~	dem65	1.3259
13	y8	~	dem65	1.3911

Figure 4.6 – Estimates of all path coefficients and loadings for Bollen's democracy model

5. Visualizing the path diagram

   In this step of the practice, the task is to plot the path diagram for Bollen's democracy model, as depicted in *Figure 4.5*. Here, we will outline the code logic, although the actual code is not included due to space constraints.

   The code processes the `MODEL_SPEC` to create graph edges that fall into three categories: first, measurement edges that are directed from latent variables to observed variables; second, structural edges that are directed from predictor variables to outcome variables within a regression equation; and third, bi-directed edges that represent residual covariance pairs, indicated by `~~` in the specification. The notebook's code attempts to render a compact diagram in two main ways: primarily through `Graphviz`, with a fallback option using `Matplotlib` if `Graphviz` is unavailable.

The next section will explain the interpretation of the empirical results obtained in the R practice for this use case.

Interpreting the empirical results

In this section, we provide a high-level discussion of the empirical results obtained from the code in the practice notebook, utilizing `semopy` estimates for Bollen's democracy model. The estimated structural equations are as follows:

$$dem60 \; = \; 1.435 \, ind60 \; + \; \varepsilon 1$$

$$dem65 \; = \; 0.507 \, ind60 \; + \; 0.816 \, dem60 \; + \; \varepsilon 2$$

The estimated structural regressions indicate significant relationships between industrialization and democracy at two time points: 1960 and 1965. The first equation shows that industrialization in 1960 (ind60) has a strong positive impact on democracy in that same year (dem60), with a coefficient of approximately 1.44. The second equation indicates that democracy in 1960 (dem60) has a positive influence on democracy in 1965 (dem65), with a coefficient of 0.82. Furthermore, industrialization in 1960 (ind60) has a direct impact on dem65, with a coefficient of 0.51. This finding suggests two pathways: a mediated pathway (ind60 → dem60 → dem65) and a direct pathway (ind60 → dem65). Thus, higher levels of industrialization are associated with increased democracy, both in the same period and in the following period.

Concluding this section, we provide the takeaway points. Bollen's political democracy model and the dataset from 75 countries in 1960 and 1965 describe a real-life use case for the application of SEM. The empirical results using `semopy` in the Python practice and the estimated model parameters highlight the stability of democratic institutions over time, as shown by the higher coefficient indicating the influence of democracy in 1960 (dem60) on democracy in 1965 (dem65) compared to the direct effect of industrialization in 1960 (ind60) on dem65. Overall, the analysis reinforces the idea that industrialization has both immediate and lagged effects.

Having discussed the fundamentals of SEM in the preceding section and its application in this section, we will now turn our attention to structural causal models in the following section.

Representing with Directed Acyclic Graphs (DAGs)

In this section, we discuss the qualitative representation of causal equation models through **Directed Acyclic Graphs** (DAGs). A DAG illustrates the causal influence structure of a structural equation model. We will first define key concepts and properties related to the DAG representation of causal graphs. Next, we will explore visualization and programmable tools that assist in constructing causal graphs and answering queries regarding causal structural analysis. Finally, we will review various categories of causal influence structures that underpin prototypical causal mechanisms, as described by the structural equations.

What is in a DAG?

In this section, we start by defining DAGs that represent causal relationships. Next, we revisit the important condition of d-separation and its use to make assertions about conditional independence between variables represented as nodes in a causal DAG.

Defining a DAG

DAGs provide a formal language for representing the structure and qualitatively depicting the causal influences between structural causal model variables. A DAG specifies the model structure by defining the variables of the graph nodes. Each node in the graph has a corresponding structural equation in the model where the node is the dependent variable, and the node's parents are the independent variables in the structural equation.

Let us begin with a definition of a causal DAG and specification of its structure.

Definition: causal DAG

A causal graph is a directed acyclic graph $G = (V,\ E)$, where $V = \{\, V_1,\ V_2, \cdots,\ V_n\}$ is a set of nodes and $E = \{\, (V_i,\ V_j) |\ 1 \leq i, j \leq n,\ i \neq j\}$ is a set of directed edges taken to mean an assumption that V_i causes V_j. We denote an edge directed from X_k to X_i as $X_k \rightarrow X_i$, and we say X_k is a **parent** of X_i or X_i is a **child** of X_k. A **directed path** comprises a sequence of directed edges, all oriented in the same direction and such that every edge points to the vertex that its successor edge points from. If there is a directed path from node X to node Y, we say Y is a **descendant** of X and X is an **ancestor** of Y. A node and all its parents constitute a **causal family**. A DAG has no **directed cycles**, a directed path that begins and ends at the same node.

Figure 4.7 shows an example DAG. In the example, A is a parent of C and C is a child of A. An example of a directed path in the example DAG is (A, C, E, G). In the DAG we have, A is an ancestor of G and G is a descendant of A. The family of the node H is the set consisting of the node and its parents, which is the set $\{F, H, H\}$. Notice the DAG has no directed cycles – that is, a directed path that begins and ends at the same node.

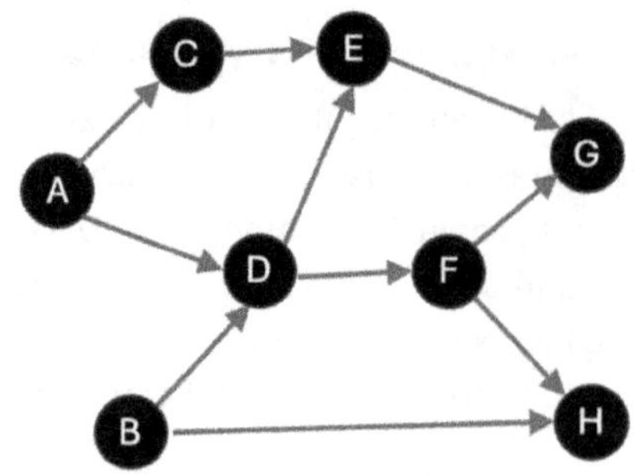

Figure 4.7 – Example DAG

A causal graph has a causal ordering, which is the topological sort of the nodes of the graph. Given a DAG, a topological sort is an ordering of its nodes such that every directed edge points from a lower-ordered node to a higher-ordered node.

In a DAG, nodes can be arranged along a horizontal line so that all directed edges point from left to right. A topological sort provides a total ordering that respects all partial orders established by the DAG, ensuring that each parent node appears before its children in the sequence. *Figure 4.8* illustrates a causal order corresponding to the DAG depicted in *Figure 4.7*.

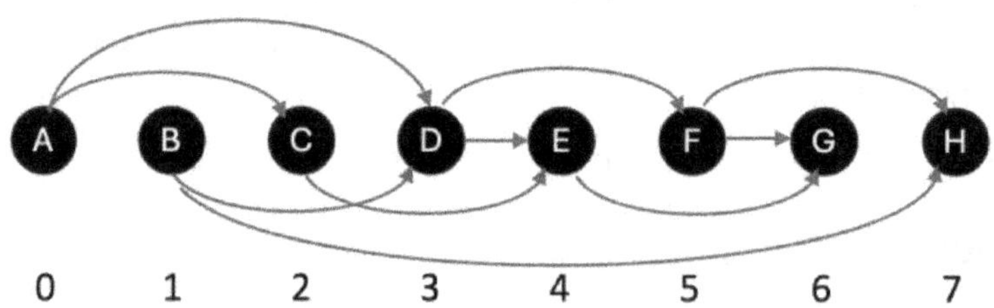

Figure 4.8—Causal ordering of the example DAG

The paths in DAGs indicate direct and indirect causal influences between model variables. Just as path analysis is central in SEM, paths in the DAG of a structural causal model also have significant interpretations and analytical implications, as we will explain next.

The paths between different nodes in the graph are essential for causal analysis and for determining conditional independencies. The following two definitions formally describe paths in a causal DAG and explain how a third set of nodes can open or block paths between a node and a set of other nodes.

Definition: paths

A path between two nodes, X and Y, in a causal graph G is a sequence of nodes that begins at X and ends at Y, where an edge in G joins every pair of consecutive nodes in the sequence regardless of the edge's directions. We call it a **causal path** if all edges along the path point in the same direction. We call it a **biasing path** if the path has directed edges with different directions. Suppose a directed edge connects X and Y. In that case, X is a **direct cause** for Y and Y is an immediate effect of X. If there is a causal path from X to Y with intermediate nodes in between, then X is an **indirect cause** of Y and Y is an indirect effect of X.

Example: causal and biased paths

The path (A, D, F, G) is an example of a causal path in the causal graph presented in *Figure 4.8*, as all edges between consecutive node pairs along this path share the same direction. Conversely, an example of a biased path in the same causal graph is (C, A, D, B, H), where the edges exhibit different directions. In the causal graph, A serves as a direct cause of C and an indirect cause of F.

Revisiting d-separation and conditional independence

In this section, we revisit the definition of d-separation, which provides a means for checking conditional independence between two variables represented by nodes of a causal DAG, as was covered previously in *Chapter 3*, on Bayesian networks. Conditional independence plays an important role in causal analysis, as we will discuss throughout this chapter.

The following definition describes the conditions in which a path between two nodes is open or blocked and defines when two sets of nodes are d-separated or d-connected by a third set of nodes, possibly empty and disjoint from the other two sets.

Definition: d-separated, d-connected

Given a set of nodes Z possibly empty in a causal graph G, we say that Z blocks a path between two nodes X and Y, not in Z if one or both of the following hold:

- The path has a node in the set Z such that the pair of edges incident to the node either point in the same direction, forming a **chain**, or point away from each other, creating a **fork**.
- The path has a **collider** node whose incident edges point opposite one another, and the collider node and all its descendants are outside the set Z.

We call the path **open** if Z does not block a path between X and Y.

If Z blocks all paths between X and Y, we say that Z d-separates X and Y, and X and Y are d-connected otherwise.

Example: d-separation

In the causal graph depicted in *Figure 4.7*, we observe that A is d-separated from E given the set {C, D}. This is because all paths from A to E are blocked by either C or D. Specifically, C is on the path (A, C, E), while D is on the path (A, D, E). The edges incident to C (or D) along the paths from A to E are oriented in the same direction, thereby forming a chain. Additionally, in the causal graph, A and B are d-connected given $\{D\}$ since D acts as a collider and is present on the path (A, D, B), which constitutes a biasing path.

Let X, Y, and Z be three disjoint sets in the DAG. We say that the set Z d-separates X and Y given Z if every path between a node in X and a node in Y is blocked by the set Z. If a set Z d-separates X and Y in the DAG, then X and Y are conditionally independent given Z in the probability distribution factored according to the DAG representation of the Bayesian network, as was covered in *Chapter 3*.

Conditional independence

The d-separation of X and Y given Z implies that X and Y are conditionally independent given Z. That is,

$$X \perp Y \mid Z \Leftrightarrow P(X,Y|Z) = P(X|Z)P(Y|Z)$$

If X and Y are d-connected, Z and Y are possibly dependent or associated.

Let us review the main points of this section. A causal graph is a DAG that illustrates the qualitative component of a structural causal model. In this graph, the nodes represent the model variables, while the directed edges indicate the causal influences between these variables. The causal graph explicitly states causal assumptions and qualitatively describes the causal mechanisms for each causal family within the DAG, which includes a node and all of its parent nodes. Furthermore, the structure of the causal DAG defines all causal and biasing paths between every pair of nodes in the graph.

In the following, we will describe how to construct causal graphs, programmatically specify them, and perform structural analysis on them.

Constructing causal graphs

In this section, we describe the categories of various variables in a DAG and how to construct causal graphs in existing software and leverage libraries to build models and query them programmatically.

To construct causal graphs, we will use the graph specification language of `DAGitty` and the `DAGitty` library (Textor, van der Zander, & Ankan, Package 'dagitty', 2022) in the R language to perform modeling and graphical analysis of structural causal models. `DAGitty` is a tool developed for causal analysis with DAGs for practitioners in the epidemiology application domain.

The endogenous variables represented by a DAG include various categories of variables. In the following, we list those categories following the `DAGitty` lexicon (Textor, Drawing and Analyzing Causal DAGs with DAGitty, 2022):

- **Exposure**: The exposure variables are the independent variables or the causes, which are the manipulated variables defining the intervention (treatment) in an observational or experimental study.
- **Outcome**: The outcome variables are the dependent variables, which are the observed effects of the exposures.
- **Adjusted**: The adjusted variables are those variables that, when we control or condition on them, we remove the confounding bias in the causal effect between the exposure and the outcome.
- **Selected**: The selected variables (**covariates**) impact the cause-effect relationship between the exposure and the outcome. Typically, the covariates define characteristics of the population sampled in an observational or experimental study, such as age, gender, and so on.

The covariates can impact the causal effect relationship between the exposure and the outcome in various ways, depending on the structure of the DAG (Gilthorpe). A covariate can be the following:

- **Confounder:** A confounder is both a) an ancestor of the exposure on a directed path not mediated by the outcome and b) an ancestor of the outcome on a directed path not mediated by the exposure
- **Mediator:** A mediator is a variable on a directed path between the exposure and the outcome, thus being a descendant of the exposure and an ancestor of the outcome
- **Proxy confounder:** A proxy confounder is a descendant of a confounder and an ancestor of either the exposure or the outcome (not both, or else it is a confounder!)

Example: causal graph

Figure 4.9 presents a causal graph created using `DAGitty`. This software features a user-friendly graphical interface and a powerful graph editor for causal modeling. The graphical representation of structural causal models in `DAGitty` employs various colors for the nodes and edges of the DAG, helping to illustrate the different categories of nodes and types of paths.

The colored nodes distinguish the various node categories. Here X is the exposure and Y is the outcome. There are two covariates I and Z. Z is a confounder (an ancestor of both the exposure and effect). I is a mediator (a node on a causal path between the exposure X and the effect I). I is also a proxy confounder (descendant of a confounder and ancestor of the outcome). There are two causal paths: $X \rightarrow Y$ and $X \rightarrow I \rightarrow Y$. There is one biasing path (concatenated edges, ignoring their directions): $X \leftarrow Z \rightarrow I \rightarrow Y$.

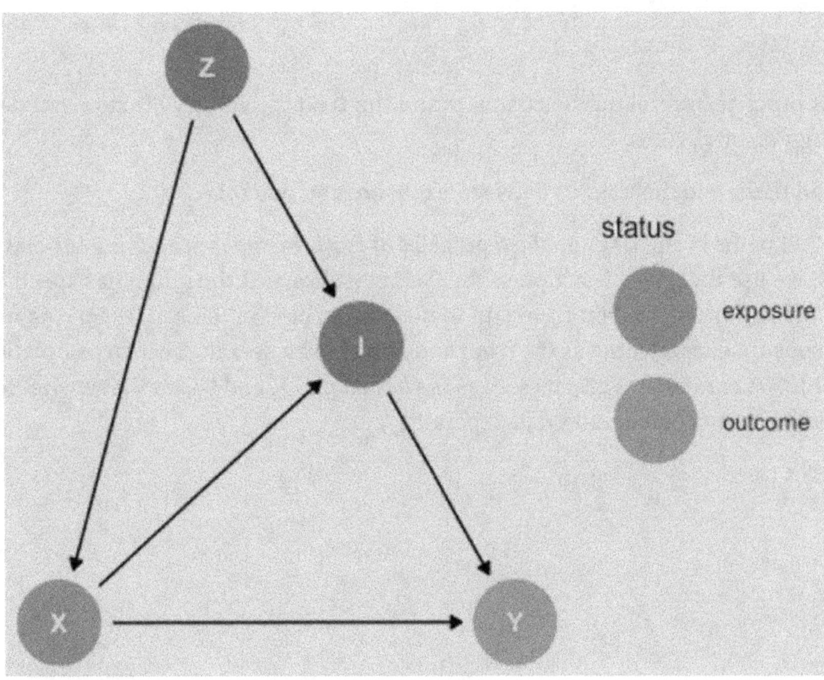

Figure 4.9 – Example DAG

In the following, we will describe how to create and analyze DAGs programmatically using the R language. `DAGitty` has a graph specification language like `Graphviz's` "dot" language.

Practice using R (DAG)

The R Markdown document `chapter4.Rmd` contains all the R code to construct the DAG in *Figure 4.9* and to run queries as shown in the following practice record. The following code will generate the DAG in *Figure 4.9*.

```
# Example: DAG with one exposure, and one outcome, and two covariates
g <- dagitty("dag{
  X -> Y
  X <- Z -> I
  X -> I -> Y
  X [exposure]
  Y [outcome]
}")
# set coordinates for graph layout
coords<-list(X=c(X=1, Z=2, I=3, Y=4),
             Y=c(X=1, Z=3, I=2, Y=1))
coords_df<-coords2df(coords)

coordinates(g)<-coords2list(coords_df)
#Plot causal graph
ggdag_status(g) + theme_dag()
```

Next, we will practice several queries using R and the `DAGitty` library (Textor, van der Zander, & Ankan, Package 'dagitty', 2022).

1. Find all paths and their status between exposure and outcome.

   To get all paths in the DAG, g, between pairs of from-to nodes specifying the start and end of paths, we use the `paths` function of the `DAGitty` library. If the graph has specified exposure and outcome nodes and no from and to nodes specified in the call, the `paths` function uses the exposure and outcome as the from and to nodes by default. The `paths` function returns a list with two components: `$paths` gives the actual paths, and `$open` shows whether each path is open (d-connected) or closed (d-separated).

   ```
   paths(g)
   ## $paths
   ## [1] "X -> I -> Y"      "X -> Y"        "X <- Z -> I -> Y"
   ##
   ## $open
   ## [1] TRUE TRUE TRUE
   ```

2. Find all causal paths between exposure and outcome.

   In the following query, we use the paths function again, with the added directive that the paths shown should only be directed (i.e., causal) paths.

```
paths(g, "X", "Y", directed= TRUE)$paths
## [1] "X -> I -> Y" "X -> Y"
```

3. Find all descendants of a given node in the graph.

   To get the descendants of node "Z", we use the descendants function, taking two arguments: the DAG, g, and the node, "Z".

```
descendants(g, "Z")
## [1] "Z" "X" "Y" "I"
```

4. Find all ancestors of a given node in the graph.

   To get the ancestors of node "Y", we use the ancestors function, taking two arguments: the DAG, g, and the node "Y".

```
ancestors(g, "Y")
## [1] "Y" "X" "Z" "I"
```

5. Find all conditional independencies.

   To get all conditional independencies in the DAG, g, we use the impliedConditionalIndependencies function, which takes one argument, namely the DAG, g. The function returns a list of conditional independence statements that must hold for every joint probability distribution compatible with the graph g.

```
impliedConditionalIndependencies(g)
## Y _||_ Z | I, X
```

6. Check if a third list of nodes d-connects between two nodes.

   To check whether it is true or false that the node "I" d-connects between node "Y" and node "Z" in the DAG, g, we use the dconnected function as follows:

```
dconnected(g, "Y", "Z", "I")
## [1] TRUE
```

7. Check if a third list of nodes d-separates between two nodes.

   To check whether it is true or false that the list of nodes "X" and "I" d-separates between node "Y" and node "Z" in the DAG, g, we use the dseparated function as follows:

```
dseparated(g, "Y", "Z", c("X","I"))
## [1] TRUE
```

To conclude this section, let us summarize what we covered. We described the types of variables in the DAG: exposure/treatment, outcome, and the covariates of type selected, adjusted, confounder, and proxy-confounder. With hands-on practice, we learned how to programmatically construct and pose queries for a DAG using R and `DAGitty`.

In the next section, we describe categories of causal structures corresponding to different causal mechanisms, standard building blocks in causal models of DAGs.

Categorizing causal influence structures

In previous sections, we covered definitions of DAGs and their use in representing the structural component of the structural causal model.

In this section, we describe five patterns of causal influence that describe common part structures of many causal DAGs. The five causal structures represent primitive causal influence, building blocks of more complex causal mechanisms in a structural causal model. Interpreting those basic mechanisms facilitates the causal analysis of more complex mechanisms in a causal DAG.

The basic five mechanisms presented in the following are 1) causation, 2) mediator, 3) confounder, 4) collider, and 5) instrument variable.

Causation is the most basic, where there is a direct influence from exposure to the outcome. All others add levels of complexity by contributing causal or biasing paths, including other covariates. The mediator is a mechanism where a covariate mediates the conduit for the causal effect. A confounder is a mechanism where an additional covariate exerts a causal influence on both the exposure and the outcome, creating a biased path and confounding the direct causal effect of the exposure on the outcome when not conditioned upon in a controlled experiment—a phenomenon known as **confounding bias**. The collider is a mechanism where an additional covariate is a common effect of both the exposure and the outcome, creating a biasing path and a "spurious" correlation when conditioned upon in a controlled experiment, a phenomenon referred to as **collider bias**. The last mechanism, the instrumental variable, uses an additional variable called an instrument to mitigate the effect of unobserved confounders. The following sections describe the five mechanisms and depict their causal graphs.

Causation

The most basic mechanism is causation, which is the starting point of any observational study investigating possible causal relations between two random variables corresponding to a probable cause interpreted generically as exposure or treatment (commonly inherited terminology from the epidemiology domain) and another variable representing the treatment's effect or outcome.

Figure 4.10 shows the causation structure where T represents a cause and Y its effect. T is the parent of Y, and Y is a child of T. The variable T can represent a treatment or exposure in the broad sense of the word, not restricted to the original use of the terms coming from the epidemiology application domain. The direction of causality implies a temporal precedence relation that the treatment or exposure must precede the outcome.

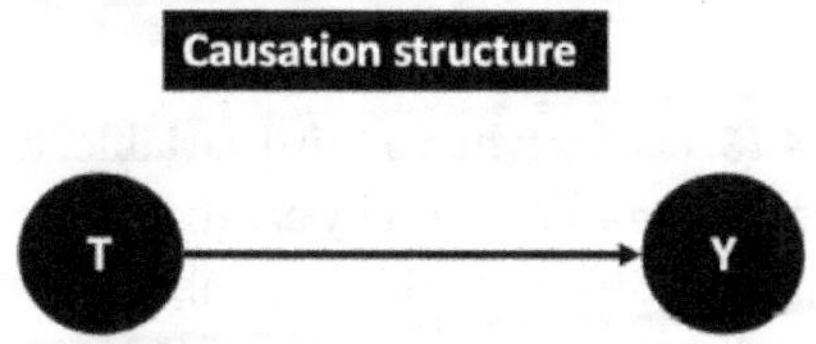

Figure 4.10 – Causation structure: treatment T directly affects outcome Y

The causation structure represents a hypothesis suggesting that two variables are causally linked. For instance, one might assume a cause-effect relationship between dieting and weight loss, intelligence and academic performance, and smoking and lung cancer, as well as regular exercise and overall well-being.

Treatment may refer to a policy decision, such as implementing a new economic policy to increase the minimum wage, and one must predict its effect on employment rates.

Mediator

The mediator mechanism is where one or more variables mediate the causal path, forming a chain from the treatment to the outcome. The mediator structure describes a causal mechanism consisting of a causal path representing a chain graph where variable M mediates the causal effect of the exposure or treatment T and the outcome Y. See *Figure 4.11*. Here, M d-separates T from Y. T and M are conditionally independent given M. Knowing M makes T irrelevant to Y. The causal effect of T on Y transitions through the mediating variable M. Here, T does not directly affect Y but instead does so through the intermediary of M. That is, T causes M, and M causes Y. The mediating mechanism may provide an added pathway by which T indirectly causes Y or explain a way in which T causes Y.

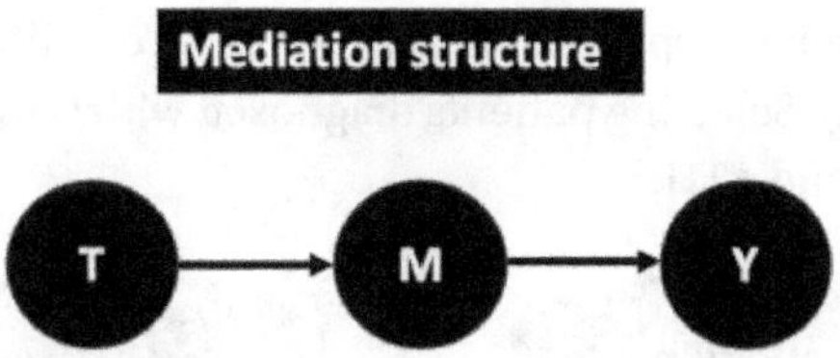

Figure 4.11 – Mediator structure: treatment T indirectly affects Y through mediating variable M

Example: mediator

Figure 4.12 presents a mediator causal chain that illustrates the relationship between exercise and happiness, with endorphin release serving as the mediator. In this causal chain, **Exercise** is identified as the treatment, **Happiness** as the outcome, and **Endorphin release** as the mediator. Exercise leads to the release of endorphins, which function as the mechanism through which exercise influences happiness.

Figure 4.12 – Example mediator causal mechanism

Collider

The collider mechanism, *Figure 4.13*, occurs when a third variable, known as a collider, is influenced by both the treatment (T) and the outcome (Y), thereby creating a biasing path between them. In this causal structure, both T and Y directly affect the third variable, C, which serves as the collider. The presence of the collider blocks the direct path between T and Y. However, conditioning on the collider variable opens up this biasing path, leading to what is referred to as **collider bias**.

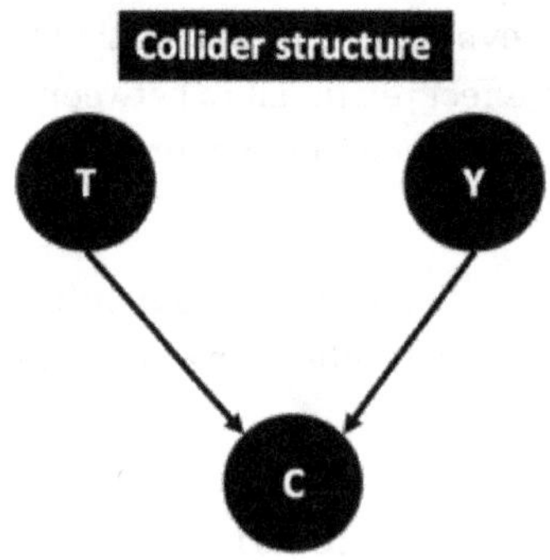

Figure 4.13 – Collider structure: controlling C makes T and Y associated

Example: collider

Figure 4.14 illustrates a collider causal mechanism in the medical field (Stanbouly, Lee, Dodson, & Chuang, 2022). Retrognathia (a small lower jaw) and adenotonsillar hypertrophy (enlarged adenoids and tonsils) both heighten the risk of narrow airways by physically obstructing them and affecting craniofacial development. **Adenotonsillar Hypertrophy** (ATH) can directly block the airway, while retrognathia decreases the available space for breathing. Both conditions can lead to issues such as **Obstructive Sleep Apnea** (OSA). Selecting patients diagnosed with OSA may create a misleading association between retrognathia and ATH.

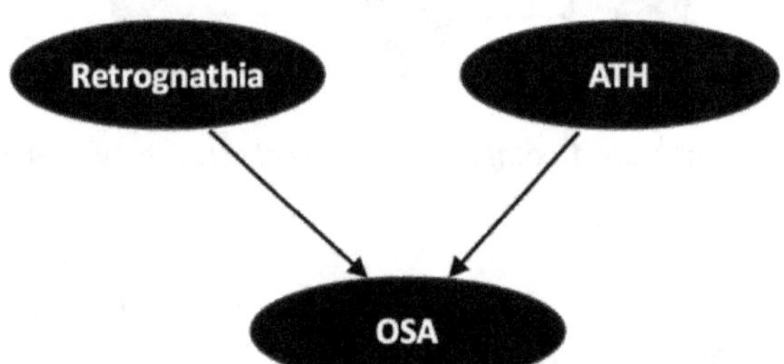

Figure 4.14 – Example collider causal mechanism

Confounder

The confounder mechanism, *Figure 4.15*, occurs when a third variable serves as a common cause for both the treatment and the outcome, creating a biasing path between them that can be blocked by controlling for this confounder. The confounder structure has a variable, C, that causally impacts both the treatment variable T and the outcome variable Y. Not knowing C creates an association between T and

Y; T and Y become correlated, a phenomenon referred to as **confounding bias**. Not knowing C opens a biasing path between T and Y, and learning C blocks that path. The confounder C d-separates T and Y.

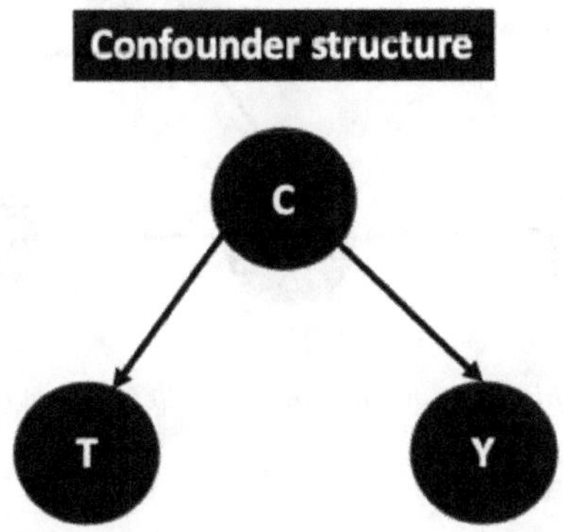

Figure 4.15 – Confounder structure: not knowing C makes T and Y associated

Example: confounder

The example provided, *Figure 4.16*, comes from the field of public health (Health, 2025). A study found a correlation between alcohol consumption and the risk of **Coronary Heart Disease** (**CHD**); however, smoking may have confounded this relationship. Smoking is known to be a risk factor for CHD, and individuals who smoke often consume more alcohol. Controlling for smoking in a study will remove the confounder bias and might demonstrate no significant association between alcohol consumption and CHD.

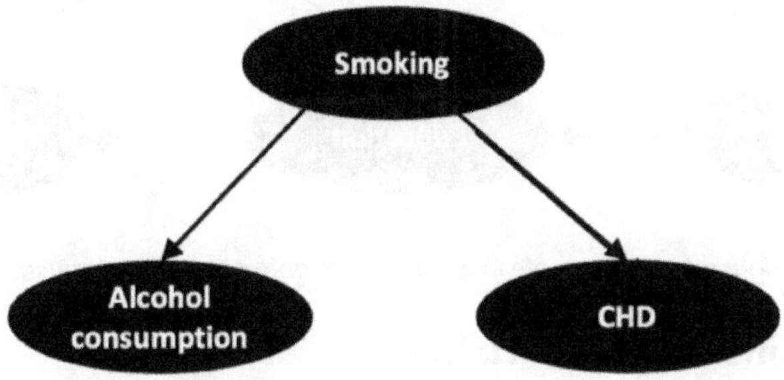

Figure 4.16 – Example confounder causal mechanism

Instrument variable

The **Instrument Variable** (**IV**) structure aims to eliminate confounding bias due to an unobserved confounder. As mentioned earlier, in a confounder structure, failing to control the confounder creates a so-called confounding bias when estimating the causal effect of the treatment T on the outcome Y. An IV allows for the estimation of the causal impact of T on Y in the case of a linear structural equation model under two conditions. First, the instrument I must be uncorrelated with the unobserved confounder C. Second, I must be correlated with the treatment T. See *Figure 4.17*.

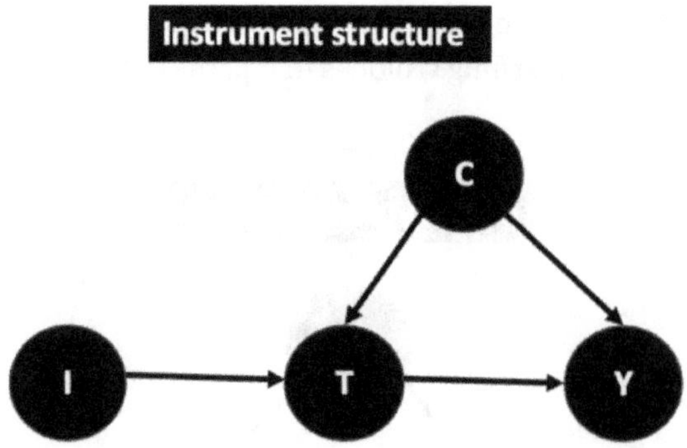

Figure 4.17 – Instrument variable structure: I correlates with T but not the unobserved confounder C

Example: instrument variable

A causal model examining the effect of education on earnings includes an unmeasured confounder representing innate ability, which influences both education and earnings. To address this confounding bias, an IV is employed to estimate the causal effect of education on earnings. This approach utilizes a third variable, referred to as an instrument, that affects education but is not correlated with unobserved factors such as innate ability that impact earnings. An example of such an IV, illustrated in *Figure 4.18*, is geographical proximity to a college, which is presumed to influence education levels without being related to innate ability.

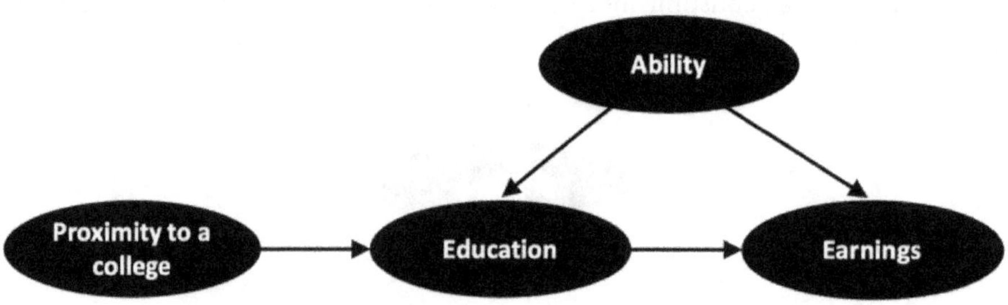

Figure 4.18 – Example instrument variable causal mechanism

Practice using R : instrument variable

In this practice, we construct the instrument structure of *Figure 4.18* and pose queries on the causal graph regarding paths between treatment (exposure) T and the outcome Y, and check that I is an IV under the linearity assumptions and the conditions mentioned in the preceding section.

The R Markdown document, `chapter4.Rmd`, contains all the R code to construct the DAG in *Figure 4.18* and to run queries as shown in the following practice record.

1. Constructing the DAG

   The first step consists of specifying the causal graph, g, and visualizing the graph, as in the following code.

```
#create graph
g <- dagitty("dag{
  T -> Y
  I -> T <- C -> Y
  T [exposure]
  Y [outcome]
}")
#Set position of nodes
coords<-list(x=c(I = 0, T = 1, C = 2, Y = 3),
             y=c(I = 0, T = 0, C = 1, Y = 0))
coords_df<-coords2df(coords)

coordinates(g)<-coords2list(coords_df)
#Plot causal graph
ggdag_status(g) + theme_dag()
```

2. Running queries on the DAG

   In the following, we run two queries using the R `DAGitty` library. The first query is to find all paths between exposure "`T`" and outcome "`Y`" in the DAG, g, and output the status (open or closed) of each path. The first query is implemented in the following code.

```
paths(g, "T", "Y")
## $paths
## [1] "T -> Y"          "T <- C -> Y"
##
## $open
## [1] TRUE TRUE
```

   The result from the first query shows there are two open paths: a causal path $T \to Y$ and a biasing path $T \leftarrow C \to Y$.

   The second query will check if "`I`" can be used as an IV to infer the total effect of an exposure, "`T`", on an outcome, "`Y`", in the presence of the latent (unobserved) confounder, "`C`". The query can be run as in the following code.

```
instrumentalVariables(g, "T", "Y" )
## I
```

The result shows that "I" is an IV.

In conclusion, let us recap what we have covered. We described five causal influence structures as standard building blocks in a causal model of a DAG: causation, mediator, confounder, collider, and instrument variable. With hands-on practice, we demonstrated how to programmatically construct an example of those structures and pose queries on their structural properties.

So far, we have discussed the qualitative aspect of a structural causal model, which includes the DAG. In the following section, we will provide a formal description of the quantitative aspect of a structural causal model and examine causal effects through linear models, utilizing path diagrams and linear regression.

Structural Causal Modeling (SCM)

Previous sections covered the qualitative aspects of structural causal models represented in DAGs. This section focuses on a quantitative description of structural equation systems.

We begin by defining structural equations for a structural causal model. Following this, we focus on linear structural models and their connection to the representation of path diagrams and path analysis. Lastly, we describe the use of regression for estimating the path coefficients in linear SEMs.

Defining a Structural Causal Model (SCM)

A **Structural Causal Model** (**SCM**) is a mathematical framework used to describe causal relationships formally within a system. An SCM consists of structural equations that describe how variables influence one another. This representation of causal relationships expands upon probabilistic graphical models, enabling formal methods for reasoning about causal effects, interventions, and counterfactuals.

A structural causal model is a generative model that quantifies the causal relationships between a set of endogenous variables, which are nodes of a causal DAG, G, specifying the value of each endogenous variable as a function of the values of the parents of the endogenous variable in G and an exogenous variable whose value is determined by mechanisms outside the model. The exogenous variables represent modeling errors, background factors, disturbances, and noise.

Formally, an SCM, denoted as (M), is represented as a 4-tuple: ($M = \langle V, U, F, P(u)\rangle$) (Pearl, 2009). The four components that specify an SCM are as follows:

- **Endogenous variables** ($V = \{X_1, \cdots, X_n\}$): This set includes observable variables whose values are determined by functions of other variables within the model
- **Exogenous variables** ($U = \{U_1, \cdots, U_n\}$): This set consists of unobserved variables whose values are influenced by factors external to the model
- **Functions** ($F = \{f_1, \cdots, f_n\}$.): Each function ($f_i \in F$) defines a structural equation for an endogenous variable (X_i), determining the value of (X_i) based on the values of its parents $pa(X_i)$ in the causal DAG (G), along with the value of the exogenous variable (U_i)
- **Probability distribution** ($P(u)$): This specifies the joint probability distribution over all the exogenous variables (U)

Important note

Typically, the causal graph does not explicitly represent the exogenous (latent) variables. When the graph does, it uses dotted lines for the edges between the latent and endogenous variables.

Each structural equation given by a function $f_i \in F$ describes a **causal mechanism** for a **causal family** consisting of a node $X_i \in V$ and the node's parents $pa(X_i)$ in the causal graph G. The structural equation for X_i is given by

$$X_i = f_i(pa(X_i), U_i), i = 1, \cdots, n \qquad (4.8)$$

The set of variables, U, is considered random in the structural causal model and given by a probability distribution $P(u)$.

We call an SCM **Markovian** if the exogenous variables are pairwise mutually independent, that is,

$$U_i \perp U_j \; \forall i, j = 1, \cdots, n, i \neq j \qquad (4.9)$$

We call an SCM **semi-Markovian** if the exogenous variables are not pairwise mutually independent.

A typical example of dependencies between exogenous variables is when the same exogenous variable appears in two structural equations, such as an unobserved confounder. We represent that dependency with a (dotted) bi-directed edge between pairs of variables.

Figure 4.19(a) shows an example of a Markovian causal model where the exogenous (unobserved) variables are mutually independent. *Figure 4.19(b)* shows a causal graph for a semi-Markovian model where the unobserved variable U_1 appears in the functions f_1, f_2 of the structural equations of variables X_1, X_2; the exogenous variable U_1 is a confounder for X_1 and X_3. *Figure 4.19(c)* represents the dependency between X_1 and X_2 created by the unobserved confounder by a bi-directional edge using a dotted line to replace the explicit reference to the unobserved variable U_1.

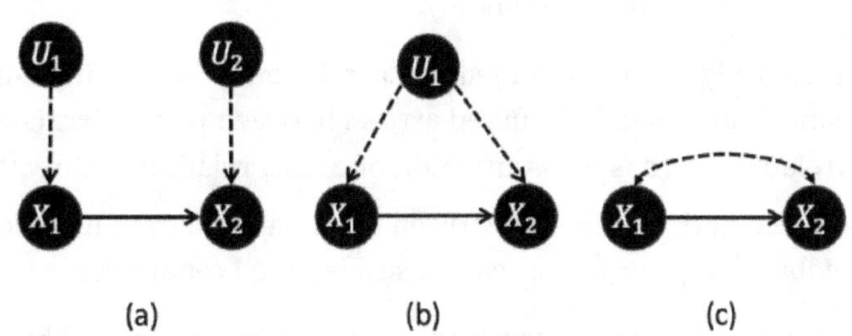

Figure 4.19— Examples of SCM's causal graph categories: Markovian (a), semi-Markovian (b), and (c)

The probability distribution of the unobserved variables $P(u)$ and the structural equations induce a joint probability distribution on the endogenous variables V. To see this, let us order the variables $V = \{X_1, \cdots, X_n\}$ in a causal (topological) order according to the DAG, G. The structural equations for the variables V following the causal ordering will be as follows:

$$X_1 = f_1(U_1),\ X_2 = f_2(X_1, U_2), \cdots, X_n = f_n(X_1, \cdots, X_{n-1}, U_n) \tag{4.10}$$

Equation 4.10 is solvable for $X_1, \cdots, X_n$, as explicit functions of $U_1, \cdots, U_n$. Hence, the probability distribution $P(u)$ and the functional mappings of *Equation 4.10* produce an implicit probability distribution $P(v)$ and define a Bayesian network having the DAG, G, with nodes, the set V, and the complete probability distribution $P(v)$. Hence, the SEM of the SCM is a generative data model of the observed variables V.

Markovian versus semi-Markovian SCM

A DAG shows a Markovian SCM structure. The graph's nodes symbolize endogenous variables, while it typically does not directly display exogenous variables. Each node in the Markovian SCM DAG is associated with a structural equation that involves a distinct unobserved (exogenous) variable, with all unobserved variables being pairwise mutually independent. In contrast, a semi-Markovian SCM is illustrated by a directed graph that features bi-directed edges between pairs of nodes, indicating mutual dependencies between the unobserved variables in the two structural equations linked to those node pairs.

In the next section, we consider the case of linear SCMs, which are syntactically equivalent to the formulation of path analysis and path diagrams.

Representing linear SCMs with path diagrams

In this section, we consider strict causal models with linear structural equations. The formulation of linear SCMs is syntactically equivalent to representing them using path diagrams and path coefficients (Spirtes, Richardson, Meek, Scheines, & Glymour).

In **path diagrams**, directed edges show assumed causal relations where a single-headed arrow points from cause to effect, and double-headed, curved arrows between pairs of variables indicate that the two variables are correlated; there is no assumption of causal relation. A directed edge from node i to node j has an associated **path coefficient**, written p_{ji}. A path diagram in which the causal flow is unidirectional (no feedback loops or reciprocal causes) is called **recursive**.

A path diagram on n variables has linear structural equations parameterized by the path coefficients that we can represent as follows:

$$X_i = \begin{cases} U_1, i = 1 \\ \sum_{j<i} p_{ij} X_j + U_i, i = 2, \cdots, n \end{cases} \tag{4.11}$$

We assume without loss of generality that the variables are standardized to mean 0 and variance 1, all the exogenous variables $U_1, \cdots, U_n$ are normally distributed and are uncorrelated with all the endogenous variables $X_1, \cdots, X_n$.

$$U_i \sim N(0,1), \sigma(U_i, X_j) = 0, \forall i, j = 1, \cdots, n, i \neq j \quad (4.12)$$

The exogenous variables can be mutually pairwise dependent, and we denote their covariances as

$$r_{ij} = \sigma(U_i, U_j) \quad (4.13)$$

We note that for standardized variables, the correlation between two variables has the same value as their covariance, and we use $\sigma(X, Y)$ to denote both the covariance between standardized variables X and Y and also the correlation between X and Y.

The variables X_j summed over on the right-hand side of *Equation 4.11* are the direct **causes** of X_i. The coefficients p_{ij} quantify the **direct effect** of X_j on X_i and can annotate the directed edges between nodes in the causal graph. The variable U_i represents the exogenous variable affecting X_i and typically means "omitted factors" determined by external mechanisms outside the model's scope. In the Markovian SCM, the variables U_i are considered random variables with a multivariate normal distribution that are pairwise mutually independent.

Relating path coefficients to variable correlations

In path analysis, the correlation between two variables can be decomposed into direct and indirect effects, as defined by the path coefficients. The total correlation represents the sum of contributions from all pathways connecting the two variables. This decomposition is based on **Wright's tracing rule** (Wright, 1934) (Wright, Correlation and causation, 1921) and is applicable to recursive linear models where causality flows in one direction (without feedback loops).

The following definition of Wright's tracing rule, commonly referred to as Wright's rule or tracing rule, is based on the version presented by Chen et al. (Chen, Pearl, & Kline, 2018). This version provides a concise characterization of the rule and is more formal than other equivalent definitions.

Definition: Wright's tracing rule

Wright's tracing rule states that the correlation, $\sigma(Y, X)$, between any pair of variables, X and Y, is equal to the sum of products of path coefficients and error correlations along non-collider routes between X and Y. Formally, let $\Pi = \{\pi_1, \cdots, \pi_k\}$ denote the non-collider routes between X and Y, and let p_i be the product of path coefficients along route π_i. Then the model-implied correlation between variables X and Y is $\sum_i p_i$.

In the following sections, we will present two examples to illustrate the tracing rule applied to two linear SCMs, M1 and M2.

Example: tracing rule model M1

We will use model M1 along with the path diagram in *Figure 4.20* to illustrate the application of the tracing rule. Let us apply Wright's rule to compute the correlation between pairs of variables. For example, consider the correlation between X_2 and X_4. There are two non-colliding paths between these variables: $X_2 \rightarrow X_4$ and $X_2 \rightarrow X_1 \rightarrow X_4$. By applying the tracing rule, we determine that $\sigma(X_2, X_4) = p_{42} + r_{12}\, p_{41}$.

Next, to compute the correlation between X_1 and X_2, we note that there is only one non-colliding path: $X_1 \rightarrow X_2$ (or equivalently, $X_2 \rightarrow X_1$). The rule yields $\sigma(X_1, X_2) = r_{12}$.

Finally, to compute the correlation between X_1 and X_5, we identify three non-colliding paths: $X_1 \rightarrow X_4 \rightarrow X_5$, $X_1 \rightarrow X_2 \rightarrow X_4 \rightarrow X_5$, and $X_1 \rightarrow X_3 \rightarrow X_5$. Thus, the rule provides us $\sigma(X_1, X_4) = p_{41}\, p_{54} + r_{12}\, p_{42}\, p_{54} + r_{13}\, p_{53}$.

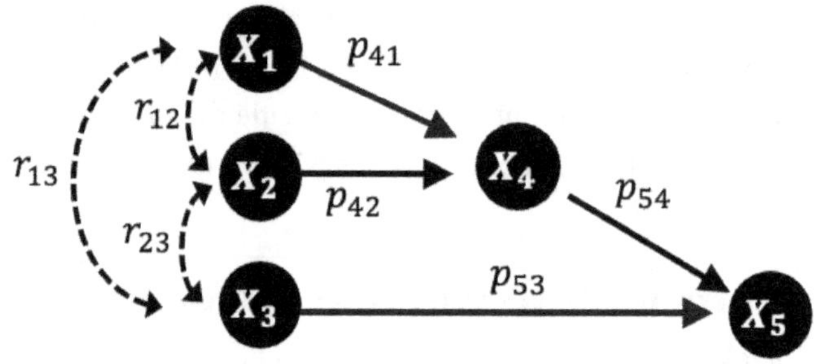

Figure 4.20 - Path diagram of linear SCM, M1

Example: tracing rule model M2

Figure 4.21 illustrates the linear SCM M2. Let us apply this tracing rule to compute the correlation between X_1 and X_2. We have two paths: $X_1 \rightarrow X_2$ and $X_1 \rightarrow X_3 \leftarrow X_2$, where only the first is a non-collider with coefficient p_{21}. Therefore, $\sigma(X_1, X_2) = p_{21}$.

Next, we compute the correlation between X_1 and X_3. There are two paths: $X_1 \rightarrow X_3$ and $X_1 \rightarrow X_2 \rightarrow X_3$. By applying the tracing rule, we find that $\sigma(X_1, X_3) = p_{31} + p_{21}\, p_{32}$.

Lastly, we will compute the correlation between X_2 and X_3. There are two non-colliding paths: $X_2 \rightarrow X_3$ and $X_2 \leftarrow X_1 \rightarrow X_3$_3. By applying the tracing rule, we determine that $\sigma(X_2, X_3) = p_{32} + p_{21}p_{31}$.

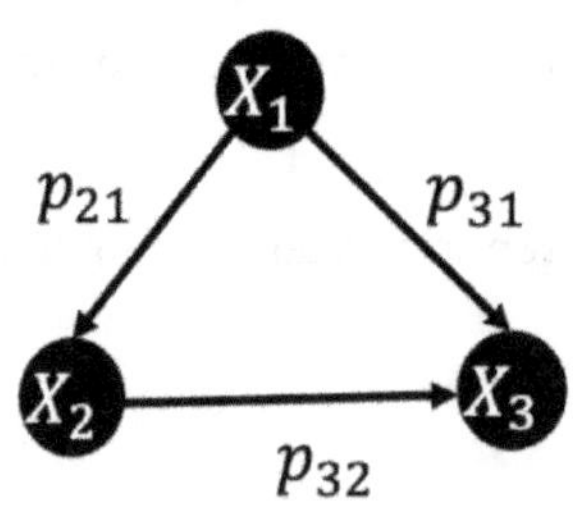

Figure 4.21 — Path diagram of linear SCM, M2

To summarize the key takeaways, Wright's tracing rule provides a method for calculating correlations between variables in a path diagram. According to this rule, the correlation between two nodes in the diagram is the total sum of the products of path coefficients along all non-colliding paths that connect the two nodes.

In the next section, we focus on the identifiability problem concerned with the question of determining the path coefficients from variable correlations.

Identifiability of causal effects in Linear Structural Causal Models (LSCMs)

In this section, we state the problem of identifiability of causal effects in **Linear Structural Causal Models (LSCMs)**.

LSCMs are represented by a mixed graph (specifically, an **Acyclic Directed Mixed Graph**, or **ADMG**). Directed edges (→) represent a direct causal effect from one variable to another, while bidirected edges (↔) are used to signify a statistical association between two variables that share a common unobserved, or latent, cause. This structure helps distinguish direct causation from correlation induced by hidden variables within the model.

To formally state the identifiability problem of LSCMs, let us make use of the following notations. Let $G = (V, D, B)$ be a directed graph with node set V, set of directed edges D, and a set of bidirected edges B. We assume the $V = \{X_1, X_2, \cdots, X_n\}$ and that the subgraph of directed edges is acyclic and topologically ordered so that $(X_i \to X_j) \in D$ implies $i < j$. The set V consists of the endogenous (observed) variables. For each $X_j \in V$, a linear structural equation corresponds:

$$X_j = \sum_{i:\, i<j} \lambda_{ij} X_i + \varepsilon_j \qquad (4.14)$$

λ_{ij} is the path coefficient representing the direct effect of X_i on X_j. ε_j is a random variable in standardized form representing an exogenous (latent or unobserved) variable.

Let X be a column vector of the variables in V, ε a column vector of the latent variables, having a normal distribution $\varepsilon \sim \mathcal{N}(0, \Omega)$. The relationships between the variables are described by a set of linear equations in the following matrix form,

$$X = (I - \Lambda)^{-T} \varepsilon \qquad (4.15)$$

Λ is an $n \times n$ upper diagonal matrix consisting of the path coefficients for the directed edges in D, $\lambda_{ij} \geq 0$, for all $i < j$, and $\lambda_{ij} = 0$ for all $i \geq j$.

It follows from *Equation 4.15* that the covariance matrix of the random vector X denoted as Σ can be expressed in terms of the path coefficients and the error covariance, as follows:

$$\Sigma = E[\boldsymbol{X}\boldsymbol{X}^T] = (I - \Lambda)^{-T} \Omega \, (I - \Lambda)^{-1} \qquad (4.16)$$

Equation 4.16 provides an algebraic equation that specifies the relation between variable covariances and LSCM parameters. The algebraic equation specifies how the variable covariances are computed from knowing the parameters of the LSCM.

An LSCM is considered **generically identifiable** (Foygel , Draisma, & Drton, 2012), (Drton & Weihs, 2015) if the causal effect parameters in Λ can be uniquely identified from the elements of Σ, based on the observed data distribution for nearly all parameter values that are consistent with the structure of the LSCM graph.

Generic identifiability can theoretically be completely solved by computer algebra, notably employing **Gröbner bases**, as demonstrated by García-Puente et al. (Garcia-Puente, Spielvogel, & Sullivant, 2010). This method has a doubly exponential computational complexity in terms of running time (Bardet & Chyzak, 2005), rendering it impractical for instances beyond four or five variables.

To illustrate the computation of the path coefficients from variable correlations, let us consider some examples.

Example: identifiability model M3

Consider the path diagram of the linear SCM, M3, depicted in *Figure 4.22*. Can the parameter λ_{XY} be uniquely determined from the probability distribution on X, Y, and Z?

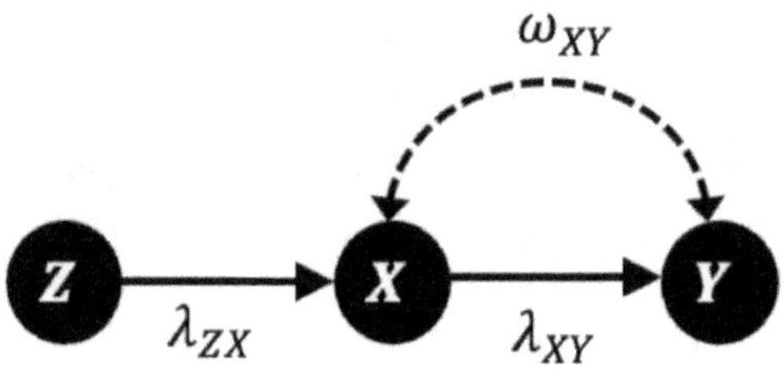

Figure 4.22 – Path diagram of linear SCM, M3

From the probability distribution, we compute the variable correlations $\sigma(X,Y), \sigma(X,Z)$, and $\sigma(Y,Z)$.

Using the tracing rule, we derive the following equations that represent the correlations among the three variables:

$$\sigma(X,Z) = \lambda_{ZX}$$

$$\sigma(Y,Z) = \lambda_{ZX}\,\lambda_{XY}$$

Provided $\sigma(X,Z) \neq 0$, we can uniquely determine λ_{XY}, solving the preceding equations, which yields

$$\lambda_{XY} = \frac{\sigma(Y,Z)}{\sigma(X,Z)}$$

We conclude that the model is generically identifiable.

Example: identifiability model M4

Consider the path diagram of the linear SCM, M4, depicted in *Figure 4.23*. Can the parameter λ_{XY} be uniquely from the probability distribution on X, Y, and Z?

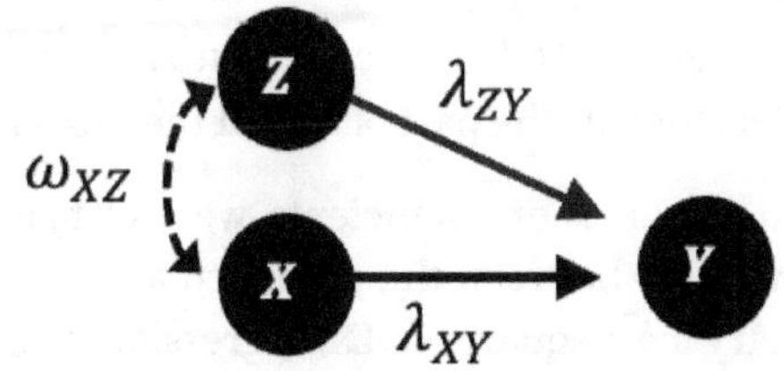

Figure 4.23 – Path diagram of linear SCM, M4

From the probability distribution, we compute the variable correlations $\sigma(X,Y)$ and $\sigma(Y,Z)$. Using the tracing rule, we derive the following equations that represent the correlations among the three variables:

$$\sigma(X,Y) = \lambda_{XY} + \omega_{XZ}\,\lambda_{ZY}$$

$$\sigma(Y,Z) = \lambda_{ZY} + \omega_{XZ}\,\lambda_{XY}$$

The preceding are two equations in two unknowns, λ_{XY} and λ_{ZY}, which can be easily solved given the variable correlations$\sigma(X,Y)$ and $\sigma(Y,Z)$. Hence, the LSCM with the mixed graph in *Figure 4.22* is generically identifiable.

The main takeaways can be summarized as follows: Generic identifiability in an LSCM refers to the ability to uniquely recover the model's causal parameters from the observed joint covariance matrix for nearly all potential parameter values. This concept is critical in causal inference, where the goal is to determine whether the observed data is sufficient to clarify the underlying causal structure and its associated parameters. The term "generic" is used because it excludes rare, degenerate cases in which the parameters may not be uniquely identifiable.

The following section considers the connection between linear structural equation models and multiple regression.

Relating LSCMs to linear regression

In this section, we begin with a description of linear regression. Next, we discuss the distinctions between regression and structural models. Finally, we give an example of a linear structural model where we estimate the model parameters using regression on simulated data.

Describing linear regression

A linear regression problem consists of a given dataset of n unit cases of observations composed, each of a set m of explanatory variables $X_1, \cdots, X_m$ and a corresponding response variable Y.

$$Y = \beta_0 + \beta_1 X_1 + \cdots + \beta_m X_m + \varepsilon \qquad 4.17)$$

The coefficients on the right-hand side of *Equation 4.17* are the partial regression coefficients, which have the following interpretation. The coefficient β_0 is the intercept, the point where the regression line crosses the Y axis. The coefficients β_i $(i = 1,\ldots,m)$ represent the partial slopes for Y. Each β_i represents the change in the expected value of Y, denoted as E[Y], for each one-unit change in X_i, while keeping all other independent variables $\{X_j : j \neq i\}$ constant. The error term, ε, also called the residual, represents the unobserved factors, or the part of Y that the model does not explain.

To fit a linear model and determine regression coefficients, we need to identify the values for the slope and intercept that minimize the error between the predicted and actual values in our dataset. The most common method is **ordinary least squares (OLS)** regression, which minimizes the **sum of the squared residuals** (**RSS**), or the differences between the observed and predicted values. This process adjusts the model's parameters (coefficients) to create the "best-fit" line for the data.

$$(\hat{\beta}_1,\cdots,\hat{\beta}_m) = \underset{(\beta_0,\beta_1,\cdots,\beta_m)\in\mathbb{R}^{m+1}}{\operatorname{argmin}} \sum_{i=1}^{n}\left(Y_i - (\beta_0 + \beta_1 X_1 + \cdots + \beta_m X_m)\right)^2 \qquad (4.18)$$

In linear regression, the coefficients estimated by OLS will be biased if the residual error is correlated with the regressor variables.

Important note

In linear regression, if the error term is correlated with one or more of the regressor (independent) variables, the OLS estimator will be biased and inconsistent. This phenomenon is known as endogeneity.

Estimating causal effect using linear regression

A primary objective in an SCM is to estimate the structural parameters representing the causal effect of a variable X on another variable Y.

In an LSCM, Y can be represented as a linear function of regressor variables (predictors) consisting of X, a set of adjustment variables, $W_1,\cdots,W_m$, and an error term U_Y, given by the linear regression model

$$Y = \alpha X + \gamma_1 W_1 + \cdots + \gamma_m W_m + U_Y \qquad (4.19)$$

subject to the **zero conditional mean assumption** stated next.

$$E[U_Y \mid X, W_1,\cdots,W_m] = 0 \qquad (4.20]$$

The zero conditional mean assumption means that the covariance between the error term and each regressor variable is zero, that the error term contains no information that is also present in the regressor variables, that for any set of values of the regressor variables, the average of the errors is zero, and that the errors between estimated and observed values of Y are, on average, centered around the regression line.

The zero conditional mean assumption is important for unbiased estimation and causal interpretation. The assumption ensures that the regression coefficients estimated by the **Ordinary Least Squares (OLS)** estimators are unbiased, meaning they are not systematically over- or under-estimating the true model coefficients. The assumption allows for a causal interpretation of the parameter α as the average causal effect: the average change in the outcome Y for a one-unit change in the cause X.

When the zero conditional zero mean assumption is violated, the OLS estimators of α will be biased and/or inconsistent, meaning the estimates will be systematically wrong and will not converge to the true values of α even with a large sample size. The estimated average causal effect results of the regression will not be reliable, and any conclusions drawn from them will be flawed.

The takeaway of this section is that linear regression models can be used to estimate the average causal effect between a cause X and outcome Y, conditioning on a set of adjustment variables and such that the zero conditional mean assumption is respected.

In the next section, we address the important topic of selecting adjustment variable sets, given a DAG representing the structural relationships between the observed (endogenous) variables.

Selecting the adjustment set

The previous section outlined the use of linear regression to estimate the **Average Causal Effect** (ACE) of cause X on outcome Y, where the predictor variables include, besides X, other adjustment variables so that the zero conditional mean assumption is respected.

In this section, we turn our focus to the subject of selecting adjustment sets. First, we state an important result that provides a graphical condition, called single-door, for finding adjustment variables that ensures the ACE can be estimated using linear regression. Next, we give two examples that illustrate the application of the single-door criterion to two linear structural causal models represented by their acyclic directed mixed graphs.

The following presents a graphical criterion for identifying the ACE using linear regression within LSCMs.

Theorem: single-door criterion (Pearl, 2009)

Let G be the directed graph representing the path diagram in which α is the path coefficient associated with the path $X \rightarrow Y$. Let G_α denote the diagram that results when the path $X \rightarrow Y$ is deleted from G. The coefficient α is identifiable if there exists a set of variables W such that (i) W contains no descendants of Y, and (ii) W d-separates X from Y in G_α. If W satisfies both conditions, then α is equal to the partial regression coefficient $\beta(Y, X|W)$, which represents the coefficient of X when regressing Y on both X and W. In this case, W is referred to as single-door admissible. Conversely, if W does not meet these conditions, then $\beta(Y, X|W)$ is not equal to α.

The theorem's intuition is to ensure that all paths between X and Y in the acyclic directed mixed graph, which represents the LSCM, are blocked by the set$W$ in the mutilated graph after removing the directed edge between X and Y.

Let us turn to some examples to illustrate how to apply the single-door criterion to select adjustment variables for estimating the ACE using linear regression.

Example: linear regression model M5

To illustrate the single-door criterion, we adopt the example from Chen, Pearl, & Kline(Chen, Pearl, & Kline, 2018). The directed graph G in *Figure 4.24* depicts an example path diagram where the path coefficient α of the path $X \to Y$ can be identified by regressing Y on X while adjusting for either Z or W. Z (or W) is not a descendant of Y, and Z (or W) d-separates X and Y as it blocks the path between them. Therefore, Z (or W) satisfies the single-door criterion, and the coefficient α can be estimated without bias by fitting either one of the following regression models with the measurement data:

$$Y = \alpha X + \gamma Z + U_Y \qquad (4.21)$$

$$Y = \alpha X + \gamma W + U_Y \qquad (4.23)$$

The zero conditional mean assumption of *Equation 4.20* holds true in both regression models mentioned previously. It is important to note that U_Y is independent of X because the path between them is blocked by either Z or W, which means they are d-separated.

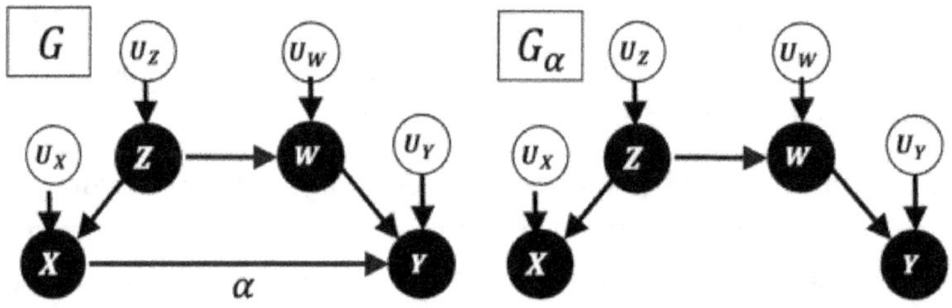

Figure 4.24 – Model M5: The path coefficient a is identifiable by regressing Y on X while adjusting for either Z or W

The next example is a modified version of the previous example that includes a bidirected edge in the graph.

Example: linear regression model M6

Figure 4.25 continues with the example after adding a correlation between the exogenous variables U_Y and U_W. In the diagram G shown in *Figure 4.25*, we can obtain an unbiased estimate of α by fitting the regression model from *Equation 4.21* while controlling for Z only and not for W. If we use W as an adjustment, an open path will form between X and Y, represented as $Y \leftarrow U_Y \to U_W \to W \leftarrow Z \to X$, due to conditioning on W and the fact that W is a collider.

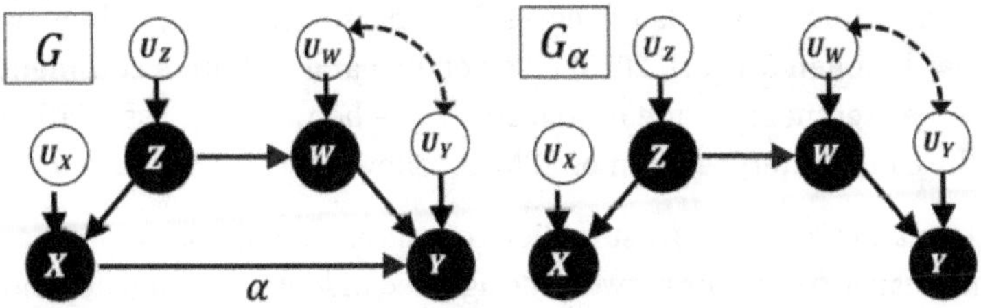

Figure 4.25 – Model M6: the path coefficient a is identifiable by regressing Y on X while adjusting for Z only

The main takeaway is that the single-door criterion establishes conditions based on d-separation in the mutilated graph. This allows for adjustments that enable the use of linear regression to obtain an unbiased estimate of the path coefficient for the direct causal link $X \rightarrow Y$ in the LSCM. This coefficient represents the average causal effect of X on Y.

In the following section, we will explore the key differences between linear regression and linear structural equations that describe the relationship between dependent and independent variables.

Differentiating between LSCMs and linear regression

In the previous sections, we discussed how linear regression is utilized to estimate parameters of LSCMs, specifically the path coefficients that represent direct causal links. This section aims to clarify the differences between a linear regression model and a linear structural equation within a structural causal model.

As previously mentioned in the context of linear regression, the objective is to achieve the best fit for a linear regression model, as described in *Equation 4.17*, based on the OLS criterion outlined in *Equation 4.18*. This process is applied to a dataset comprising sample records of value tuples from the independent variables and their corresponding dependent variable.

The linear structural equation model, represented by *Equation 4.19*, depicts a causal relationship between two variables, conditioned on a set of other variables known as adjustment variables. The aim is to estimate the causal effect parameter in the structural equation using OLS regression.

Although structural equations are syntactically equivalent to linear regression, the semantics are different. The designation of dependent and independent variables in regression is merely statistical; it represents the best fit between the dependent and independent variables based on minimizing the mean square error between the predicted and actual values of the dependent variable. Structural equations incorporate causal influence assumptions that have interpretations grounded in *intervention* or manipulation. The structural equations represent assumptions about being a model of the world. They represent a *generative model* of the observable datasets with testable implications that data could confirm or disconfirm.

The distinction between linear models of regression and structural equations is a much-debated topic. Bollen and Pearl (Bollen & Pearl, 2013) give detailed and elaborate comments on that subject. We can summarize the arguments made in that reference between the associational interpretation of regression and the causal interpretation of SEM as follows:

- A structural equation represents assumptions of causal links and structural parameters of causal effects rather than associations between designated independent and dependent variables and regression coefficients for best predictors fitting empirical data.
- The error terms in the structural equations represent inherent factors or disturbances generated by stochastic processes independent of the causal mechanisms of the LSCM. In regression, the errors represent the residuals between the dependent variable's actual and predicted values based on the estimated regression parameters.

Wrapping up this section, let us recap. A linear regression model is syntactically equivalent to a linear structural equation in LSCMs. Linear regression is used to estimate the causal effect parameters of the structural equations. A linear regression equation mathematically represents the statistical association between a dependent variable and one or more independent variables. It uses a linear function to model this relationship, allowing for the *prediction* of the dependent variable's value based on the independent variables. A linear structural equation represents assumptions of causal effect relations, and the parameters of the structural equation represent the average change in the dependent variable, representing the effect of an *intervention* of a unit change in an independent variable representing the cause.

In the following section, we offer an overview of issues related to the use of linear regression for estimating causal effects in linear structural models.

Potential biases and failures from regression

In LSCMs, regression analysis is utilized to estimate the average causal effect parameter by fitting a linear equation that expresses the outcome variable Y as a function of the causal variable X and a set of adjustment variables. For further details, refer to *Equation 4.19*. However, challenges such as endogeneity and multicollinearity can introduce significant biases and failures, making it difficult or even impossible to draw valid causal inferences.

Issues from endogeneity

Endogeneity arises when an explanatory variable in a regression model is correlated with the error term. This correlation results in biased and inconsistent estimates when employing OLS regression. Common sources of endogeneity include **Omitted Variable Bias** (**OVB**), simultaneity or reverse causality, and measurement error.

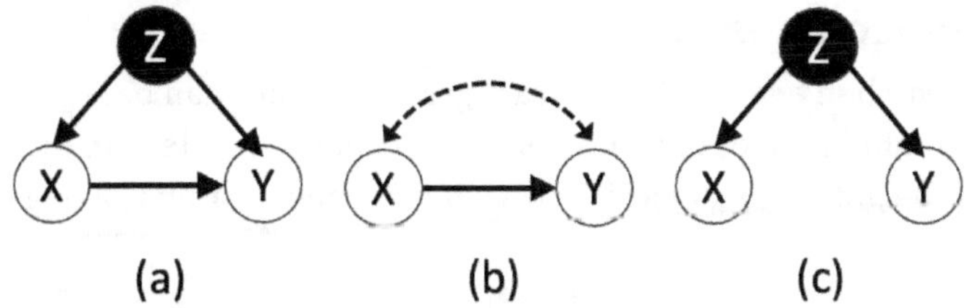

Figure 4.26 – Causal graphs of endogeneity issues for linear regression (a) omitted variable; (b) simultaneity/inverse causality; (c) measurement error

Omitted Variable Bias (OVB)

OVB occurs when an unobserved variable Z affects both the independent variable X and the dependent variable Y; refer to *Figure 4.26(a)*. If Z is excluded from the regression model, its influence becomes part of the error term, resulting in a correlation between X and the error term. As a result, the OLS regression of Y on X will yield a scatter plot in which the residuals are not randomly distributed around the regression line. Instead, they will exhibit a pattern associated with the omitted variable Z, which leads to biased estimates of the causal effect of X on Y. For example, a researcher may overlook a person's natural ability or work ethic when analyzing the relationship between wages and education. Since ability likely influences both the level of education attained and the wages earned, the estimated coefficient for education will be biased, potentially exaggerating the true causal effect of education on wages.

Simultaneity/reverse causality

Simultaneity/reverse causality occurs when the presumed effect (dependent variable, Y) actually influences the presumed cause (independent variable, X), contrary to the expected direction (Y causes X, and not X causes Y). See *Figure 4.26(b)*. A simple scatterplot and OLS regression of Y on X will capture both effects simultaneously, leading to a biased estimate of the effect in one direction (e.g., X on Y). The data points will reflect a complex equilibrium rather than a single causal path. As an example, take the relationship between police force size and crime rates: more police might reduce crime, but cities with high crime rates also tend to hire more police. Both variables are determined in a feedback loop.

Measurement error

Measurement error occurs when an observed variable X inaccurately measures an unobserved variable Z, which in turn influences a dependent variable Y. See *Figure 4.26(c)*. The measurement error associated with Z creates a correlation between X and the error term when regressing Y on X. As a result, OLS estimates become biased and inconsistent, particularly exhibiting **attenuation bias**, where the estimated effect is skewed toward zero. For instance, in a study examining the effect of physical exercise on weight loss, significant errors in reporting exercise levels (e.g., individuals overstating their activity) can distort the estimated relationship between exercise and weight loss, resulting in biased outcomes.

Issues from multicollinearity

Multicollinearity arises when two predictors in a regression model can be represented as an exact or approximate linear combination of one another. This scenario may lead to unstable and imprecise coefficient estimates, heightened standard errors, and erroneous or misleading outcomes.

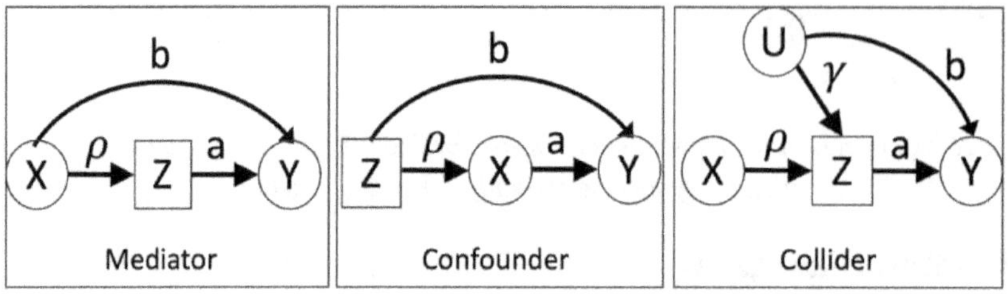

Figure 4.27 – Three causal structures for linear regression illustrating collinearity between the predictor variables X and Z

Schisterman et al. (Schisterman, Perkins., Mumford, Ahrens, & Mitchell, 2017) presented three structural scenarios, illustrated in *Figure 4.27*, in which a high correlation can occur between a given variable Z and the exposure X. These scenarios include mediators, confounders, and colliders. They evaluated the consequences of increasing correlation ρ between the variable Z and the exposure X, focusing on the bias and variance related to the total effect of exposure X on the outcome Y. This evaluation was conducted using both unadjusted and adjusted models. The linear regression models for each of the three causal structures are described by *Equation 4.24* for the model without adjustment and by *Equation 4.25* for the model with an adjustment variable.

$$Y = \beta_1 X + \varepsilon \qquad (4.24)$$

$$Y = \beta_2 X + \beta_3 Z + \varepsilon \qquad (4.25)$$

Schisterman et al. (Schisterman, Perkins., Mumford, Ahrens, & Mitchell, 2017) derived closed formulas for the expected value of the regression parameter β_1 in the unadjusted model (*Equation 4.24*) and the parameter β_2 in the adjusted model (*Equation 4.25*). They also provided formulas for the expected bias and expected variance of these estimates, which depend on the path coefficients and the sample size.

For the mediator model, the total causal effect of X on Y is equal to $b + a\,\rho$, which is estimated without bias by the coefficient $\beta_1 = b + a\,\rho$ in the unadjusted model, while estimated by the coefficient $\beta_2 = b$ in the adjusted model with a bias equal to $-a\,\rho$.

The confounder model indicates that the total causal effect of X on Y is represented by the value a. In the unadjusted model, this effect is estimated with bias through the coefficient $\beta_1 = a + b\,\rho$, resulting in a bias of $b\,\rho$. In contrast, the adjusted model estimates the effect using the coefficient $\beta_2 = a$, which does not involve any bias.

For the collider model, the total causal effect of X on Y is represented by $a\,\rho$, which is estimated without bias by the coefficient $\beta_1 = a\,\rho$ in the unadjusted model. In contrast, for the adjusted model, the coefficient β_2 is calculated as $-\ (b\,\gamma\,\rho)/(1-\rho^2)$,, which introduces a bias of $-\ (b\,\gamma\,\rho)/(1-\rho^2), -a\,\rho$. In this scenario, conditioning on the collider Z creates collider bias between the unobserved variable U and the exposure X. It is important to note that the correlation ρ can be either positive or negative, which can subsequently influence the degree of bias in the effect of interest.

The formulas for the expected variances for the three collinearity models given in Schisterman et al. (Schisterman, Perkins., Mumford, Ahrens, & Mitchell, 2017) demonstrate that high correlation from collinearity can inflate standard errors, which reduces the statistical power and reliability of the model, making it harder to conclude that a variable has a significant effect on the dependent variable.

The three collinearity structures illustrate that the causal relationship between the variables is crucial for valid inference. This relationship should be the primary consideration when determining the total effect of exposure X on outcome Y and when selecting which covariates to include in the regression formulation.

In summary, endogeneity results in fundamentally biased estimates, which completely invalidate causal claims. Meanwhile, multicollinearity renders individual coefficient estimates unreliable and unstable, obstructing accurate causal interpretation. Addressing both issues is essential for achieving robust structural causal modeling.

In the following section, we are going to practice using R with examples that illustrate the effect of adjustments or the variables used in the regression model when estimating the structural equation coefficients. The goal of the practice is to acquire skills for implementing linear regression models for testing causal assumptions and enforcing learned theories.

R practice: LSCMs and regression

In this practice session, we will apply linear regression to estimate causal effects within two distinct linear structural causal models. This exercise will demonstrate that the estimation of the path coefficient corresponding to the ACE will be unbiased if the regression includes variables from an adjustment set that meets the single-door criterion. Conversely, if we fail to make the necessary adjustments, linear regression will produce a biased estimate of the ACE.

During this practice, you will receive a walk-through of implementing these tasks in R:

1. Specifying the structural equation model and plotting the path diagram
2. Generating a synthetic dataset
3. Finding adjustment sets
4. Estimating the path coefficient

We will use the ggdag library (Barrett, 2023) to represent the path diagrams, which are the causal graphs for the LSCM. The `dagify` function in the ggdag package allows us to specify `DAGitty` DAGs using the syntax of the `lavaan` package (Rosseel, 2012), which assigns the following operators to define relationships in a structural equation model:

```
~ "regressed on"
~~ "covaries with".
```

We will be utilizing the `DAGitty` package in R to find adjustment sets to ensure the identification of the path coefficient parameter representing the ACE from the measurement records in the synthetic dataset.

The entire interactive R code session for this practice is found in the *R Markdown* document `chapter4.Rmd` in the book's GH repo.

LSCM M7

Figure 4.28 shows the structural causal model M7, consisting of three nodes corresponding to the variables Z, T, and Y, two directed edges $Z \to Y, T \to Y$, and a bidirectional edge $T \leftrightarrow Z$, indicating a correlation between the exogenous variables acting on the Z and T variables. *Figure 4.28* annotates the directed edges with the path coefficients. The figure does not show the exogenous unobserved variables explicitly.

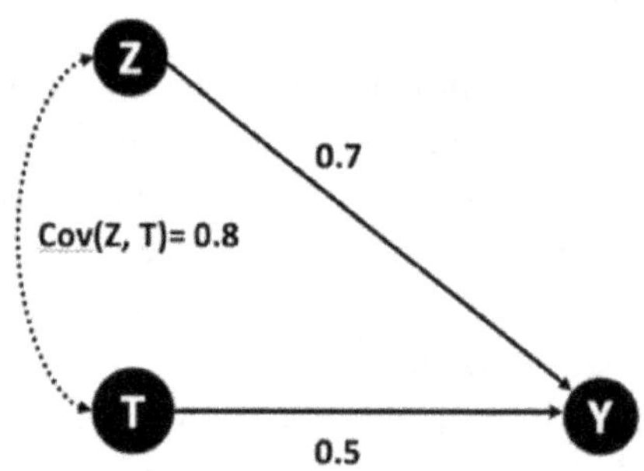

Figure 4.28 – Path diagram for the linear structural causal model M7

We'll begin our first task next.

Specifying the structural equation model and plotting the path diagram

The following code snippet first defines the graph representation of the structural causal model, `g1`. The first line in the specification says that `Y` is regressed on `T` and `Z`, and the second line says that `T` and `Z` are correlated. The rest of the code will generate the path diagram for the linear SCM, M4, depicted in *Figure 4.28*.

```
g1 <- dagify(
  Y ~ T + Z,
  T ~~ Z
)
```

```
# set coordinates for graph layout
coords<-list(X=c(Z = 1, T = 1, Y = 2),
             Y=c(Z = 2, T = 1, Y = 1))
coords_df<-coords2df(coords)

coordinates(g1)<-coords2list(coords_df)
#Plot causal graph
ggdag(g1)+theme_dag()
```

We'll follow up with our second task next.

Generating a synthetic dataset

The following code will create a dataset for the random variables `Z`, `T`, and `Y`. The linear equation of the SCM uses the path coefficients and the covariance from the path diagram in *Figure 4.28*. To simulate the covariance between the variables `Z` and `T`, we use the `mvtnorm` library in R. The code uses the `rnorm(n)` function to generate `n` random numbers from a standard normal distribution (mean=0, sd=1) to simulate an error term, representing the unobserved factors in the linear equation model. The code generates a dataset stored in `model.data`, which is a data frame of `n` records, where `n` is assigned a large number to improve the estimation accuracy.

```
# simulate data (linear model)
n <- 1e4
# we will generate correlated, normal random variables with the program
rmmvnorm in the mvtnorm library
out <- rmvnorm(n, mean = c(0, 0),
               sigma = matrix(c(1,.8, .8, 1), 2, 2))
Z <- out[,1]
T <- out[,2]
Y <- 0.5 * T + 0.7 * Z + rnorm(n)
model.data <- data.frame (
  Z <- Z,
  T <- T,
  Y <- Y
)
```

We'll move on to our third task next.

Finding adjustment sets

To determine the adjustment sets for the direct effect of the exposure `"T"` on the outcome `"Y"`, we utilize the DAGitty `adjustmentSets` function, which outputs `Z` as the adjustment variable, as anticipated. It is evident that `Z` blocks the pathway between T and Y and meets the single-door criterion.

```
print( adjustmentSets(g1, "T", "Y"))
## { Z }
```

Next, we get to the fourth and final task.

Estimating the path coefficient

In this task, we will estimate the path coefficient for the path $T \rightarrow Y$ using linear regression. That path coefficient corresponds to the average causal effect of X on Y. To perform linear regression on the dataset, we use the `lm` function in R, which takes as an argument the linear model and outputs the best fit of linear model coefficients concerning the dataset, in the sense of the least mean square of residual errors between the model and the actual data. The `lm` function returns, for every coefficient in the regression model, an estimated value and a standard error of that estimate, and other statistics: the **t-value** and the **p-value**. The t-value is a derived statistic, the ratio of the estimate over the standard deviation. The p-value measures the statistical significance of this t-statistic – the smaller the p-value, the more unlikely the **null hypothesis** (i.e., that the relationship between the variables could have happened by chance alone). For the purpose of this practice, we only use the `coef` function on the `lm` output to get the estimated coefficients, which outputs the intercept and the regression coefficients of the respective independent variables.

First, we conduct linear regression without including any adjustment variables, as shown in the following code.

```
coef(lm(Y ~ T))
## (Intercept)           T
## 0.004530807 1.060607636
```

The result shows that the intercept is nearly zero, as expected, since the variables are standardized, and the path coefficient is estimated to be approximately 1.06, which is far off from the true value of 0.5, indicating estimation bias.

Second, we conduct the linear regression while using `Z` as an adjustment variable, as shown in the following code.

```
coef(lm(Y ~ T + Z))
## (Intercept)           T           Z
## 0.003432434 0.503902193 0.703071340
```

The code output shows that when we incorporate the adjustment variable `Z` in the regression model, we now get estimated path coefficients, which are pretty accurate. The estimated coefficient we get is 0.5039 for the direct causal effect of `T` on `Y`, compared to the actual value of 0.5. We also get an estimate of 0.703 for the path coefficient between the adjustment variable `Z` and the outcome `Y`, compared to the actual value of 0.7. The results now indicate an unbiased estimate of the direct causal effect of `T` on `Y`. Remember, the estimated path coefficients are themselves random variables, so you will not get the same values for the estimates shown here on different runs of the `lm` function.

Let us now proceed to practice on our next model.

LSCM M8

Figure 4.29 shows the structural causal model M8 corresponding to a confounder C causally influencing treatment and outcome. The figure shows the path diagram annotated with the values of the path coefficients. In the following practice using R, we will perform four consecutive tasks, as we have done previously for the model M7.

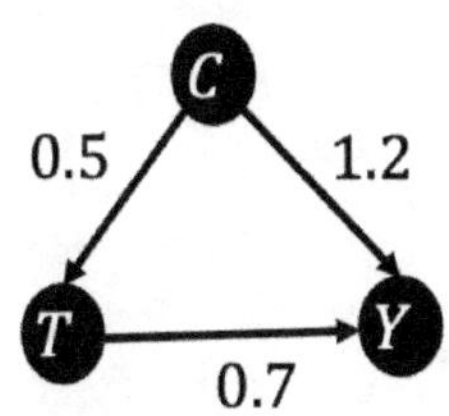

Figure 4.29 – Path diagram for the linear structural causal model M8

We'll begin with our first task next.

Specifying the structural equation model and plotting the path diagram

The following code snippet first defines the graph representation of the structural causal model, g. The first line of the specification indicates that `T` is regressed on `C`, while `Y` is regressed on both `T` and `C`. The rest of the code will generate the path diagram for the LSCM, M8, similar to the graph depicted in *Figure 4.29*, without the path coefficient annotations.

```
#create graph
g <-dagify(T~C,Y~T+C)
#Set position of nodes
coords<-list(x=c(T = 0, C = 1, Y = 2),
             y=c(T = 0, C = 1, Y = 0))
coords_df<-coords2df(coords)
coordinates(g)<-coords2list(coords_df)

#Plot causal graph
ggdag(g)+theme_dag()
```

We'll follow up with our second task next.

Generating a synthetic dataset

The following code simulates the structural equations using the linear model and the path coefficients as per *Figure 4.30*. The confounder `C` is modeled as a random variable with the standard normal distribution of zero mean and standard deviation equal to one. The code output is a data frame consisting of `n` records of the dataset for simulated observation on the three model variables `C`, `T`, and `Y`.

```
# simulate data (linear model)
n <- 1e4
```

```
C <- rnorm(n)
T <- 0.5 * C + rnorm(n)
Y <- 0.7 * T + 1.2 * C + rnorm(n)
model.data <- data.frame (
  Z <- Z,
  T <- T,
  Y <- Y
)
```

We'll move on to our third task next.

Finding adjustment sets

We use the `DAGitty` library's `adjustmentSets` function in the following code to obtain the adjustment sets for estimating the direct causal effect of the exposure `"T"` on the outcome `"Y"`. As expected, there is only a singleton set with the confounder `C` as the adjustment variable.

```
print( adjustmentSets(g, "T", "Y"))
## { C }
```

Next, we get to the fourth and final task.

Estimating the path coefficient

As before, we will first estimate the path coefficient for the path $T \rightarrow Y$ by regressing `Y` on `T` without adjustment for the confounder `C`, which represents the case of an unobserved confounder.

```
coef(lm(Y ~ T))
## (Intercept)          T
##  0.01319587  1.18046752
```

The code output shows an estimated value of the path coefficient equal to 1.18 compared to the true value of 0.7, indicating an estimation bias, referred to in this case as a confounder bias.

Second, we conduct linear regression while using `C` as an adjustment variable, as shown in the following code.

```
coef(lm(Y ~ T + C))
## (Intercept)          T           C
##  0.01444396  0.68187555  1.23169505
```

The code output shows that, when adjusting for the confounder C, we attain estimated path coefficients that are quite satisfactory compared to their actual values. The estimated coefficient for the direct path from $T \rightarrow Y$ is 0.6818 compared to the actual value of 0.7. The estimated coefficient for the direct path $C \rightarrow Y$ is 1.2316 compared to the actual value of 1.2. Again, remember the estimated path coefficients are random variables, meaning the estimated coefficients obtained from different code executions can be different.

Let us conclude this section by summarizing the takeaway points. Applying linear regression to estimate the parameters of linear structural causal models requires careful considerations to address challenges such as endogeneity and multicollinearity. However, when selecting appropriate control variables for adjustments, linear regression provides a powerful vehicle for estimating the ACEs from observational data. For linear regression to converge to the true causal effect parameters in an LSCM, the parameters must be identifiable, and the regression model must be properly specified to reflect that identification strategy.

In the next section, we will focus on the causal interpretation of Bayesian networks, which is based on the framework of structural causal modeling.

Interpreting causal Bayesian networks

In this section, we describe the causal interpretation of Bayesian networks. We begin with a discussion on association and causation. Next, we explain the connection between causality and intervention and define the "do" operator. Then, we present the difference between intervention conditioning and observational conditioning in Bayesian networks. Following this, we demonstrate examples of Bayesian network models.

Association and observational questions

Researching statistical association is a fundamental topic across various research fields. A primary objective of biomedical research is to identify connections between exposure and outcome. In medicine, one significant area of interest is examining the relationship between a drug (the exposure) and the subsequent adverse or beneficial events (the outcomes) that may occur.

Observational questions in statistics and machine learning often include inquiries such as, "Is there a relationship between a patient's pre-existing conditions and their hospital recovery time?" or "How strong is the relationship between a company's advertising budget and its sales figures?" These questions rely on the analysis of observational data, allowing for the prediction of outcomes or the identification of correlations.

Observational questions seek to determine the probability of a variable Y based on observing another variable X and involve evaluating the conditional probability $P(Y|X)$, which can be calculated using the joint probability $P(Y,X)$ and Bayes' rule.

$$P(Y|X) = \frac{P(Y,X))}{P(X)}$$

Observational questions between two observed variables, say X and Y, arise when the variables are associated. Association means that knowing Y influences our belief about X, and knowing X influences our belief about Y. That is,

$$P(X|Y) \neq P(X)$$

and

$$P(Y|X) \neq P(Y)$$

The Pearson correlation coefficient measures the association between the variables X and Y,

$$\rho(X,Y) = \frac{Cov(X,Y)}{\sqrt{Var(X)Var(Y)}} = \frac{E[(X - \mu_X)(Y - \mu_Y)]}{\sqrt{E(X - \mu_X)^2]\,E[(Y - \mu_Y)^2]}}$$

The association relation is symmetric. The correlation between X and Y is the same as between Y and X,

$$\rho(X,Y) = \rho(Y,X)$$

The majority of statistics aim to determine whether two variables are correlated, the strength of that correlation, and whether chance accounts for the observed relationship. Statistics fundamentally evaluate the influence of randomness in this correlation. The p-value indicates the improbability of the observed link occurring by chance. Thus, statistical analysis alone cannot provide evidence of a causal relationship, which we'll discuss next.

Intervention and causal questions

Although the association can be observed, causality cannot be established. While association is symmetric or bidirectional, causation, on the other hand, has only one direction represented by a directed edge in the causal DAG.

Intuitively, causality implies a causal ordering between the variables, reflecting a temporal precedence relation between the variables. The causes must occur before their effects can happen.

Causal inference focuses on interventional questions: "Would the effect of Y change if we intervened and changed X?"

The following is an example of an intervention question: "What would happen to a person's income *if we forced them* to receive training?" In that question, the intervention is "forcing someone to receive training," and the outcome is "changing a person's income."

Other examples of intervention questions include the following:

- "Will reducing class sizes by a third increase student test scores?"
- "Does providing free legal aid decrease the rate of homelessness among low-income tenants?"

To investigate causation, we must operationalize the intervention operation to represent a physical or a hypothetical, intentional manipulation of the data-generative Bayesian network model.

The intervention operation transforms a pre-intervention data-generative model into a post-intervention generative model, akin to an experiment in which we set the value of a variable X for the entire population, rather than just for a subset that naturally possesses that value. This process enables us to simulate what a **Randomized Controlled Trial** (**RCT**) might entail, without actually conducting the experiment, which may be impractical or unethical.

The following section explains the "do" operator to represent unit intervention and define conditioning under intervention.

Unit intervention – the do operator

Intervention in a structural causal model on a variable X, denoted as $do(X = x)$ or simply $do(X)$, signifies a modification of the causal mechanism associated with the functional equation for X. This mechanism is superseded by the equation $X = x$, which requires that X take on a specific value, x. Consequently, this intervention eliminates all preexisting causal influences that would have determined X's value in the model prior to it. Additionally, it modifies the causal graph by removing all directed edges leading into X, thereby severing the causal relationships between X and all its parent variables, denoted as $pa(X)$.

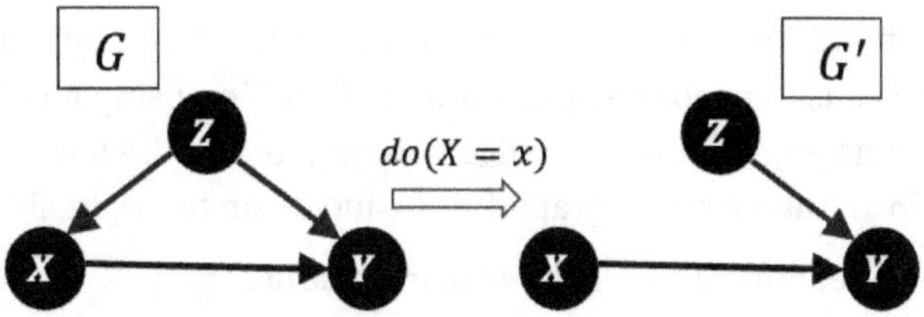

$Z = f_1(U_Z)$
$X = f_2(Z, U_X)$
$Y = f_3(X, Z, U_Y)$

$Z = f_1(U_Z)$
$X = x$
$Y = f_3(x, Z, U_Y)$

Figure 4.30 – The unit intervention do(X=x) changes the causal model on the left to the causal model on the right

Figure 4.30 presents an illustrative example in which the causal model on the left shows a direct causal influence of treatment X on outcome Y. Confounder$Z$ influences both the treatment and the outcome, creating a biasing path between X and Y. The modified causal model on the right illustrates the impact of a unit intervention $do(X = x)$. In this model, the structural equation for X is replaced with the equation $X = x$, and the directed edge $Z \rightarrow X$ is removed from the causal graph.

We use the **do operator** notation to indicate the intervention or manipulation of network variables. To represent the distribution of variables Y when we intervene on other variables X, we write $P(Y|do(X))$ and call that distribution **intervention conditioning**, which in general is different from the conditional probability,$P(Y|X)$, which is the probability of Y when we observe or see X. We refer to the latter conditioning as **observation conditioning.**

In general, the probability of conditioning by intervention $P(Y|do(X))$ will be different from that of conditioning by observation $P(Y|X)$:

$$P(Y|do(X)) \neq P(Y|X)$$

To gain intuition about the causal interpretation of Bayesian networks and the difference between observation and intervention, let us consider a simple BN example representing a joint distribution $P(X, Y)$ on two variables X and Y. We can model the BN by either one of the two graphs in *Figure 4.31*. In both models, we obtain the same result for the probability of Y conditioned on seeing X.

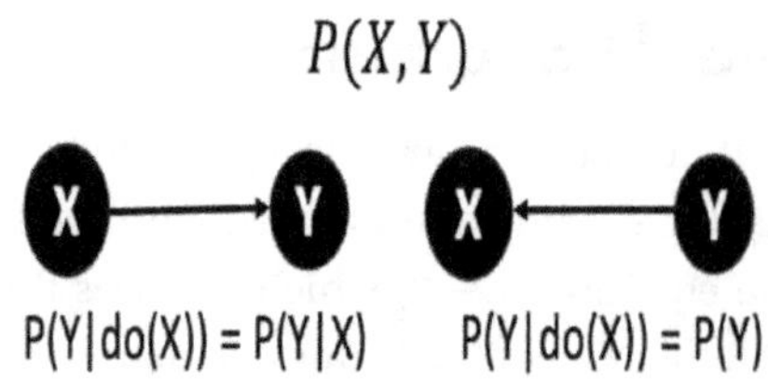

Figure 4.31 – Conditioning by intervention

If we interpret the directed edge in the Bayesian network models of *Figure 4.31* as representing causal influence, we obtain different results regarding the probability of Y when conditioned on performing intervention on X. This difference arises from the semantics of unit intervention. Intervening on X entails overriding the causal mechanism that affects X and imposing a specific value for it. This action is analogous to removing the directed edge pointing to X in the left-side graph of *Figure 4.31*. This alteration results in an **intervention graph** for the model on the right side of *Figure 4.31*, where Y and X are disconnected, leading to $P(Y \mid do(X))$ being distinct from $P(Y \mid X)$.

In the following, we present two examples of intervention conditioning for causal inference in real-world scenarios.

Example: treatment and recovery

A pharmaceutical lab is investigating whether a new drug enhances recovery. Observational data has been collected from recovery rate records provided by doctors who prescribe the new drug to patients with more severe illnesses. Statistical analysis indicates that, on average, the drug is associated with poorer recovery rates, likely due to unobserved confounders related to factors such as illness severity, age, and other variables.

The pharmaceutical company then assigns a team of data scientists to construct a structural causal model that accounts for the confounders, which is illustrated in the causal graph shown in *Figure 4.30*. The team employs the do operator to compute the intervention conditioning $P(recovery \mid do(drug = 1))$, which mimics a situation in which the drug is given to everyone at random, much like an RCT. The confounder causal model reveals a more nuanced effect of the new drug, indicating that its impact on recovery rates can be positive, negative, or neutral, depending on the patients' covariates.

Example: decision-making in public health policy

A decision-maker wants to evaluate the impact of a city-wide vaccination program on disease rates (for example, COVID-19).

The challenge: simply observing a lower disease rate after the program starts,

$$P(disease \mid vaccination = 1) < P(disease \mid vaccination = 0)$$

is insufficient, as additional variables (confounders) may influence the situation, including seasonal variations, public health initiatives, or behavioral modifications.

SCMs help disentangle the vaccination's effect from confounding factors to design better, more effective policies with predictable outcomes.

By using an SCM, which visualizes causal relationships, decision-makers can estimate the impact of a policy intervention on the desired outcome,

$$P(disease \mid do(vaccination = 1))$$

enabling them to refine the policy before large-scale implementation.

In the next section, we state the main goal of causal inference and outline how the intervention conditioning with the do operator can be evaluated.

The goal: from intervention to observation

The primary objective of causal inference is to determine whether a causal effect is identifiable—that is, whether it can be estimated from observational data.

Although RCTs are widely regarded as the most reliable method for collecting intervention data, conducting physical experiments to obtain this data is not always feasible.

As you will learn in *Chapter 9*, we can utilize the rules of **do-calculus**, developed by Judea Pearl, to translate an interventional distribution $P(Y)|do(X)$ into an expression that relies solely on observational probabilities, without using the do operator.

This is achieved through a set of three mathematical rules that manipulate probability expressions containing the do operator based on the structure of an underlying causal graph (DAG). The rules provide a formal, algorithmic framework to do the following:

- Determine identifiability: Assess whether a specific causal quantity can be estimated using the available observational data and the presumed causal structure.
- Derive the identification formula: If the effect is identifiable, the rules can be applied to create a formula that includes only standard conditional probabilities. This formula can then be estimated using traditional statistical machine learning methods on observed data.

The three rules of do-calculus are complete, meaning that if a causal effect is identifiable, a sequence of applications of the rules will successfully transform the causal expression into an observational one. This allows researchers to interpret associations as causation when the underlying assumptions encoded in the causal graph are met.

In the next section, you will practice using R with a hands-on coding example to experiment with the intervention conditioning example discussed earlier, in *Figure 4.31*.

R practice: intervention conditioning

The goal of this practice is to demonstrate how to use the R software to derive an expression for intervention conditioning. Intervention conditioning determines the identifiability of causal effect, and the R software uses the dosearch library (Tikka, Hyttinen, & Karvanen, 2021).

In the following, we query the two causal graphs of *Figure 4.31* for the intervention conditioning $P(Y|do(X))$.

The *RMarkdown* document 'chapter4.Rmd' includes the entire code to run for the example.

```
data1 <- "P(x,y)"
query1 <- "P(y|do(x))"
graph1 <- "
    x -> y
  "
dosearch(data1, query1, graph1)
## p(y|x)
data2 <- "P(x,y)"
query2 <- "P(y|do(x))"
graph2 <- "
    y -> x
  "
dosearch(data2, query2, graph2)
## p(y)
```

To conclude this section, let us highlight the key takeaways:

- The do operator represents a mathematical intervention that sets a specific value for the variable being intervened upon, effectively overriding the causal mechanisms that typically determine that variable's value
- Graphically, it involves deleting all incoming edges to the intervened variable in the causal graph
- Intervention conditioning $P(Y\} \mid do(X))$ is fundamentally different from observation conditioning $P(Y|X)$
- It allows us to answer "what if?" causal questions and simulate randomized controlled trials

The following section will explain the probabilistic and causal interpretations of Bayesian networks and formally define causal Bayesian networks.

Bayesian networks: probabilistic and causal interpretations

In this section, we will explain the probabilistic and causal interpretations of BNs. First, we will review the probabilistic interpretation. Then, we'll describe the causal interpretation based on unit intervention concerning a subset of variables.

Probabilistic interpretation

In this section, we focus on the probabilistic interpretation of the graphical model in a BN.

A BN represents the factorization of a joint distribution P across a set of variables, $V = \{X_1, \cdots, X_n\}$, which is illustrated by a DAG, G. In this context, the parents of the variable X_i are denoted as $pa(X_i)$. The following equation illustrates how P is factored according to the DAG:

$$P(x) = \prod_{i=1}^{n} P(x_i \,|x_{pa(X_i)}) \qquad (4.26)$$

In this context, $P(x)$ represents the joint probability of the value assignments of variables$V$ corresponding to the value tuple x;x_i, which indicates the value assignment of variable X_i within that tuple, and $x_{pa(X_i)}$ signifies the value assignments of the parents ofX_i, denoted as $pa(X_i)$ in tuple x.

The probabilistic interpretation of a Bayesian network is as follows:

- A directed edge between two nodes corresponds to probabilistic dependence
- The lack of a directed edge between two nodes means conditional independence
- Every node conditioned on its parents is independent of all its non-descendants
- The network is a generative model of all probability distributions that factor according to its DAG
- A Bayesian network enables the computation of the conditional probability of Y upon observing $X, P(Y|X)$, or the probability of X when seeing $Y, P(X|Y)$, for any two disjoint subsets of the network variables
- There can be multiple DAGs that encode the same independence relations in the probability distribution represented by the Bayesian network

The following section describes the causal interpretation of the Bayesian network.

Causal interpretation

As noted earlier, probabilistic relations between variables are associative, and the direction of probabilistic influence is reversible. Causal relations have a unique directionality that goes from cause to effect, such as temporal relations that are ordered from past to present.

A causal interpretation of a Bayesian network enables us to formulate causal queries that go beyond those under the probabilistic interpretation. Queries under the causal interpretation seek to estimate the causal effects of interventions. Interventions imply structural perturbations of the data generative model.

The causal interpretation of a BN implies the following:

- A directed edge from node X to node Y means X is a direct cause of Y
- A node X has an indirect causal influence on the descendants of X
- Conditioning by intervention via the do operator enables us to compute the data generative model under changed conditions, where we fix the values of certain variables

The following definition offers a formal interpretation of the data generative model for a causal BN. The standard representation of the probability distribution of a BN, as illustrated in *Equation 4.26*, is considered the degenerative case of the causal interpretation when no intervention occurs.

Definition 4.5: causal Bayesian Network (Lauritzen, 2000)

Let $BN = (G, P)$ be a Bayesian network representing a joint probability distribution P over variables $V = \{X_1, \cdots, X_n\}$ that factors according to the DAG, G. Let $pa(X_i)$ denote the parent of X_i in G. We say the BN is *causal for unit interventions at* $B \subseteq V$ if it holds for any $A \subseteq B$ that,

$$P(x|\text{do}(x_A^*)) = \prod_{v \in V \setminus A} P(x_v|x_{pa(v)})|_{x_A = x_A^*} \quad 4.27)$$

Here, $P(x|\text{ do}(x_A^*))$ denotes the conditioning of x for unit intervention on variables in set A, forcing their values to equal x_A^*. In *Equation 4.29,* we use x_v to denote the value assignment of variable v in the value tuple x, and $x_{pa(v)}$ to represent the value assignment of the parents of $v, pa(v)$, in tuple x.

Equation 4.27 is the factored representation of the joint probability of the BN after trimming the factors for the intervention variables and after assigning the set values for the variables subjected to the intervention. Thus, we can rewrite *Equation 4.27* in the equivalent form:

$$P(x|\text{do}(x_A^*)) = \frac{\prod_{v \in V} P(x_v|x_{pa(v)})}{\prod_{v \in A} P(x_v|x_{pa(v)})}|_{x_A = x_A^*} \quad (4.28)$$

To contrast the intervention conditioning with the observation conditioning, we write the conditional probability based on observing x_A^* under the Bayes' rule as follows:

$$P(x|x_A^*) = \frac{\prod_{v \in V} P(x_v|x_{pa(v)})}{\sum_{v \in V \setminus A} \prod_{v \in V} P(x_v|x_{pa(v)})}|_{x_A = x_A^*} \quad (4.29)$$

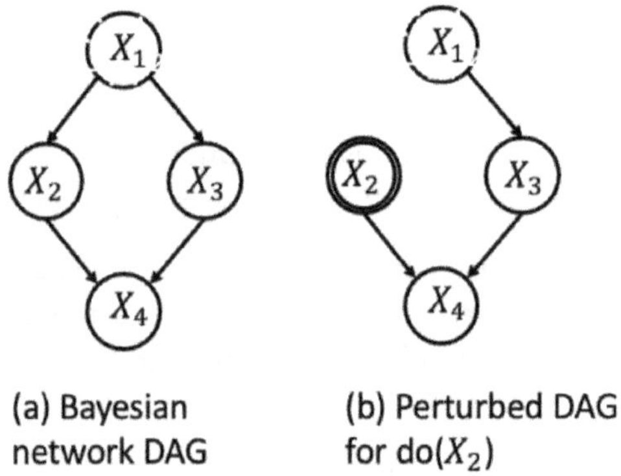

Figure 4.32 – A unit intervention on nodes in the DAG results in a perturbed graph discarding all edges from the nodes' parents

Example: intervention conditioning

To illustrate the effect of the do operator and unit intervention, let us consider the BN-directed acyclic graph of *Figure 4.32(a)*. Let us compute the intervention conditioning, the probability of X_4 conditioning on unit intervention on X_2, written $P(X_4|do(X_2 = x_2^*))$. An intervention fixing the value of variable X_2, represented as $do(X_2 = x_2^*)$, corresponds to the perturbed graph of *Figure 4.32(b)*, deleting the directed edge $X_1 \rightarrow X_2$. We can write the joint probability conditioning on $do(X_2 = x_2^*)$ from the perturbed graph as follows:

$$P\left(X_{1,}, X_2, X_{3,}, X_4 \middle| do(X_2 = x_2^*)\right) = P(X_{3,}|X_{1,})P(X_4|x_2^*, X_{3,})P(X_{1,}) \quad (4.30)$$

To obtain the probability of X_4 conditioning on $do(X_2 = x_2^*)$, we need to marginalize over X_3, X_1:

$$P(X_4|do(X_2 = x_2^*)) = \Sigma_{x_3,x_1}\left(P(x_{3,}|x_{1,})P(X_4|x_2^*, x_{3,})P(x_{1,})\right) \quad (4.31)$$

Let us compute the joint probability under intervention conditioning in *Equation 4.30* for the following numerical data for the Bayesian network **Conditional Probability Tables (CPTs)** specified in *Figure 4.33*.

X1	P(X1)
0	0.2
0.8	0.8

X1	X3	P(X3|X1)
0	0	0.35
0	1	0.65
1	0	0.55
1	1	0.45

X2	X3	X4	P(X4|X2.X3)
0	0	0	0.9
0	0	1	0.1
0	1	0	0.55
0	1	1	0.45
1	0	0	0.65
1	0	1	0.35
1	1	0	0.75
1	1	1	0.25

Figure 4.33 – Conditional probabilities for the intervention conditioning example

It is a straightforward exercise to verify the numerical results presented in *Figure 4.34*. The table on the left side of the figure displays the joint probability under intervention conditioning, as derived from *Equation 4.30* with ($x_2^* = 1$). Meanwhile, the table on the right side of the figure shows the probability of (X_4) under intervention conditioning, based on *Equation 4.31* with ($x_2^* = 1$). This probability was obtained by marginalizing over (X_1) and (X_3) from the joint distribution listed in the left table.

X1	X2	X3	X4	P(X1,X3,X4|do(X2=1))
0	1	0	0	0.0455
0	1	0	1	0.0245
0	1	1	0	0.0975
0	1	1	1	0.0325
1	1	0	0	0.286
1	1	0	1	0.154
1	1	1	0	0.27
1	1	1	1	0.09

X4	P(X4|do(X2=1))
0	0.2
1	0.6

Figure 4.34 – Joint probability and the probability of X4 under the intervention do(X2=1) for the intervention conditioning example demonstrates the impact of the intervention on the distribution of the variables involved

We also encourage you to compute the conditional probability $P(X_4|(X_2 = x_2^*))$ using Bayes' rule. Note that the result will differ from the distribution obtained through intervention conditioning.

To conclude this section, let us highlight the key takeaways:

- The do operator represents a mathematical intervention that sets a specific value for the variable being intervened upon, effectively overriding the causal mechanisms that typically determine that variable's value
- Graphically, it involves deleting all incoming edges to the intervened variable in the causal graph
- Intervention conditioning is fundamentally different from observation conditioning
- It allows us to answer "what if?" causal questions and simulate randomized controlled trials

In the following section, we describe the primary distinctions between SEM and SCM.

Comparing SCM versus SEM

Throughout this chapter, we examined two structural modeling frameworks: **Structural Equation Modeling (SEM)** and **Structural Causal Modeling (SCM)**. While both SEM and SCM serve as tools for analyzing relationships between variables, they differ significantly in their focus, graphical representation, goals, and techniques. In this section, we provide a summary of the key distinctions between SCM and SEM, as outlined in *Table 4.1*.

Functionality	Structural Causal Modeling	Structural Equation Modeling
Focus	Identifying and analyzing causal relationships between variables.	Estimating predictive and associative relationships between variables
Graphical representation	Causal graphs—directed edges represent direct causal links; bi-directed edges represent correlations between exogenous variables.	Path diagrams—directed edges represent regression relationships (prediction); bi-directed edges represent covariances or associations between variables.
Technique	Interventions, conditional independence and d-separation in DAGs	Multivariate multiple regression, path analysis, confirmatory factor analysis
Goal	Causal effect analysis, what-if analysis, counterfactual reasoning, causal explanation, causal attribution	Testing theoretical models on how latent variables are measured by observed variables, estimating the direct and indirect relationships between observed and latent variables

Table 4.1 – Comparative analysis of SCM and SEM

SEM aims to test hypothesized relationships between latent (unobserved) and observed (measured) variables through two main components: the measurement model, which examines how latent variables are indicated by observable measures, and the structural model, which explores the relationships among these latent variables. SEM provides advantages in testing theoretical models, estimating both direct and indirect relationships, accounting for measurement error, and analyzing complex systems with multiple independent and dependent variables, thereby offering a more comprehensive approach compared to traditional regression methods.

SCM offers a framework for understanding how variables influence one another and the effects of interventions, which is essential for distinguishing causation from correlation. By using causal graphs with directed edges (→) to represent direct relationships among observed variables, SCM effectively illustrates causal links, such as the relationship where X causes Y. Furthermore, bi-directed edges (↔) indicate the existence of unobservable variables that may jointly affect observed variables, thereby emphasizing correlations that are pertinent to causal inquiries.

Interventions, conditional independence, and d-separation within DAGs are some of the most important methods in SCM. These methods facilitate visual representations of causal relationships, determinations of conditional independence among variables, and adjustments for confounding factors to better estimate causal effects from observational data. Interventions, demonstrated through variable manipulation, are instrumental for testing causal hypotheses, exemplified by randomized controlled trials (RCTs). Conditional independence suggests two variables remain unaffected by one another when conditioned on a third. The d-separation illustrates the graphical rules needed to dictate which variables should be controlled or adjusted for to achieve unbiased causal effect estimates.

The objectives of SCM encompass **causal effect analysis, what-if analysis, counterfactual reasoning, causal explanation**, and **causal attribution**. Causal effect analysis quantifies relationships, allowing assessment of outcomes, such as the impact of advertisements on sales. What-if analysis examines potential outcomes resulting from varying conditions, while counterfactual reasoning—focused on hypothetical scenarios—explores outcomes that could have occurred under different circumstances. Causal explanation aims to provide understandable accounts for events using causal effect analysis and counterfactuals, while causal attribution seeks to identify the extent to which outcomes can be traced back to specific events or conditions, determining factors contributing to outcomes, such as the success of a marketing campaign.

Summary

In this chapter, you learned about fundamental topics related to two interrelated frameworks that represent complex relationships between multiple variables: **Structural Equation Modeling** (**SEM**) and **Structural Causal Modeling** (**SCM**).

The topics covered in SEM provided you with insights into techniques such as path analysis, confirmatory factor analysis, and multivariate regression. You can apply these techniques to estimate and test complex causal and correlational hypotheses regarding multivariate relationships.

The section on SCM introduced you to several structural-based methods, including intervention conditioning, d-separation, and conditional independence. Mastering these tools equips you with the ability to analyze assumptions about potential causal relationships. Additionally, you can evaluate paths between cause and effect, allowing you to block paths of association that are separate from the key causal relationship of interest.

In *later chapters* of the book, we will expand on the concepts you learned about in this chapter and explore the identification and estimation of causal effects in greater depth.

References

- Amani, V. J. (2020). *The effect of customer satisfaction on their loyalty by mediating the mental image of the brand and trust using structural equations in the Banking Industry (Case study: Pasargad Bank). Journal of Internet Banking and Commerce*, 25(4).
- Bardet, M., & Chyzak, F. (2005). *On the complexity of a gröbner basis algorithm. Retrieved from Algorithm Seminar*: `https://algo.inria.fr/seminars/summary/Bardet2002.pdf`
- Barrett, M. (2023, May 29). *Package 'ggdag'*. Retrieved from `https://cran.r-project.org/web/packages/ggdag/ggdag.pdf`
- Bollen, K. A. (1989). *Structural Equations with Latent Variables*. John Wiley & Sons.
- Bollen, K., & Pearl, J. (2013). *Eight Myths About Causality and Structural Equation Models. In S. Morgan, Handbook of Causal Analysis for Social Research* (pp. 301-328). Springer.
- Brannick, M. (2020, December). *Path Analysis*. Retrieved from `http://faculty.cas.usf.edu/mbrannick/regression/Pathan.html`
- Chen, B., Pearl, J., & Kline, R. (2018, April 5). *Graphical Tools for Linear Path Models*. Retrieved from `https://ftp.cs.ucla.edu/pub/stat_ser/r469.pdf`
- Drton, M., & Weihs, L. (2015, April). *Identifiability of Linear Structural Equation Models by Ancestor Decomposition*. Retrieved from arXiv:1504.02992 [stat].

- Foygel , R., Draisma, J., & Drton, M. (2012). *Half-trek criterion for generic identifiability of linear structural equation models. The Annals of Statistics*, 1682–1713.
- Garcia-Puente, L. D., Spielvogel, S., & Sullivant, S. (2010). *Identifying Causal Effects with Computer Algebra*. Proceedings of the *26th Conference of Uncertainty in Artificial Intelligence*, UAI (pp. 193–200). AUAI Press.
- Gilthorpe, M. (n.d.). *Covariate Roles in DAGs*. Retrieved November 27, 2023, from `https://dagitty.net/learn/graphs/roles.html`
- Healey, J. F. (2012). *Statistics: A Tool for Social Research*. Belmont: Wadsworth Cengage Learning.
- Health, F. o. (2025). *Biases and Confounding*. Retrieved from Epidemiology: `https://www.healthknowledge.org.uk/public-health-textbook/research-methods/1a-epidemiology/biases`
- Huntington-Klein, N. (2020). *Causal Diagrams Cheat Sheet*. Retrieved from `https://dagitty.net/learn/primer/Dagitty-Cheat-Sheet.pdf`
- Igolkina, A. A., & Meshcheryakov, G. (2020). *semopy: A Python Package for Structural Equation Modeling. Structural Equation Modeling: A Multidisciplinary Journal*, 952–963.
- Kim, S. (2015). *ppcor: An R Package for a Fast Calculation to Semi-partial Correlation Coefficients. Commun Stat Appl Methods, Commun Stat Appl Methods.*
- Kline, R. (2011). *Principles and Practice of Structural Equation Modeling* (Third Edition). New York: THE GUILFORD PRESS.
- Kyono, T. (2010). *Commentator: A Front-End User-Interface Module for Graphical and Structural Equation Modeling*. Los Angeles: Cognitive Systems Laboratory Department of Computer Science, UCLA.
- Lauritzen, S. (2000). *Causal Inference from Graphical Models*. Retrieved from `https://web.math.ku.dk/~richard/BSc/Lauritzen.pdf`.
- lavaan.org. (n.d.). *Lavaan model syntax* (conceptual reference). Retrieved from `https://lavaan.ugent.be/`
- M Parascandola, M. a. (2001). *Causation in epidemiology*. J Epidemiol Community Health, 905–912.
- Pearl, J. (2009). *Causality: Models, Reasoning and Inference*. New York, NY, USA: Cambridge University Press.
- Rosseel, Y. (2012). *lavaan: An R Package for Structural Equation Modeling*. Journal of Statistical Software, 1 - 36.
- Schisterman, E. F., Perkins., N. J., Mumford, S. L., Ahrens, K. A., & Mitchell, E. M. (2017). *Collinearity and Causal Diagrams: A Lesson on the Importance of Model Specification*. Epidemiology, 47-53.
- Schumacker, R., & Lomax, R. (2010). *A Beginner's Guide to Structural Equation Modeling*. Third Edition. New York, N.Y.: Routledge Taylor & Francis Group.
- Spirtes, P., Richardson, T., Meek, C., Scheines, R., & Glymour, C. (n.d.). *Using Path Diagrams as a Structural Equation Modelling Tool*. Retrieved 12 1, 2023, from `https://www.cs.cmu.edu/~tom/10-702/smr11.pdf`

- Stanbouly, D., Lee, K. C., Dodson, T. B., & Chuang, S.-K. (2022). *What is Collider Bias and Why Should We Care?* Journal of Oral and Maxillofacial Surgery.
- StataCorp. (2025). *STATA Structural Equation Modeling Reference Manual Release 19*. Retrieved from `https://www.stata.com/manuals/sem.pdf`
- Textor, J. (2022, October 13). *Drawing and Analyzing Causal DAGs with DAGitty*. Retrieved from `https://dagitty.net/manual-3.x.pdf`
- Textor, J., van der Zander, B., & Ankan, A. (2022, October 13). *Package 'dagitty'*. Retrieved from `https://cran.r-project.org/web/packages/dagitty/dagitty.pdf`
- Tikka, S., Hyttinen, A., & Karvanen, J. (2021). *Causal Effect Identification from Multiple Incomplete Data Sources: A General Search-Based Approach*. Journal of Statistical Software, 1–40.
- Wright, S. (1921). *Correlation and causation. Journal of Agricultural Research*, 557–585.
- Wright, S. (1934). *The Method of Path Coefficients. Annals of Mathematical Statistics*, 161-215.
- Zelterman, D. (2015). *Applied Multivariate Statistics with R*. New York: Springer.

Join our community on Discord

Join our community's Discord space for discussions with the author and other readers: `https://packt.link/hYCWN`

5

Relational Database Models

In this chapter, we introduce a relational database framework designed for representing and reasoning with probabilistic graphical models. This framework expands the standard representation of relational databases by incorporating the novel concept of weighted relations. We will describe how to express inference tasks in probabilistic graphical models in the relational database framework using the languages of relational algebra and **structured query language** (SQL), both of which are relationally complete.

We will explain various types of dependencies present in relational databases and highlight the desirable properties of acyclic databases, which enhance the efficiency of inference tasks. This chapter will teach you how to represent **Bayesian networks** (BNs) as relational databases and how to formulate inference tasks as relational queries. Additionally, we will address the problem of decision-making under uncertainty and its graphical model representation in influence diagrams. We will demonstrate how the relational database framework can be applied to represent and reason with influence diagrams.

To reinforce the concepts introduced in this chapter, we will use illustrative examples and provide hands-on programming demonstrations in the practice sections.

In this chapter, we're going to cover the following topics:

- Functions, relations, and databases
- Acyclic database schemes
- Bayesian network as a relational database
- Decision-making under uncertainty
- Influence diagram as a relational database

Technical requirements

To run the Python code, we recommend installing Python 3 following the installation and setup guide at `https://realpython.com/installing-python/`. Then, install the PyCharm **integrated development environment** (IDE). You can download and install the community version of PyCharm for your operating system from `https://www.jetbrains.com/pycharm/`.

The Python code in this chapter is for a Python project named `Chapter 5` that you can download from `https://github.com/PacktPublishing/Causal-Inference-with-Bayesian-Networks/tree/main/Chapter_05`

The code requires that you install the package for the Python SQL toolkit and object-relational mapper, **SQLAlchemy**: `https://www.sqlalchemy.org/#:~:text=SQLAlchemy%20is%20the%20Python%20SQL,power%20and%20flexibility%20of%20SQL`.

The examples in this chapter implement the models in the world's most advanced open source relational database, PostgreSQL, which you can install from `https://www.postgresql.org`. We highly recommend also installing the PostgreSQL web-based GUI tool pgAdmin, from `https://www.pgadmin.org/download/`.

You can run the **RMarkdown** document included in the `Chapter 5` code folder from PyCharm or use the RStudio IDE. To run it from within PyCharm, you will need to install the R plugin as per the documentation at `https://www.jetbrains.com/help/pycharm/r-plugin-support.html`.

You will be required to install various R libraries to run the examples of this chapter, and you can do that from within the **RStudio** IDE. The required packages to install are the following:

- **relations**: `https://cran.r-project.org/web/packages/relations/index.html`
- HyperG: `https://cran.r-project.org/web/packages/HyperG/index.html`

Functions, relations, and databases

This section will teach you the general definitions of functions, relations, and databases. We will introduce you to a particular type of relation, called a weighted relation, with applications for representing the graphical models of BNs and influence diagrams throughout the chapter.

We will begin this section with the definition of relations and **functional dependencies** (**FDs**). After that, we will define the terms relational database and database scheme. Following this, we will define a particular type of relation, called a weighted relation, and explain its potential for representing and solving problems in many domains. Lastly, we will describe three operators on relations that are essential for stating queries on relations to manipulate data and solve problems.

Representing relations

We define a relation, $r(X_1, \cdots, X_n)$, over a set of n attributes, $X_1, \cdots, X_n$, as a subset of the Cartesian product of the value domains of respective n attributes. Formally, that looks as follows:

$$r(X_1, \cdots, X_n) \subseteq dom(X_1) \times \cdots \times dom(X_n)$$

We use $dom(X_i)$ to denote the value domain of attribute X_i, which is assumed to be a finite set of discrete values. We call the set of attributes $X = \{X_1, \cdots, X_n\}$ the **scheme** for the relation, r(X). A relation, r(X), consists of a set of value tuples that are a subset of the Cartesian product of the domains of the attributes in the relation's scheme. The tuples constitute the rows in a table representing the relation; each tuple is a mapping from attributes to values in the attributes' respective domains. For a tuple , t, and a set of attributes, Z, we use $t[Z]$, which means t restricted to domain Z, or the components oft for the attributes in Z.

Let us give an example of relations and learn how to programmatically specify them using R and the library of the `relations` package(Meyer, et al., 2025) (Meyer, et al., 2016). First, we will describe the relations, and then we will implement them using R.

Example: relations

Let x, y, z, and w be attributes with the following domains:

$$dom(x) = \{1,2\};\ dom(y) = \{1,2,3,4\},;\ dom(z) = \{1,2,4\};\ dom(w) = \{1.2.3\}$$

We define two relations: $r_1(x, y, z)$ and $r_2(x, z, w)$. *Figure 5.1* shows tables with the tuples for each relation.

r1(x, y, z)			r2(x, z, w)		
x	y	z	x	z	w
1	2	1	1	2	3
1	3	2	2	4	1
1	4	2	1	2	2
2	1	4	1	4	2

Figure 5.1 – Example relations: $\boldsymbol{r_1(x, y, z)}$ ***and*** $\boldsymbol{r_2(x, z, w)}$

Practice using R

The code snippets shown here are part of the R Markdown `relations.rmd` file included in the `Chapter 5` folder in this book's GH repository.

Let's define the $\boldsymbol{r_1(x, y, z)}$ relation:

```
r1 <-
  data.frame(
    x = c(1, 1, 2, 1),
    y = c(2, 3, 1, 4),
    z = c(1, 2, 4, 2)
  )
r1 <- as.relation(r1)
relation_table(r1)
```

Running the preceding code will print the following table:

```
x y z
1 2 1
1 3 2
1 4 2
2 1 4
```

Let's now define the $\boldsymbol{r_2(x, z, w)}$ relation:

```
r2 <-
  data.frame(
    x = c(1, 2, 1, 1),
    z = c(2, 4, 2, 4),
    w = c(3, 1, 2, 2)
  )
r2 <- as.relation(r2)
relation_table(r2)
```

Running the preceding code will print the following table:

```
X Z W
2 4 1
1 2 2
1 4 2
1 2 3
```

So far, we have defined the basic concept of relations. In the following section, we extend the concept of relations to include a special attribute that we call the weight attribute. We will describe the concept of weighted relations next.

Weighted relations

The concept of weighted relations enables modeling and reasoning tasks in graphical models, including BNs and influence diagrams (El Fattah, et al., 2000).

The weight attribute enables using relations to represent uncertainty in graphical models of BNs. The weight can also represent the cost function in optimization problems, such as in model-based diagnosis (El Fattah, et al., 1995), as well as utility functions in decision-making with influence diagrams.

Definition: weighted relation

A weighted relation is a pair, (r, w), where r is a relation and w is a function defined on the scheme of r. That is, for each tuple $t \in r$, there is a weight assigned whose value is $w(t)$. A weighted relation, (r, w), can be represented by a single relation whose scheme is $R \cup \{W\}$, where R is the scheme of r and W is a weight variable whose value is determined by the weight function, w.

Example: weighted relation

A **joint probability distribution** (**JPD**) on a set of attributes is a weighted relation where the weight is a probability function on the value tuples of the attributes such that the sum of the weight over all tuples must equal 1. A JPD on the attributes of the $r_1(x, y, z)$ relation from the previous example is represented by the weighted relation shown in *Figure 5.2*.

x	y	z	w
1	2	1	0.3
1	3	2	0.2
1	4	2	0.1
2	1	4	0.4

Figure 5.2 – Weighted relation representing a probabilistic relation where the weight, w, is the JPD over the scheme of relation r1

The concept of a weighted relation plays a crucial role in embellishing relational databases with application-specific semantics. In doing so, we can leverage the power of relational algebra and expressions in the SQL language of relational databases to perform probabilistic inference tasks with BNs and influence diagrams. We will use the notion of a weighted relation throughout this chapter and in *Chapter 7*.

Operating on relations

In this section, we will describe three main operators on relations: the **select operator**, the **project operator**, and the **join operator**. They make up part of the operations defined in the **relational algebra theory** introduced by Codd (Codd, 1970). We will illustrate the three operators with examples. Then, in a follow-up practice section, we will demonstrate how to run the examples programmatically in R using the `relations` library by Meyer and Hornik (Meyer, et al., 2016).

The select operator

The select operator is a unary operator that selects a subset of rows of the tuples from the relations satisfying a specified logical condition on an attribute in the relation schema. Let r be a relation on a scheme, R. Let A be an attribute in R and c(A) be a selection criterion on an attribute, A, in R. Applying the $\sigma_{c(A)}(r)$ select operator returns a new relation, $r'(R) = \{t \in r | c(t_A)\}$, consisting of the subset of tuples of r satisfying the selection criterion, c(A). We express this as follows:

$$\sigma_{c(A)}(r) \stackrel{\text{def}}{=} \{t \in r | c(t_A)\} \quad (5.1)$$

Let's look at an example of a select operator.

We'll return to the $r_1(x, y, z)$ relation from the previous example. Consider the selection operation of selecting tuples with $x = 1$ and $y < 3$. We can apply two consecutive select operations, as follows:

$$\sigma_{y<3}\left(\sigma_{x=1}\, r_{y<3}\left(\{x, y, z\}\right)\right) \quad (5.2)$$

The result is the$r_1'(\{x, y, z\}) = \{(1,2,1)\}$ relation, which has only one tuple. The result does not depend on the order of invoking the select operator.

The project operator

The project operator is a unary operator that returns another relation on a specified subset of attributes after removing duplicate rows. Let r be a relation on a scheme, R. Let X be a subset of R. The projection of $r(R)$ onto X, written $\pi_X(r)$, returns a new relation, $r'(X) = \{t(X)|t \in r\}$. We express this as follows:

$$\pi_X(r) \stackrel{\text{def}}{=} \{t(X)|t \in r\} \quad (5.3)$$

Let's look at an example of a project operator.

Consider the r_1 relation from the previous example. The projection operation on the x and y variables, denoted as $\pi_{\{x,y\}}r_1(x,y,z)$, produces the following relation, $r_1'(x,y)$, which is represented by the set of (x, y) tuples: $\{(1,2),(1,3),(1,4),(2,1)\}$.

The join operator

The join operator is a binary operator that takes two relations, $r(R)$ and $s(S)$, and returns a new relation on the combined set of attributes of the two relations containing a set of tuples, t, such that t restriction on R is in$r$ and the restriction on S is in s. We express the join operator on the two relations as follows:

$$r \bowtie s \stackrel{\text{def}}{=} \{t|t[R] \in r \land t[S] \in s\} \quad (5.4)$$

Alternatively, the join of two relations returns tuples on the union set of attributes of the two relations, specified by the combined tuples from the two relations, restricted to the subset where the elements of the shared attributes have the same value assignments.

Example: join operator

To illustrate the working of the join operator, let us consider the join of the two relations depicted in *Figure 5.1*, expressed as $r_1(x,y,z) \bowtie r_2(x,z,w)$. *Figure 5.3* depicts the join operation on the two relations, making it easy to verify the result.

r1(x,y,z)

x	y	z
1	2	1
1	3	2
1	4	2
2	1	4

r2(x,z,w)

x	z	w
1	2	3
2	4	1
1	2	2
1	4	2

r1 (x,y,z)⋈ r2 (x,z,w)

x	y	z	w
2	1	4	1
1	3	2	2
1	4	2	2
1	3	2	3
1	4	2	3

Figure 5.3 – The join operator on relations r1 and r2

The following explains how the join table in *Figure 5.3* corresponds to the definition provided in Equation 5.4. The schema of relation $r1$ is $R = (X,Y,Z)$, and the schema of relation $r2$ is $S = (X,Z,W)$, which results in a join relation schema of (X,Y,Z,W). We will show that every tuple, t, in the join relation meets the requirement that $t[R]$ is in $r1$ and $t[S]$ is in $r2$.

The first tuple in the join relation is $t = (2,1,4,1)$, which results in $t[R] = (2,1,4)$, corresponding to the fourth tuple in the $r1$ table, and $t[S] = (2,4,1)$, which corresponds to the second tuple in the $r2$ table. We can similarly verify that the join definition holds for all other tuples in the join relation.

The join table contains five rows, while the cross product of both relations results in 16 combinations. This indicates that not every tuple in the $r1$ relation can be paired with a tuple in the $r2$ relation to produce valid tuples in the join relation. For instance, the first tuple in the $r1$ table has $(x = 1, z = 1)$, which does not match any tuples in the $r2$ table. Conversely, the second and third tuples in the $r1$ relation, which have $(x = 1, z = 2)$, can join with the first tuple in the $r2$ table, yielding the second and third rows of the join table, as illustrated in *Figure 5.3*.

In the following practice section, we demonstrate how to define relations and run operations on them using R.

Practice using R

The code snippets shown here are part of the R Markdown `relations.rmd` file included in the `Chapter 5` folder in this book's GH repository.

Invoking the select operator

The following code selects the tuples of r_1 with $x = 1$ and $y < 4$:

```
relation_table(relation_selection(relation_selection(r1, x==1), y<4))
```

The selection operation outputs the following relation:

```
x y z
1 2 1
1 3 2
```

Invoking the project operator

The following code projects the r_1 relation on the subset of attributes, $\{x. y\}$:

```
relation_table(relation_projection(r1, c("x", "y")))
```

The projection operation outputs the following relation:

```
x y
2 1
1 2
1 3
1 4
```

Invoking the join operator

The following code performs the $r_1(x, y, z) \bowtie r_2(x, z, w)$ join operation:

```
relation_table(r1 %|><|% r2)
```

Running this code will produce the following table:

```
x y z w
2 1 4 1
1 3 2 2
1 4 2 2
1 3 2 3
1 4 2 3
```

To recap, a relation is defined over a scheme, which consists of a set of attributes that have values in finite domains represented by sets of discrete values. The relation is made up of a collection of value tuples, where each tuple maps attributes to their corresponding values within the respective domains. The tuples in a relation form a subset of the Cartesian product of all the domains associated with the attributes. A weighted relation builds upon a standard relation by adding an attribute known as weight, which is a function mapping on the relation scheme and carries domain-specific semantics. This weight attribute can denote probability distributions in BNs or represent cost or utility functions in optimization problems, as well as in modeling with influence diagrams.

In the remainder of this section, we define relational databases, describe various types of dependencies, and introduce relational operators used in relational algebra.

Relational databases

In this section, we formally define the concepts of relational databases and FDs.

Relational databases are the most common framework for representing structured data in the real world. The formal basis for relational database systems is provided by the relational data model(Codd, 1970).

In the relational model, a database is a collection of relational tables (or simply relations). The database language SQL is now widely accepted as the standard query language for relational database systems. SQL can be used to interact with a database or can be embedded within another programming language to interface with a database. Additional programming extensions have transformed SQL into a complete database programming language.

Relational tables can represent the following:

- **Discrete probability distributions:** A map from input values to probability distributions over a discrete random variable, for example, the result of a weather forecast given the actual weather.
- **Constraints:** The set of allowable combinations of input and output values. For example, the output is 1 if the inputs have an odd number of 1s.

Definition: relational database

A **relational database** on a database scheme, $\mathbf{R}$, is a collection of relations, $\{r_1, r_2, \cdots, r_p\}$. Each relation, r_i, on a relation scheme, R_i, written $r_i(R_i)$, is a set of value tuples, $\{t_1, t_2, \cdots, t_p\}$, for the attributes in R_i from their respective domains. We use $t[X]$ (or simplyt_X) to denote the value assigned by a tuple, t, to the subset of attributes $X \subseteq R_i$. A relation, $r_i(R_i)$, **key** is a minimal subset, K_i, of R_i that uniquely identifies a tuple within the relation. A **foreign key** is a set of attributes within one relation that matches the key of some (possibly the same) relation. A relation scheme may contain more than one key; the keys explicitly listed are called **designated keys**. We distinguish one of the designated keys as the **primary key**.

A relational database implements the data model serving an application in some domain. The data model must faithfully represent the dependencies underlying the information relevant to the application. The database design must efficiently encode the data model, adhering to normalization rules and the database's normal forms. The following section will explain two forms of dependencies in relational databases, **functional** and **multivalued**. Later, we will discuss another type of dependency: **join**.

Dependencies in relational databases

First, we will describe functional dependencies.

The following definition of functional dependencies is a generalization of the relation keys. Informally, a **functional dependency**, or **FD**, means the values of a tuple on one set of attributes uniquely determine the values on another set of attributes.

Definition: functional dependency

We say that a relation, $\mathrm{r(R)}$, satisfies the FD $X \to Y$, where X and Y are subsets of R. For every X value, x, there corresponds only one value for Y; that is, the set $\{t(Y)|t \in \mathrm{r} \wedge t(X) = x\}$ has at most one tuple. A database scheme, **R**, represents the FDs $G = \{X \to Y \mid some\ \mathrm{R} \in \mathbf{R}\ embodies\ X \to Y\}$. $\mathbf{R}$ completely characterizes a set of FDs, F, if $F \equiv G$.

A JPD on a set of attributes, X, is an example of the FD X→P, where each value tuple, x, for X corresponds to only one value, p, of the probability attribute P.

Another form of dependency is the **multivalued dependency** (**MVD**). An MVD is a generalization of the FD. Every FD is an MVD, but the converse is not true.

We write $X \twoheadrightarrow Y$ to denote an MVD in a relation, r(R), between two disjoint subsets, X and Y, of the scheme R If $X \twoheadrightarrow Y$ in a relation, r(R), we say that the subset of attributes Y of scheme R is multi-dependent on subset X of scheme R, or that attributes X multi-determine attributes Y.

Informally, an MVD on a scheme, R, implies that the relation on R can be decomposed into two relations on subsets of R. Specifically, R can be partitioned into three disjoint subsets, X, Y, and Z, such that the relation on R is the join of the projection of the R relation on XUY and the projection of R on XUZ. This informal characterization is a theorem stated in the next section. The formal characterization of MVD is as follows.

Definition: multivalued dependency (Maier, 1983)

A relation, $r(R)$, satisfies the multivalued dependency (MVD) $X \twoheadrightarrow Y$, where X and Y are two disjoint subsets of the scheme R, under the following condition. Let $Z = R - X - Y$ denote the remaining attributes. Whenever two tuples t_1 and t_2 in r have the same values for attributes in X, swapping their Y-values while leaving their Z-values unchanged must produce two tuples that also belong to r. Because Y and Z are interchangeable, the MVD $X \twoheadrightarrow Y$ implies the complementary MVD X↠Z. See *Figure 5.4*.

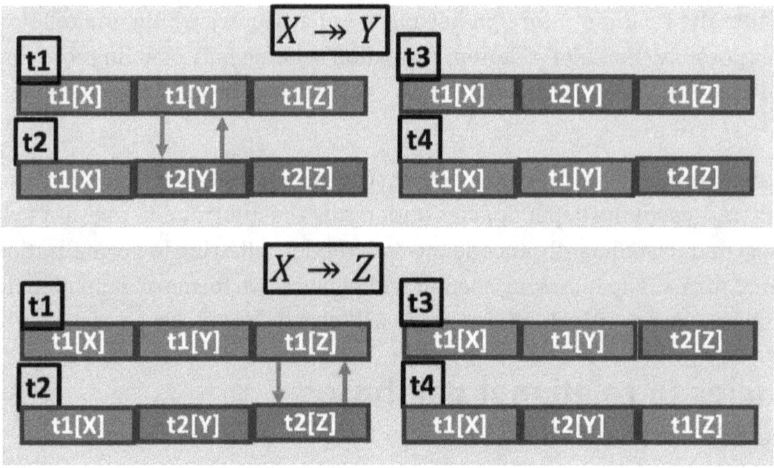

Figure 5.4 – MVD explained

Example: functional and multivalued dependencies

Figure 5.5 shows an instance of a relation on attributes X, Y, Z, and W with five tuples. The relation satisfies several FDs: $X\,Z \rightarrow W$. All pairs of tuples that agree on X, Z (first and fourth, second and third) have the same value for W.

The relation in *Figure 5.5* also satisfies the MVD $X \twoheadrightarrow Y$. The first and second tuples agree on X. If we swap the value of the Y attribute in the pair of tuples, keeping the value of other attributes in the tuples unchanged, we produce two tuples that also hold in the relation, which are the third and fourth tuples. Also, it follows that $X \twoheadrightarrow Z\,W$.

X	Y	Z	W
0	0	1	0
0	1	0	1
0	0	0	1
0	1	1	0
1	0	1	1

Figure 5.5 – Relational table satisfying multiple FDs

Let's summarize the main points from this section. Relational databases consist of relations on subsets of attributes forming the database scheme. The database specifies designated and foreign keys for the relational tables to ensure the integrity of the data model. A relation is a subset of the Cartesian product of the value domain of all attributes of the relation's scheme. Relations embody dependencies that include FDs and MVDs. Understanding the dependencies inherent in the data model is crucial for implementing appropriate and efficient database schemas. A weighted relation is a relation that has a special attribute named a weight attribute, to be handled specially by relational algebra operators depending on the semantics of the application domain. Instances of relational operators include the select operator, the project operator, and the join operator, with the join operation generally being the most computationally expensive.

In the next section, we will describe acyclic databases. Acyclic databases have many desirable structural properties that carry over to BN models. For example, learning about database representation in hypergraphs and join trees will serve as building blocks for inference algorithms in BNs, which will be covered *later in* of the book.

Acyclic database schemes

In the previous section, we formally described relational databases and characterized FDs and MVDs.

This section will teach you essential concepts relating to relational databases that will equip you with in-depth knowledge about structural properties with parallels that carry over to BNs and related graphical models.

We will begin this section by describing **join dependencies** (JDs) in relational databases, extending the MVDs concept. After that, we will define local and global consistency in relational databases and explain their connection. Following this, you will learn about hypergraphs and their use for representing the database schema structure. Lastly, we will formally define acyclic relational schemes and their representation in acyclic hypergraphs and associated join trees.

Join dependencies

Informally, the JD condition means that the decomposition of a relation, R, into n smaller relations by projecting the relation onto n subsets of R produces a lossless join in the sense that joining the decomposed relations recovers or reproduces the original relation. Alternatively, everything is preserved by decomposing the relation into smaller tables, and the relation can be reassembled by joining the relations in the decomposition without losing any information. The following is a formal definition of the JD

Definition: join dependency (Beeri, et al., 1981)

Let r be$a$ relation over a set of attributes, U, and let $R_1, \cdots, R_n$ be a set of relation schemes over U. Each R_i is a subset of U and $\bigcup_{i=1}^{n} R_i = U$. The relation $r(U)$ satisfies the JD expressed syntactically as $\bowtie (R_1, \cdots, R_n)$ if r decomposes losslessly onto $R_1, \cdots, R_n$. When we project the relation r onto the n smaller subsets of attributes $\{R_i\}$ and then join the results, we get the relation $r(U)$ back without any loss. This looks as follows:

$$r = \pi_{R_1}(r) \bowtie \cdots \bowtie \pi_{R_n}(r) \quad (5.5)$$

To simplify the notation, we will adopt the convention of referring to a relation, $r(R)$, by its scheme, R, the scheme being the set of attributes over which the relation is defined. Adopting this convention should not result in any confusion, as the intended meaning will be apparent from the context.

The MVD is a special case of the JD. The MVD X↠Y about R is equivalent to the JD $\bowtie ((X\,Y), (X\,Z))$, where $Z = R - X - Y$. The following theorem formally states the connection between the MVD and JD, or lossless decomposition.

Theorem: multivalued dependency (Maier, 1983)

Let r be a relation on scheme R, and let X, Y, and Z be subsets of R such that Z = R - (X Y). Relation r satisfies the MVD $X \twoheadrightarrow Y$ if, and only if, r decomposes losslessly onto the relation schemes R1 = X Y and R2 = X Z.

Example: join dependency, multivalued dependency

The relation $XYZW$ in *Figure 5.5* satisfies the MVD $X \twoheadrightarrow Y$ and can be decomposed into two relations, XY and XZW, obtained by a projection operation on the respective schemes. It is easy to verify that the join of the decomposed relations, XY and XZW, shown in *Figure 5.6*, produces the relation $XYZW$.

X	Y
0	0
0	1
1	0

X	Z	W
0	1	0
0	0	1
1	1	1

Figure 5.6 – Relation XYZW satisfies the JD $\bowtie (\boldsymbol{XY}, \boldsymbol{XZW})$

To recap, the JD generalizes the MVD, which involves multiple schemas instead of just two. The JD allows the decomposition of a relation into n relations on smaller subsets of the relation scheme.

The following section will describe local and global consistency and explain their connection.

Local and global consistency

In this section, we will formally define local and global consistency in relational databases and illustrate the concepts using an example.

Definition: Local Consistency

We say the n relations $R_1, \cdots, R_n$ are locally join consistent if for each i and j every tuple in R_i has a matching tuple in R_j. Alternatively, every tuple in R_i can be extended by a tuple in R_j, thus there are no dangling tuples between R_i and R_j. Formally, this is written as follows:

$$\pi_{R_i}(R_i \bowtie R_j) = R_i \quad (5.6)$$

The equation also holds when i and j are reversed.

Definition: global consistency

We say the n relations $R_1, \cdots, R_n$ are globally join consistent if they satisfy the JD condition; that is, every tuple in a local relation R_i can be extended by matching tuples in other relations (no dangling tuples). Every tuple in a relation R_i is consistent with a tuple in the global relation R, which is the join of all the local relations. Formally, for all values of i, we have the following:

$$\pi_{R_i}(R_1 \bowtie \cdots \bowtie R_n) = R_i \quad (5.7)$$

It is clear from the definitions that global consistency implies local consistency. When local consistency implies global consistency, the output of the join of two relations is always as large as the input relations, thus producing a monotone join where the number of tuples following every join will grow monotonically.

In general, local consistency does not imply global consistency, as the following example of local and global consistency demonstrates.

Example: local-global consistency

Figure 5.7 shows three relations, XY, YZ, and ZX, over the X, Y, and Z attributes. The three relations are binary and locally consistent. However, the join of the three relations is empty, meaning there is no globally consistent tuple for XYZ. The example can be found in the `relations.rmd` R Markdown file.

X	Y
0	1
1	3
2	5

Y	Z
1	2
3	8
5	4

Z	X
2	1
8	2
4	0

Figure 5.7 – Relations are locally consistent but not globally consistent

In the next section, you will learn about hypergraphs and their use for representing JDs.

Representing join dependencies with hypergraphs

Hypergraphs provide a formalbasis for representing a JD. A hypergraph is just like an undirected graph with the distinction that the edges are arbitrary subsets of nodes instead of just pairs of nodes (doubleton subsets), as in a familiar undirected graph.

Definition: hypergraph

A hypergraph is a pair, $\mathcal{H} = (U, \mathcal{E})$, where U is a finite set of vertices and $\mathcal{E}$ is a collection of subsets of U called hyperedges, such that $\bigcup_{e \in \mathcal{E}} e = U$.

A hypergraph gives a graphical representation of a JD, $\bowtie (R_1, \cdots, R_n)$:

- The hypergraph nodes represent the attributes, U, and the set union of all the n schemes, $\bigcup_{i=1}^{n} R_i$

The relation schemas $R_1, \cdots, R_n$ are the hyperedges

Example: hypergraph

Let us consider the JD given by $\bowtie (ABC, ACE, ABFG, ABD)$ in Baixeries (Baixeries, 2017), which we will use as an example throughout this section. *Figure 5.8* shows the hypergraph for the JD schemas. The hypergraph has the following:

- The nodes $\{A. B. C. D, E, F, G\}$ are the members in the set union of the schemas
- The hyperedges are the schemas

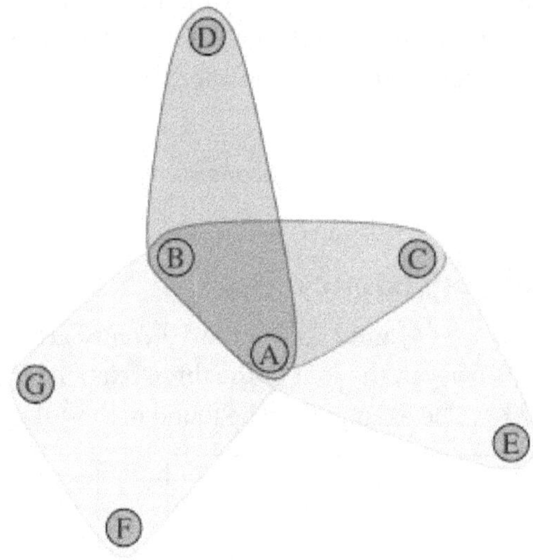

Figure 5.8 – Hypergraph for the relation schemas: **ABC, ACE, ABFG, ABD**

Practice using R

The notebook in the `acyclicJD.rmd` R Markdown file has the full code for this section's examples. The R code requires the `HyperG` and `gRbase` libraries.

```
hg1 <- hypergraph_from_edgelist(
  list(c("A","B","C"),c("A","C","E"),c("A","B","F","G"),c("A","B","D")))
plot(hg1)
summary(hg1)
```

The output includes the hypergraph depicted in *Figure 5.8* and the following summary description:

```
Hypergraph on 7 vertices with 4 hyperedges
Hyper-edge orders:
[1] 3 3 4 3
The hypergraph is connected.
     does not have empty hyper-edges
     does not have loops
     does not have isolated vertices
```

In summary, a hypergraph gives a formal tool for representing the structure of JDs among relation schemas.

The following section covers the characterization of acyclic JDs and their representation by join trees.

Acyclic join dependencies and join trees

This section defines acyclic JDs, a specialization of JDs. Here, we consider a specific class of acyclic schemes called **alpha-acyclic**, or **α-acyclicity**. Acyclic database schemes have multiple desirable properties, and we will describe the main ones. In the following chapters, you will leverage what you learn in this section to understand inference algorithms in graphical models of BNs.

We will begin by defining acyclic JDs using the representation in hypergraphs that you learned about in the previous section. Then, we will describe a graph-based method for checking that a given JD is acyclic. Following this, we will describe another representation of an acyclic JD as join trees. We will conclude with a theorem that gives the main properties of schemas with acyclic JDs.

Definition: acyclic join dependency

An acyclic JD is a JD whose hypergraph is acyclic. There are many notions of acyclicity; see Fagin (Fagin, 1983). What we refer to here as acyclic is the notion of α-acyclicity.

There is a well-known test for checking whether a hypergraph is α-acyclic, known by the name **GYO reduction**, named by the initials of the authors who arrived at that test independently (**Graham, Yu, and Oszoyoglu**).

Definition: GYO-reduction (Brault-Baron, 2014)

A hypergraph, $H = (U, \mathcal{E})$, is α-acyclic if repeatedly applying the following GYO reduction on H results in an empty hypergraph:

- **Included edge removal:** Delete a hyperedge that is a subset of, or contained within, another hyperedge
- **Singleton vertex removal:** Eliminate a node that belongs to only one hyperedge

Example: GYO-reduction

Let us apply GYO reduction to the hypergraph of *Figure 5.8* to check whether the hypergraph is acyclic:

Nodes D and E have a degree of 1, so we can delete them.

Now, we have two hyperedges, AB and AC, each contained in the ABC hyperedge, so we can eliminate both hyperedges, AB and AC.

Node C now has a degree of 1, so we remove C.

The AB hyperedge is now contained in the $ABFG$ hyperedge, so we delete the AB hyperedge.

We are left with one hyperedge with all its nodes having a degree of 1; we can delete all of them.

We end up with an empty hyperedge, concluding that the hypergraph is acyclic.

A hypergraph is acyclic if, and only if, it has the running intersection property. This property is a key characterization of hypergraph acyclicity and is equivalent to others, such as the existence of a join tree or a chordal and conformal graph representation. Next, we will define the running intersection property.

Definition: running intersection property

A JD $\bowtie$ **R** has the **running intersection property** if there is an ordering, $R_1, \cdots, R_n$, of **R** such that for every $j > 1$, there exists an $i < j$, such that the following applies:

$$R_j \cap (R_1 \cup \cdots \cup R_{j-1}) \subseteq R_i \quad (5.8)$$

That is, the intersection of each R_j with the union of the previous R_i is contained in one of them.

The running intersection property and the existence of a join tree are equivalent characterizations of acyclicity in hypergraphs. Next, we will define the representation of acyclic hypergraphs as join trees.

Theorem: join tree (Beeri, et al., 1983)

A hypergraph, $\mathcal{H} = (U, \mathcal{E})$, is acyclic if, and only if, it has a join tree, $JT(\mathcal{H})$. A $JT(\mathcal{H})$ is a tree whose vertices are the hyperedges of $\mathcal{E}$ and that satisfies the running intersection property that whenever an attribute in U occurs in two tree nodes, $e_1, e_2 \in \mathcal{E}$, then the attribute occurs in each tree node on the unique path linking e_1 and e_2 in $JT(\mathcal{H})$. A join tree is not unique, as multiple distinct trees can satisfy the running intersection property for the same set of hyperedges.

Example: join tree

Figure 5.9 shows a join tree for the JD $\bowtie (ABC, ACE, ABFG, ABD)$ of the previous example, with the hypergraph shown in *Figure 5.8*. The nodes of the join tree are the schemas (which are attributes constituting a **clique**). The edges of the join tree are labeled with the intersection of every pair of connected nodes, and those labels are called the **separators**. The nodes are ordered according to their level or depth, beginning from the root and moving toward the leaves.

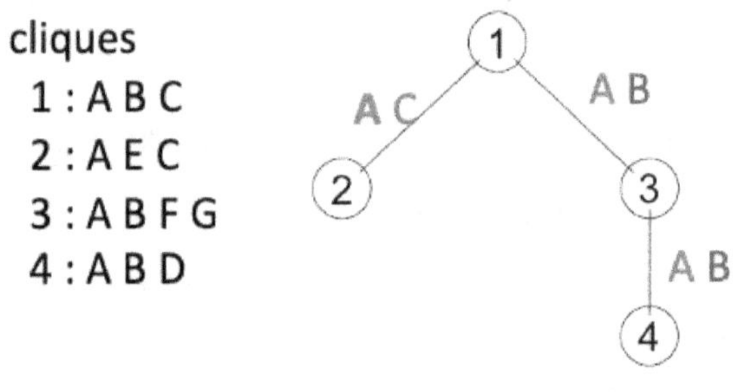

Figure 5.9 – Join tree for an acyclic database of four relations, which are the cliques – the nodes are the cliques; the edges are the separators, which are the shared attributes between a node and its parent

Figure 5.10 shows the running intersection property of ordering the cliques in the join tree. The third column gives the intersection of a clique and the union of previous cliques; the fourth column shows the index of the preceding clique that the intersection is a subset of. The intersection of C_2 and C_1 is the separator set $\{A. C\}$, a subset of C_1. The intersection of C_3 and the union of previous cliques, $C_1 \cup C_2$, is the separator set $\{A. B\}$, a subset of C_1. Lastly, the intersection of C_4 and the union of previous cliques, $C_1 \cup C_2 \cup C_3$, is the separator set $\{A, B\}$, a subset of C_3.

j	C_j			i
1	{A, B, C}			
2	{A, E, C}	$C_2 \cap C_1$	$\{A.C\} \subseteq C_1$	1
3	{A, B, F, G}	$C_3 \cap (C_1 \cup C_2)$	$\{A.B\} \subseteq C_1$	1
4	{A, B, D}	$C_4 \cap (C_1 \cup C_2 \cup C_3)$	$\{A, B\} \subseteq C_3$	3

Figure 5.10 – Clique ordering satisfies the running intersection property

In the following practice section, we will programmatically implement the previously discussed hypergraph example using R. We will generate its representation as a join tree to verify our worked example from this section.

Practice using R

The code first maps the hypergraph to an equivalent undirected graph and then computes a join tree for the graph using the `jTree` method from the `gRbase` library. Plotting the join tree produces what is shown in *Figure 5.9*.

Here is the code:

```
g <- as.graph(hg1)
jt <- jTree(g)
jt
```

The following is the output from running this code:

```
cliques
  1 : A B C
  2 : A E C
  3 : A B F G
  4 : A B D
separators
  1 :
  2 : A C
  3 : A B
  4 : A B
parents
  1 : 0
  2 : 1
  3 : 1
  4 : 3
```

The following section describes why acyclic schemes are beneficial.

Advantages of the acyclic database scheme

Acyclic databases have several desirable properties—see Maier (Maier, 1983) and Beeri et al. (Beeri, et al., 1983).

The two main ones are as follows:

- **Local consistency implies global consistency**: Whenever relations for these schemes are pairwise consistent, they are globally consistent.
- **Sequential monotone joins are guaranteed to exist**: An ordered sequence of join operations can be established, leading to **monotone joins**, where the number of tuples in the consecutive joins does not decrease. These monotone joins enhance query performance by ensuring that no intermediate solutions need to be discarded to arrive at the complete solution.

A join tree representation of an acyclic database scheme provides a recipe for a monotone sequential join. The sequential join performs a pairwise join on the relations of the schemes in reverse order, from the leaves to the root of the join tree.

Having a monotone sequential join expression entails the following:

- No intermediate join has more tuples than the final join
- At any one time, we only need to keep one intermediate join

Concluding this section on acyclic database schemes, let us highlight the main points. Hypergraphs provide a formal representation of the structure of the database scheme. A relational database schema is considered acyclic if the hypergraph that corresponds to it is acyclic. A formal method called GYO reduction can determine whether a hypergraph is acyclic. If a hypergraph is acyclic, it is a hypertree and represented as a join tree. A join tree has the running intersection property that whenever an attribute occurs in two tree nodes, the attribute occurs in each tree node on the unique path linking the two nodes. Acyclic database schemes ensure that pairwise local consistency is equivalent to global consistency. A join tree gives a recipe for monotone sequential joins where the number of tuples is non-decreasing, and we only need to keep one intermediate join at any one time.

You will learn how to use a relational database formulation to represent graphical models in BNs in the following section. The benefit of this approach is that we can use the well-established theories and tools of SQL to help us with reasoning tasks in both deterministic and probabilistic graphical models.

Bayesian network as a relational database

In this section, you will learn how a relational database with weighted relations, as outlined in the previous section, can be utilized to represent BNs.

In a previous section titled "weighted relations," we discussed the concept of a weighted relation. In that section, we provided an example of how to represent a joint probability distribution (JPD)as a weighted relation, where the weight assigns a probability value between 0 and 1 to each tuple, ensuring that the sum of these probabilities across all tuples equals 1.

This section builds upon the basic model of representing a JPD using a single weighted relation, evolving into a more intricate representation in the form of a probabilistic database. The probabilistic database decomposes the JPD into several weighted relations, reflecting the structure found in Bayesian networks (BNs).

In this section, you will do the following:

- Explore the formal foundations for representing BNs using the language of relational databases
- Learn how to implement BNs in a **PostgreSQL** database through practical examples
- Learn how to conduct probabilistic queries using SQL on the BN database, illustrated with examples

We will begin with formal definitions of the representation of BNs in the language of relational databases, and we will illustrate all concepts with a running example of a simple BN. After that, you will learn how to use an open source database management system, specifically PostgreSQL, to implement a relational database schema. Lastly, we will demonstrate how to programmatically use Python to perform probabilistic queries on the BN database and get the posterior probabilities conditioned on observed values of network variables.

Representing a BN as a relational database

This section shows how to represent a BN as a relational database. First, we formally define a database schema for representing a BN. Then, we present an example BN model and show how we map the qualitative and quantitative description of the BN to a relational database using the concept of weighted relations introduced earlier in this chapter.

Definition: Bayesian network database schema

Let BN=(G,P) be a BN where G is a **directed acyclic graph(DAG)** whose nodes are the set of n variables, $X = \{X_1, \cdots, X_n\}$. Also, P is the conditional probability table specifying, for each value tuple, t, of the variable X_i and all its parents, pa(X_i), the probability, $P(t_{X_i}|t_{pa(X_i)})$, of X_i with the value t_{X_i}, given the value assignments $t_{pa(X_i)}$ of all its parents in the DAG, G. We represent a BN as a database with a weighted relation for each node. The weighted relation, r_i, for node X_i is defined on the $\{X_i\} \cup pa(X_i)$ scheme and consists of the tuples, t, whose weight is the probability $P(t_{X_i}|t_{pa(X_i)})$. We define a primary key for each table, uniquely identifying each value tuple and its weight. We use foreign keys for an X_i variable's relation to reference the primary keys of the tables for all its parents to specify the value assignments of X_i's parents.

To illustrate the database representation of a BN based on the concept of weighted relations and the BN schema, let us work on an example in detail. We will use this example BN throughout this section.

Example: BN model M1

Figure 5.11 shows an example BN model, M1. The model consists of three variables: a, b, and c. The domain of each variable is a Boolean: True or False. The model has a DAG, a chain representing a causal mechanism from a to c mediated by variable b.

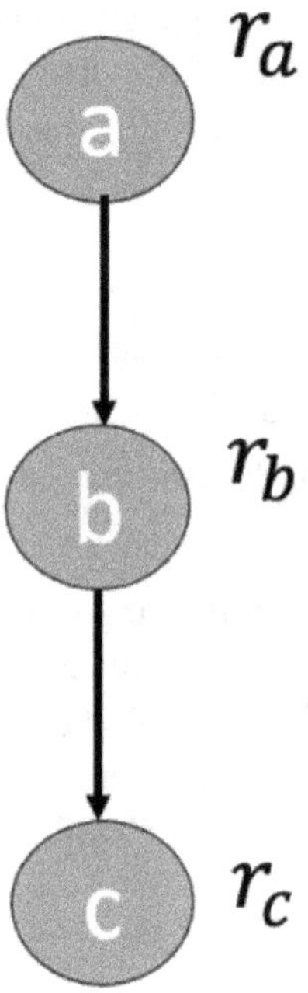

id	a	p_a
1	FALSE	0.75
2	TRUE	0.25

id	a_id	b	p_b_a
1	1	FALSE	0.2
2	1	TRUE	0.8
3	2	FALSE	0.9
4	2	TRUE	0.1

id	b_id	c	p_c_b
1	1	FALSE	1
2	2	FALSE	0.25
3	2	TRUE	0.75
4	3	FALSE	1
5	4	FALSE	0.25
6	4	TRUE	0.75

Figure 5.11 – Relational tables representing a BN model, M1

The tables shown in *Figure 5.11* next to each node in the network represent a weighted relation where the weight variable represents the probability that the node's variable takes a specific value conditioning on the value tuple of its parents. The id column of each table represents a primary key for the relation represented by the table. We define three relations, $r_a, r_b,$ and r_c, for the respective network variables, $a, b, and\ c$. Each relation's table gives the value tuples of the relation and the corresponding weight for the value tuple, which is the probability of the value assigned to the variable given the value assignment of all its parents. The variable a is a root node with no parent, so the weight, p_a, is simply the prior probability of a. There are two rows in the r_a table indicating that variable a takes the value $True$ with a probability of 0.75 and $False$ with a probability of 0.25. The table for the r_b relation of variable b has one additional column for a foreign key, `a_id`, which points to a value assignment of a in table r_a. The first two rows of table r_b give the conditional probability of b given $a = False$, and the last two rows give the probability of b given $a = True$.

Similarly, table r_c has a foreign key that references the primary key of table r_b, and the weight p_c_b for the r_c relation is the conditional probability of c conditioning on the value assignment of its parent variable, b, from its r_b table. Table r_c discards tuples where the probability is 0. This exclusion of zero-probability tuples results in space savings, which can be significant for large networks.

In the following section, we will explain how to implement the database schema for the BN example in a real-world relational database and how to perform this task programmatically using Python and SQL.

Implementing BNs in relational databases

In this section, we describe how to implement the relational database model of the BN. The database mapping of the BN model consists of two phases: qualitative and quantitative specification. We will begin by detailing the qualitative specification and then follow up with the quantitative specification.

The qualitative specification of the BN consists of designing the database schema represented by an **entity-relation** (**ER**) diagram; the entities are the database tables, and the relations are the connections between the tables.

The quantitative specification consists of populating the tuples of the weighted relational tables of the database, which assigns the probability for each BN variable conditioned on the value assignment of all its parents.

In the following practice, we demonstrate how to map the example BN of *Figure 5.11* into a relational database. Specifically, you will learn how to create a PostgreSQL database for the BN model and how to do it programmatically using Python. We will also draw the ER model for the BN database using the **pgAdmin** GUI tool.

Practice using Python and PostgreSQL

You can find the entire code for this practice in the `bndb.jpynb` Jupyter notebook and the HTML version of the notebook, `bndb.html`, which can be found in the `Chapter 5` folder in this book's GH repository. In the following subsections, we explain the code snippets and their outputs. We have broken down the code into a sequence of steps.

Step 1: importing the required libraries and the database login username and password

`DB_USERNAME` and `DB_PASSWORD` are the database server login and password to be imported from a `config.py` file. We need to import the following libraries from the `sqlalchemy` package:

```
from config import DB_USERNAME, DB_PASSWORD
from sqlalchemy import create_engine, text
from sqlalchemy.exc import SQLAlchemyError
```

Step 2: creating an engine object

You first need to create a new PostgreSQL database named `m1_db` using pgAdmin. You can find a step-by-step guide on how to create a new PostgreSQL database on this website: `https://www.tutorialsteacher.com/postgresql/create-database`.

You need to specify the following information to access the `m1_db` database on your PostgreSQL server; fill in the placeholders for `username` and `password` with your specific username and password for your installation of the PostgreSQL server.

`db_string` is the URL required to create the `engine` object. We will use the `engine` object next to invoke SQL statements. For more information on SQLAlchemy and PostgreSQL, see the documentation on the website: `https://docs.sqlalchemy.org/en/14/dialects/postgresql.html`.

Here is the code:

```
username = DB_USERNAME
password = DB_PASSWORD
hostname = "localhost"
port = "5432"
database_name = "m1_db"
# Create the engine
db_string = f"postgresql+psycopg2://{username}:{password}@{hostname}:{port}/
{database_name}"
engine = create_engine(db_string)
```

Step 3: Specifying the database schema and populating the tables for the BN nodes

The following code snippet creates the tables for representing the nodes of the BN model, M1, depicted in *Figure 5.11*. To create the tables, we use the `connection` object and write SQL statements to create the database tables. The following SQL first establishes the schema for the table using **`CREATE TABLE`**, the qualitative specification part of the BN model. Next, we use the **`INSERT INTO`** statement to instantiate the table with the proper data, the quantification specification part of the BN model.

The following is the Python code for specifying the table of variable a:

```
try:
    with engine.begin() as connection:
        connection.execute(text('''
            CREATE TABLE IF NOT EXISTS public.a
            (
                id INTEGER NOT NULL PRIMARY KEY,
                a boolean NOT NULL,
                p_a REAL NOT NULL
            )
        '''))
        connection.execute(text('''
            INSERT INTO a(id, a, p_a)
            VALUES (1, false, 0.75), (2, true, 0.25)
        '''))
except SQLAlchemyError as e:
    print(f"An error occurred: {str(e)}")
```

The following is the Python code for specifying the table of variable b:

```
try:
    with engine.begin() as connection:
        connection.execute(text('''
```

```
            CREATE TABLE IF NOT EXISTS public.b
            (
                id INTEGER NOT NULL PRIMARY KEY,
                a_id INTEGER NOT NULL,
                FOREIGN KEY (a_id) REFERENCES a(id),
                b boolean NOT NULL,
                p_b_a REAL NOT NULL
            )
        '''))
        connection.execute(text('''
            INSERT INTO b(id, a_id, b, p_b_a)
            VALUES (1, 1, false, 0.2), (2, 1, true, 0.8),
            (3, 2, false, 0.9), (4, 2, true, 0.1)
        '''))
except SQLAlchemyError as e:
    print(f"An error occurred: {str(e)}")
```

Lastly, the following is the Python code for the database table of variable `c`:

```
try:
    with engine.begin() as connection:
        connection.execute(text('''
            CREATE TABLE IF NOT EXISTS public.c
            (
                id INTEGER NOT NULL PRIMARY KEY,
                b_id INTEGER NOT NULL,
                FOREIGN KEY (b_id) REFERENCES b(id),
                c boolean NOT NULL,
                p_c_b REAL NOT NULL
            )
        '''))
        connection.execute(text('''
            INSERT INTO c(id, b_id, c, p_c_b)
            VALUES (1, 1, false, 1),
            (2, 2, false, 0.25),(3, 2, true, 0.75),
            (4, 3, false, 1),
            (5, 4, false, 0.25), (6, 4, true, 0.75)
        '''))
except SQLAlchemyError as e:
    print(f"An error occurred: {str(e)}")
```

The following code shows how to query the database to fetch all records in each of the three tables of the BN database model.

The following is the code and the results for fetching the records of the relation for variable a:

```
try:
    with engine.begin() as connection:
        result = connection.execute(text("SELECT * FROM a")).fetchall()
        for row in result:
            print(row)
except SQLAlchemyError as e:
    print(f"An error occurred: {str(e)}")
```

The output looks as follows:

```
(1, False, 0.75)
(2, True, 0.25)
```

The following is the code and the results for fetching the records of the relation for b:

```
try:
    with engine.begin() as connection:
        result = connection.execute(text("SELECT * FROM b")).fetchall()
        for row in result:
            print(row)
except SQLAlchemyError as e:
    print(f"An error occurred: {str(e)}")
```

The output looks as follows:

```
(1, 1, False, 0.2)
(2, 1, True, 0.8)
(3, 2, False, 0.9)
(4, 2, True, 0.1)
```

The following is the code and the results for fetching the records of the relation for c:

```
try:
    with engine.begin() as connection:
        result = connection.execute(text("SELECT * FROM c")).fetchall()
        for row in result:
            print(row)
except SQLAlchemyError as e:
    print(f"An error occurred: {str(e)}")
```

The output looks as follows:

```
(1, 1, False, 1.0)
(2, 2, False, 0.25)
(3, 2, True, 0.75)
(4, 3, False, 1.0)
(5, 4, False, 0.25)
(6, 4, True, 0.75)
```

Step 4: Drawing the ER diagram

Now that we have created the BN database, we can visualize the schema by drawing the ER diagram using the **pgAdmin** ER diagram tool. For instructions on using the tool, see the documentation on the website: `https://www.pgadmin.org/docs/pgadmin4/development/erd_tool.html`.

Figure 5.12 shows the ER diagram for the database scheme of the BN model, M1. Note that the chart indicates the **one-to-many relationship** between table A and table B and between table B and table C. This means that one record in table A is linked to many records in table B, with the same value assigned to variable a—similarly, one record in table B shares the b variable's value with many records in table C.

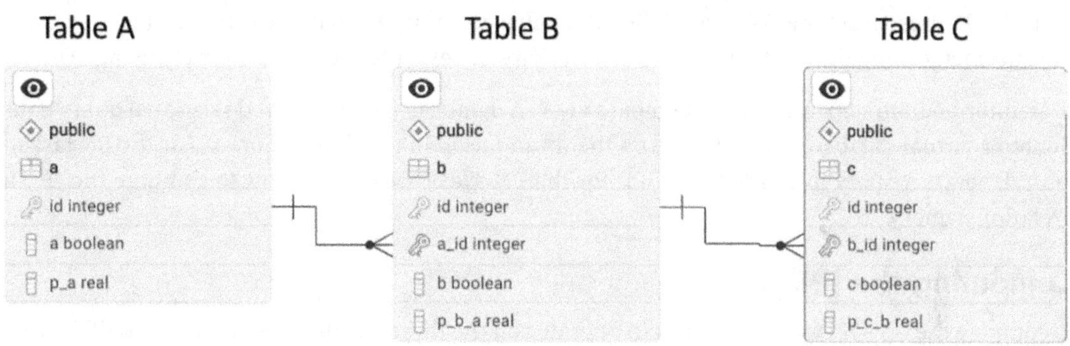

Figure 5.12 – ER diagram of the database scheme for the BN model, M1

To recap, we showed in this section how to programmatically create a database representation of a BN model using SQL and Python with implementation in PostgreSQL, the world's most advanced open source database.

In the next section, we will explain how to use SQL with the BN database representation to perform BN tasks of probabilistic queries.

Probabilistic queries of the Bayesian network using SQL

This section explains how to use SQL to perform probabilistic queries on the BN represented in relational databases. First, we will give the formal definition of the BN tasks, and after that, we will illustrate, with the example of the BN model M1, how to perform the BN tasks using SQL queries on the BN relational database model.

Definition: BN tasks

Two of the most common BN tasks are **posterior distribution** computation and computation for the **most probable explanation** (MPE):

- The posterior distribution task is to compute the $P(X|e)$ distribution for unobserved variables, X, given observations or evidence, e

The MPE task is to find an assignment, $x^* = (x_1^*, \ldots, x_n^*)$, such that $x^* = \arg\max_x \prod_{i=1}^{n} P(x_i | x_{pa_i}, e)$

Example: BN tasks

Consider a BN that represents the relationship between diseases and symptoms. The posterior distribution task is to update the probability of having a specific disease based on the observation of a particular set of symptoms. The MPE task is to identify the state of diseases that explain the symptoms and have the highest posterior probability.

Having defined two main BN tasks, let us now demonstrate how to perform them, formulating them as relational queries using relational algebra operators and implementing them using SQL.

The relational database formulation of the BN enables a direct approach to performing the two main BN tasks. Representing the BN as a relational database enables us to express the BN tasks elegantly and succinctly using the relational operators and directly execute them declaratively using SQL.

As an intermediate step, we create a **database view**. A database view is a stored relational query equivalent to a virtual database table created on the fly and used in relational queries. For the BN tasks, a central view is a query to compute the JPD for the BN. Next, we explain how to compute the JPD as a relational query.

Computing the JPD

To compute the BN's JPD, we apply the join operator on the three weighted relations, $r_a, r_b, and\ r_c$, and we write the following:

$$jpd = r_a \bowtie r_b \bowtie r_c \quad (5.9)$$

The join operator treats the weight attribute specially. The semantics of the weights as probabilities in the database representation of the BN implies that the weight attribute of the join is the product of the weights of all relations in the join.

The following code in Python contains the SQL to create the JPD view:

```
try:
    with engine.begin() as connection:
        connection.execute(text('''
            CREATE OR REPLACE VIEW JPD
            AS
            SELECT a.a, b.b, c.c, a.p_a * b.p_b_a * c.p_c_b AS prob
            FROM a
```

```
            JOIN b ON a.id = b.a_id
            JOIN c ON c.b_id = b.id
        '''))
        print('Created view JPD.')
except SQLAlchemyError as e:
    print(f"An error occurred: {str(e)}")
```

The output indicates that the view creation succeeded:

```
Created view JPD.
```

Figure 5.13 shows how you can use the pgAdmin GUI to run a simple `SELECT` query on the JPD view to get all the rows in the JPD of the BN.

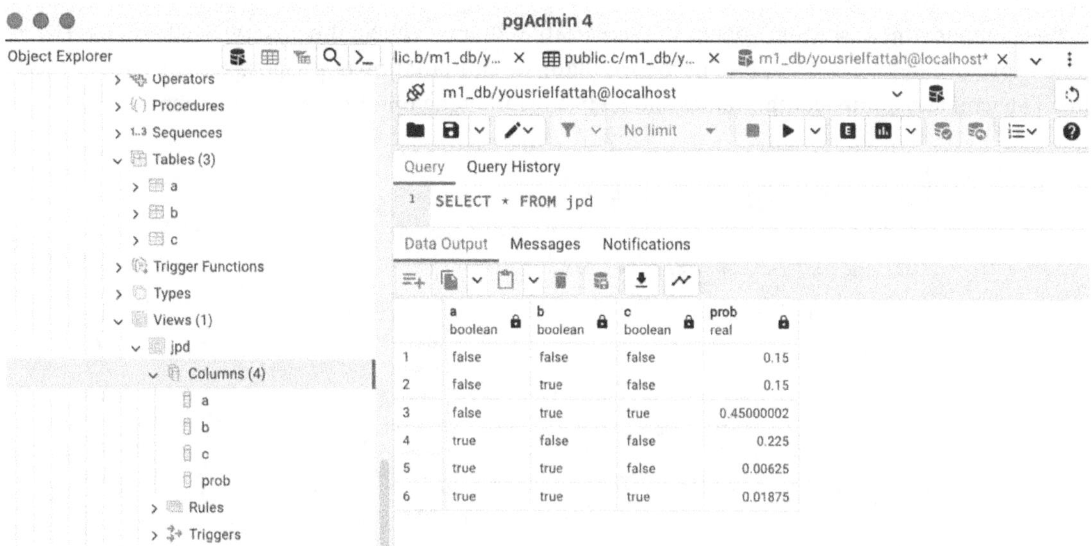

Figure 5.13 – All rows of the joint probability view in the BN database model

You can also fetch all the JPD rows using Python as follows:

```
try:
    with engine.begin() as connection:
        result = connection.execute(text("SELECT * FROM JPD")).fetchall()
        for row in result:
            print(row)
except SQLAlchemyError as e:
    print(f"An error occurred: {str(e)}")
```

The following is the resulting output:

```
(False, False, False, 0.15)
(False, True, False, 0.15)
(False, True, True, 0.45000002)
(True, False, False, 0.225)
(True, True, False, 0.00625)
(True, True, True, 0.01875)
```

Once we have the JPD view, we can easily perform the BN tasks by writing other queries on the JPD view, as we shall demonstrate next.

Computing the posterior distribution

Let us compute the posterior probability for variables a and b, given the observed evidence $c = true$. The expressiveness of SQL enables us to formulate the computation of the posterior distribution task in an elegant and compact manner, as shown in the following one-liner SQL statement:

```
try:
    with engine.begin() as connection:
        print('getting posterior probability of a, b given c is true')
        result = connection.execute(text('''
            SELECT a, b, Prob / (SELECT SUM(Prob) FROM JPD WHERE c = true)
            as prob1
            FROM JPD
            WHERE c = true
        '''))
        for row in result:
            print(row)
except SQLAlchemyError as e:
    print(f"An error occurred: {str(e)}")
```

The following is the output from executing the posterior distribution query:

```
getting posterior probability of a, b given c is true
(False, True, 0.96)
(True, True, 0.04)
```

Figure 5.14 explains the posterior probability query, a compound SQL statement with a parent query and an inner subquery. The subquery, enclosed between parentheses, computes the sum of the joint probabilities conditioning on the observed evidence, $c = true$, which amounts to the marginal probability of the evidence, $c = true$. The parent query selects the value tuples for attributes a and b. It divides the probability attribute by the sum from the subquery, which results in the posterior probability for the value tuple of a and b per the Bayes formula.

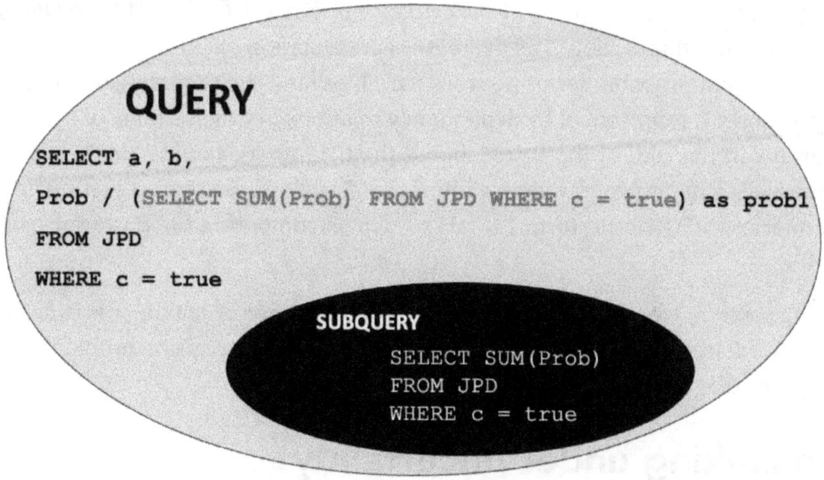

Figure 5.14 – SQL query to compute the posterior probability for the BN model, M1

In the next section, we demonstrate how to solve the MPE task for the BN example.

Computing the most probable explanation

We will continue with the example BN model, M1, of *Figure 5.11* and compute the MPE for the observed evidence, $c = true$. We can do this using the following code in Python using a one-liner SQL statement, which includes a subquery in the `WHERE` clause that determines the maximum probability overall value tuples of the joint distribution view JPD:

```
try:
    with engine.begin() as connection:
        print('getting most probable explanation of a, b given c is true')

        result = connection.execute(text('''
            SELECT a, b
            FROM JPD
            WHERE c = true
            AND prob = (SELECT Max(prob) from JPD)
        '''))

        for row in result:
            print(row)
except SQLAlchemyError as e:
    print(f"An error occurred: {str(e)}")
```

The following is the output from executing the MPE query:

```
getting most probable explanation of a, b given c is true
(False, True)
```

The output says that the MPE for the $c = True$ evidence is that $a = False$ and $b = True$. Let's summarize the key points from this section. The database representation of a BN has relational tables representing the conditional probabilities of network variables, and the probabilistic influence between the network variables is represented by dependency relations between a table of a network variable and all its parents in the DAG of the BN. Mapping the BN into a database can be automated using Python and the SQLAlchemy programmable interfaces. The database representation enables the use of relational queries in SQL to perform BN tasks, such as computing the posterior probability and finding the MPE.

In the following section, we will transition to the important topic of making decisions under uncertainty, where you will learn about graphical models for representing decision problems using decision trees and influence diagrams.

Decision-making under uncertainty

Decision-making under uncertainty underlies most problem-solving in the real world. The outcome of most real-life decisions is more or less uncertain, requiring us to make future decisions based on contingencies, and we need to think through all possible courses of action or decision paths that yield the most beneficial or preferred consequences. Decision-making is required to solve complex problems in professional fields in engineering, medicine, marketing, the military, government, international relations, and diplomacy:

- Government leaders make complex foreign policy decisions regarding treaties, sanctions, or military interventions, considering national security, economic impacts, and the potential for conflict
- Doctors make decisions about courses of treatment for patients and consider resource allocation trade-offs during crises such as pandemics, accounting for uncertainty
- Engineers must choose design options and parts while keeping safety in mind and taking risks into account
- Marketers decide whether to enter or exit a market based on advertising campaigns, pricing strategies, and which marketing medium to use
- The military uses decision-making to develop courses of action and weighs suitability, feasibility, and risk to achieve mission success

With this brief introduction to the importance of decision-making under uncertainty, let us now explain how the problem can be represented to enable computing a decision policy and performing decision analysis. The following section provides an overview of the decision analysis paradigm.

Explaining the decision analysis paradigm

Under the assumption of a rational decision-maker, making decisions under uncertainty is formulated under the **expected utility theory** (EUT) framework. This theory integrates probability theory and utility theory to model rational choices when outcomes are uncertain:

- **Probability theory**: This provides the mathematical framework to quantify and analyze the risks and likelihood of different chance events

- **Utility theory**: This is used to incorporate an individual's subjective preferences and risk attitudes (e.g., risk-averse, risk-neutral, or risk-seeking) into the decision model, typically via a utility function

Keeney and Raiffa (Keeney, et al., 1976) prescribe a paradigm of decision analysis that encompasses various processes that we will review in the following sections, which will serve to provide a useful background regarding representing and analyzing problems of decision-making under uncertainty. The first step in Keeney and Raiffa's paradigm is structural analysis, which we will discuss next.

Structural analysis

The goal of this step is to qualitatively describe the decision-making problem by ascertaining the decision-making sequence and pinpointing the information that conditions those decisions. The outcome of the structural analysis is typically represented as a decision tree, as illustrated in *Figure 5.15*. The decision tree consists of nodes where the choice of a branch is under the control of the decision-maker (these nodes are represented by squares in *Figure 5.15*) and nodes that are not fully under the decision-maker's control (represented by circles in *Figure 5.15*). We will refer to these two types as **decision nodes** and **chance nodes**.

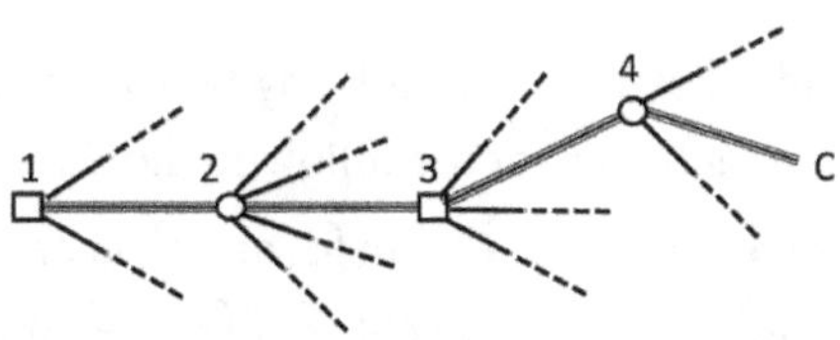

Figure 5.15 – Decision tree schematic: Nodes 1 and 3 are decision nodes; nodes 2 and 4 are chance nodes; branches at a decision node represent possible choices (actions); branches at a chance node represent possible outcomes (events)

There are two structural categories of decision problems: symmetric and asymmetric. An **asymmetric decision problem** can be revealed by its decision tree structure when the number of paths (scenarios) is smaller than the cardinality of the Cartesian product of the state spaces of all chance and decision variables in the context of decision analysis. The distinction between symmetric and asymmetric decision trees lies in the structure of the paths, the available variables at different points, and the possible scenarios they depict.

Following structural analysis, the second step in Keeney and Raiffa's paradigm is uncertainty analysis, which we will discuss next.

Uncertainty analysis

The decision-maker assigns a probability to each branch that originates from a chance node. The probability assignment is derived from multiple sources that should be fused and verified for internal consistency. The sources include the following:

- The subjective judgment of the decision-maker and domain experts, calibrated to eliminate biases and personal idiosyncrasies
- Statistical estimation models based on experiential and historical data

The third step in Keeney and Raiffa's paradigm is utility or value analysis, which we will discuss next.

Utility or value analysis

This phase of decision analysis focuses on clarifying the decision-maker's preferences among various scenarios associated with different paths in a decision tree. The decision-maker utilizes a real-valued utility function (U) to assign a real value, $u_i = U(C_i)$, for the consequence, C_i, of every i-th decision scenario corresponding to a path from the root to a leaf in the decision tree. A utility function quantifies the level of preference or desirability associated with each consequence of a decision scenario (a path in the decision tree). The preference relation entailed by the utility function must respect the following constraints under the assumption of the decision-maker's rationality.

In rational decision-making, the **preference relation** ($\succ$) defined over the set of consequences corresponding to all decision scenarios must adhere to the rational preference axioms as posited by von Neumann and Morgenstern (von Neumann, et al., 1944), which are formally stated in the following definition.

The axioms that define what constitutes a rational and consistent preference ordering are given by the following rational preference constraints:

- **Completeness:** An individual can always rank any two alternatives (prefers C_1 to C_2, C_2 to C_1, or is indifferent). That is, exactly one of the following holds: $C_1 \succ C_2$, $C_1 \succ C_2$, or $C_1 \sim C_2$.
- **Transitivity:** If C_1 is preferred or indifferent to C_2, and C_2 is preferred or indifferent to C_3, then C_1 must be preferred or indifferent to C_3. That is, if $C_1 \succcurlyeq C_2$ and $C_2 \succcurlyeq C_3$, then $C_1 \succcurlyeq C_3$
- **Continuity:** There exists a probability where an individual is indifferent between a certain middle outcome and a lottery of the best and worst outcomes. That is, if $C_1 \succcurlyeq C_3 \succcurlyeq C_2$, then there exists a probability, p, such that C_3 is equally preferred to a lottery with probability p for C_1 and probability $1 - p$ for C_2.
- **Independence:** Adding the same third option to two existing choices does not change the original preference order between a lottery with probability p for C_1 and probability $1 - p$ for C_3, and a lottery with probability p for C_2 and probability $1 - p$ for C_3.

Utility functions are inherently subjective, meaning that different decision-makers may assign distinct functions to the same decision problem, reflecting their individual preference structures. The numerical values of the utility function assigned to consequences must reflect their relative ordering of preference. It is the relative values of the utility functions assigned to consequences and not the absolute values that matter. Utility is typically defined up to a linear transformation (multiplying by a positive constant and adding another constant), meaning the specific zero point and scale are not individually meaningful; only the relative ordering and differences matter. The following theorem formally states the connection between the utility function and the preference relation.

If a preference relation ($\succ$) satisfies the rational preference axioms, then there exists a function, $U: C \rightarrow \mathbb{R}$, such that the following apply:

- $U(C_1) > U(C_2)$ if, and only if, $C_1 \succ C_2$. That is, C_1 is preferred over C_2.
- $U(C_1) = U(C_2)$ if, and only if, $C_1 \sim C_2$. That is, C_1 is equally preferred to C_2.

Here, C_1 and C_2 are any two members of set C. Set C contains the consequences of every decision scenario, corresponding to a path in the decision tree.

In rational choice theory, decision-makers are assumed to behave in a way that maximizes their expected utility, choosing the action that leads to the highest numerical value.

To illustrate that making a choice between decision options is akin to selecting between lotteries in a decision tree, consider the scenario where you must decide whether to take your umbrella before leaving home for various activities. The weather forecast indicates that it will be sunny with probability p_1, cloudy with probability p_2, and rainy with probability p_3. Each path in the decision tree corresponds to a specific outcome that represents the consequences of either taking or leaving the umbrella, depending on whether the weather turns out to be sunny, cloudy, or rainy. For each of the six possible outcomes, you, as the decision-maker, are tasked with assigning a utility value—a real number—to reflect your preference order of these consequences.

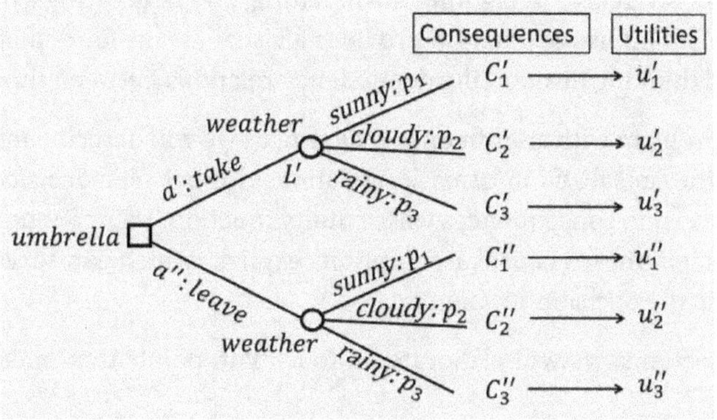

Figure 5.16 – A decision tree illustrating that choosing between decision options a' and a" is equivalent to choosing between the expected utility of outcomes of lotteries L' and L"

As depicted in *Figure 5.16*, you must assign numbers to the consequences of every scenario (such as u_i' to C_i' and u_i'' to C_i'') in a way that makes the choice between a' and a" the same as the choice between lotteries L' and L''. In other words, the choice between a' and a'' is based on the order of the expected results of lotteries L' and L'', for example.

$$a' \succ a'' \Leftrightarrow (\Sigma_i p_i' u_i' > \Sigma_i p_i'' u_i'') \quad (5.10)$$

To sum up, the assignment of utility numbers to outcomes must ensure that the maximization of expected utility forms the foundation for identifying the decision-maker's optimal course of action.

Following the utility or value analysis, the fourth and last step in Keeney and Raiffa's paradigm is optimization analysis, which we will discuss next.

Optimization analysis

The goal of the optimization analysis step is to determine the strategy that provides the **maximum expected utility (MEU)**. The decision strategy will determine the best course of action at each decision node based on information about what happened and what actions were taken along the path that led to that node in the tree. For representation with decision trees, optimization analysis is typically performed using the backward recursion method of dynamic programming, which Keeney and Raiffa call the averaging-out-and-folding-back method. For representation with influence diagrams, which we will describe in the following section, optimization analysis can utilize the structural parameters of the DAG that represents the influence diagram to improve the efficiency of the analysis.

Now that we have reviewed decision problems and the processes involved in decision analysis, let's transition to the next section, which focuses on the significance of influence diagrams in this field. We will utilize influence diagrams to demonstrate our database framework and perform optimization analysis.

Using influence diagrams for decision analysis

In this section, we describe the importance of using influence diagrams for decision analysis; to help understand why we should care about the subject.

Influence diagrams are invaluable in eliciting, formulating, and structuring a decision problem. The DAG representation of an influence diagram provides a visual means for depicting the elements of a decision problem and thinking through the dependency relations between those elements.

Modeling a decision problem with an influence diagram, as we will describe in an upcoming section, requires articulating the causal and information relationships between decisions and chance events and the valuation of decision consequences with a utility function. Using influence diagrams to structure the decision problem makes causal assumptions explicit, which can serve in later revisions or updates in formulating the decision problem.

In the following subsections, we will elaborate on four main points that make influence diagrams important.

Influence diagrams are a compact representation of complex decision problems

Influence diagrams are a compact representation of complex decision problems that structurally define causal and information dependencies between variables representing events, decisions, and utility functions.

The DAG of the influence diagram will be a much more compact representation of **symmetric decision problems**, which contain much fewer nodes than those depicted in the associated decision tree. This distinction is because influence diagrams focus on the dependency structure between variables and do not explicitly represent the decision scenarios with respect to the possible value instantiations of those variables. The decision tree's paths (trajectories) show decision scenarios where each path has the same set of variables (decision or chance), but the edges are different because the value instantiations of the variables are different. This means that the size of the tree grows exponentially as the number of variables and the size of their value domains increase.

For **asymmetric decision problems**, decision trees are asymmetric, making them less expansive than in the symmetric case, as the tree paths will differ in their length. This trait makes decision trees advantageous for depicting and visualizing the contextual dependencies of various scenarios in asymmetric decision problems.

For asymmetric decision problems, influence diagrams introduce dummy states for chance and decision variables as a workaround, which makes the asymmetry hidden while adding to the computational complexity. To address asymmetric decision problems without introducing dummy states, alternative approaches to influence diagrams have been developed in this active research area. The main approaches include valuation networks (Shenoy, 1992) (Shenoy, 2000) (Bielza, et al., 1999), asymmetric influence diagrams (Smith, et al., 1993) (Covaliu, et al., 1995), and decision circuits (Bhattacharjya, et al., 2007) (Bhattacharjya, et al., 2008) (Shachter, 1990).

Let's now look at the second point.

Influence diagrams ensure a consistent specification of the probability model

Influence diagrams necessitate the specification of conditional probabilities for chance nodes based on the values assigned to their direct parents. The DAG in the influence diagram illustrates the factorization of the joint distribution of the probability model, which generates all the conditional probabilities. Any required conditional probabilities can be derived from a consistent joint distribution, similar to the process used in BNs.

Because influence diagrams use modular conditional probability tables (CPTs), the same probability only needs to be defined once rather than repeated in many different scenarios. This makes the model easier to build and maintain. It also helps keep the probabilities consistent across the entire analysis. In contrast, decision trees often require the modeler to re-enter and rebalance probabilities at many different branches, which increases the risk of inconsistency.

Decision trees also require events to be arranged in a fixed time sequence. As a result, they may require probabilities in a form you do not originally have. For example, if a decision is made after a test result is observed, the tree needs the posterior probability—that is, the probability of the true state given the test result—at the chance nodes after that decision. Converting inputs into this form and checking that all probabilities remain consistent must usually be done manually, which can take considerable time and can introduce errors.

Now we will look at the third point.

Influence diagrams leverage structure-based algorithms for efficient computation

The main advantage of influence diagrams is that they allow the use of structure-based algorithms for optimization analysis. Structure-based algorithms can leverage the topology of the DAG representing the influence diagram to make time-space trade-offs to alleviate computational complexity. Examples of structure-based algorithms are those based on join tree clustering and variable elimination methods, which you will learn about in later chapters in this book.

Optimization analysis with decision trees using dynamic programming is constrained by the ordering of the variables in the tree and can suffer from the "curse of dimensionality." Decision trees expand exponentially as the depth increases, rendering them disadvantageous for complex decision problems involving many variables.

The fourth and final point will be discussed next.

Influence diagrams excel at showing the value of information

Influence diagrams excel at showing the **value of information** (VoI) by visually representing how resolving uncertainty about certain chance events can impact the expected utility. Quantifying the VoI is crucial in determining the maximum cost of acquiring information sources. *Figure 5.17* shows a textbook example of a toy problem involving decision-making under uncertainty. The figure presents an influence diagram illustrating the decision of whether to take an umbrella when going out. You do not know whether there will be rain. To resolve the uncertainty about the weather, you are offered a weather forecast service at a cost. The VoI is computed by calculating the increase in the expected MEU of the "umbrella utility" based on the added weather forecast information. The VoI in this case is equivalent to the expected purchase utility.

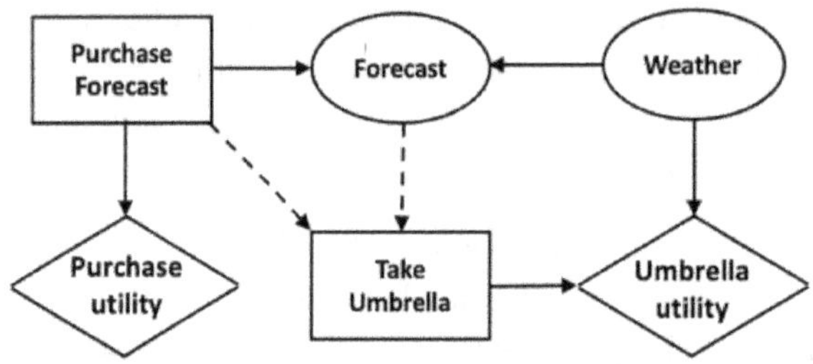

Figure 5.17 – Influence diagram of the umbrella decision example, where the value of the forecast information gives an upper bound on the cost of the forecast purchase

In a more significant real-life scenario, consider decision analysis in medical diagnosis. The influence diagram in *Figure 5.18* explicitly shows how a potential new test (information node) influences a belief about the patient's disease state (chance node), which in turn influences the choice of treatment (decision node) and ultimately the patient's health outcome (utility node). This visualization reveals whether a cheap, imperfect test provides significant value by reducing uncertainty about the disease and justifying its cost, or whether a more expensive test is necessary to obtain crucial information, thereby directly affecting the value of the test result.

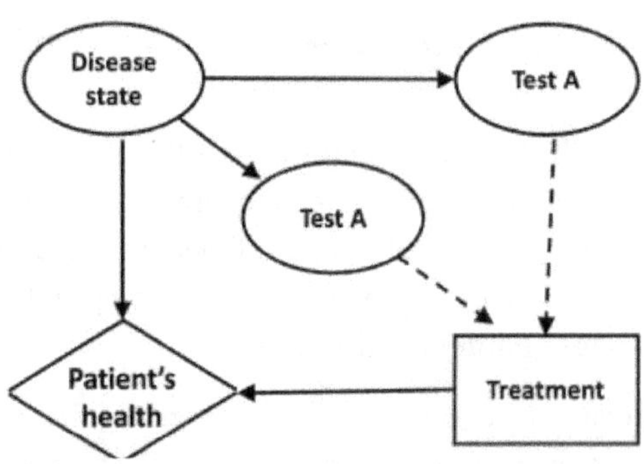

Figure 5.18 – Influence diagram of medical diagnosis based on multiple tests before treatment. The VoI helps determine the trade-off cost versus benefit of test A and test B as information sources

To conclude this section, here is the key takeaway: influence diagrams support all stages of decision analysis, from structural analysis to optimization analysis, and provide a tool for computing not only the decision that yields the MEU but also the VoI (like lab tests in medical diagnosis). This process establishes an upper bound on the cost of obtaining additional sources of information and helps determine the best information-gathering strategy before making the decision.

In the following section, we present a formal description of influence diagrams and specify the graphical icons depicting the elements of the influence diagram.

Describing influence diagrams

In this section, we describe the graphical elements of an influence diagram. As stated earlier, an influence diagram represents a problem of decision-making under uncertainty. Generally speaking, influence diagrams are extensions of the graphical models of BNs that include, in addition to the chance nodes of the BNs, two special types of nodes: decision nodes and utility nodes.

An influence diagram is a DAG whose nodes are of three types:

- **Chance nodes:** Represent uncertain events on random variables just like the nodes in a regular BN. An influence diagram uses circles or ellipses for drawing chance nodes.
- **Decision nodes:** Represent the actions or decisions that are variables under the decision-maker's control. An influence diagram uses rectangles or boxes to draw decision nodes.
- **Utility nodes:** Represent the decision-maker's assessment of the monetary value as a deterministic function of the decisions taken and the realized random events. An influence diagram uses a diamond shape to draw utility nodes.

The edges between nodes in the influence diagram have different interpretations depending on whether the pair of nodes linked by the edges are of the same type, either as chance nodes or decision nodes, or are of mixed types: chance node, decision node, and utility node.

Directed edges between node pairs indicate influences of two categories:

- **Informational influences** correspond to edges **directed to a decision node**; the nodes that the edges emanate from are the events whose outcome the decision-maker will know when making the decision
- **Conditioning influences** correspond to edges **directed to a chance or a utility node**; the nodes that the edges emanate from are the variables whose value will condition the probability assignment to the chance node or the valuation of the utility function to the utility node

Figure 5.19 depicts four types of relationships between chance and decision nodes in an influence diagram.

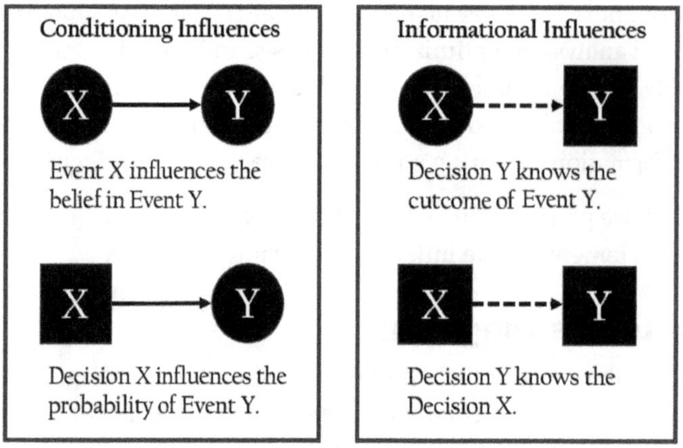

Figure 5.19 – Types of relationships in an influence diagram: Circles represent events and boxes represent decisions; directed edges use solid lines for conditioning influence and dotted lines for informational influence

Next, we will give a formal definition of an influence diagram.

An influence diagram has two components, one qualitative and another quantitative. A DAG, $G = (V, E)$, is the qualitative component. The nodes, V, represent three types of variables: chance nodes for events, S, decision nodes for decisions, D, and utility nodes for utilities, U. The quantitative component contains the following:

- **Conditional Probability Tables (CPTs):** For each chance node $X_i \in S$, a CPT specifies $P(X_i|pa(X_i))$ for every combination of parent states.
- **Utility functions:** For every utility node $U_i \in U$, a utility function $u_{i:}$ assigns a real-value to each combination of its parent states.

The utility functions must reflect the decision-maker's preferences by ranking decision scenarios based on their desirability, in accordance with the axioms of rational preference. This allows for the optimization analysis of the decision problem to be framed as the task of computing the decision policy that corresponds to the MEU.

The preceding definition of an influence diagram is general and covers various decision-making models. Decision networks add two conditions to the influence diagram. Decision networks are related to decision trees. Decision trees are intuitive and popular decision-making models. Decision networks form a compact and concise representation of a decision problem. Decision trees and decision networks present equivalent views of the same decision-making problem.

Let's now define decision networks and decision trees.

A **decision network** is an influence diagram respecting two conditions:

- **The single decision-maker condition:**

  The single decision-maker condition requires the influence diagram to yield a directed path containing all the decisions

- **The no-forgetting condition:**

  The no-forgetting condition imposes that each decision node and its direct predecessors (parents) influence all successor decision nodes

The two conditions guarantee that an ID is equivalent to a decision tree whose branches represent sequences of decisions and events forming realizations of all possible scenarios. Every leaf node is associated with a valuation of the utility function for the series of chance events and decisions along the path from the root node to the leaf node of the decision tree.

The following sections use the terms "influence diagram" and "decision network" interchangeably.

To conclude this section on influence diagrams, let's recap the key points. Influence diagrams are graphical models designed for decision-making under uncertainty, enhancing BNs by incorporating decision and utility nodes. Decision networks are a type of influence diagram that meets the criteria of having a single decision-maker and adhering to the no-forgetting condition. Decision trees, on the other hand, illustrate every possible realization of the outcome scenarios entailed by a decision network. In general, decision networks serve as compact and concise representations of decision problems, while decision trees tend to be more intuitive and easier to understand and interpret.

In the next section, we will formally state the MEU problem, which is the optimization problem of computing the MEU and the associated MEU decision policies for general decision networks.

Stating the MEU problem

There are two main tasks in reasoning with influence diagrams. One task is calculating the MEU, and the other is identifying the **MEU decision policies**.

To formally state the MEU problem, consider an influence diagram with chance nodes, S, decision nodes, D, and an additively decomposable utility function, u(x), relative to $Q_1, \cdots, Q_t, Q_i \subseteq X, X = S \cup D$, which is defined as follows:

$$u(x) = \sum_{i \in I} u_i(x_{Q_i}) \quad (5.11)$$

Here, $I = \{1, \cdots, m\}$ is a set of indices denoting the utility nodes in the decision network, the subset of variables $\{Q_i\}$ is the parents of the utility nodes, and x is an instantiation of all the chance and decision variables. The utility function must adhere to the rational preference axioms described in the previous section to ensure that the decision-maker's optimal course of action corresponds to the MEU.

We define a **decision function** for every decision node, $X_i \in D$, as a mapping, $\delta_i: x_{pa_i} \to x_i$, where x is an instantiation of all the chance and decision variables. The set of all the decision functions for a decision node, X_i, denoted by Δ_i, is called the **decision function space** for X_i. The Cartesian product, as follows, is called the **policy space**:

$$\Delta = \prod_{X_i \in D} \Delta_i \quad (5.12)$$

Given a policy, $\delta = \{\delta_{X_i} | \forall X_i \in D\} \in \Delta$, a probability, P_δ, can be defined over the chance nodes, S, and the decision nodes, D, as follows:

$$P_\delta(x) = \prod_{X_i \in S} P(x_i | x_{pa(X_i)}) \prod_{X_i \in D} P_{\delta_i}(x_i | x_{pa(X_i)}) \quad (5.13)$$

Here, $P(x_i|x_{pa(X_i)})$ for $X_i \in S$ is given in the specification for the influence diagram, while $P_{\delta_i}(x_i|x_{pa(X_i)})$ is given by δ_i, as follows:

$$P_{\delta_i}(x_i|x_{pa(X_i)}) = \begin{cases} 1, & \text{if } \delta_{X_i}(x_{pa(X_i)}) = x_i \\ 0, & \text{otherwise} \end{cases} \quad (5.14)$$

The expected utility under policy δ, denoted by $E[u|\delta]$, is a function of the policy δ, defined as follows:

$$g(\delta) = E[u|\delta] = \sum_{i\in I} u_i\,(x_{Q_i})P_\delta(x_{Q_i}) \quad (5.15)$$

The maximum of $g(\delta)$ over all the possible policies is the MEU of the influence diagram:

$$\text{MEU} \stackrel{\text{def}}{=} \max_{\delta\in\Delta} g(\delta) \quad (5.16)$$

An MEU policy is a policy, δ^*, that achieves the MEU:

$$\delta^* = \operatorname*{argmax}_{\delta\in\Delta} g(\delta) \quad (5.17)$$

Having stated the MEU problem, let us review the key points. We gave a mathematical derivation of the MEU problem for decision networks. The MEU problem comprises two tasks: one is computing the MEU, and the other is determining the MEU policies. The MEU problem addresses the optimization analysis phase in the decision analysis paradigm.

In the following section, we describe a relational database approach for representing influence diagrams and decision-making under uncertainty.

Influence diagram as a relational database

This section will describe a relational database approach to representing influence diagrams, specifically decision networks. The database representation of the influence diagram comprises a qualitative specification, building the database schema, and a quantitative specification, populating the database with data.

We will begin with the qualitative component of the database specification, which maps the DAG of the influence diagram to a relational database schema.

Qualitative specification

The qualitative specification takes as input the DAG representing the influence diagram and outputs a database schema.

We create a topological sort of the influence diagram's DAG and process every node (v) in increasing order:

If v is a chance node, create a weighted relation represented by a table that includes the following:

- One column for the primary key
- If $|pa(v)| > 0$, the table will include foreign keys that reference the primary key of each table corresponding to the nodes $x \in pa(v)$
- One column for the value of v

- One column for the weight attribute, which represents the probability of v taking a specific value, conditioned on the value tuple of its parent tables

If v is a decision node, create a weighted relation represented by a table that consists of the following:

- One column for the primary key
- One column will be designated for the value of v

- If v is a **utility node**, create a weighted relation represented by a table that consists of the following:
    - One column for the primary key
    - The table will include foreign keys that reference the primary key of each table corresponding to the nodes $x \in pa(v)$
    - One column for the weight attribute, which is the utility value as a function of the value tuple of its parent tables

After creating the database schema for the influence diagram using the procedure outlined here, the next step is to populate the database tables according to that schema. This process of filling the database tables is known as the quantitative specification step, which will be explained next.

Quantitative specification

The quantitative specification fills the database tables with the relevant data from the influence diagram. For all tables, while inserting data, assign a unique identifier for the primary key, for example, an integer that orders the rows in the tables.

To execute the quantification specification, create a topological sort of the influence diagram's DAG and process every node, v, in increasing order according to d:

- If v is a **chance node**, these are the steps:
    - If v has no parents in the influence diagram's DAG, then for each value, x, in v's domain, dom(v), with a non-zero probability, p, insert a record in v's table where you set the value of v to x and the probability to p
    - If v has parents in the influence diagram's DAG, for each value, x, in v's domain, dom(v), with non-zero probability, p, conditioned on the value tuple, t, assigned to pa(v), insert a record in v's table where you set the value of v to x, the probability to p, and the foreign keys to the primary keys of the respective records of pa(v) where pa(v)=t
- If v is a **decision node**, for each value, x, in v's domain, dom(v), insert a record in v's table where you set the value of v to x
- If v is a **utility node**, for each value tuple, t, of v's parents, pa(v), insert a record in v's table where you set the value of v to the value, x, of the utility function for pa(v)=t and the foreign keys to the respective primary keys of the records of pa(v) where pa(v)=t

Let's summarize the key points of this section on representing influence diagrams as relational databases. The database representation of influence diagrams includes both qualitative and quantitative specifications. The qualitative specification procedure in this section mapped the influence diagram's

DAG to a relational database schema. The quantitative specification procedure populated the database with the influence diagram data, respecting the influence diagram's data model.

In the next section, we will describe a use case for an influence diagram that we will use as a running example for demonstrating the database framework for representing and reasoning with influence diagrams.

Describing a use case of an influence diagram

In this section, we describe a use case of an influence diagram representing a hypothetical decision problem modeling a medical decision problem borrowed from Shenoy (Shenoy, 2009). The problem consists of a decision problem where a doctor is deciding how to help people who might have disease D. Disease D causes a pathological state called P, which then manifests as symptom S. The doctor checks first to see whether the patient has symptom S. Based on this observation, they either administer treatment to the patient (for D and P) or refrain from doing so. The physician's utility function is contingent upon their decision to provide or withhold treatment, the presence or absence of disease D, and the existence or nonexistence of pathological condition P. There is a 10% chance that people will contract disease D. 80% of people with D have a condition that causes P. Conversely, 15% of patients devoid of D possess P. 70% of people diagnosed with P show symptoms. 20% of patients who do not have P exhibit symptom S. We presume that D and S are conditionally independent in a probabilistic framework, contingent upon P. The doctor's utility function is shown in *Table 5.1*.

Physician's utility (u)		**States**			
	Has pathological state (p)		No pathological state (~p)		
	Has disease (d)	No disease (~d)	Has disease (d)	No disease (~d)	
Acts	Treat (t)	10	6	8	4
	Not treat (~t)	0	2	1	10

Table 5.1 – Utility function of the decision network use case (Shenoy, 2009)

Figure 5.20 depicts the DAG for the influence diagram of the hypothetical medical diagnosis problem as described in Shenoy (Shenoy, 2009).

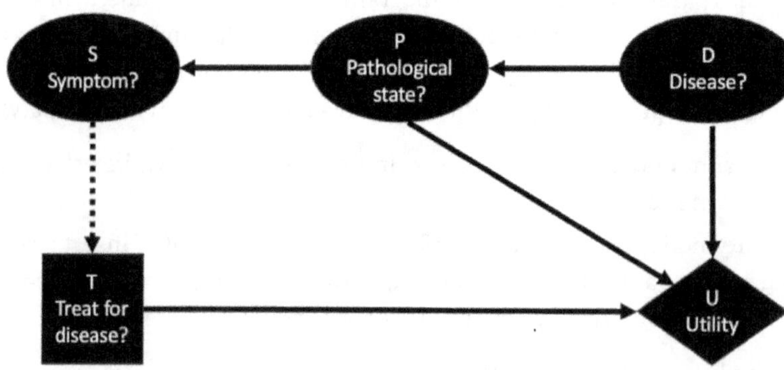

Figure 5.20 – Influence diagram of the medical decision example

The conditional probability functions of the influence diagram from the data described in the Shenoy medical decision problem are presented in *Table 5.2*.

D	Pr(D)
D	0.1
~d	0.9

D	Pr(P=p|D)	Pr(P=~p|D)
d	0.8	0.2
~d	0.15	0.85

P	Pr(S=s|P)	Pr(S=~s|P)
P	0.7	0.3
~p	0.2	0.8

Table 5.2 – Conditional probability functions of the decision network use case

To flesh out the fictitious use case with a real-life example, consider a use case where the disease is prostate cancer (Lucia, et al., 2008), the symptom is an elevated **prostate-specific antigen** (**PSA**), the pathological state is the stage of cancer as measured by the **Gleason score** (**GS**), and the treatment is either surgery or radiation. The utility of the prostate cancer example would be the value of life extension traded off with risks and side effects, which will depend on the treatment decision, the pathological state, and the disease. A rational treatment decision policy will be one that maximizes the expected value of the doctor's utility function.

In the next section, we will use the influence diagram example to demonstrate the application of the qualitative and quantitative specification procedures outlined in the section titled *Influence Diagram as a Relational Database.*

Representing the influence diagram example as a relational database

In this section, we provide a comprehensive guide to demonstrate how to implement both qualitative and quantitative specifications to represent the influence diagram for the use case within a relational database.

The demonstration is split into two parts. The first part is on implementing the qualitative specification on the influence diagram example, applying the mapping procedure from the previous section to map the DAG of the influence diagram to a database schema. The second part implements the quantitative specification procedure, populating the database tables with the influence diagram's data. In the practice sections, we demonstrate the implementations programmatically using Python and the open source **relational database management system** (**RDBMS**) PostgreSQL.

We will first see part 1 of the demonstration of mapping the influence diagram's DAG to a database schema.

Part 1: implementing the qualitative specification

In this section, we map the influence diagram's graphical structure to a relational database schema.

Applying the procedure in the previous section for qualitative specification, which consists of mapping the influence diagram to a database schema, will yield the database schema depicted in *Figure 5.21*. The implementation is done programmatically using Python and the PostgreSQL RDBMS. The highlights of the programming steps will be described in the following Practice section.

The database schema shown in *Figure 5.21* has five tables, one for each node in the DAG of the influence diagram of *Figure 5.20*. The relational table T corresponds to the decision node in the DAG. The S, D, and P relational tables correspond to the S, D, and P chance nodes, respectively, in the DAG. The relational table U corresponds to the utility node U in the DAG. As was described in the previous section, all the chance and decision variables in the decision network of the use case are Booleans, having either a true or false value.

The one-to-many relationships that connect the tables and the definitions of the primary and foreign keys impose required constraints respecting the DAG structure of *Figure 5.20*.

IIt is important to notice that the database schema does not link the S relation to the T relation to represent the informational influence indicated in the influence diagram's DAG in *Figure 5.20* by a dotted directed edge. The informational link will be established through the relational query used to compute the MEU policy, which will define the informational influence as the result of the MEU-related query demonstrated in part 3 of the database implementation.

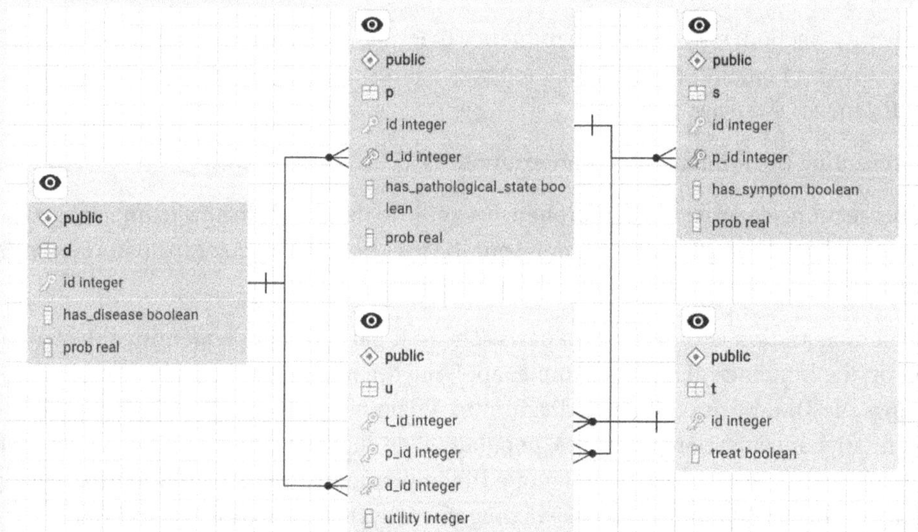

Figure 5.21 – Database schema for the influence diagram of the decision network scenario

The following practice section provides a walk-through of the process of building the database schema for the influence diagram using Python and PostgreSQL.

Practice using Python and PostgreSQL

The complete code for building the database schema for the influence diagram example can be found in the `medid.ipynp` Jupyter file.

Creating a new PostgreSQL database and programmatically using SQLAlchemy to create engine and connection objects follows the same steps as before in building the schema for the BN example, so we skip repeating the details here. Please refer to the section titled *Implementing BNs in relational databases*.

The following code snippets describe the implementation of the qualitative specification of the influence diagram. The result of executing the code is building the database schema in PostgreSQL, whose ER model was depicted in *Figure 5.21*.

Each of the following steps creates a relational table representing a node in the influence diagram's DAG and links the table to other tables corresponding to the node's parents in the DAG, except for tables corresponding to decision nodes.

To build the schema, the nodes must be processed according to a topological sort ordering of the influence diagram's DAG; specifically, the database schema should create the relational tables in five steps, following this order: D, P, S, T, U. The following sections describe each of the five steps.

We begin with the first node in the ordering: the chance node for the disease, denoted as D.

Step 1: specifying the D table

The following code creates the table schema for the chance variable representing the disease in the decision network of the medical diagnosis use case:

```
try:
    with engine.begin() as connection:
        # disease (D) table
        connection.execute(text('''
            CREATE TABLE IF NOT EXISTS public.d
            (
                id INTEGER NOT NULL PRIMARY KEY,
                has_disease boolean,
                prob REAL NOT NULL
            )
        '''))
except SQLAlchemyError as e:
    print(f"An error occurred: {str(e)}")
```

Next, we create the schema for the second node in the ordering, which is the pathological state chance node, denoted as P.

Step 2: specifying the P table

The following code creates the table schema for the chance variable representing the pathological state in the medical diagnosis example. To show that the probability assignment of P depends on the conditioning D variable, the table has a foreign key that points to the D table:

```
try:
    with engine.begin() as connection:
        # has pathological state (P) table
        connection.execute(text('''
            CREATE TABLE IF NOT EXISTS public.p
```

```
            (
                id INTEGER NOT NULL PRIMARY KEY,
                d_id INTEGER,
                FOREIGN KEY (d_id) REFERENCES d(id),
                has_pathological_state boolean NOT NULL,
                prob REAL NOT NULL
            )
        '''))
except SQLAlchemyError as e:
    print(f"An error occurred: {str(e)}")
```

Next, we create the schema for the third node in the ordering, which is the symptom chance node, denoted as S.

Step 3: specifying the S table

The following code creates the table schema for the chance variable representing the symptom in the decision network of the medical diagnosis use case. The table has a foreign key referencing the P table to implement the dependency of the probability assignment on the conditioning P variable:

```
try:
    with engine.begin() as connection:
        # has symptoms (S) table
        connection.execute(text('''
            CREATE TABLE IF NOT EXISTS public.s
            (
                id INTEGER NOT NULL PRIMARY KEY,
                p_id INTEGER,
                FOREIGN KEY (p_id) REFERENCES p(id),
                has_symptom boolean NOT NULL,
                prob REAL NOT NULL
            )
        '''))
except SQLAlchemyError as e:
    print(f"An error occurred: {str(e)}")
```

Continuing, we create the schema for the fourth node in the ordering, which is the treatment decision node, denoted as T.

Step 4: specifying the T table

The following code creates the table schema for the decision variable representing whether to treat in the decision network of the medical diagnosis use case:

```
try:
    with engine.begin() as connection:
```

```
        # decision whether to treat (T) table
        connection.execute(text('''
            CREATE TABLE IF NOT EXISTS public.t
            (
                id INTEGER NOT NULL PRIMARY KEY,
                treat boolean NOT NULL
            )
        '''))
except SQLAlchemyError as e:
    print(f"An error occurred: {str(e)}")
```

Finally, we create the schema for the fifth and final node in the ordering, which is the doctor's utility node, denoted as U.

Step 5: specifying the U table

The following code creates the table schema for the utility function in the decision network of the medical diagnosis use case. The table has three foreign keys representing the function's dependency on the T, P, and D variables:

```
try:
    with engine.begin() as connection:
        # utility (U) table
        connection.execute(text('''
            CREATE TABLE IF NOT EXISTS public.u
            (
                t_id INTEGER NOT NULL,
                p_id INTEGER NOT NULL,
                d_id INTEGER NOT NULL,
                PRIMARY KEY (t_id, p_id, d_id),
                FOREIGN KEY (t_id) REFERENCES t(id),
                FOREIGN KEY (p_id) REFERENCES p(id),
                FOREIGN KEY (d_id) REFERENCES d(id),
                utility INTEGER NOT NULL
            )
        '''))
except SQLAlchemyError as e:
    print(f"An error occurred: {str(e)}")
```

To recap, we mapped the influence diagram of the medical diagnosis example to a relational database schema. We demonstrated how to programmatically build the database schema for the qualitative specification of the influence diagram in a PostgreSQL database using Python.

Now that we have built the schema for the influence diagram's relational database, we will demonstrate in part 2 how to populate the database table with data from the medical decision use case.

Part 2: implementing the quantitative specification

In this section, we perform the quantification task of the relational database tables using the quantification data of the influence diagrams shown in *Table 5.1* and *Table 5.2*.

Applying the procedure in the previous section for quantitative specification will fill the database tables with the influence diagram's data. Executing the quantification task will result in all relational tables in the influence diagram database being filled with data records, as shown in *Figure 5.22*.

D (Diseae?)

id	has_disease	prob
1	TRUE	0.1
2	FALSE	0.9

P (Pathological state?)

id	d_id	has_pathological_state	prob
1	1	TRUE	0.8
2	1	FALSE	0.2
3	2	TRUE	0.15
4	2	FALSE	0.85

S (Symptoms?)

id	p_id	has_symptom	prob
1	1	TRUE	0.7
2	1	FALSE	0.3
3	2	TRUE	0.2
4	2	FALSE	0.8
5	3	TRUE	0.7
6	3	FALSE	0.3
7	4	TRUE	0.2
8	4	FALSE	0.8

T (Treat for diseae?)

id	treat
1	TRUE
2	FALSE

U (Utility)

t_id	p_id	d_id	utility
1	1	1	10
1	1	2	6
1	2	1	8
1	2	2	4
2	1	1	0
2	1	2	2
2	2	1	1
2	2	2	10

Figure 5.22 – Database tables for the influence diagram of the medical decision use case

The quantification task is done programmatically using Python and the PostgreSQL RDBMS. The highlights of the programming steps will be described in the following practice section.

Practice using Python and PostgreSQL

The complete code for the quantitative specification of the influence diagram is included in the `medid.ipynp` Jupyter file.

As with the qualitative specification, the database tables are also processed according to a topological sorting of the DAG representing the influence diagram; specifically, the database tables are populated with data in the following order: table D, then P, then S, then T, then U. The following sections describe each of the five steps of filling each table with data.

First, we populate the database table for the first node in the ordering, the chance node D, which denotes disease.

Step 1: quantifying the D table

The following code creates the table schema for the chance variable representing the disease in the medical diagnosis example. The code populates the table with the data according to the D table specification in *Table 5.2*:

```
try:
    with engine.begin() as connection:
        # disease (D) table
        connection.execute(text('''
            INSERT INTO d(id, has_disease, prob)
            VALUES (1, true, 0.1), (2, false, 0.9)
        '''))
except SQLAlchemyError as e:
    print(f"An error occurred: {str(e)}")
```

Next, we populate the database table for the second node in the ordering, which is the pathological state chance node, denoted as P.

Step 2: quantifying the P table

The following code creates the table schema for the chance variable representing the pathological state in the medical diagnosis example. The table has a foreign key referencing the D table to implement the dependency of the probability assignment on the conditioning D variable. The code populates the table with the data according to the P table specification in *Table 5.2*:

```
try:
    with engine.begin() as connection:
        # has pathological state (P) table
        connection.execute(text('''
            INSERT INTO p(id, d_id, has_pathological_state, prob)
            VALUES (1, 1, true, 0.8), (2, 1, false, 0.2),
            (3, 2, true, 0.15), (4, 2, false, 0.85)
        '''))
except SQLAlchemyError as e:
    print(f"An error occurred: {str(e)}")
```

Next, we populate the database table for the third node in the ordering, which is the symptom chance node, denoted as S.

Step 3: quantifying the S table

The following code creates the table schema for the chance variable representing the symptom in the medical diagnosis example. The table has a foreign key referencing the P table to implement the dependency of the probability assignment on the conditioning P variable. The code populates the table with the data according to the S table specification in *Table 5.2*:

```
try:
    with engine.begin() as connection:
        # has symptoms (S) table
        connection.execute(text('''
                INSERT INTO s(id, p_id, has_symptom, prob)
                VALUES (1, 1, true, 0.7), (2, 1, false, 0.3),
                       (3, 2, true, 0.2), (4, 2, false, 0.8),
                       (5, 3, true, 0.7), (6, 3, false, 0.3),
                       (7, 4, true, 0.2), (8, 4, false, 0.8)
            '''))
except SQLAlchemyError as e:
    print(f"An error occurred: {str(e)}")
```

Continuing, we populate the database table for the fourth node in the ordering, which is the treatment decision node, denoted as T.

Step 4: quantifying the T table

The following code creates the table schema for the decision variable representing whether to treat. The table lists the elements of the binary decision variable domain in this example:

```
try:
    with engine.begin() as connection:
        # decision whether to treat (T) table
        connection.execute(text('''
                       INSERT INTO t(id, treat)
                       VALUES (1, TRUE), (2, FALSE)
            '''))
except SQLAlchemyError as e:
    print(f"An error occurred: {str(e)}")
```

Finally, we populate the database table for the fifth and final node in the ordering, which is the doctor's utility node, denoted as U.

Step 5: quantifying the U table

The following code creates the table schema for the utility function. The table has three foreign keys representing the function's dependency on the T, P, and D variables. The code populates the table with the data shown for the U table in *Table 5.1*:

```
try:
    with engine.begin() as connection:
        # utility (U) table
        connection.execute(text('''
            INSERT INTO u(t_id, p_id, d_id, utility)
            VALUES (1,1,1,10), (1,1,2,6), (1,2,1,8),
                   (1,2,2,4), (2,1,1,0), (2,1,2,2),
                   (2,2,1,1), (2,2,2,10)
        '''))
except SQLAlchemyError as e:
    print(f"An error occurred: {str(e)}")
```

We have created the influence diagram's database schema and populated all tables with its data, which concludes the mapping of the influence diagram's representation from the DAG to the relational database.

In the following section, we will address reasoning tasks with the relational database representation of the influence diagram example. These tasks will involve implementing relational queries in SQL for computing the MEU and the associated decision policy.

Part 3: reasoning with the influence diagram example database

In this section, we address reasoning with the influence diagram database. We will demonstrate how to compute the MEU, with *Equation 5.16*, and derive the MEU policy, with *Equation 5.17*. We will demonstrate how to express the influence diagram reasoning tasks as relational queries implemented programmatically using Python and the PostgreSQL RDBMS.

The initial reasoning task involves calculating the JPD of all chance variables within the influence diagram. This joint probability is represented as a relational view in the database, which is essential for formulating queries to compute decision trees, evaluate the MEU, and determine the MEU decision policy.

We will begin with the task of computing the JPD next.

Computing the JPD

The JPD is a relational view, a virtual relational table, that is calculated as a join operation that combines the three weighted relations that represent all the chance variables in the influence diagram database: D, P, and S. The weight attribute in the join relation is the product of the weights of the three joined relations. We discussed an example of how to compute a JPD in the practice section on BNs earlier in this chapter.

The following Python code demonstrates how to create the JPD view programmatically in the influence diagram database:

```
try:
    with engine.begin() as connection:
        connection.execute(text('''
        CREATE OR REPLACE VIEW JPD
            AS
            SELECT d.has_disease, p.has_pathological_state, s.has_symptom,
                d.prob * p.prob * s.prob AS prob
            FROM d
            JOIN p ON p.d_id = d.id
            JOIN s on s.p_id = p.id
        '''))
        print('Created view JPD.')
except SQLAlchemyError as e:
    print(f"An error occurred: {str(e)}")
```

Having computed the JPD, we will now proceed to our first core reasoning task, which is computing the decision tree.

Computing the decision tree

Computing the decision tree provides the overall picture of all courses of action and corresponding expected utilities. In our use case, computing the decision tree involves calculating all value tuples for the chance variable (the symptom, S), the decision variable (treat, T); and the resulting expected utility (U). Those value tuples correspond to branches in the decision tree, representing course-of-action scenarios.

The following Python code produces a table with three columns: `has_symptom`, `treat`, and `sum_utility`. Each row of the table corresponds to a tuple consisting of an observed value of the chance variable, s, the decision, t, and the expected utility for the corresponding observation-decision value pair:

```
try:
    with engine.begin() as connection:
        result = connection.execute(text('''
            WITH tot AS (
                SELECT has_symptom, SUM(prob)
                AS sum_prob FROM jpd GROUP BY has_symptom
            ),
            prob_data AS
            (
                SELECT jpd.has_pathological_state,
                       jpd.has_symptom, jpd.has_disease,
                       jpd.prob / tot.sum_prob as prob1,
```

```
                    p.id as p_id, d.id as d_id
                FROM jpd
                JOIN tot ON jpd.has_symptom
                            tot.has_symptom
                JOIN d ON d.has_disease = jpd.has_disease
                JOIN p ON p.has_pathological_state
                          jpd.has_pathological_state
            )

            SELECT
              prob_data.has_symptom, t.treat,
              SUM(prob_data.prob1 * u.utility) as sum_utility
            FROM prob_data
            JOIN u ON (u.p_id = prob_data.p_id
                       AND u.d_id prob_data.d_id)
            JOIN t ON (t.id = u.t_id)
            GROUP BY prob_data.has_symptom, t.id;
        '''))

    for row in result:
        print(row)

except SQLAlchemyError as e:
    print(f"An error occurred: {str(e)}")
```

The following section explains the logic behind the SQL query in the preceding code:

- The `WITH` clauses create two temporary views or subqueries referenced later in the query.
- The first subquery, `tot`, computes the marginal probability of `has_symptom` by grouping the `jpd` table by the `has_symptom` column and calculating each group's sum of the `prob` column.
- The second subquery, `prob_data`, joins the `jpd` table with the `tot` subquery and the tables for the non-observed chance variables (`d` and `p`). It calculates a new column, `prob1`, by dividing `jpd.prob` by `tot.sum_prob`. The subquery also selects a few more columns for later use. The `prob1` column is the joint probability of `has_disease` and `has_pathological_state` conditioning on `has_symptom`. We are conditioning on `has_symptom` because this information influences the treatment decision.
- The `SELECT` clause in the body section of the query is used to get information from the `prob_data` temporary view created in the earlier step and join it with the `u` (utility) and `t` (treat) tables. The join is based on conditions specified in the `ON` clause. It also calculates `SUM(prob_data.prob1 * u.utility)`, the main calculation that this query performs. The results are finally grouped by `prob_data.has_symptom` and `t.id`.

The whole SQL statement essentially divides the `prob` value in the `jpd` table by the sum of `prob` values for each symptom, joins these values with utilities from the utility table, `u`, as per the associated pathological state and disease, and then sums up the multiplied result for each combination of symptom and treatment.

Executing the code to compute the decision tree results in four records. Each record has the values for the `has_symptom` attribute, the `treat` decision, and `sum_utility`, in that order. The results mentioned correspond to the decision tree illustrated in *Figure 5.23*. This decision tree indicates that expected utilities are higher when treatment aligns with the presence of symptoms.

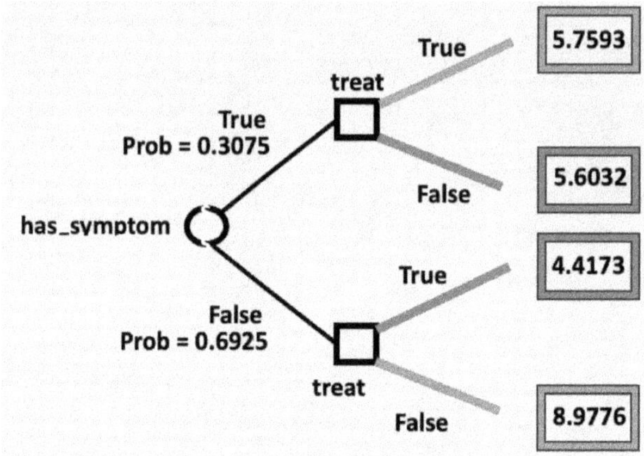

Figure 5.23 – Decision tree representing the expected utility for the four decision scenarios resulting from the chance event of the patient having symptoms and the doctor deciding whether to treat

The `medid.ipynp` Jupyter notebook contains additional code for evaluating the MEU and determining the MEU policy for the influence diagram example. The logic of that code is similar to the one explained previously for the decision tree code. Due to a lack of space, the details of the code for the following reasoning tasks are not listed here, but we will briefly describe the tasks performed and display the calculated results in the following sections.

Next, we will review the task of evaluating the MEU.

Finding the MEU policy

The code for finding the MEU policy includes SQL statements designed to calculate the treatment decision that maximizes the expected utility for two scenarios: when a symptom is present (`has_symptom = True`) and when it is absent (`has_symptom = False`).

The code will generate two records that reflect the MEU decision policy for the two possible scenarios: one where the symptom is present and one where it is absent. These scenarios correspond to the green-colored sections in the decision tree depicted in *Figure 5.23*.

The output from executing the code consists of two records, each with four fields: `has_symptom`, `treat`, MEU, and probability. Each record corresponds to a scenario involving a symptom and decision, the MEU, and the probability of the scenario, which reflects the likelihood of `has_symptom` occurring.

The code output is as follows:

```
(False, False, 8.977617133408785, 0.6925, 6.2169998220769696)
(True, True, 5.759349569678307, 0.3075, 1.771000020138798)
```

This section concludes the review of the code for identifying the MEU policy. Next, we will discuss the code used to evaluate the MEU for the influence diagram.

Evaluating the MEU

The code for this step computes the probability for each decision scenario based on whether the symptom is present (`has_symptom = True`) or absent (`has_symptom = False`). The probability of `has_symptom` is determined by marginalizing the JPD, which involves summing the JPD over all chance variables except for `has_symptom`. In relational database terminology, this process projects the weighted relation JPD onto `has_symptom`. The probabilities for the two scenarios are shown in *Table 5.3* under the `sum_prob` column.

The code computes the expected utility for each scenario under the MEU policy. This policy determines the decision that maximizes the expected utility conditioning on the value of the chance variable, `has_symptom`, for each scenario.

The first five columns in *Table 5.3* contain the results from the code of the previous reasoning task on determining the MEU policy.

The first column presents the values of the `has_symptom` variable, which correspond to distinct scenarios. The second column indicates the treatment decision that maximizes the expected utility, conditioned on the value of `has_symptom`. The third column, labeled `sum_utility`, provides the MEU conditioned on the value of `has_symptom`. The fourth column shows how likely each scenario is to happen. Finally, the fifth column, titled `total_utility`, reflects the product of the expected maximum utility for the scenario that occurred and the probability of that scenario.

Summing the values in the `total_utility` column results in the MEU value generated by the code, which is `7.98799984`.

has_symptom	treat	sum_utility	sum_prob	total_utility = sum_utility * sum_prob
FALSE	FALSE	8.977617133	0.6925	6.21699982
TRUE	TRUE	5.75934957	0.3075	1.77100002
				MEU = 7.98799984

Table 5.3 – MEU is computed as the weighted sum of the expected utility over all scenarios under the MEU policy

This concludes our exploration of reasoning with influence diagrams for the influence diagram example using the relational database framework.

Let's summarize the lessons learned from the demonstration on the influence diagram example. The representation of an influence diagram as a relational database within a framework based on weighted relations allows for the utilization of powerful relational algebra methods. This approach enables the declarative formulation of queries for reasoning tasks related to influence diagrams. Practically, the database framework facilitates the implementation of these reasoning queries using SQL, along with the optimization methods that are standard in the database industry. The demonstration, accompanied by hands-on practice using the influence diagram example, highlights the benefits of adopting the database framework for both representation and reasoning with influence diagrams.

Summary

You have acquired knowledge and learned skills in three key areas in this chapter. First, we introduced the concept of weighted relations, which extends a standard relational database by adding a special weight attribute. This extension enables the framework to incorporate probabilistic databases, BNs, and influence diagrams. Second, we learned about inference tasks, which can run efficiently in acyclic databases by taking advantage of their significant properties: pairwise local consistency is equivalent to global consistency, and sequential monotone joins are guaranteed to exist. Third, we saw that inference tasks in BNs and influence diagrams can be expressed declaratively in relational algebra and implemented programmatically in SQL, which was demonstrated through hands-on practice exercises throughout the chapter.

The knowledge and skills gained in this chapter empower you to develop BN or influence diagram models that can be applied to specific problems in your field. Additionally, you will be able to create a working prototype using open source RDBMSs, such as PostgreSQL.

The upcoming chapters in the book will build on the concepts introduced in this chapter. We will look at graph algorithms that use join trees to make inference tasks such as belief propagation and updates more efficient. Specifically, *Chapter 8* will detail elimination-based algorithms that incorporate the concept of weighted relations.

References

- Bacchus F. and Grove A. *Graphical models for preferences and utility*. [Conference]. Uncertainty in AI (UAI-95). 1995. pp. 3–10.
- Baixeries Jaume. *A Formal Context for Acyclic Join Dependencies*. [Conference]. Lecture Notes in Computer Science. [s.l.]: Springer, 2017. pp. 563–572.
- Beeri C. [et al.] *Properties of Acyclic Databases*. [Conference]. STOC '81: Proceedings of the thirteenth annual ACM symposium on Theory of computing. Milwaukee Wisconsin USA: Association for Computing Machinery, 1981.
- Beeri Catriel , Fagin Ronald and Maier David. *On the Desirability of Acyclic Database Schemes*. [Journal]. Journal of the ACM (JACM). 1983. 3: Vol. 30. pp. 479–513.
- Bernstein P. A. and Goodman N. *The power of natural semijoins*. [Journal]. SIAM Journal on Computing. 1981. 4: Vol. 10. pp. 751–771.

- Bhattacharjya D. and Shachter R. D. *Evaluating influence diagrams with decision circuits* [Conference]. [s.l.]: Proceedings of the 23rd Conference on Uncertainty in Artificial Intelligence (UAI), 2007.
- Bhattacharjya D. and Shachter R. D. *Sensitivity analysis in decision circuits* [Conference]. [s.l.]: Proceedings of the 24th Conference on Uncertainty in Artificial Intelligence (UAI), 2008.
- Bielza C. and Shenoy P. P. *A comparison of graphical techniques for asymmetric decision problems.* [Journal]. Management Science. 1999. pp. 1552–1569.
- Brault-Baron Johann. *Hypergraph Acyclicity Revisited.* [Online]. arXiv. 2014. `https://arxiv.org/abs/1403.7076`.
- Codd E. F. *A relational model of data for large shared data banks.* [Journal]. Communications of the ACM. 1970. 6: Vol. 13. pp. 377–387.
- Covaliu Z. and Oliver R. M. *Formulation and solution of decision problems using decision diagrams.* [Journal]. Management Science. 1995. 12: Vol. 41. pp. 1860–1881.
- El Fattah Yousri and Dechter Rina. *Diagnosing tree-decomposable circuits.* [Conference]. IJCAI'95: Proceedings of the 14th international joint conference on Artificial intelligence. Montreal Quebec Canada: [s.n.], 1995.
- El Fattah Yousri and Peot Mark A. *A compositional structured query approach to automated inference.* [Conference]. KR'00: Proceedings of the Seventh International Conference on Principles of Knowledge Representation and Reasoning. San Francisco, CA, United States: Morgan Kaufmann Publishers Inc., 2000. pp. 213–224.
- El Fattah Yousri. *Structured modeling language for automated modeling in causal networks.* [Conference]. Proceedings of the Sixteenth International Joint Conference on Artificial Intelligence (IJCAI-99). Stockholm, Sweden: [s.n.], 1999. pp. 1108–1114.
- Fagin R. *Degrees of Acyclicity for Hypergraphs and Relational Database Schemes.* [Journal]. Journal of the Association for Computing Machinery. 1983. 3: Vol. 30. pp. 514–550.
- Keeney L and Raiffa H. *Decisions with Multiple Objectives: Preferences and Value Trade-Offs* [Book]. New York: Wiley, 1976.
- Kochenderfer Mykel J. *Decision Making Under Uncertainty: Theory and Application* [Book]. Cambridge, Massachusetts: The MIT Press, 2015.
- Koller D and Pfeffer A. *Probabilistic frame-based systems.* [Conference]. Proceedings of the Fifteenth National Conference on Artificial Intelligence (AAAI- 98). 1998. pp. 580–587.
- Lucia M.S. [et al.] *Pathologic characteristics of cancers detected in the Prostate Cancer Prevention Trial: Implications for prostate cancer detection and chemoprevention.* [Journal]. Cancer Prevention Research. 2008. 3: Vol. 1. pp. 167–173.
- Maier D. *The Theory of Relational Databases.* [Book]. Rockville, Maryland: Computer Science Press, 1983.
- Marchette David J. *Package 'HyperG'.* [Online]. October 12, 2022. `https://cran.r-project.org/web/packages/HyperG/HyperG.pdf`.
- Meyer David, Hornik Kurt and Buchta Christian. *Package 'relations'.* [Online]. relations: Data Structures and Algorithms for Relations. July 23, 2025. `https://cran.r-project.org/web/packages/relations/relations.pdf`.

- Meyer David and Hornik Kurt. *Good Relations with R*. [Online]. July 1, 2016. January 1, 2024. `https://cran.r-project.org/web/packages/relations/vignettes/relations.pdf`.
- Nielsen T. D. and Jensen F. V. *Representing and solving asymmetric decision problems*. [Journal]. International Journal of Information Technology & Decision Making. 2003. 2: Vol. 2. pp. 217–263.
- *pgAdmin Development Team ERD Tool*. [Online]. pgAdmin 4 documentation. 2025. January 10, 2024. `https://www.pgadmin.org/docs/pgadmin4/development/erd_tool.html`.
- Shachter R. D. and Bhattacharjya D. *Solving Influence Diagrams: Exact Algorithms*. [Online]. February 15, 2011. `https://doi.org/10.1002/9780470400531.eorms0808`.
- Shachter R. D. *Dynamic programming and influence diagrams*. [Journal]. IEEE Transactions on Systems, Man and Cybernetics. 1990. 2: Vol. 20. pp. 365–379.
- Shenoy P. P. *Valuation network representation and solution of asymmetric decision problems*. [Journal]. European Journal of Operational Research. 2000. pp. 579–608.
- Shenoy P. P. *Valuation-based systems for Bayesian decision analysis*. [Journal]. Operations Research. 1992. pp. 463–484.
- Shenoy Prakash P. *Decision Trees and Influence Diagrams*. [Book Section]. Encyclopedia of Life Support Systems, Optimization and Operations Research, Vol. 4 / book auth. Derigs U. (ed.). [s.l.]: UNESCO-EOLSS, 2009.
- Smith J. E., Holtzman S. and Matheson J. E. *Structuring conditional relationships in influence diagrams*. [Journal]. Operations Research. 1993. 2: Vol. 41. pp. 280–297.
- Ullman Jeffrey D and Ron Fagin. *Acyclic Hypergraphs*. [Online]. July 2016. January 21, 2024. `https://databasetheory.org/sites/default/files/2016-07/RonFagin-ullman.pdf`.
- von Neumann John and Morgenstern Oscar. *Theory of Games and Economic Behavior*. [Book]. [s.l.]: Princeton University Press, 1944.

6

Join Tree Clustering

This chapter introduces join tree (junction tree) clustering as a practical method for transforming a Bayesian network's directed acyclic graph (DAG) into a tree-decomposable structure that supports efficient, exact inference. You will learn the key graph transformations—moralization and triangulation—along with the core structural ideas behind chordal graphs, simplicial nodes, perfect elimination orderings, and maximum cardinality search (MCS). Additionally, you will discover how to identify maximal cliques, assemble them into a join tree that satisfies the running intersection property (RIP), and understand why this tree structure enables scalable inference. Throughout, hands-on R examples show how to construct and query graphs, verify d-separation and separation, and implement the algorithms for tree clustering from start to finish using common graph libraries.

In this chapter we're going to cover the following main topics:

- Graph representation of Bayesian networks
- Moralization
- Triangulation
- Computing join tree representation

Technical requirements

The R code in this chapter is in a folder named `R-chap6` that you can download from the GitHub URL: `https://github.com/PacktPublishing/Causal-Inference-with-Bayesian-Networks/tree/main/Chapter_06`

To run the R code provided in the RMarkdown document, you'll need to have the following:

- R Runtime Environment: An installation of R, a free software environment for statistical computing and graphics. It can be downloaded from the **Comprehensive R Archive Network (CRAN)**.
- **RStudio**: An integrated development environment for R. While not mandatory, RStudio enhances the user experience. It can be downloaded and installed via RStudio's download page.
- **RMarkdown Package**: This can be installed from within R or RStudio by running `install.packages("rmarkdown")`.

R package requirements

The code requires installing the following R packages, which you can do from within R or RStudio. You can use the `install.packages` function in R to do this. Each package must be installed separately. The following lines of code show how to install these packages:

```
# R command to install packages
install.packages("gRbase")
install.packages("ggm")
install.packages("igraph")
install.packages("RBGL")
install.packages("sets")
```

Please note that if you're using a version of RStudio that is not up to date, you may need to update it to ensure compatibility with the latest versions of the packages above.

Graph representation of Bayesian networks

This section begins with formal definitions related to directed and undirected graphs and their structural properties. Next, we describe how directed acyclic graphs give us a formal language for describing the factored probability models of Bayesian networks and answering conditional independence queries between model variables. To ensure you grasp these concepts, we provide an example Bayesian network on fault diagnosis of a combinatorial circuit, which we use in our practice section using R. This example will serve as a guiding thread throughout this chapter, supporting your learning process and helping you apply these concepts in practice.

Directed and undirected graphs

The structure of Bayesian network models is described using directed and undirected graphs. The graphs represent the probabilistic influence between the Bayesian network variables and describe the Bayesian network factored representation of its complete probability distributions on the network variables. The graph structure can be used to derive structure-based algorithms for inference tasks and answering conditional independence queries.

Undirected graphs are the fundamental data structure underlying all the structure-based algorithms for inference with BNs. In this section, we give the mathematical notations and preliminaries on directed and undirected graphs that we will use and build on as we progress in the following sections.

Let us begin with definitions of undirected graphs.

Definition: undirected graph

An undirected graph G is a pair (V, E) where V is a finite set of $|V|$ elements called nodes or vertices, and E is a set of $|E|$ doubletons, sets of node pairs called edges. Given $x \in V$, we define:

$$adj(x) = \{y \in V | \{x, y\} \in E\} \quad (6.1)$$

as the set of nodes **adjacent** to x. The set $adj(x)$ is also called the **neighbors** of x, denoted $ne(x)$, or the **boundary** of x, denoted $bd(x)$. The set consisting of a node x and its boundary $bd(x)$ is called the **closure** of x:

$$cl(x) = \{x\} \cup bd(x) \quad (6.2)$$

A **path** of length n between two nodes $x, y \in V$ is an ordered set of distinct nodes $[v_0, \cdots, v_{i-1}, v_i, \cdots, v_n], v_0 = x,\ v_n = y$, such that $v_i \in adj(v_{i-1}), i = 1, \cdots, n$. A **cycle** of length n is a path of length n that begins and ends at the same node. We say a graph is **connected** if, for each pair of distinct nodes $x, y \in V$, there is a path from x to y.

Directed graphs differ from undirected graphs in that the edges are directed. In an undirected graph, as stated in the above definition, an edge is a pair of nodes represented by a doubleton set $\{x, y\}$, and an edge $\{x, y\}$ is the same as $\{y, x\}$. In directed graphs, the edges are directed and thus specified as ordered tuples of node pairs, so an edge (x, y) is not the same as (y, x). We write $x \to y$ to denote the directed edge from x to y to emphasize the directionality.

Now, for definitions on directed graphs.

Definition: directed graph

Formally, a directed graph (**digraph**) G is a pair (V, E) where V is a finite set of $|V|$ elements called nodes or vertices, and E is a set of $|E|$ edges, ordered pairs of nodes (x, y) denoted as $x \to y$. The **out-degree** of a node x is the number of edges pointing away from x, and the **in-degree** of a node y is the number of edges pointing into y. A **directed path** of length n in a digraph G is a chain from x to y, an ordered set of distinct nodes $[v_0, \cdots, v_{i-1}, v_i, \cdots, v_n], v_0 = x,\ v_n = y$, such that $v_{i-1} \to v_i, i = 1, \cdots, n$. A **directed cycle** of length n is a path of length n that begins and ends at the same node. A **directed acyclic graph (DAG)** is a digraph having no directed cycle.

We note that every directed graph has an **underlying graph**, an undirected graph with the same nodes and edges but with the directionality of the edges ignored.

Subgraphs are derived from graphs directed or undirected by including only subsets of nodes. The following is the definition of a subgraph, also called a section graph.

Definition: subgraph, section graph

A subgraph $G(A)$ of a graph $G = (V, E)$ is the graph $(A, E(A))$ whose nodes are a set A and whose edges are a subset of edges $E(A) \subseteq E$ such that each edge in $E(A)$ is an edge $e \in E$ between two nodes both of which fall in the set A.

In the following, we define ancestral sets and ancestral graphs. Ancestral sets and ancestral graphs are defined only for directed acyclic graphs. Ancestral graphs and subgraphs play roles in analyzing graph separations and graph decomposition.

Definition: ancestral sets, ancestral graphs

Given a DAG, $G = (V, E)$ and a node x, the **ancestors** of x, $ancestors(x)$, is the set of all nodes y such that there is a directed path from y to x in G. The **ancestral set** for a subset of nodes $A \subseteq V$ is the union of A and ancestors of all nodes in A.

$$ancestralSet(A) = A \cup \bigcup_{x \in A} ancestors(x) \quad (6.3)$$

Given a DAG, $G = (V, E)$, the ancestral graph for a subset of nodes $A \subseteq V$ is the subgraph of G for the ancestral set of A:

$$ancestralGraph(A; G) = G\big(ancestralSet(A)\big) \quad (6.4)$$

Definition: complete set, complete graph

Let $G = (V, E)$ be an undirected graph, and $S \subseteq V$ be a subset of nodes. S is a complete set or a **clique** in G if every pair of distinct nodes x and y in S corresponds to an edge $\{x, y\} \in E$. The subgraph $G(S)$ is a complete graph if S is a complete set.

In the next section, we delve into the structural representation of Bayesian networks using directed acyclic graphs.

Directed acyclic graphs of Bayesian network

In *Chapter 3*, we introduced Bayesian networks as a generative model of complete probability distributions on network variables. This model encapsulates all the quantification of probability distributions on the network variables, respecting the conditional independencies dictated by the probabilistic relationships between network variables. These conditional independencies are not just theoretical concepts but are crucially and formally represented by the graphical structure of the model.

A directed acyclic graph, $G = (X, E)$, represents the factored representation of a Bayesian network of a complete probability distribution over a set of network variables $X = \{X_1, X_2, \cdots, X_n\}$ as follows. Given a node $X_i \in X$, we write $pa(X_i)$ to denote the set of all parents of the node in the directed graph, G. The set $pa(X_i)$ specifies all the variables having a direct probabilistic influence on the variable X_i, meaning the probability distribution of X_i is a function of the value assignment tuple for the set $pa(X_i)$. We write that functional dependency as $P\big(X_i \big| pa(X_i)\big)$, which denotes the conditional probability distribution (CPD) of X_i as a function of the values of $pa(X_i)$.

The factored representation of the Bayesian network of the complete probability distribution can be written as the product of the conditional probability distribution of every node in the graph conditioned on the value of all its parent variables as follows:

$$P(X_1, \dots, X_n) = \prod_{i=1}^{n} P\big(X_i \big| pa(X_i)\big) \quad (6.5)$$

Throughout this chapter, we will delve into the practical application of Bayesian networks using the circuit diagnosis example from *Chapter 3*. This example will vividly illustrate how graphical representations and structural properties can be used to model and solve real-world problems. The circuit example is described again here for self-containment.

Example: circuit model

Figure 6.1 presents the circuit diagnosis example from *Chapter 3*. This example describes a combinational circuit that outputs the exclusive NOR of its inputs. In this chapter, we are primarily concerned with the model structure.

The circuit model has ten variables; six of them, X, Y, Z, D, E, and F, are binary signals representing Boolean variables, the inputs, and outputs of the logic gates forming the circuit's four components. In comparison, the other four variables, which are $A1, A2, A3$, and $A4$, are assumption variables representing the working status of each of the four components. The status of a component can either be normal or abnormal. The diagnosis problem is to compute the probabilities of the working status of the four components given the observed input outputs on the circuit components. We will revisit the inferencing task in the next chapter, but for this chapter, we are focused on the structural aspects of the model's graphical representations.

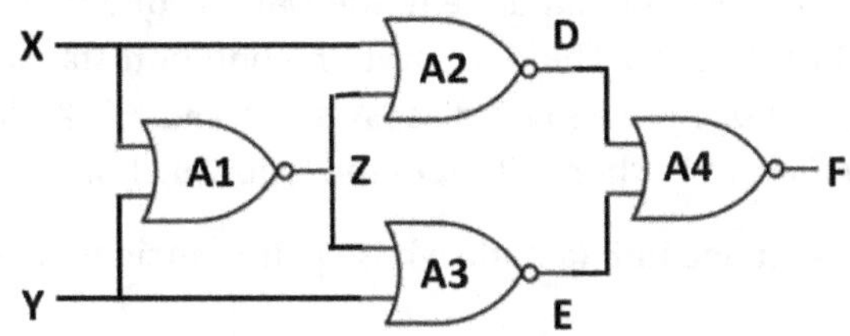

Figure 6.1—Combinatorial digital circuit

Figure 6.2 shows the DAG representing the structure of the circuit BN model. The BN model represents a complete probability distribution over the set of ten variables represented by the nodes of the DAG. The DAG defines a causal family for each variable consisting of the variable and all its parents, and the BN model specifies a conditional probability distribution for each variable as a function of the values of all its parents. The DAG has ten nodes and 12 edges. The DAG is sparse and has at most three parents per family.

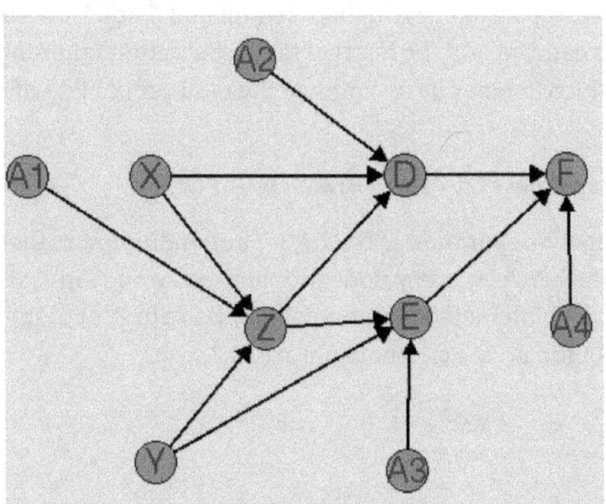

Figure 6.2—Directed acyclic graph for the circuit BN model

The factored representation of the BN model for the DAG of *Figure 6.2* is written as follows:

$$P(X,Y,Z,D,E,F,A1,A2,A3,A4) = P(A1)P(A2)P(A3)P(A4)P(X)P(Y)$$

$$P(Z|A1,X,Y)P(D|A2,X,Z)P(E|A3,Y,Z)P(F|A4,D,E) \quad (6.6)$$

As we conclude this introductory section on the graphical representation in Bayesian networks, revisiting the central concept of conditional independence is crucial. The graphical representation provides us with a topological test for checking the conditional independence in the probability model of a Bayesian network by verifying d-separation in the DAG representing the factored probability model of the BN. We will review d-separation and conditional independence next.

D-separation and conditional independence

In this section, we delve into the practical application of conditional independence in Bayesian networks. As you may recall from *Chapter 3*, a node in the DAG is conditionally independent of all its non-descendants given all its parents. The d-separation condition states that given three disjoint subsets of nodes in the DAG (X,Z,Y), we say Z d-separates X and Y if Z blocks all undirected paths between a node in X and a node in Y where either of the following holds:

- There is a non-collider node in Z (a node whose pair of incident edges are not both pointing toward the node);
- There is a collider node (a node whose pair of incident edges are both pointing toward the node), and the node and all its descendants are not in Z.

This concept is not just theoretical but has real-world applications. Every d-separation in the DAG is a conditional independence in the model. The condition of d-separation states that for any three disjoint sets of nodes, X, Y, and Z, in the DAG, if Z d-separates X and$Y$ in the DAG, then X and Y are conditionally independent given Z in the model. We write:

$$\text{d-separate}(X,Z,Y) \Leftrightarrow X \perp Y \mid Z \quad (6.7)$$

to denote that X and Y are d-separated by Z in the DAG and that X and Y are conditionally independent given Z in the model. This concept is at the heart of the factorization inherent in the probability model of a Bayesian network, which enables us to compute the joint probability of X and Y given Z using the decomposable function.

$$\text{d-separate}(X,Z,Y) \Leftrightarrow \big(P(X,Y|Z) = P(X|Z)P(Y|Z)\big) \quad (6.8)$$

Let us recall the condition of d-separation in the DAG. The condition says that X and Y are d-separated by Z if all undirected paths between every node in X and every node in Y are blocked by Z. A path is blocked by Z if either: 1) the path includes a non-collider node in Z, or 2) the path has a collider node and that node and none of the node's descendants are in Z.

Example: d-separation

Figure 6.2 of the DAG of our circuit BN has the following d-separation: $(E \perp X \mid Y, Z)$. The following is a list of undirected paths between E and X, and in parentheses, the explanation of why the path is being blocked:

```
X -> Z -> E               (blocked by Z)
X -> Z <- Y -> E          (blocked by Y)
X -> D <- Z -> E          (blocked by Z)
X -> D <- Z <- Y -> E     (blocked by either Y or Z)
X -> D -> F <- E          (blocked by collider node F)
```

Our next step is to put the d-separation concept into practice. We will programmatically verify the d-separation we just discussed in the upcoming section. This practical application will not only solidify our understanding of this concept but also demonstrate its relevance in our circuit Bayesian network example.

Let's embark on a rewarding practice session that uses R to implement and reinforce the concepts we've learned. This journey of learning and implementing coding with R will keep you engaged and motivated. By the end of this session, you'll have gained practical experience in coding with R and learned how to leverage powerful graph libraries. In the next section, we'll further enhance our skills in programming practice using R, where we'll learn how to programmatically construct a directed acyclic graph and perform various graph queries, such as computing ancestral sets and ancestral graphs.

Practice using R

The R code in our practice throughout the chapter will require that we install and import the following libraries.

```
library(gRbase)
library(ggm)
library(igraph)
library(RBGL)
library(sets)
```

The code was executed using the following versions of the library packages mentioned above:

```
gRbase_2.0.3
ggm_2.5.2
igraph_2.2.2
RBGL_1.76.0
sets_1.0-25
```

Directed Acyclic Graph

We begin by specifying the DAG for the circuit example BN.

To do that, we use the `ggm::DAG()` function.

We specify the DAG as a sequence of formulas. Each formula defines a node of the graph prefixed by the tilde symbol, followed by all the node's parents separated by the plus symbol.

The DAG returns the adjacency matrix (AMAT), which is an nxn matrix where n is the number of nodes.

The element in row i and column j in AMAT equals one if there is an edge from node i to node j and equals zero otherwise.

The names of the rows and columns of the adjacency matrix are the nodes of the DAG.

If order = TRUE, the adjacency matrix is permuted to have the nodes listed in topological order where parents precede children in the order, with the resulting adjacency matrix being upper triangular.

```
G <- ggm::DAG(F ~ D+E+A4, D ~ X+Z+A2, E ~ Y+Z+A3, Z ~ X+Y+A1,order=TRUE)
print(G)
##     A1 A3 Y A2 X Z E D A4 F
## A1  0  0  0  0 0 1 0 0  0 0
## A3  0  0  0  0 0 0 1 0  0 0
## Y   0  0  0  0 0 1 1 0  0 0
## A2  0  0  0  0 0 0 0 1  0 0
## X   0  0  0  0 0 1 0 1  0 0
## Z   0  0  0  0 0 0 1 1  0 0
## E   0  0  0  0 0 0 0 0  0 1
## D   0  0  0  0 0 0 0 0  0 1
## A4  0  0  0  0 0 0 0 0  0 1
## F   0  0  0  0 0 0 0 0  0 0
```

We use the function `ggm::plotGraph()` to plot the DAG from the adjacency matrix:

```
ggm::plotGraph(G)
```

Figure 6.2 shows the visualization of the DAG for the circuit BN model resulting from the execution of the `ggm::plotGraph` function. Notice there are four DAG families corresponding to the nodes "Z," "S," "E" and "F," which are the outputs of the four gates in the circuit. We use the term a DAG family to mean a child node in the DAG and all its parents. We do not use the term DAG family for the root nodes with no parents, which are all the other nodes in the DAG.

We use the `as()` function to coerce the AMAT, G, to be represented as an *igraph*:

```
dag <- as(G,"igraph")
```

We use the `igraph::V()` function to get a list of all vertices (nodes) of the `dag`:

```
igraph::V(dag)
## + 10/10 vertices, named, from 9cca248:
##  [1] A1 A3 Y A2 X Z E D A4 F
```

We use the `igraph::E()` function to return a list of all edges in the `dag`:

```
igraph::E(dag)
## + 12/12 edges from 9cca248 (vertex names):
##  [1] A1->Z A3->E Y ->Z Y ->E A2->D X ->Z X ->D Z ->E Z ->D E ->F D ->F A4->F
```

Next, we demonstrate how to use the `gRbase` library functions to get the list of DAG families and how to list the family for a given node in the DAG.

Getting DAG families—nodes and their parents

With the `gRbase::vpar()` function, we efficiently retrieve a comprehensive list of all families in the DAG, i.e., nodes and their parents in the DAG. To identify the family for a specific node, such as "F," we simply use the $ operator to return the node's family from the list:

```
vpardag <- gRbase::vpar(dag)
vpardag$F
## [1] "F"  "E"  "D"  "A4"
```

Next, we demonstrate how to get the ancestral set and the ancestral graph for a given DAG and a subset of nodes. Recall the ancestral set for a given input DAG and the input subset of nodes is the union of the input subset and all the ancestor nodes for every node in the input subset; the ancestral graph is the subgraph of the DAG on the ancestral set.

Ancestral sets and ancestral graphs

The following code demonstrates how to get the ancestral set in the DAG, `dag`, for the subset of the two nodes "D" and "E." We use the `gRbase::ancestralSet()` function to return the ancestral set in the DAG, `dag`, and the two-node subset.

```
ancestralSet(c("D","E"),dag)
## [1] "A1" "A3" "Y"  "A2" "X"  "Z"  "E"  "D"
```

The following code demonstrates how to get the ancestral graph in the DAG, `dag`, for the subset of the two nodes "D" and "E." We use the `gRbase::ancestralGraph()` function to return the ancestral graph in the DAG, `dag`, of the two-node subset, which is the subgraph for the ancestral set of the two-node subset.

```
gRbase::ancestralGraph(c("D","E"),dag)
## IGRAPH d384aa6 DN-- 8 9 --
## + attr: name (v/c), label (v/c)
## + edges from d384aa6 (vertex names):
## [1] A1->Z A3->E Y ->Z Y ->E A2->D X ->Z X ->D Z ->E Z ->D
```

Next, we demonstrate how to use the *igraph* library to check if in a DAG, G, and three disjoint node subsets (X, Z, Y), Z d-separates X and Y.

d-separation

We use `ggm::dSep(G,X,Y,Z)` to determine if two sets of nodes, X and Y, are d-separated by a third set Z of nodes in a directed acyclic graph G. The function takes four arguments. The first argument, G, is a Boolean matrix representing the adjacency matrix of a directed acyclic graph. Note that the function does not check if graph G is acyclic; that is, a DAG. Also, if any of the three sets X, Z, or Y is a singleton set, we can input the individual node rather than a set.

The following code is an example of checking that the set {"Y", "Z"} d-separates the node "E" and the node "X" in the DAG of the circuit example:

```
ggm::dSep(G, "E", "X", c("Y","Z"))
## [1] TRUE
```

With the above definitions on graphs and the practice session using R, we conclude this section.

Now, we have reached the end of this section on the graph representation of Bayesian networks. Let us recap what we have learned. We covered the formal definitions of directed and undirected graphs, along with related concepts such as subgraphs, ancestral sets, ancestral graphs, complete sets, and complete graphs. We will use these definitions to build upon and introduce graph-based algorithms in the following sections. We revisited the connection between directed acyclic graphs and the specification of Bayesian networks. Additionally, we presented an example of a Bayesian network model for a circuit diagnosis application and described its directed acyclic graph. Finally, we examined the critical property of conditional independence in the Bayesian network probability model and the connection with the topological condition of d-separation in the associated directed acyclic graph illustrated by an example on our circuit BN. The last part of this section demonstrated in hands-on programming practice using R how to use graph libraries in R to specify the BN directed acyclic graph, visualize the DAG, perform basic queries on the DAG structure, and verify d-separation between two sets of nodes, conditioning on a separator third set.

In the next section, we describe the step in transforming the DAG of the BN model to an undirected graph called a moral graph, which constitutes the first step initiating the process toward building a join tree representation from the DAG representing the BN. Mapping the DAG to a corresponding moral graph is called moralization, which is the topic of the following section.

Moralization

In this section, we begin by defining what is in a moral graph and then proceed to a detailed, step-by-step description of the moralization process. This process takes the directed acyclic graph of a Bayesian network and produces an associated undirected graph called the moral graph. We illustrate this process with the moralization of our circuit BN example, ensuring that you understand each step of the process.

Next, we revisit the critical topic of d-separation and present a theorem that formalizes the connection between graph separation in the moral graph and d-separation in the directed acyclic graph. This theorem is a cornerstone in understanding the relationship between these two concepts, emphasizing its importance in the field. We then describe a graph-based algorithm called *d_separates* for checking the d-separation based on the theorem and illustrate the working of the algorithm using the DAG of our circuit BN. We follow with a section where we demonstrate the implementation of the *d_separates* algorithm as a function in R and test the function on our circuit BN example. Separation in an undirected graph is a crucial topic pertinent to this chapter, as we will learn in the coming sections the duality between graph decomposition and graph separation; graph decomposition is what tree clustering is all about.

We continue with a section demonstrating how to implement the *d_separates* algorithm programmatically using R. This section walks you through the steps of employing various functions in graph libraries to implement the algorithm as an R function. We then test this function on our circuit BN model, showcasing its practical application in a real-world scenario. We conclude this section with a practice session using R that reinforces the concepts related to the moralization of the BN-directed acyclic graph using our circuit BN, further emphasizing the real-world relevance of these concepts.

What is in a moral graph?

The structure of the moral graph, an undirected graph $G = (V, E)$, is of utmost importance. It is produced by moralizing a corresponding directed acyclic graph $G_1 = (V, E_1)$. For each node y in V, every doubleton subset of nodes in the parents of y must be joined by an edge in the moral graph G. This ensures that every pair of parents in the directed acyclic graph is connected in the moral graph, a crucial aspect that goes beyond the labeling of this operation as marrying the parents or moralizing the directed acyclic graph.

Formally, an undirected graph $G = (V, E)$ is the moral graph for a directed acyclic graph $G_1 = (V, E_1)$ if it satisfies the following two conditions:

- Every pair of nodes x and y sharing a child in the DAG, G_1, corresponds to an undirected edge $(x, y) \in E$ in the moral graph G.
- Every directed edge $(x, y) \in E_1$ in the DAG, G_1, has a corresponding undirected edge $(x, y) \in E$.

Figure 6.3 illustrates the moral graph, which is an undirected graph that reflects the Bayesian Network (BN) model.

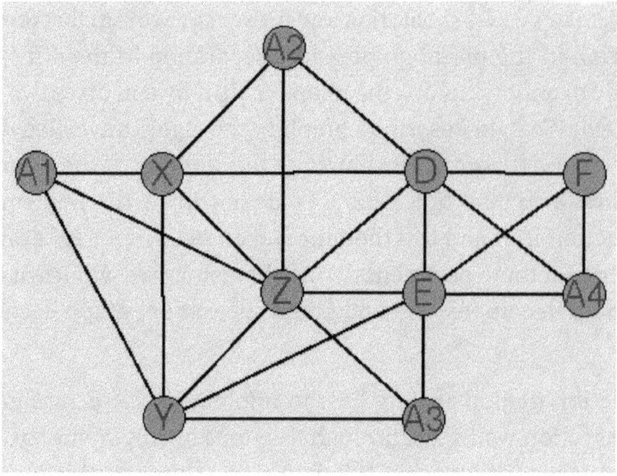

Figure 6.3—Moral graph for the circuit BN model

The figure features additional edges that connect the parents of nodes from the corresponding directed acyclic graph in *Figure 6.2*. Specifically, the moral graph introduces 12 new edges: A1--Y, A1--X, A1--Z, A3--Y, A3--Z, A3--E, A2--X, A2--Z, A2--D, E--A4, D--A4, and A4--F.

The next section introduces a theorem of profound significance in Bayesian network modeling. This theorem, which establishes the link between graph separation in the moral graph and d-separation in the directed acyclic graph, is a cornerstone in understanding the structure of a Bayesian network and the conditional independence in its probability model. Its practical relevance and applicability in Bayesian network modeling are truly invaluable.

Graph separation

The section begins with the definition of graph separation, which sets the foundation for algorithms to verify d-separation. Next, we present a theorem that links the d-separations in the Bayesian network DAG and separator sets in the corresponding moral graph.

Definition: graph separator

Let G be an undirected graph, and let the triplet (X, Z, Y) be pairwise mutually disjoint subsets of nodes in G where $X \cup Z \cup Y = V$ and every path in the subgraph $G(V)$ between every node in X and every node in Y contains a node in Z. Z is said to be a **separator** of $G(V)$ concerning (X, Z, Y).

To illustrate the concept of a graph separator, we consider the moral graph of our circuit BN in *Figure 6.3*. It is easy to verify that the node subset {"D","E","Z"} is a separator of the moral graph concerning ({"A4","F"}, {"D","E","Z"}, {"A2", "X", "A1", "Y", "A3"}).

The following theorem defines the connection between separation in the moral graph and d-separation in the DAG.

Theorem: d-separation

Let X, Y, Z be three pairwise mutually disjoint subsets of nodes in a DAG G. Let anG be the ancestral graph for the union set, $X \cup Y \cup Z$. That is, $anG = G(ancestralSet(X \cup Y \cup Z))$. Let m_anG be the moral graph of the ancestral graph anG. Then, X and Y are d-separated by Z in G if and only if Z separates X and Y in the moral ancestral graph m_anG.

The following algorithm in pseudocode can be described based on the theorem.

Algorithm: d_separates

Input: directed acyclic graph $G = (V, E)$; X, Y, Z are disjoint subsets of V

Output: *True* or *False*.

```
1. Ensure X, Y, and Z are mutually pairwise disjoint.
2. Get the ancestral graph anG in G for the set X ∪ Y ∪ Z
3. Moralize the ancestral graph m_anG = moralize(anG)
4. Remove all nodes in Z from m_anG
5. If X and Y are connected, return False; else, return True
```

The following example illustrates the `d_separates` algorithm for checking d-separation using our circuit BN model.

Example: d_separates

Let us apply the `d_separates` algorithm to the d-separation example we had earlier in our circuit directed acyclic graph. In the example, we check if the node pair "E" and "X" are d-separated by the set {"Y," "Z"}. *Figure 6.4* shows the steps applying the `d_separates` algorithm. The ancestral graph shown on the left is the subgraph of the circuit DAG of *Figure 6.2* for the ancestral set of $\{E, X, Y, Z\}$. The undirected graph in the middle is the moralized ancestral graph. After removing the set $\{Y, Z\}$ from the undirected graph in the middle, we have a disconnected graph on the right where there is no path between "E" and "X". Hence the d-separation query returns True.

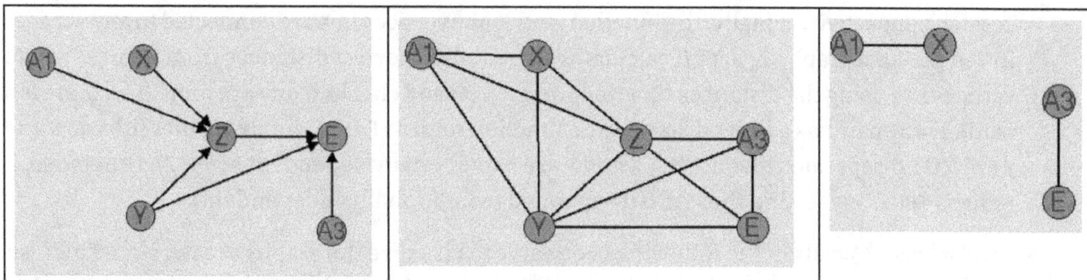

Ancestral graph- G *(ancestralSet({E, X, Y, Z})*	Moral Ancestral Graph	Moral Ancestral Graph with *Y* and *Z* removed

Figure 6.4—Steps of the d_separates algorithm for checking the set {"Y," "Z"} d-separates between the node "E" and the node "X" in the DAG of the circuit example

In the following section, we present a step-by-step guide for implementing the d_separates algorithm. We begin by illustrating how to convert it into an R function. Next, we will demonstrate its application in testing d-separation within a directed acyclic graph (DAG) by examining graph separation in the moral ancestral graph.

D_Separates function implemented using R

In this section, we demonstrate the implementation of the d_separates algorithm described in the previous section. Specifically, we write a function d_separates in R that takes a DAG and three disjoint subsets (X, Z, Y) of nodes in the DAG and returns True or False depending on whether Z d-separates X and Y.

The d_separates function tests conditional independence between disjoint sets of vertices in a given directed acyclic graph (DAG). This test relies on the concept of d-separation in graph theory. The igraph and gRbase libraries are used for the graph data structures and some specific operations on the graph, respectively.

As a first step, the d_separates function verifies whether the input sets X, Y, and Z are disjoint, meaning they share no common elements. This check is pivotal, as the concept of d-separation is applicable only to disjoint subsets of vertices in the graph.

The function then performs the following series of operations:

- Ancestral Graph: This operation generates the ancestral graph anG of the dag restricted to the vertices in X, Y, and Z. The ancestral graph of a set of vertices is the subgraph induced by the vertices and their ancestors (vertices from which there is a directed path).
- **Moralization**: This step turns the ancestral graph anG from the previous step into an undirected moral ancestral graph m_anG by adding an undirected edge between every pair of parents in each family of anG that are not already connected and then removing the direction from all edges.
- Removal of Z vertices: This step creates an undirected graph m2_anG by removing all vertices in set Z from the moral ancestral graph m_anG. The idea is to eliminate the influence of these vertices on the rest of the graph to enable checking if Z is a separator.
- Test for Connectivity: Finally, the function tests if any vertices in X are connected to any vertices in Y in the final graph m2_anG. It calculates all pairwise shortest distances from vertices in X to vertices in Y using the distances function from *igraph* and checks if any are finite (i.e., there is a path). If any pair has a finite distance, the function returns FALSE, implying that subsets X and Y are NOT d-separated given Z (so, X and Y are conditionally dependent given Z). Otherwise, it returns TRUE, suggesting they ARE d-separated (conditionally independent).

The ggm::plotGraph() calls in the following code serve to visualize the graphs at each step. This visualization helps us understand the transformations that occur in the graph throughout the computation.

```
library(igraph) # for the graph data structures
library(gRbase) # for the ancestralGraph and moralize functions

d_separates <- function(dag, X, Y, Z) {
```

```
  if (!is.vector(X)) { X <- c(X) }
  if (!is.vector(Y)) { Y <- c(Y) }
  if (!is.vector(Z)) { Z <- c(Z) }
  # Ensure X, Y, and Z are disjoint
  if (length(intersect(X, Y)) > 0 || length(intersect(Y, Z)) > 0 ||
      length(intersect(X, Z)) > 0)
  {
    stop("X, Y, and Z must be disjoint sets")
  }
  # 1. Get the ancestral graph
  anG <- gRbase::ancestralGraph(union(X, union(Y, Z)), dag)
  ggm::plotGraph(anG)

  # 2. Moralize the ancestral graph
  m_anG <- gRbase::moralize(anG)
  ggm::plotGraph(m_anG)
  # 3. Remove all nodes in Z from m_anG
  m2_anG <- delete_vertices(m_anG, Z)
  ggm::plotGraph(m2_anG)
  # 4. Test if X and Y are connected in m2_anG
  are_connected <- any(sapply(X, function(x) {
    # If the minimum distance
    # from x to any node in Y is gnot infinity, they are connected.
    min(igraph::distances(m2_anG, v = x, to = Y)) != Inf
  }))
  return(! are_connected)
}
```

The following code runs the `d_separates` function for testing d-separation on the example we presented earlier:

```
d_separates(dag, "E", "X", c("Y","Z"))
## [1] TRUE
```

The graphs plotted at each of the three steps in the `d_separates` code are those depicted in *Figure 6.4*.

We now end this section by emphasizing that with the above implementation of the *d_ separates* algorithm as an R function, you have learned how to use various functions in the R graph libraries to:

- Get the ancestral graph for a node subset in a directed acyclic graph
- Moralize a directed acyclic graph
- Check if a given subset of nodes in an undirected graph is a separator set for two other node subsets

Walking you through all the steps of implementing the `d_separates` function empowers you to implement the d-separation testing algorithm with working code in R, reinforcing your understanding and confidence in the topic. Those programming skills are an invaluable stepping stone to implementing structure-based algorithms for specific tasks requiring graph-based processing in Bayesian network applications.

In the upcoming practice section, we will provide you with a hands-on demonstration of how to apply the concepts we've covered. Using R, we will show you how to: obtain the moral graph for our circuit example, represent the moral graph as an undirected graph object in the `gRbase` and `igraph` R libraries, print and plot the moral graph using functions from those graph libraries, and obtain the moral subgraphs for the DAG families.

Practice using R

In this practice session, we will demonstrate how to utilize the R graph libraries to: 1) moralize the directed acyclic graph (DAG) of our circuit Bayesian network (BN) model from the previous R practice session; 2) print and plot the undirected moral graph; and 3) obtain the node set of the DAG families along with their respective moral subgraphs.

Let us begin with the first part of this practice: getting the moral graph for the DAG of the circuit BN.

Moralization of the DAG

We use the `gRbase::moralize(G)` function to create the moral graph for an input DAG, G, specified by an adjacency matrix. We run that function on the DAG of our circuit BN, given by the adjacency matrix `G` we created in the earlier practice. The function will return an adjacency matrix corresponding to the undirected graph representing the moral graph of the DAG of our circuit BN:

```
morG <- gRbase::moralize(G)
print(morG)
##    A1 A3 Y A2 X Z E D A4 F
## A1  0  0 1  0 1 1 0 0  0 0
## A3  0  0 1  0 0 1 1 0  0 0
## Y   1  1 0  0 1 1 1 0  0 0
## A2  0  0 0  0 1 1 0 1  0 0
## X   1  0 1  1 0 1 0 1  0 0
## Z   1  1 1  1 1 0 1 1  0 0
## E   0  1 1  0 0 1 0 1  1 1
## D   0  0 0  1 1 1 1 0  1 1
## A4  0  0 0  0 0 0 1 1  0 1
## F   0  0 0  0 0 0 1 1  1 0
```

The moral graph is an undirected graph where all the parents of every node of the DAG are connected. We leave it as an exercise to verify that fact by comparing the adjacency matrix of the moral graph, `morG`, with the input DAG G, G. As an example, you can see that the parents of the node 'Z' in the DAG, `G`, which are the nodes 'X,' 'Y,' and 'A1' are all connected in the `morG` matrix, i.e., having the value 1 in the `morG` cells for every pair of nodes of the parents of 'Z.' Also, note the symmetry of the adjacency matrix since the graph is undirected.

Let us move on to the second task, which involves printing and plotting the undirected moral graph. This will allow us to view and visualize the structure of the moral graph.

Print and plot the undirected moral graph

To write the nodes and edges of the moral undirected graph `morG`, we need to map the moral graph represented as an adjacency matrix `morG` to a graph object of nodes and edges by coercing the adjacency matrix to a graph object of type `igraph` to become an `igraph` object. Then, we can use the print function to output the list of nodes and edges of the `igraph` object. The first line of the R code below uses the `as(morG, "igraph")` function to coerce the adjacency matrix `morG` to become an `igraph` object, `m_ug`. The second line prints the `igraph` object `m_ug`, which will output the moral graph's list of nodes and edges:

```
m_ug <- as(morG, "igraph")
print(m_ug)
## IGRAPH ba8831c UN-- 10 22 --
## + attr: name (v/c), label (v/c)
## + edges from ba8831c (vertex names):
##  [1] A1--Y  A1--X  A1--Z  A3--Y  A3--Z  A3--E  Y --X  Y --Z  Y --E  A2--X
## [11] A2--Z  A2--D  X --Z  X --D  Z --E  Z --D  E --D  E --A4 E --F  D --A4
## [21] D --F  A4--F
```

To visualize the moral undirected graph, we use the `ggm::plotGraph(morG)` from the `ggm` library, which takes the graph represented by its adjacency matrix as input. The result is the plot shown in *Figure 6.3*.

```
ggm::plotGraph(morG)
```

So far, we have practiced the first two tasks that demonstrate how to programmatically get the moral graph for the specified directed acyclic graph for our circuit BN, output the nodes and edges of the moral undirected graph, and visualize the moral graph.

Now, let's proceed to the third and final task of this tutorial. We will explore the node sets of the directed acyclic graph (DAG) families and their corresponding subgraphs within the moral undirected graph.

Getting the moral subgraphs for the DAG families

As we've seen in our previous exercises, the DAG families, `vpardag`, are crucial. They contain a list of all families in the DAG, i.e., nodes and their parents. This concept is fundamental to our understanding of the graph. Remember, retrieving the family for a specific node, such as 'F,' is as simple as using the $ operator to return the node's family from the list.

In the following code, we find the subgraph of the DAG family of the node "F" in the moral undirected graph. The first two lines in the code below retrieve the list of nodes for the DAG family of the node "F" and print them. The third line of code retrieves the subgraph `ugF` for the DAG family of the node "F" in the moral graph given by the `igraph` object `m_ug`. We then print the subgraph `m_ug`, which lists the nodes that are the family of the node "F" and the set of its undirected edges. Note that all elements of the DAG family are connected in pairs in the moral graph, forming a complete set.

```
vpardag <- gRbase::vpar(dag)
nodes <- vpardag$F
print(nodes)
## [1] "F"  "E"  "D"  "A4"

ugF <- subgraph(m_ug, vpardag$F)
print(ugF)
## IGRAPH 799254c UN-- 4 6 --
## + attr: name (v/c), label (v/c)
## + edges from 799254c (vertex names):
## [1] E --D  E --A4 D --A4 E --F  D --F  A4--F
```

As we wrap up this section on moralization, let's reflect on its practical application. First, we defined moralization as a graph transformation step, where we map the directed acyclic graph to an undirected graph. This is done by adding undirected edges between every pair of unconnected parents sharing a child node in the directed graph and dropping the directionality of all directed edges afterward. Second, we introduced the crucial concepts of graph separation. We demonstrated how d-separation in the DAG representation of the BN translates to graph separation in the moral ancestral graph. We also described a graph algorithm for checking d-separation. Third, in a practical R section, we showcased code and examples to reinforce the concepts learned throughout this section on moralization. We also provided an implementation of the `d_separates` algorithm, further solidifying the practical application of these concepts.

The following section explores the crucial triangulation step, which follows the moralization step and is a vital prerequisite before the final step in the algorithm of join tree clustering.

The following section will explain the triangulation process, which is a complex procedure that transforms an undirected graph into a triangulated graph. In this triangulated graph, every cycle with a length greater than 3 contains a chord—an edge that connects two nonconsecutive nodes within the cycle. By triangulating the moral undirected graph of the Bayesian network, we obtain a graphical representation with several beneficial properties. These properties facilitate the decomposition of the graph into a collection of irreducible subgraphs. Additionally, triangulation allows for a tree-decomposable representation of the Bayesian network, which will be discussed in the final section of this chapter.

Triangulation

Graph triangulation, a crucial intermediate step in tree clustering, is not just a theoretical concept. It has direct and practical applications in data and computer science, especially in the moralization of the directed acyclic graph representing a Bayesian network, which is essential for facilitating efficient inference and reasoning in probabilistic models. This practicality emphasizes the value of understanding and mastering this concept in your work.

Let us overview the content of this section. We begin with a section on what's in a triangulated graph, giving formal definitions of triangulation and triangulated graphs.

Next, we consider ordered graphs, undirected graphs whose nodes are assigned a total ordering. We present definitions in the context of ordered graphs in general and ordered triangulations. The definitions include the width of an ordering, the induced width, simplicial nodes, and perfect ordering.

Next, in a subsection on triangulating ordered graphs, we take a step-by-step approach. We describe an algorithm for triangulating an ordered graph and guide you through implementing it as a function using R. This structured approach ensures you understand each part of the process and feel confident in your learning.

Next, we explain a method for arranging a graph using a maximum cardinality search and a process for creating a maximum cardinality order for an undirected graph. We also demonstrate the implementation of the maximum cardinality ordering as a function using R.

We conclude with a section on practice using R. This is not just about theory but about putting your knowledge into action. You will actively participate in a hands-on session, running coding examples, reinforcing the learned concepts, and making them stick. This interactive approach ensures a deeper understanding and retention of the material.

What is in a triangulated graph?

Triangulating an undirected graph is a complex process that involves adding edges to the graph in a way that it becomes decomposable into one or more complete graphs. A triangulated graph is one in which an edge joins two nonconsecutive vertices for every cycle of length greater than 3. The following is the formal definition originally due to Berge (Berge, 1967).

Definition: triangulated graph

A graph G is triangulated if, for every cycle, a sequence of adjacent nodes beginning and ending at the same node,$\mu = [v_0, \cdots, v_{i-1}, v_i, \cdots, v_n, v_0]$, n > 3 there is an edge of G joining two nonconsecutive vertices of μ; such edges are called **chords** of the cycle. Thus, a graph is triangulated if and only if the graph has no cycles of length greater than three without a chord. If a graph is not triangulated, it can be made by adding edges, so-called **fill-ins**.

It is always possible to add many fill-in edges to triangulate a graph. We do not want to add more fill-ins than are necessary. Here, we define minimal triangulation, which means irredundant fill-ins.

Definition: minimal triangulation

Given an undirected graph $G = (V, E)$, a triangulated graph $G_T(V, E \cup E')$ is created by adding fill-ins E' to the edges E. A minimally triangulated graph is a triangulated graph whose fill-in edges are such that removing any fill-ins makes the graph untriangulated.

Ensuring that the minimal triangulation corresponds to the minimum fill-ins is highly desirable. The following defines minimum triangulation.

Definition: minimum triangulation

Minimum triangulation is the graph triangulation with the smallest fill-ins. The minimum triangulation is unique.

Computing the minimum triangulation, a problem known to be NP-hard (Yannakakis, 1981), is of significant importance in graph theory and computer science. Although finding the minimum triangulation (the smallest number of added edges) is NP-hard, computing a minimal triangulation (a triangulation where removing any added edge makes the graph non-triangulated) can be done efficiently. Minimal triangulation typically begins with specifying a total ordering of the graph. Following the determined order, vertices are processed one by one in reverse, connecting all unconnected lower-indexed neighbors to form a complete set.

It is important to note the distinction between minimum and minimal triangulation of undirected graphs.

A minimal triangulation is a local optimum where no proper subset of edges makes the graph chordal. A minimum triangulation is the global optimum with the fewest total fill edges. While all minimum triangulations are minimal, the converse is not true; there can be exponentially many minimal triangulations. While finding a minimal triangulation can be done in polynomial time, finding the minimum triangulation is NP-hard. A common algorithm for computing minimal triangulation is based on the maximum cardinality ordering computed by the **maximum cardinality search** (MCS), which will be described in subsequent sections.

In tree clustering for inference with Bayesian networks, the moral graph is first triangulated to form a chordal graph, which is then used to construct a join tree, also called a junction tree or clique tree. An important structural parameter is the **treewidth**, which is the maximal clique size minus 1. The efficiency of inference with tree clustering is dictated by the tree width. Selecting an ordering that produces a minimal triangulation with low treewidth is crucial for reducing computation time.

Because the size of the largest clique in a triangulation depends on the ordering, finding the optimal node ordering that minimizes the treewidth is NP-hard.

Minimal triangulations aim to create small separators and preserve sparse graphs, which is essential for decreasing the complexity of inference in Bayesian networks. This topic will be further discussed in *Chapter 7*, which focuses on probabilistic inference using join tree clustering.

To conclude this section, let's summarize the main takeaways. We defined a triangulated graph as one in which every cycle of length greater than 3 contains a chord, which is an edge connecting two nonconsecutive nodes. We also explained that it is possible to triangulate any graph by adding edges known as fill-ins to create a chordal graph. We defined minimal triangulation as a form of triangulation where the removal of any fill-ins renders the graph untriangulated. Additionally, we noted that the problem of finding the minimum triangulation with the fewest fill-ins is NP-hard. Lastly, we mentioned that the minimum triangulation of a graph is not necessarily unique.

Let us move on to the next section on ordered graphs, graphs with complete ordering of their nodes. We will present several definitions related to ordered graphs that we will use next to describe the triangulation of ordered graphs and formalize concepts related to triangulated ordered graphs.

Ordered graphs

In this section, we introduce the concept of ordered graphs and present preliminaries and definitions that we will use to describe the *triangulate ordered graph* algorithm. The concept of ordered graphs is fundamental in graph theory and has practical applications in various fields, such as network design and optimization. Understanding this concept is, therefore, crucial for anyone working in these areas.

Given an undirected graph $G = (V, E)$, we define the ordered graph, a pair (G, d), where d is an ordering, a bijection $V \leftrightarrow \{1,2,..,|V|\}$, assigning an order to each node in V. For example, consider a graph with four nodes, A, B, C, and D. An ordering of this graph could be $\{A = 1, B = 2, C = 3, D = 4\}$, which assigns the numbers 1, 2, 3, and 4 to the nodes A, B, C, and D, respectively.

Assigning a complete ordering on the graph nodes splits the neighbors of every node into two sets: the set of lower neighbors, the nodes that appear earlier in the ordering, and the set of higher neighbors, the nodes that appear later in the ordering. This distinction is crucial in the context of the *triangulate ordered graph* algorithm, as it allows us to determine the sequence in which nodes are processed.

Definition: lower and higher neighbors

Let $G = (V, E)$ be a graph and d be the total ordering of the nodes V of the graph.

For each node $x \in V$, we denote the set of higher neighbors, an ordered set of nodes adjacent to x, and follow x in the ordering d as,

$$adj^{+}(x; G, d) = adj(x) \cap \{y | d(y) > d(x)\} \quad (6.9)$$

Similarly, for each node $x \in V$, we define the set of lower neighbors as an ordered set of nodes adjacent to x and precedes x in the ordering d as:

$$adj^{-}(x; G, d) = adj(x) \cap \{y | d(y) < d(x)\} \quad (6.10)$$

A structural parameter of an ordered triangulated graph is the induced width of its ordering. The induced width of a triangulated graph concerning a given ordering is determined as the maximum node width over all nodes in the ordering. The node width is simply the cardinality of its lower neighbors. We formally define the concepts as follows.

Definition: node width, ordered induced width

The node **width of a node** x in an ordered triangulated graph (G, d) is the cardinality of its adjacent nodes preceding x.

$$w(x; G, d) = |adj^{-}(x; G, d)| \quad (6.11)$$

In an ordered triangulated graph (G, d), the ordered induced width over the ordering d is the maximum width over all nodes.

$$w(G, d) = \max_{x \in V} w(x; G, d) \quad (6.12)$$

The induced width of a triangulated graph is the minimum ordered induced width of the ordered triangulated graph over all orderings. We state this formally in the following definition.

Definition: induced width

The induced width of a graph G is the minimal ordered induced width of the ordered triangulated graph over all graph orderings, d.

$$w^{*} = \min_{d} w(G_T, d) \quad (6.13)$$

We call a node in a graph simplicial if the closure set of the node, the node, and all its neighbors, together form a clique. We state this formally in the following definition.

Definition: simplicial node

Given an undirected graph $G = (V, E)$, we call a node $x \in V$ **simplicial** if the closure of $x, cl(x)$, is a clique. In other words, x and all its adjacent nodes, $adj(x)$, in the graph, form a clique, i.e., if x has a complete neighbor set, and thus every pair of nodes in $cl(x)$ is an edge in E.

Every triangulated graph has an ordering called perfect ordering, **simplicial ordering** or **perfect elimination ordering (PEO)**. We define that ordering next.

Definition: perfect elimination ordering

Given a triangulated graph $G = (V, E)$, we call the ordering d of the nodes of the triangulated graph a **perfect ordering** if the nodes are ordered such that every node $x \in V$ is a simplicial node in the subgraph of G formed from x and all its lower neighbors, $adj^{-}(x, d)$. In other words, a perfect ordering is an ordering that arranges the nodes in such a way that every node is a simplicial node in its 'local' subgraph. This concept is essential in the context of the triangulate graph algorithm, as it allows us to identify an ordering that minimizes the induced width of the graph, thereby potentially improving the algorithm's efficiency.

Alternatively, d is a perfect ordering if, for every x in V, the subgraph $G(cl(x, d))$ is a complete graph. Here, $cl(x, d)$ is the closure of x with its lower neighbors:

$$cl(x, d) = \{x\} \cup adj^{-}(x, d) \quad (6.14)$$

With this, we conclude this section on ordering and ordered graphs. In summary, ordering the nodes on an undirected graph engenders partitions of adjacent nodes to those that fall earlier in the ordering versus those later. The notations we introduced in this section are more than just theoretical; they will play a crucial role in making the definitions of triangulating a graph in the following section more precise and applicable in real-world scenarios.

In the following section, we can formally define the *triangulate ordered graph* algorithm using the above definitions.

Triangulating ordered graphs

This section presents an algorithm for triangulating an ordered undirected graph. The ordering of the graph is not just a detail but a crucial factor that makes the triangulation of the graph a tractable task. Note that the ordering plays a pivotal role in influencing the number of fill-in edges and the resulting induced graph width. We will go over a method for ordering the graph nodes later.

Triangulating an ordered graph follows a logical sequence. We process the nodes in reverse order, which is a key step in the algorithm. This order allows us to handle the nodes with the most connections first, simplifying the process. For each node, we examine its preceding adjacent nodes. If a pair of nodes in the preceding adjacent set is not connected, then we add an edge between the pair so that all the preceding adjacent nodes form a clique (a complete set). These added edges, known as fill-ins, are what transform the input graph into a triangulated graph. Here's a pseudo-code to illustrate this process:

Algorithm: Triangulate Ordered Graph

Input: Ordered graph (G, d); $G = (V, E)$

Output: Triangulated graph $G_T = (V, E \cup E')$

Initialize: $E' \leftarrow \emptyset$

For every node x in V from last to first do:

```
    S_x ← adj⁻(x; G_T, d)
    For y in S_x:
        For z in S_x:
            If d(z) > d(y) and {z,y} ∉ E ∪ E' :
                E' ← E' ∪ {z,y}
                G_T ← (V, E ∪ E')
Return G_T
```

The following theorem states the computational complexity of the *triangulate ordered graph* algorithm.

Theorem: minimal triangulation

Given a graph $G = (V, E)$ and an orderingd of the nodes in V. The algorithm *triangulate ordered graph* computes minimal triangulation in $O(n\, w^2)$ time, with $n = |V|$ and w being the induced width of the ordering.

Proof

The width w of the ordering determines the size of the cliques. Because each node adds at most $O(w^2)$ edges (connecting neighbors) and there are n nodes, the complexity is $O(n\,w^2)$.

Now, let's bring the theory to life in a practical setting. The following section provides the R code for implementing the *triangulate ordered graph* algorithm, demonstrating its real-world application. This practical demonstration will help you see the direct relevance and usefulness of the algorithm in your work as software developers, data scientists, and researchers.

Triangulate function implemented using R

The following function takes an undirected graph and an ordering of graph nodes as input. It outputs a triangulated graph constructed by adding fill-in edges to the input graph to make it chordal. A graph is chordal (or triangulated) if each of its cycles of four or more nodes has a chord, an edge joining two nodes that are not adjacent in the cycle. An equivalent definition is that any chordless cycle has at most three nodes.

The code below implements the *triangulate ordered graph* algorithm as an R function named `triangulate`. Please note that this function requires the `igraph` package in R.

This function takes an undirected graph and an ordering of the graph's nodes as input and returns a minimally triangulated graph as output. Here is an overview of the steps of what the function does.

The `triangulate` function takes in two arguments:

- `ug`: An undirected graph, which is checked and stopped if it's a directed graph. The function uses `igraph's` `is_directed` function to perform the check.
- `Ordering`: An ordering of the graph's nodes.

The function code block begins with an initialization step.

Initialization: It then makes an exact copy of the input graph `ug`, saving it in the variable `ug_triangulated`. It also reverses the ordering of the graph's nodes using the `rev` function, saving the result in `r_ordering`.

After the initialization step, the function code implements its main logic, which executes an iteration within a for loop.

Main Logic: The function loops through each node in the reversed ordering. On each iteration:

1. It checks if the current node (`vi_name`) is in the list of names of vertices in the triangulated graph. If it is:
2. It identifies the nodes before the current node in the ordering (`nodes_before_vi`)
3. It identifies the neighbors of the current node in the original graph (`all_neighbors_vi`) and finds the intersection (`neighbors_of_vi`) with `nodes_before_vi`
4. It then explores each pair of neighbors in `neighbors_of_vi`, checking whether an edge exists between them in the triangulated graph
5. If no such edge exists and the vertices are unique, it adds this edge to the triangulated graph

Final Output: Once it finishes iterating over all nodes, the function returns the resultant graph `ug_triangulated`, now a triangulated version of the original undirected input graph ug.

Here is the entire code of the function `triangulate`:

```
triangulate <- function(ug, ordering) {
  if(igraph::is_directed(ug)) stop("The graph must be undirected.")

  ug_triangulated <- ug
  r_ordering <- rev(ordering)

  for(i in seq_along(r_ordering)) {
    vi_name <- r_ordering[i]
    print(paste("Processing node: ", vi_name))

    if (vi_name %in% V(ug_triangulated)$name) {
      nodes_before_vi <- r_ordering[seq(i+1, length(r_ordering))]
      # Fetch neighbors from the original graph 'ug'
      all_neighbors_vi <- neighbors(ug, vi_name)$name
      neighbors_of_vi <- intersect(all_neighbors_vi, nodes_before_vi)

      print(paste("Neighbors of ", vi_name, " : ",
            paste0(neighbors_of_vi, collapse = ", ")))

      for(j in seq_along(neighbors_of_vi)) {
        for(k in seq(j+1, length(neighbors_of_vi)))  {
          nodes = c(neighbors_of_vi[j], neighbors_of_vi[k])
          nodes = nodes[order(match(nodes, r_ordering))]

          node1 <- nodes[1]
          node2 <- nodes[2]

          if(node1 %in% V(ug_triangulated)$name & node2 %in% V(
            ug_triangulated)$name & node1 != node2
          ){
            node1_vertex <- V(ug_triangulated)[node1]
            node2_vertex <- V(ug_triangulated)[node2]
            if(length(E(ug_triangulated)[node1_vertex %--% node2_vertex])== 0) {
              print(paste("Adding edge between", node1, "and", node2))
              ug_triangulated <- igraph::add_edges(ug_triangulated,
                                                   c(node1, node2))
            }
```

```
                }
            }
        }
    }
}

    return(ug_triangulated)
}
```

As the triangulate R function described above requires total ordering of the nodes of the input undirected graph, we need to emphasize the crucial role of a method for generating an ordering based on a heuristic. One such heuristic is based on maximum cardinality search (MCS), which we describe in the following section.

Maximum cardinality ordering

The following describes an algorithm known as maximum cardinality search (MCS), which is used for ordering the nodes of an undirected graph.

The MCS algorithm is a linear time algorithm that orders the nodes of an undirected graph in n steps, where n is the number of the graph nodes. The algorithm orders the nodes from 1 to n by visiting the graph nodes in such a way that the node with the most already visited neighbors is selected for visitation each time, with ties being broken arbitrarily. The algorithm is a basis for determining whether a graph is chordal.

The following code provides a pseudocode description of the maximum cardinality search algorithm, which is used to compute a maximum cardinality ordering:

Algorithm: maximum cardinality search

Input: undirected graph $G = (V, E)$

Output: maximum cardinality ordering, d.

Initialize: weight function $w(v) = 0 \ \forall v \in V$; unordered nodes $U \leftarrow V$.

For $i = 1$ to $|V|$:

$x \leftarrow \underset{v}{\text{argmax}}\, w(v)$

$d(x) \leftarrow i$

$U \leftarrow U - \{x\}$

For $y \in adj(x; G) \wedge y \in U$

$w(y) = w(y) + 1$

Return d.

An important theorem about the MCS algorithm is that its ordering is a perfect elimination ordering if the graph is chordal, i.e., it is triangulated.

It is a theorem (Carneiro, 2018) that the MCS algorithm produces the *perfect ordering* (*simplicial ordering*) if and only if the input graph is triangulated. Therefore, the MCS algorithm can be used to check if the input graph is a triangulated graph.

The MCS algorithm is highly efficient, requiring time and space linear in the graph size. We state this in the following theorem.

Theorem: maximum cardinality search

Given a graph $G = (V, E)$, the complexity of the MCS algorithm is $O(n + m)$ both in time and space, with $n = |V|$ and $m = |E|$.

Proof

The MCS algorithm processes each node once and traverses each edge at most twice, once for each node pair connected by an edge, making its time complexity $O(|V| + |E|)$. The algorithm has $O(|V| + E|)$ space complexity to store the graph's adjacency matrix and the auxiliary data structure for managing the weights.

The following section presents an implementation of the MCS ordering algorithm in R. We will utilize the computed MCS ordering to triangulate the moral graph, employing the `triangulate` R function described previously, which we developed in R.

Max_Cardinality_Ordering function implemented using R

In this section, we describe an implementation of the MCS algorithm as a function in R named `max_cardinality_ordering`. The function takes an undirected graph and an optional first node in the ordering. It returns a total order based on the maximum cardinality search (MCS), which visits the vertices in such an order that every time the vertex with the most already-visited neighbors is visited. Ties are broken arbitrarily.

Here is a walkthrough of the `max_cardinality_ordering` function, which performs a maximum cardinality search on an undirected graph:

Initial checks

The function begins by ensuring that the input graph ug is undirected using the `igraph::is_directed` function. It will stop with an error message if the input graph is directed.

Adjacency list creation

Next, the adjacency list `adj_list` of the undirected graph ug is created using the `as_adj_list` function of the `igraph` package. An adjacency list represents a graph as a list of edges for each vertex, providing quick and easy access to all vertices connected to a given vertex.

Initialize node variables

The function then initializes various variables, including:

- The total number of nodes `n` in the graph, calculated using the `vcount` function of the `igraph`

- An empty character vector order_vec to hold the resulting order of nodes
- A weights variable initialized to a zero vector

Apply first node

If a starting node x is supplied and exists in the graph, it is added as the first node in the order_vec, and its weight is set to NA so that it won't be considered in the upcoming rounds.

Calculating weights and reordering

The function then begins a for loop that iterates from the second node to the nth node in the ordering. For each iteration:

- It gets the node selected in the last round and increases the weight of its neighbors in the adjacency list by 1.
- It then identifies the nodes not yet added to the order_vec (remaining nodes).
- From these remaining nodes, it selects the node with maximum weight and adds it to the order_vec. The weight of this added node is then set to NA, so it won't be considered for the upcoming selections.

Return result

Finally, the function returns order_vec as a named vector, with names as "Node_" followed by their order in the final ordering.

This function performs a maximum cardinality search, aiming to minimize the fill-in when using the ordering for triangulating the graph. This fill-in corresponds to the edges that must be added to the graph (transforming it into a chordal graph).

The following is a listing of the entire code for the max_cardinality_ordering function. Please note that this function requires the igraph package in R. The implementation handles ties when multiple unordered nodes have the same maximum weight by selecting the first one. In the following code, the R function which.max() is used to return the index of the vector's first occurrence of the maximum value in a vector.

```
max_cardinality_ordering <- function(ug, x = NULL) {
  # Ensure the graph is undirected
  if(igraph::is_directed(ug)) stop("The input graph must be undirected.")

  # Construct adjacency list
  adj_list <- as_adj_list(ug)

  # Number of nodes, and create an empty named vector to hold the result
  n <- vcount(ug)
  order_vec <- character(n)

  # Initialize node weights to 0 and set the names as node names
```

```
  weights <- rep(0, n)
  names(weights) <- names(adj_list)

  # If a starting node is provided, and exists in the graph
  if(!is.null(x) && x %in% names(adj_list)) {
    order_vec[1] <- x   # Set the starting node as the first node
    weights[x] <- NA   # So it won't be selected next time
  }

  # For each node from 2 to n
  for(i in 2:n) {
    x <- order_vec[i-1]  # The node selected in the last round
    # Increase the weight of its neighbors
    if (!is.na(x)) weights[adj_list[[x]]] <- weights[adj_list[[x]]] + 1

    # From those nodes not already in the order,
    # select a node with maximum weight and add it to the order
    remaining_nodes <- setdiff(names(weights), order_vec)
    max_weight_node <- remaining_nodes[which.max(weights[remaining_nodes])]
    order_vec[i] <- max_weight_node
    weights[max_weight_node] <- NA   # So it won't be selected next time
  }

  # Return the final dictionary as a named vector
  return(setNames(order_vec, paste0("Node_", 1:n)))
}
```

With this description of the maximum cardinality ordering, we have the toolbox needed to triangulate the moral graph of the Bayesian network. We can proceed to the next practice section using R, where we will learn how to use all the functions we developed and apply them to the running example of our combinatorial circuit BN.

Practice using R

In this hands-on practice, we will build upon the knowledge and skills we gained in the previous practice. We will continue working with our circuit BN example, validating and experimenting with the implemented R functions for ordering and triangulating an undirected graph. Here's a quick overview of what we'll be doing. First, we'll calculate a maximum cardinality ordering for the moral undirected graph `m_ug` from the previous R practice. We'll use our implemented function `max_cardinality_ordering` to compute an ordering of the graph `m_ug` that starts with the node "A1." Then, we'll use the ordering computed in the previous step to triangulate the graph `m_ug` using our implemented function `triangulate`, which will return a triangulated graph `Tmg`.

Generating maximum cardinality ordering of the moral graph

The following generates a maximum cardinality ordering of the nodes of the moral graph that begins with the node "A1":

```
ordering <- max_cardinality_ordering(m_ug,"A1")
print(ordering)
## Node_1 Node_2 Node_3 Node_4 Node_5 Node_6 Node_7 Node_8 Node_9
Node_10
##    "A1"    "Y"    "X"    "Z"    "A3"    "E"    "D"    "A2"    "A4"
"F"
```

Figure 6.5 shows the moral graph of our circuit example with the MCS ordering of the graph nodes.

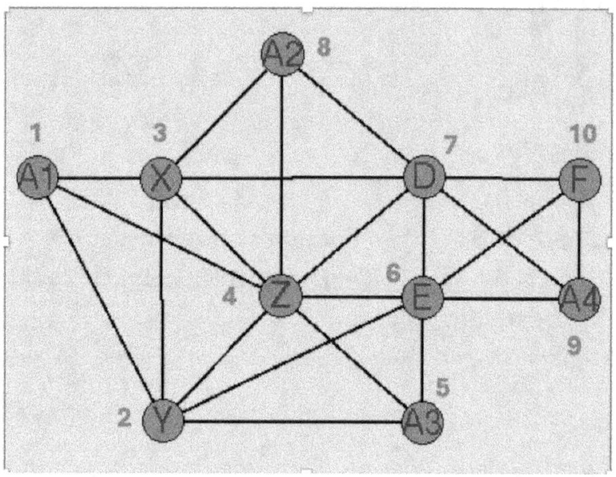

Figure 6.5—Moral graph with the MCS ordering computed by the max_cardinality_ordering function using R

Figure 6.5 adds labels beside every node, reflecting the total ordering output of the `max_cardinality_ordering` function. The ordering of the ten nodes from 1 to 10 is as follows: A1, Y, X, Z, A3, E, D, A2, A4, F. It is instructive to verify that the ordering is a maximum cardinality ordering. For example, the ordering starts with A1. The weights of its unordered neighbors, Y, X and Z, are incremented by 1 from the initial zero. Y, X, and Z all have a max weight of 1, so Y is chosen second. The weights of Y's unordered neighbors are incremented by 1, resulting in the nodes X, Z, E, and A3 having weights of 2, 2, 1, and 1, respectively; therefore, since X and Z have the maximum weights, X is selected as the third node in the ordering. You now follow the logic. The point of this explanation is for you to understand how the maximum cardinality search (MCS) works and also see that there can be multiple maximum cardinality orderings for the nodes of the same graph.

This concludes the section on maximum cardinality ordering, and we are now ready to proceed with triangulating the moral graph. This process takes as input the ordered moral graph based on the maximum cardinality ordering that we have computed.

Triangulating the ordered moral graph

The `triangulate` function, which implements the *triangulate ordered graph* algorithm we discussed earlier, is used in the following code. We apply this function to a practical example—our circuit. By inputting the moral undirected graph and the MCS ordering of the graph, we obtain the triangulated moral graph as output. The function `triangulate` is designed to print messages that detail the execution of each algorithm step. This feature is especially beneficial because it enables you to track the execution and understand the logic behind the triangulation algorithm.

```
TmG <- triangulate(m_ug, ordering)
## [1] "Processing node:  F"
## [1] "Neighbors of  F  :  E, D, A4"
## [1] "Processing node:  A4"
## [1] "Neighbors of  A4  :  E, D"
## [1] "Processing node:  A2"
## [1] "Neighbors of  A2  :  X, Z, D"
## [1] "Processing node:  D"
## [1] "Neighbors of  D  :  X, Z, E"
## [1] "Adding edge between E and X"
## [1] "Processing node:  E"
## [1] "Neighbors of  E  :  A3, Y, Z"
## [1] "Processing node:  A3"
## [1] "Neighbors of  A3  :  Y, Z"
## [1] "Processing node:  Z"
## [1] "Neighbors of  Z  :  A1, Y, X"
## [1] "Processing node:  X"
## [1] "Neighbors of  X  :  A1, Y"
## [1] "Processing node:  Y"
## [1] "Neighbors of  Y  :  A1"
## [1] "Processing node:  A1"
## [1] "Neighbors of  A1  :  "
```

Let's highlight the workings of the triangulation algorithm based on the printed output. The algorithm proceeds in reverse ordering, processing each node in sequence. When processing a node, it finds the adjacent nodes that precede it in the ordering and checks if they are mutually pairwise connected, i.e., forming a complete set, and if not, it fills in edges between pairs of nodes that are not connected. To follow along, node F has its earlier neighbors already forming a complete set; no fill-in edges are added. Next, nodes A4 and then A2 also have their earlier neighbors as complete sets. Next, for node D, its earlier neighbors X, Z, and E do not form a complete set; therefore, a fill-in edge is added to connect X and E. The rest of the process continues with no further fill-ins. A visualization of the triangulated graph in the next section will deepen the understanding of how the triangulation algorithm works.

The following sections will use various utility functions to print, view, and analyze the computed triangulated graph. The following practice is designed to cement and reinforce the graph concepts presented earlier and to show how to leverage existing functions in R graph libraries in `gRbase` and `igraph`.

Printing the triangulated graph's nodes and edges

The following prints the triangulated graph by listing the vertex names and their incident edges:

```
print(TmG)
## IGRAPH 5a84492 UN-- 10 23 --
## + attr: name (v/c), label (v/c)
## + edges from 5a84492 (vertex names):
##  [1] A1--Y  A1--X  A1--Z  A3--Y  A3--Z  A3--E  Y --X  Y --Z  Y --E  A2--X
## [11] A2--Z  A2--D  X --Z  X --D  Z --E  Z --D  E --D  E --A4 E --F  D --A4
## [21] D --F  A4--F  X --E
```

Verifying the computed graph is triangulated and chordal

The following code verifies that the triangulated moral graph is triangulated:

```
gRbase::is.triangulated(TmG)
## [1] TRUE
```

To confirm that the triangulated graph is chordal, we convert the triangulated moral graph to the `igraph` type and then use the `is_chordal` function:

```
TG <- as(TmG, "igraph")
igraph::is_chordal(TG)
## $chordal
## [1] TRUE
##
## $fillin
## NULL
##
## $newgraph
## NULL
```

Plotting the triangulated graph

The following line of code plots the triangulated moral graph of our circuit example, depicted in *Figure 6.6*:

```
ggm::plotGraph(TmG)
```

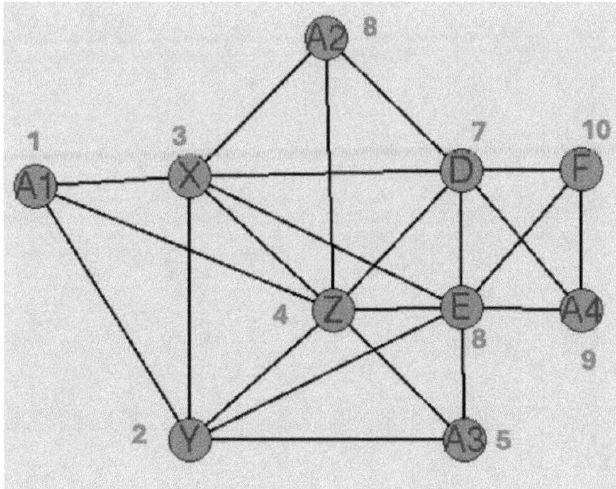

Figure 6.6—Triangulated moral graph of ordering computed by the triangulate function using R; comparing with the moral graph we note the added edge between the node pair {E, X}

Figure 6.6 includes labels next to each node in the graph that indicate the node rank in the MCS (Maximum Cardinality Search) ordering: A1, Y, X, Z, A3, E, D, A2, A4, F. The labels are added manually to the output of the plot from the `ggm::plotGraph` in the previous step. This is done to explain the induced width of the triangulated graph.

The following table lists the nodes in this order, along with their corresponding width, which counts the number of adjacent nodes that appear earlier in the ordering. The induced width for the triangulated graph is 4, representing the maximum width across all nodes.

Ordering	A1	Y	X	Z	A3	E	D	A2	A4	F
Width	0	1	2	3	2	4	3	3	2	3

Table 6.1: MCS ordering and node widths showing induced width of the triangulated graph

Generating a perfect ordering of the triangulated graph

Let us compute a perfect elimination ordering (simplicial ordering) of the nodes of the triangulated graph using a maximum cardinality search, a practical application of graph theory. This will allow us to compare it to the ordering we generated earlier using our implemented `max_cardinality_ordering` function on the moral graph prior to its triangulation. To generate the perfect ordering for the triangulated moral graph, which is a chordal graph, we use the `gRbase::mcsMAT()` function, which takes the adjacency matrix of an undirected graph as input and returns a perfect order of the nodes if they exist. The function checks for the decomposability of the input undirected graph.

```
# convert to an adjacency (association) matrix
matTG <- as_adjacency_matrix(TG, type = "both", attr = NULL, edges = FALSE,
names = TRUE, sparse = FALSE)
print(matTG)
##    A1 A3 Y A2 X Z E D A4 F
```

```
## A1  0  0 1  0 1 1 0 0  0 0
## A3  0  0 1  0 0 1 1 0  0 0
## Y   1  1 0  0 1 1 1 0  0 0
## A2  0  0 0  0 1 1 0 1  0 0
## X   1  0 1  1 0 1 1 1  0 0
## Z   1  1 1  1 1 0 1 1  0 0
## E   0  1 1  0 1 1 0 1  1 1
## D   0  0 0  1 1 1 1 0  1 1
## A4  0  0 0  0 0 0 1 1  0 1
## F   0  0 0  0 0 0 1 1  1 0

gRbase::mcsMAT(matTG)
##  [1] "A1" "Y"  "X"  "Z"  "E"  "A3" "D"  "A2" "A4" "F"
```

Let's compare the MCS (Maximum Cardinality Search) ordering of the moral graph both before and after triangulation. The following table has two rows; the first row provides the MCS ordering before triangulation and the second after triangulation. The maximum cardinality ordering of the triangulated graph is noted as 'perfect' as it corresponds to the perfect elimination ordering.

MCS	A1	Y	X	Z	A3	E	D	A2	A4	F
Perfect	A1	Y	X	Z	E	A3	D	A2	A4	F

Table 6.2: MCS vs. perfect elimination ordering after triangulation

We observe that the two orderings in the table differ only in the ranking of the two nodes, A3 and E. Note that the ordering of the triangulated moral graph differs from that of the moral graph due to the fill-in edge (E, X), which makes the moral graph chordal. Because of that added edge, the MCS ordering on the triangulated graph would rank the node E before A3.

Finding the simplicial nodes of the triangulated graph

Let us find the simplicial nodes of the triangulated graph. Recall that a node in an undirected graph is simplicial if its boundary is complete.

```
gRbase::simplicialNodes(TG)
## [1] "A1" "A3" "A2" "A4" "F"
```

It is easy to verify that the five nodes—A1, A3, A2, A4, and F—are simplicial nodes. In an undirected graph, a node is defined as simplicial if its adjacent nodes (neighbors) form a complete set, meaning that all nodes in this set are mutually pairwise connected. For instance, the adjacent set of A1 in *Figure 6.6* is {X, Y, Z}, which is clearly a complete set. Additionally, it is straightforward to confirm that only these five nodes in the graph are simplicial. In contrast, Y is not simplicial because its neighbors, {A1, X, Z, E, A3}, do not form a complete set.

Verifying the neighbors of a simplicial node is a complete set

The following code verifies that the adjacent nodes to a simplicial node are a complete set by verifying that their subgraph in the triangulated graph is complete:

```
igraph::neighbors(TG, "A1")
## + 3/10 vertices, named, from 5a84492:
## [1] Y X Z
gRbase::is.complete(subgraph( TG, unlist(neighbors(TG, "A1"))))
## [1] TRUE
```

With this, we conclude our R practice.

Concluding this section on triangulation, let's review the key points. Triangulation can be achieved by adding fill-in edges, ensuring that every cycle of length greater than 3 has a chord—an edge that connects two nonconsecutive nodes within the cycle. We implemented the *triangulate ordered graph* algorithm in R, which processes the nodes from last to first and connects adjacent nodes earlier in the order if they are not already connected. In every triangulated (chordal) graph, there exists a simplicial node, where its adjacent nodes (neighbors) form a clique. Additionally, a triangulated graph possesses a perfect elimination ordering (also known as simplicial ordering), wherein each node is simplicial relative to the subgraph formed by the preceding nodes in the order.

In the next section, we will demonstrate that a triangulated graph can be decomposed into maximal prime subgraphs with pairwise connected nodes constituting maximal cliques that can be organized in a join tree subject to the running intersection property.

Computing join tree representations

This section addresses the crucial final step of tree clustering, which involves computing a join tree representation of the Bayesian network. This representation plays a pivotal role in our understanding and application of Bayesian networks.

The aim of tree clustering is to decompose the inference problem for Bayesian network tasks into smaller, manageable problems on subnetworks. This method allows us to address the inference problem locally for each subnetwork and subsequently propagate the local solutions through the join tree, ultimately constructing a global solution. This process enhances the efficiency of computations.

The section is organized as follows. First, we state the general problem of graph decomposition and provide formal definitions to lay the foundation for subsequent sections. Next, we describe the concept of maximal prime subgraph decomposition, a natural step forward from graph decomposition. We follow with a section on finding maximal cliques in an ordered triangulated graph and discuss its connection with the maximal prime subgraph decomposition. We then move on to the section on computing a join tree, which is the primary goal of this chapter. By the end of this section, you will gain a deeper understanding of how to achieve a tree-decomposable network representation of a Bayesian network. To make these concepts more concrete, we conclude with a hands-on practice section where we apply and experiment with the learned concepts to our running circuit BN example using R.

Now that we've covered the overall map of this section, let's transition to the first subsection, graph decomposition. This will set the foundation for understanding the topic of decomposable structure and equip you with the knowledge to confidently navigate the world of Bayesian networks.

Graph decomposition

This section walks you through the fundamental concepts that set the foundation for understanding the concept of graph decomposition.

We begin by formally defining what makes a graph decomposable or prime. We follow with an explanation of graph decomposition and the definition of separator sets, with particular reference to the decomposability of chordal graphs.

What exactly makes a graph either decomposable or prime?

This section focuses on determining whether a given undirected graph is decomposable or prime, which is particularly significant in graph theory and computer science. In the context of Bayesian networks, graph decomposability enables the topology of the graphical model to be exploited to develop structure-based algorithms. These algorithms come with computational complexity guarantees, evaluating the worst-case complexity in time and space in terms of the network structural parameters, as is the case of the tree clustering method.

Graph decomposition is a concept that mirrors the graph separator we discussed earlier. A graph is decomposable if its nodes are partitioned into three sets (X, Z, Y). One of these sets, Z, the separator, is a complete set where every pair of nodes is connected, and every node in the separator is on a path between a node in X and a node in Y.

The following formally defines decomposable and prime graphs.

Definition: decomposable graph, prime graph

An undirected graph $G = (V, E)$ is decomposable if there exists a triple set (X, Z, Y) of pairwise disjoint subsets forming a partition of V, where $X \cup Z \cup Y = V$, such that:

- Z is a complete set, that is, a set of nodes, all of which are pairwise adjacent in G.
- Z separates X and Y, meaning every path between a node in X and a node in Y contains a node in Z.
- Z is called a **separator set**. If G is chordal and not a complete graph, it must possess at least one such separator.
- Under the above conditions, a graph G can be decomposed into $G(X \cup Z)$ and $G(Z \cup Y)$, meaning G is equivalent to the union of the two subgraphs. The decomposition is called **proper** if $X \neq \emptyset$ and $Y \neq \emptyset$. A graph is **prime** if no proper decomposition exists, meaning the graph is irreducible. A prime graph is a complete graph. *Figure 6.7* depicts a decomposition (X, Z, Y) of a reducible graph, G. Here, X represents the node set for the subgraph on the left, Y represents the node set for the subgraph on the right, and Z is the node set for the subgraph in the middle. Z is a complete set constituting a clique separator.

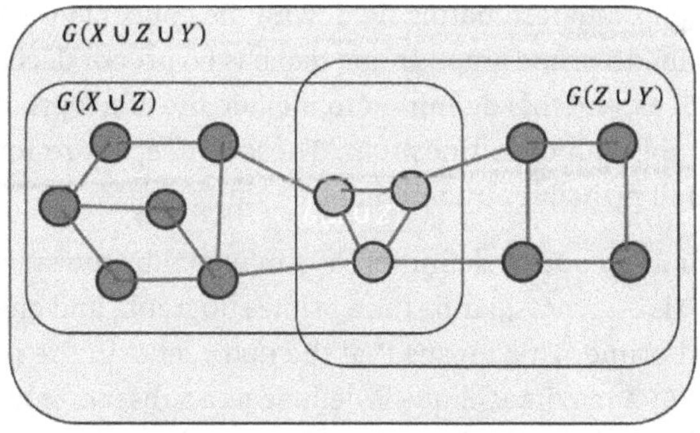

Figure 6.7—Graph G is decomposable to two subgraphs **G(X U Z)** *and* **G(Z U Y)**

Based on the work of Dirac in 1961, chordal graphs are **recursively decomposable.** A proper chordal graph $G = (V, E)$ with a separator Z is decomposable into smaller chordal subgraphs:

- $G_1 = G(X \cup Z)$
- $G_2 = G(Z \cup Y)$

where (X, Z, Y) is a three-way partition of V. The smaller subgraphs G_1 and G_2, which are chordal, can undergo further decomposition into even smaller chordal subgraphs. This leads to a recursive decomposition process. The recursion continues for G_1 and G_2 until a subgraph becomes a complete graph (a clique), which cannot be further partitioned by a separator. This recursive decomposition is illustrated by a tree-decomposable structure known as a join tree or clique tree, where the tree's nodes represent maximal cliques and the edges signify the clique separators. The process of computing a **join tree** will be described in the following sections.

Let's wrap up and summarize the takeaway points. The significance of decomposable graphs within the framework of a Bayesian network cannot be overstated. By breaking down the Bayesian network into smaller subgraphs, we can draw inferences about it more effectively. This approach allows us to simplify the inference problem of the entire network by successively decomposing it into smaller subproblems, thereby improving the efficiency of our computations.

Having defined the general concepts of graph decomposability, we will now move on to the next section, where we will explore a more refined definition of decomposability. In particular, you will learn about maximal prime subgraphs—these are prime subgraphs that cannot be expanded by adding more nodes—as well as the concept of p-decomposition, which refers to the method of decomposing a graph into a collection of maximal prime subgraphs.

Maximal prime subgraph decomposition

Be ready to explore the fascinating world of graph decomposition in greater depth. We are about to build on the concepts discussed in the previous section and examine the crucial concept of maximal prime subgraph decomposition. This concept is essential and closely related to the computation of a join tree, which we will discuss in the upcoming sections.

Graph theory is a complex and fascinating field, with the concepts we are about to explore being no exception. We have defined a prime graph as one that has no proper decomposition or lacks a complete separator. Now, we will extend this definition to include the concepts of a maximal prime subgraph and maximal prime subgraph decomposition. These concepts are foundational and essential for understanding the graph principles underlying tree clustering.

Let's begin with a clear and concise definition of a maximal prime subgraph. A maximal prime subgraph, denoted as $G(U)$ is one that qualifies as a prime subgraph, and any subgraph $G(X)$ that expands $G(U)$ is not considered prime. This means that the node set U in the prime subgraph $G(U)$ forms a maximal clique within G. A maximal clique is defined as a subset of nodes that cannot be enlarged by adding more nodes while still preserving the properties of a clique.

The following is a formal definition of a maximal prime subgraph.

Definition: mp-subgraph

Let $G = (V, E)$ be an undirected graph and let $G(U)$ be a subgraph of G on the subset of nodes $U \subset V$. A subgraph $G(U)$ is a **maximal prime (mp-) subgraph** of G if $G(U)$ is prime and $G(X)$ is reducible for all X with $U \subset X \subseteq V$, that is, the node subset X contains a clique separator for $G(X)$.

As an example, the subgraph $G(\{A1, X, Y, Z\})$ is an mp-subgraph in our circuit example triangulated graph depicted in *Figure 6.6*.

The following defines the p-decomposition and p-separator for a reducible graph.

Definition: p-decomposition, p-separator

Given a reducible graph G, we call the triple (X, Z, Y) a **p-decomposition** and Z a **p-separator** if the triple (X, Z, Y) is a proper decomposition of G into the subgraphs $G' = G(X \cup Z)$ and $G'' = G(Z \cup Y)$ such that the mp-subgraphs of G' and G'' are pairwise different and are all mp-subgraphs of G.

For our circuit example triangulated graph in *Figure 6.6*, the set $Z = \{D, E\}$is a p-separator and the triple (X, Z, Y) where $X = \{A4, F\}$ and $Y = \{ A1, A2, A3, X, Y, Z\}$ that decompose the graph into the subgraphs $G' = G(\{A4, D, E, F\})$ and $G'' = G(\{A1, A2, A3, D, E, X, Y, Z\})$.

The following defines the maximal prime subgraph decomposition of a reducible graph, which underlies the representation in join trees, which will be described in the following sections. A graph's maximal prime subgraph decomposition is a decomposition of the graph into the smallest subgraphs into which the graph can be decomposed.

Here is an operational definition of the mp-subgraph decomposition.

Definition: maximal prime subgraph decomposition

A maximal prime subgraph decomposition of a reducible graph G is the successive p-decompositions of G that derives all the mp-subgraphs of G. The successive p-decompositions is a binary tree, a practical, and visual representation, whose root node is a p-decomposition given by a triple (X, Z, Y) of G into G' = $G(X \cup Z)$ and G" = $G(Z \cup Y)$) are decomposed further until all subgraphs are the mp-subgraphs of G which constitute the leaves of the binary tree.

Figure 6.8 depicts the binary tree that represents the successive p-decompositions of the triangulated moral graph, G, based on the circuit example shown in *Figure 6.6*. The mp-subgraph decomposition concludes with six complete graphs, labeled as G_i for $i = 1, \cdots, 6$. The node sets of these graphs correspond to the maximal cliques of the triangulated graph, which will form the nodes of a join tree, as will be detailed in the next section.

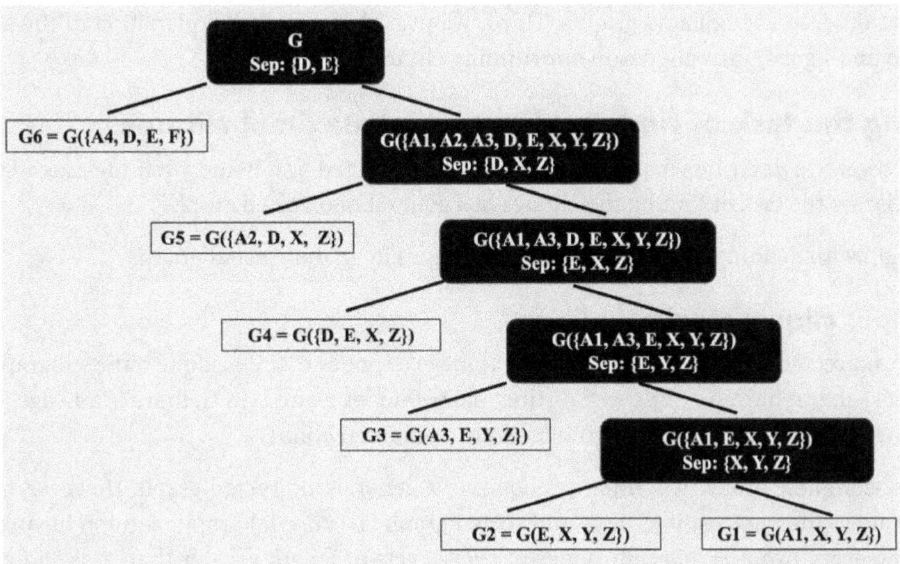

Figure 6.8—Successive p-decompositions of the triangulated graph of the circuit BN as a binary tree; nodes show the decomposable subgraphs and the p-separators, and the leaves are the mp-subgraphs

Let us recap what we covered in this section. We defined the mp-subgraph as a subgraph that is prime, and all subgraphs on supersets of the maximum prime subgraph nodes are not prime. We defined p-decomposition as a decomposition (X, Z, Y) of an undirected graph G into two subgraphs $G(X \cup Z)$ and $G(Z \cup Y)$ such that the mp-subgraphs of the two subgraphs form a partition of the mp-subgraphs of G. The separator Z of a p-decomposition is called a p-separator. A maximum prime subgraph decomposition of a graph G is a successive p-decomposition that bottoms out with all the mp-subgraphs of G.

The following section will explore the problem of finding the maximal cliques in a triangulated graph. This task is a theoretical exercise and a crucial step for computing a join tree. The next section will elucidate the connection between the maximal cliques in the triangulated graph and the maximal prime subgraph decomposition of the triangulated graph. Understanding this connection is key to applying these concepts in real-world scenarios.

Finding the maximal cliques

In a previous section, we learned that a triangulated graph is a chordal graph where every cycle of length four or more has a chord. We also defined a simplicial node as one whose neighbors form a clique. We also learned that a chordal graph has a perfect elimination ordering. A chordal graph contains at least one simplicial node.

According to a theorem established by Dirac in 1961, a non-complete chordal graph must have at least two non-adjacent simplicial nodes. In our circuit example, we illustrated in the R practice that its chordal graph includes five simplicial nodes: A1, A3, A2, A4, and F.

This section is organized as follows. First, we define the task of finding cliques and maximal cliques in general undirected graphs. Second, we describe a linear-time algorithm for computing maximal cliques for ordered triangulated graphs. Third, we provide a detailed explanation of the algorithm applied to finding maximal cliques in our running circuit example.

Defining the task of finding cliques and maximal cliques

In this section, we describe what cliques are in an undirected graph and what the maximal cliques are. We discuss the task of finding the cliques in a general undirected graph.

Let us begin with defining cliques and maximal cliques in an undirected graph.

Definition: clique, maximal clique

Given an undirected graph $G = (V, E)$, we call a subset of nodes $C \subseteq V$ a clique if the subgraph $G(C)$ is a complete subgraph where, for every distinct pair of nodes x and y in C, there is an edge $\{x, y\} \in E$. We call a clique not contained in any other clique a **maximal clique**.

It is an NP-complete problem to find the cliques of a general undirected graph. However, as we will demonstrate in the next section, if an undirected graph is a chordal graph, a graph having a chord connecting every two non-adjacent nodes in every cycle of length greater than 3, then finding the cliques can be computed in linear time in the number of cliques.

In the following section, we will describe an algorithm for finding the maximal cliques of a triangulated graph with MCS ordering. Using our running circuit example, we will show that the nodes of the maximal prime subgraph decomposition of the circuit Bayesian network triangulated graph, as discussed in the previous section, are the same as the maximal cliques that we shall compute by the algorithm, the details of which we will provide in the next section.

Finding maximal cliques in triangulated graphs

In this section, we present a linear-time algorithm designed to find all maximal cliques within a triangulated graph. The algorithm takes an ordered triangulated graph with nodes ordered in perfect elimination ordering as input and produces a list of all maximal cliques as output.

The algorithm begins by initializing the set of maximal cliques, referred to as `maxCliques`, as an empty list. The steps of the algorithm are as follows: it processes the nodes in reverse order, from last to first, within a for loop. During the processing of each node, it collects a set that includes the node and its earlier neighbors, thereby forming a clique, since the input graph is chordal. It then checks if any existing clique in the `maxCliques` list is a superset of the newly formed clique. If no such superset is found, we append the newly formed clique to the `maxCliques` list. Once the for loop concludes, the `maxCliques` list will contain all the maximal cliques.

Algorithm: finding maximal cliques

Input: An ordered triangulated graph, $G_T = (V, E)$, with perfect elimination ordering d of $V = (v_1, v_2, \cdots, v_n)$

Output: All the maximal cliques in G_T.

Initialize: maxCliques ← [ ]

For i = n down to 1:

```
    C ← {v_i} ∪ adj⁻(v_i; G_T, d)
    isMaxClique = TRUE
    For each C' ∈ maxCliques:
        If (C' ⊇ C):
        isMaxClique = FALSE
        break
    If isMaxClique:
         maxCliques.append(C)
Return maxCliques
```

The algorithm indicates that there can be, at most, n cliques, where n represents the cardinality of the node set V. The following presents a theorem regarding the algorithm for identifying maximal cliques.

Theorem: maximal cliques

A chordal graph with n vertices and m edges has at most n maximal cliques that can be identified in $O(n + m)$ time using a Perfect Elimination Ordering (PEO) generated via Maximum Cardinality Search (MCS).

To clarify how the algorithm for *finding maximal cliques* works and to deepen our understanding of maximal cliques in triangulated graphs, we will take a step-by-step approach using our circuit example. We will apply the algorithm to the ordered triangulated graph that we computed in the previous sections.

Application use case on the Circuit BN

In this section, we will illustrate the algorithm step by step to demonstrate how it identifies maximal cliques.

Example: maximal cliques

Let us compute the maximal cliques for the triangulated graph of our circuit Bayesian network model depicted in *Figure 6.6*. We start with the last node, F, which is node #10, in the MCS (Maximum Cardinality Search) ordering. Node F forms the clique {E, D, A4, F} with its earlier neighbors. We append this clique to the `maxCliques` list, which is initially empty. The next node, #9, is A4, forming the clique {E, D, A4} with earlier neighbors. This clique is a subset of the clique {E, D, A4, F} already in the `maxCliques` list, so we do nothing. The next node, #8, is A2, which forms the clique {X, Z, D, A2} with its earlier neighbors. This clique is appended to the `maxCliques` list since it is not a subset of any other element in the `maxCliques`. Next, node #7 is D, which forms the clique {X, Z, E, D} with its earlier neighbors. This clique is appended to the `maxCliques` list since it is not a subset of any other set in the list. Then, node #6 is A3, forming the clique {Y, Z, E, A3} with its earlier neighbors. This clique is appended to the `maxCliques` list because it is not a subset of any other set in the list. Moving on, node #5 is E, which forms the clique {Y, X, Z, E} with its earlier neighbors. This clique is appended to the `maxCliques` list since it is not a subset of any other set in the list. Next, node #4 is Z, forming the clique {A1, Y, X, Z} with its earlier neighbors. This clique is appended to the `maxCliques` list since it is not a subset of any other set in the list. Next, node #3 is X, forming the clique {A1, Y, X} with its earlier neighbors. Since the new clique is a subset of the clique {A1, Y, X, Z} already in `maxCliques`, we do nothing. Node #2 is Y, forming the clique {A1, Y} with its earlier neighbors. The current clique is a subset of the clique {A1, Y, X} in `maxCliques`, so we do nothing. Finally, node #1 is A1, which has no earlier neighbors, forming a singleton {A1}. The singleton is a subset of the clique {A1, Y, X} in `maxCliques`, so we do nothing. The set `maxCliques` at the end of executing the algorithm will thus have six maximal cliques: {{A1, Y, X, Z}, {Y, X, Z, E}, {Y, Z, E, A3}, {X, Z, E, D}, {X, Z, D, A2}, {E, D, A4, F}}. The six maximal cliques are the nodes of the maximal prime subgraph decomposition depicted in *Figure 6.8* of the previous section.

Figure 6.9 shows a table summarizing the above steps. The first column is the MCS in reverse ordering. The second column is the ordered node label. The third column is the clique consisting of the node and its earlier neighbors. The fourth column lists the maximal cliques. The fifth column is the index of the maximal clique in the list, `maxCliques`.

MCS #	Node	Clique	Maximal Clique	Maximal Clique #
10	F	{E, D, A4, F}	{E, D, A4, F}	6
9	A4	{E, D, A4}		
8	A2	{X, Z, D, A2}	{X, Z, D, A2}	5
7	D	{X, Z, E, D}	{X, Z, E, D}	4
6	A3	{Y, Z, E, A3}	{Y, Z, E, A3}	3
5	E	{Y, X, Z, E}	{Y, X, Z, E}	2
4	Z	{A1, Y, X, Z}	{A1, Y, X, Z}	1
3	X	{A1, Y, X}		
2	Y	{A1, Y}		
1	A1	{A1}		

Figure 6.9—Finding maximal cliques for the circuit example

It is important to note that the maximal cliques computed by the algorithm shown in *Figure 6.9* are precisely the same as the node sets of the six mp-subgraphs derived by the p-decompositions shown in *Figure 6.8*. We highlight this note as follows.

Maximal Cliques and Maximal Prime Subgraph Decomposition

Comparing *Figure 6.8* and *Figure 6.9*, we conclude that the six maximal cliques computed by the algorithm 'Finding Maximal cliques' are precisely the same as the complete sets, which are the nodes of the six maximal prime subgraphs decomposition of the triangulated graph of our running circuit example.

We conclude this section with a summary. We introduced the concepts of cliques and maximal cliques in undirected graphs. Finding cliques in a general undirected graph is known to be NP-complete. However, if the graph is chordal or triangulated, identifying the maximal cliques becomes more manageable and can be accomplished in linear time relative to the size of the graph. We then explored the algorithm for finding maximal cliques to compute the maximum cliques in a chordal graph. This algorithm, which processes the nodes in reverse order, serves as a crucial tool in graph theory. It collects cliques and retains only those that are not subsets of cliques already discovered. The maximal cliques represent complete sets corresponding to the nodes of the maximal prime subgraph decomposition of the triangulated graph. The number of maximal cliques in a graph is at most equal to the number of nodes in the graph and is typically much less than the total number of graph nodes.

The following section explores the practical application of constructing a join or junction tree, representing the p-decomposition of a triangulated graph. This representation is not just a theoretical concept but a valuable tool in computer science. The nodes of a join tree are the maximal cliques, and the edges are the separator sets respecting the running intersection property, making it a desirable if not the primary data structure for exact inference in Bayesian networks (BNs), as we will describe in *Chapter 7*.

Computing a join tree

In the previous section, we explored the practical application of computing the maximal cliques in a triangulated graph. When structured as a join tree, these maximal cliques play a pivotal role as a tree-structured representation of the triangulated graph's maximal prime subgraph decomposition. This concept is not only crucial but also highly relevant in the realms of graph theory and computational algorithms, particularly in optimizing search processes and improving the efficiency of various computational tasks.

This section is meticulously structured to guide you through the process. We present an algorithm for constructing a join tree from the ordered list of maximal cliques. We then provide a comprehensive example to illustrate the algorithm's functionality. Next, we define the running intersection property the join tree must adhere to, using our running circuit example for clarity. We conclude with a hands-on practice section in R, where we will compute the p-decomposition and the structure of the join tree for the triangulated graph of the circuit Bayesian network (BN). This practical exercise will not only validate our results but also provide you with a solid understanding of the algorithm that was applied earlier.

Next, let us begin with the construction of a join tree.

Building a join tree

This section takes you through the step-by-step process of building a join tree for a given triangulated graph. The process begins by determining all the maximal cliques in the triangulated graph described in the previous section. As demonstrated in the previous section, the algorithm orders the maximal cliques to conform with the MCS ordering of the triangulated graph. The maximal cliques are arranged according to the order of their highest ordered variable within the cliques, based on the MCS ordering. This systematic approach ensures that you understand each step clearly and can confidently apply it in your own work.

The *building a join tree* algorithm starts by picking the maximal clique, which is the first one in the list, and calling it the root of the join tree. This root node serves as the starting point for the tree's construction. Each subsequent step builds upon the previous one, leading to the complete join tree.

Then, the algorithm processes the remaining cliques in a for loop. When processing the clique C_i, it searches for a clique C_j ordered before C_i with the most significant size separator set $S_{i,j}$ which is the set intersection $S_{i,j} = C_i \cap C_j$. The algorithm assigns the clique C_j to be the parent of the cliques C_i.

Algorithm : building a join tree

Input: maximal cliques, $[C_1, \cdots, C_m]$ ordered conforming with the MCS ordering

Output: Join tree

Initialize: Make C_1 the root of the tree.

For i = 2 to m:

> Find a clique C_j, earlier in the ordering $(j < i)$ with the maximum separator $S_{i,j} = C_i \cap C_j$. That is, a separator $S_{i,j}$ such that $|S_{i,j}| \geq |S_{i,k}|, \forall j, k < i$. If there is more than one clique with maximum size separator, select one arbitrarily from them.
>
> Make the clique C_j the parent of C_i and annotate the edge with the separator set $S_{i,j}$.

The following is a formal statement of the time complexity of building a join tree from ordered maximal cliques.

Theorem: building a join tree

Building a join tree (or clique tree) from ordered maximal cliques based on perfect elimination ordering (PEO) of a chordal graph $G(V, E)$ can be achieved in linear time $O(|V| + |E|)$.

Proof

The maximum cliques are ordered $(C_1, C_2, \cdots, C_k)$ based on the perfect elimination ordering (PEO), such that the running intersection property (RIP) is satisfied. The algorithm connects each clique C_i with a preceding clique C_j $(j < i)$ that shares the maximum size separator with C_i. Because the graph is chordal, the total number of nodes across all maximal cliques, $\sum_{i=1}^{k} |C_i|$, is $O(|V| + |E|)$. The construction of the tree via parent selection (connecting to the clique with the highest intersection) can be performed efficiently within this bound.

Let us work on an example of applying the build a join tree algorithm step by step. Let us do that in the following example.

Example: building a join tree

Let us build the join tree for our running circuit example. *Figure 6.9* gives the maximal cliques ordered from 1 to 6. We apply the *building join tree* algorithm described earlier. The root node is the clique #1. Clique #2 is a child node to the root node, i.e., node #1 is the parent of node #2. We annotate the edge between node #2 and node #1 with the separator set $\{Y, X, Z, E\} \cap \{A1, Y, X, Z\}$, which equals the set $\{Y, X, Z\}$. The set intersection between clique #3 and clique #1, $\{Y, Z, E, A3\} \cap \{A1, Y, X, Z\}$, gives a separator set of size 2. The set intersection between clique #3 and clique #2, $\{Y, Z, E, A3\} \cap \{Y, X, Z, E\}$ gives a separator set of size 3. Therefore, we make clique #3 node #3 in the join tree, make node #2 the parent of node #3, and annotate the edge between them with the separator set $\{Y, Z, E\}$. Next, clique #4, $\{X, Z, E, D\}$, will be node #4 in the tree, and we need to find which of the three previous cliques, has the maximum size separator set with clique #4. The set intersection between clique #4 and clique #2 gives the maximum separator $\{X, Z, E\}$, so we make node #2 the parent of node #4 and annotate the edge with the separator set. Clique #5, $\{A2, X, D, Z\}$ is node #5; it has the maximum separator, $\{X, D, Z\}$, with node #4, which becomes the parent for node #5 and we annotate the edge between the two nodes by the separator set. Finally, node #6 is clique #6, $\{F, E, A4, D\}$, which has the maximum separator set, $\{E, D\}$, with node #4, so we make node #4 the parent of node #6, and we annotate the edge with the separator set.

Figure 6.10 illustrates a join tree for the example, with the nodes (represented as ovals) indicating the maximal cliques and the edges (shown in boxes) representing the separators. Remember that the separators, or sepsets, are the intersections of the two cliques connected by the edge.

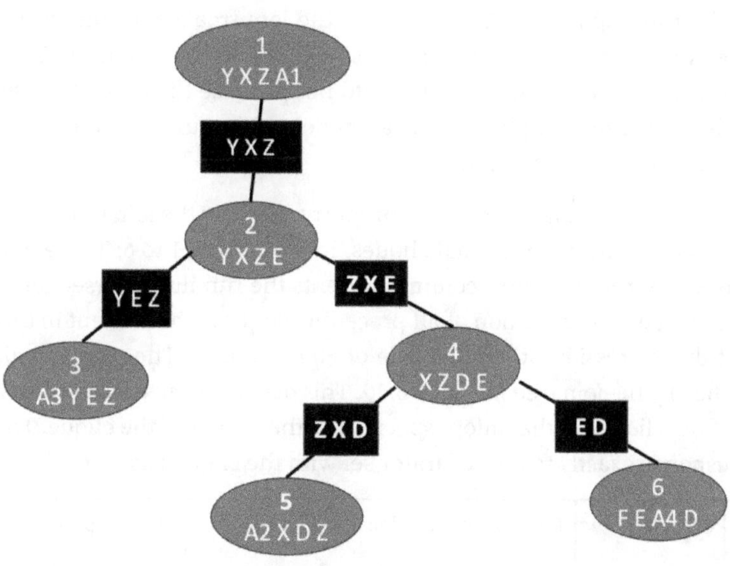

Figure 6.10: A Join Tree for the ordered triangulated graph of the circuit example

From the algorithm and the example above, we remark that the join tree connects the maximal cliques such that nodes that share the most variables are joined by an edge annotated with the set of their shared variables, called the separator set or sepset. That aspect of the construction of the tree implies that for any pair of nodes in the join tree that share a variable, that variable will be present all along the path between the node pair. This is called the **running intersection property** (RIP), which we will define next.

Consider the two nodes, #5 and #6, in *Figure 6.10* as one example of the RIP. They share the D variable, which is present along the path 5-4-6 connecting them. Another example is the pair of nodes #3 and #6; they share the E variable, which is included in all the sepsets along the path 3-2-4-6 connecting them.

The following is the formal definition of RIP.

Definition: running intersection property

The running intersection property (RIP) is an ordering of the set of maximal cliques $C_1, C_2, \cdots, C_m$ such that for every $j > 1$, there exists an $i < j$ such that:

$$C_j \cap (C_1 \cup \cdots \cup C_{j-1}) \subseteq C_i \ (6.15)$$

That is, the intersection of each with the union of the previous is contained in one.

The running intersection property (RIP) in a junction tree is not merely a property; it is the necessary and sufficient condition that enables the use of the join tree for exact inference in Bayesian networks. This inference relies on the efficient belief propagation algorithm, which is based on message passing, as will be explained in *Chapter 7*. The RIP ensures that local consistencies between neighboring cliques lead to global consistency throughout the entire probability distribution represented by the tree. It guarantees that when marginalizing the joint distribution of a clique over the variables it shares with a neighboring clique (the separator), the resulting marginal probabilities remain consistent.

By maintaining the running intersection property, the join tree algorithm ensures that inference queries on sets of variables (contained within a single clique) are accurate. Without the running intersection property, local message passing would not produce the correct marginal probabilities, rendering exact inference on complex graphical models with multiply connected structures either computationally intractable or incorrect.

Figure 6.11 illustrates the running intersection property (RIP) for the join tree of our example circuit BN. The first two columns list the maximal cliques, indexed from 1 to 6. These maximal cliques act as the nodes of the join tree. The third column presents the running intersection, defined as the intersection of each clique with the union of all preceding cliques. The content in this column reflects the evaluation of the expression on the left side of *Equation 6.15*. The results indicate that the RIP condition is satisfied by the join tree in *Figure 6.10*. This occurs because each running intersection is a subset of one of the cliques in the union, specifically the parent of the clique. The intersection for each clique corresponds exactly to the separator set with the clique's parent.

#j	***clique*** C_j	Running Intersection	Sepset
1	{Y, X, Z, A1}		
2	{Y, X, Z, E}	$C_2 \cap C_1 = \{Y, X, Z\}$	$\{Y, X, Z\} \subseteq C_1$
3	{A3, Y, E, Z}	$C_3 \cap (C_1 \cup C_2) = \{Y, E, Z\}$	$\{Y, E, Z\} \subseteq C_2$
4	{X, Z, D, E}	$C_4 \cap (C_1 \cup C_2 \cup C_3) = \{X, Z, E\}$	$\{X, Z, E\} \subseteq C_2$
5	{A2, X, D, Z}	$C_5 \cap (C_1 \cup C_2 \cup C_3 \cup C_4) = \{X, D, Z\}$	$\{X, D, Z\} \subseteq C_4$
6	{F, E, A4, D}	$C_6 \cap (C_1 \cup C_2 \cup C_3 \cup C_4 \cup C_5) = \{E, D\}$	$\{E, D\} \subseteq C_4$

Figure 6.11—Illustration of the join tree running intersection property (RIP) for the circuit example

So far, we described how to find the maximal cliques from the ordered triangulated graph and build a tree-structured representation of the graph called a join tree. With that, we are ready to proceed to a practice session to illustrate the maximal prime subgraph decomposition and how we build the join tree programmatically using R.

Practice using R

In this practice you will learn how to programmatically implement the steps for computing a join tree from a chordal graph. The steps in this practice will implement the last steps of the process of join tree clustering. By the end of this practice, the code will produce the join tree for our circuit example Bayesian network (BN) as depicted in *Figure 6.10*, which will be used for probabilistic inference—a topic explored in *Chapter 7*.

This practice consists of two sections described as follows:

- Maximal prime subgraph decomposition: In this section, you will utilize R and the `gRbase` library to obtain the p-decomposition illustrated in *Figure 6.8*. Additionally, you will create plots of the maximal prime subgraphs (`mp-subgraphs`) corresponding to the triangulated graph of *Figure 6.6* of our circuit Bayesian network (BN) example.
- Computing join tree structure: In this section, you will utilize R and the gRbase library to map the triangulated graph of our circuit BN example of *Figure 6.6* to a join tree structure as depicted in *Figure 6.10*.

The code in this practice will reference the adjacency matrix of the triangulated graph computed in the previous session on our running example of the circuit diagnosis BN.

Let us begin with the maximal prime subgraph decomposition of the triangulated graph.

Maximal prime subgraph decomposition

To get the maximal prime subgraph decomposition (MPD) of the triangulated graph, which is a process of breaking down a graph into its largest, most connected subgraphs, we apply the function `gRbase::mpdMAT()` to the triangulated graph adjacency matrix object, `TmG`. We set `x` to the triangulated graph's MPD. We extract each clique from `x$cliques`, convert it to a vector using `unlist()`, create the subgraph with `igraph::subgraph()`, and then plot it using `ggm::plotGraph()`. There are six maximal prime subgraphs corresponding to the six cliques. *Figure 6.12* shows the plots of all six maximal prime subgraphs.

```
x <- gRbase::mpdMAT(TmG)
TG <- as(TmG, "igraph")
for(i in 1:length(x$cliques)){
  y <- unlist(x$cliques[i])
  g <- igraph::subgraph(TG, y)
  print(ggm::plotGraph(g))
}
## $tkp.id
## [1] 7
```

```
##
## $igraph
## IGRAPH 2e0584b DN-- 4 6 --
## + attr: name (v/c), edge.arrow.mode (e/n)
## + edges from 2e0584b (vertex names):
## [1] A1->Y A1->X A1->Z Y ->X Y ->Z X ->Z
##
## $tkp.id
## [1] 8
##
## $igraph
## IGRAPH 9dc38c1 DN-- 4 6 --
## + attr: name (v/c), edge.arrow.mode (e/n)
## + edges from 9dc38c1 (vertex names):
## [1] Y->X Y->Z Y->E X->Z X->E Z->E
##
## $tkp.id
## [1] 9
##
## $igraph
## IGRAPH 2fdaf23 DN-- 4 6 --
## + attr: name (v/c), edge.arrow.mode (e/n)
## + edges from 2fdaf23 (vertex names):
## [1] A3->Y A3->Z A3->E Y ->Z Y ->E Z ->E
##
## $tkp.id
## [1] 10
##
## $igraph
## IGRAPH b8bd792 DN-- 4 6 --
## + attr: name (v/c), edge.arrow.mode (e/n)
## + edges from b8bd792 (vertex names):
## [1] X->Z X->E X->D Z->E Z->D E->D
##
## $tkp.id
## [1] 11
##
## $igraph
## IGRAPH 161ac96 DN-- 4 6 --
## + attr: name (v/c), edge.arrow.mode (e/n)
## + edges from 161ac96 (vertex names):
```

```
## [1] A2->X A2->Z A2->D X ->Z X ->D Z ->D
##
## $tkp.id
## [1] 12
##
## $igraph
## IGRAPH 645e0a5 DN-- 4 6 --
## + attr: name (v/c), edge.arrow.mode (e/n)
## + edges from 645e0a5 (vertex names):
## [1] E ->D  E ->A4 E ->F  D ->A4 D ->F  A4->F
```

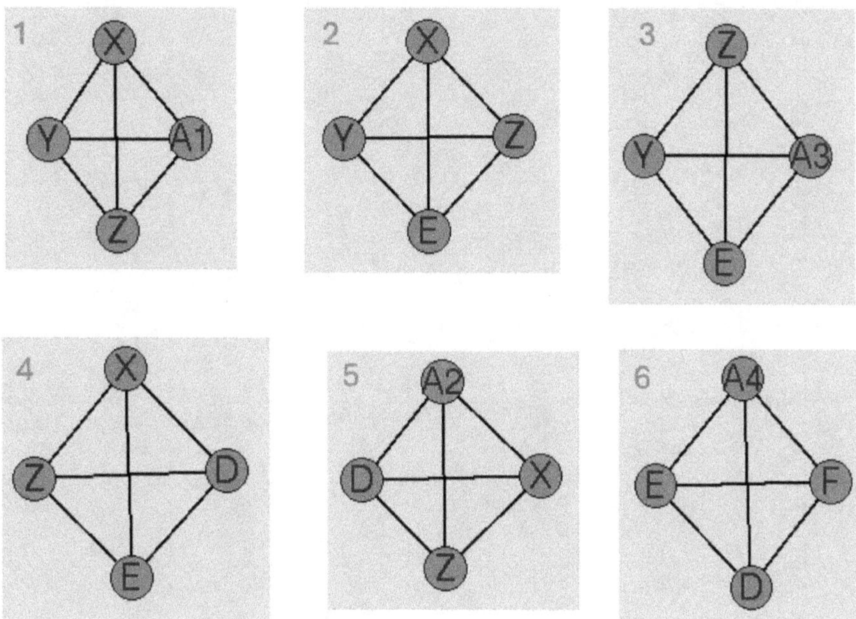

Figure 6.12—Maximal Prime Subgraph decomposition of the triangulated graph of the circuit BN model

Having explored the `mp-decomposition` of the triangulated graph for the circuit Bayesian Network (BN) example, we will now move to the final section to validate the previously derived join tree structure, as illustrated in *Figure 6.10*. Next, we will examine the results obtained by applying the methods from the *gRbase* library.

Computing the join tree of a structure

To compute the join tree structure for our circuit Bayesian network (BN) example, we use the `gRbase::rip` function from the *gRbase* package in R. This function generates a Running Intersection Property (RIP) ordering of the cliques for an undirected graph. However, it is important to note that the `gRbase::rip` function requires the graph to be chordal (triangulated); otherwise, a valid ordering cannot be produced.

In the following code, we run the `gRbase::rip` function with the input being the triangulated graph `TmG`, which we computed earlier in a previous practice session.

```
gRbase::rip(TmG)
```

The output from running the code specifies the join tree in three main aspects: 1) an ordered list of cliques based on the perfect ordering of the nodes in the triangulated graph, `TmG`; the cliques represent the nodes of the join tree, 2) an ordered list of separator sets that make up the edges of the tree, and 3) the parent of each node in the tree is indexed according to the ordering of the cliques. The separator sets correspond to the intersections between pairs of cliques that are joined by an edge in the tree.

```
## cliques
##   1 : Y X Z A1
##   2 : Y X Z E
##   3 : A3 Y E Z
##   4 : X Z D E
##   5 : A2 X D Z
##   6 : F E A4 D
## separators
##   1 :
##   2 : Y X Z
##   3 : Y E Z
##   4 : X Z E
##   5 : X D Z
##   6 : E D
## parents
##   1 : 0
##   2 : 1
##   3 : 2
##   4 : 2
##   5 : 4
##   6 : 4
```

The join tree obtained from the R practice matches the join tree illustrated in *Figure 6.10*. This figure was created using the graph algorithms detailed throughout this chapter.

Concluding this section, let's summarize the main takeaways. First, we described an algorithm for computing an ordered list of maximal cliques in a triangulated graph. Second, we detailed an algorithm for constructing a join tree from the ordered list of maximal cliques. Third, we demonstrated using R and graph libraries how to compute the maximal cliques and the join tree from the triangulated graph of our circuit Bayesian network example.

Summary

Let's summarize what was covered in this chapter, its significance, and what we will explore next.

This chapter explains how join tree (junction tree) clustering converts a Bayesian network's directed acyclic graph (DAG) into a tree-decomposable structure to support efficient, exact inference. The process proceeds through moralization (making the graph undirected and marrying co-parents), triangulation (adding fill-in edges to make the graph chordal), and then constructing a join tree from the maximal cliques of the triangulated graph. Key structural ideas include using maximum cardinality search (MCS) to obtain a good ordering, identifying simplicial nodes and perfect elimination orderings in chordal graphs, and ensuring the join tree satisfies the running intersection property (RIP) for consistent inference.

The concepts and methods discussed in this chapter on tree clustering are significant because they explain how to efficiently compute a join tree structure for a Bayesian network. Constructing a join tree is important, as it allows inference problems in Bayesian networks to be divided into smaller subproblems on smaller subnetworks, enhancing computational efficiency and facilitating time-space tradeoffs. The hands-on programming practice throughout the chapter equips you with the experience and knowledge necessary to implement graph-based algorithms in your applications using R.

In the next chapter on, we will delve deeper into how the join tree representation, which we have developed and understood in this chapter, sets the data structure for performing exact inference on various inference tasks in Bayesian networks using belief propagation. The structure-based algorithm of tree clustering allows us to ensure computational complexity guarantees for performing inference tasks, a key aspect in data science and machine learning.

References

- Berge C. *Some Classes of Perfect Graphs* [Book Section]. Graph Theory and Theoretical Physics / book auth. Harary F. (Ed.). New York: Academic Press, 1967.
- Berry A., Blair J.R.S. and Heggernes P. *Maximum Cardinality Search for Computing Minimal Triangulations* [Online]. 2 28, 2024. `https://scispace.com/pdf/maximum-cardinality-search-for-computing-minimal-50eu8e3wun.pdf`.
- Berry Anne, Pogorelcnik Romain, and Simonet Genevieve. *An Introduction to Clique Minimal Separator Decomposition* [Journal]. Algorithms . 2010. 2: Vol. 3. pp. 197-215.
- Carneiro M. *Maximum Cardinality Search and Chordal Graphs* [Online]. 9 25, 2018. `https://www.cs.cmu.edu/~janh/courses/411/18/lec/03-chordal-mcs.pdf`.
- Darwiche A. Bayesian Networks [Book Section] // Handbook of Knowledge Representation. - [s.l.] : Elsevier, 2008.
- Dechter J. *Bucket elimination: A unifying framework for reasoning* [Journal]. Artificial Intelligence. 1999. Vol. 113. pp. 41–85.
- Dechter R. and El Fattah Y. *Topological parameters for time-space tradeoff* [Journal]. Artificial Intelligence. 2001. Vol. 125. pp. 93–118.

- D'Ambrosio Bruce. *Inference in Bayesian Networks* [Journal]. AI Magazine. 1999. 2 : Vol. 20. pp. 21-36.
- Garey M.R. and Johnson D.S. Computers and Tractability- A Guide to the Theory of NP-Completeness [Book]. - New York : W.H. Freeman and Company, 1979.
- Kjaerulff Uffe. *Triangulation of Graphs: Algorithms Giving Small Total State Space* [Online]. March 1, 1990. 2 27, 2024. `http://cse.unl.edu/~choueiry/Documents/Kjaerulff-TR-1990.pdf`.
- Lauritzen S. L. & Spiegelhalter D. J. *Local Computations with Probabilities on Graphical Structures and their Application to Expert Systems* [Journal]. J. R. Statist. Soc. B . 1988. 2: Vol. 50. pp. 157-224.
- Leimer Hanns-Georg. *Optimal decomposition by clique separators* [Journal]. Discrete Mathematics . 1993. Vol. 113. pp. 99- 123.
- Mano M. M. *Computer Engineering: Hardware Design* [Book]. [s.l.]: Prentice Hall, 1988.
- Olesen Kristian and Madsen Anders. *Maximal prime subgraph decomposition of Bayesian networks* [Journal]. IEEE transactions on systems, man, and cybernetics. 2002. Vol. 32. pp. 21-31.
- Pearl J. *Probabilistic Reasoning in Intelligent Systems: Networks of Plausible Inference* [Book]. [s.l.] : Morgan Kaufmann, 1988.
- Rose Donald J. *Triangulated Graphs and the Elimination Process* [Journal]. Journal Of Mathematical Analysis And Applications. 1970. pp. 597-609.
- Wu D. *Maximal prime subgraph decomposition of Bayesian networks: A relational database perspective* [Journal]. International Journal of Approximate Reasoning. - 2007. Vol. 46. pp. 334–345.
- Yannakakis M. *Computing the minimum fill-in is NP-complete* [Journal]. SIAM J. Alg. Disc. Math. 1981. pp. 77-79.

Join our community on Discord

Join our community's Discord space for discussions with the author and other readers: `https://packt.link/hYCWN`

7

Probabilistic Inference with Join Tree Clustering

This chapter builds on the results from *Chapter 6*. It continues with inference by belief propagation with a join tree representation of the probabilistic graphical models of Bayesian and **Markov networks (MN)**. The algorithm for inference with join tree clustering and the building of the clique tables are formally described. We demonstrate with examples how to perform probabilistic inference tasks step-by-step by belief propagation with a join tree. In a practice section using Python, we show in detail how to perform inference tasks with join tree clustering on a practical fault diagnosis in a digital circuit example. We also walk through the implementations in Python using the `pgmpy` library, a powerful tool for probabilistic graphical models.

In this chapter, we're going to cover the following main topics:

- Probabilistic graphical models
- Inference tasks and inference algorithms
- From Markov models to join trees
- Belief propagation with join trees

Technical requirements

The Python code in this chapter is in a folder named `Python-chap7` that you can download from the GitHub URL: `https://github.com/PacktPublishing/Causal-Inference-with-Bayesian-Networks/tree/main/Chapter_07/inference`

To run the Python code in our *Jupyter* notebook, please follow the steps below:

1. Install Python.
2. You must have Python 3 installed on your system. If you do not already have it, you can follow this guide from Real Python to install Python: `https://realpython.com/installing-python/`
3. Install JupyterLab.

4. Once you install Python, you can install JupyterLab, which provides a Jupyter notebook interface. You can follow the installation instructions from the official Jupyter documentation: `https://jupyter.org/install`.
5. If you are using `pip`, run the following command in your terminal:

```
pip install jupyterlab
```

If you are using `Conda`, run:

```
conda install -c conda-forge jupyterlab
```

6. Install necessary libraries.

You will also need to install the following Python libraries:

- `pgmpy`: A Python library for Probabilistic Graphical Models. You can follow this guide to install it: `https://pgmpy.org/started/install.html`
- `numpy`: A library for the Python programming language, adding support for large, multi-dimensional arrays, and matrices, along with a large collection of high-level mathematical functions to operate on these arrays.
- `pygraphviz`: A Python interface to the `Graphviz` graph layout and visualization package.

Use the following commands to install these libraries using `pip`:

```
pip install pgmpy numpy pygraphviz
```

If you are using `Conda`, you can install the packages by running:

```
conda install -c conda-forge pgmpy numpy pygraphviz
```

7. Launch the notebook.
8. Once you have installed all dependencies, you can open your terminal (*Mac/Linux*) or Command Prompt (*Windows*), navigate to the directory using `cd path/to/your/directory` containing the downloaded Jupyter Notebook `chapter7.ipynb`, and run the command:

```
jupyter lab
```

9. The above command will launch JupyterLab in your web browser. From the JupyterLab interface, you can open the Jupyter notebook file and run the Python code.

Please ensure all these steps are carried out in the above order to avoid errors or issues. We included documentation of the notebook in *docx* and *HTML* format in the chapter's folder on the book website. The documents show the entire Python code and the output results you should get upon running the code.

Probabilistic graphical models

This section provides a step-by-step guide to understanding the representation of probabilistic graphical models, focusing on Bayesian and **Markov networks (MN)**. We begin by reviewing the definition of Bayesian networks, then proceed to a formal description of Markov networks. To ensure a thorough understanding, we illustrate these concepts with a digital circuit example. This example, which we'll refer to throughout, demonstrates the application of the belief propagation inference algorithm in fault diagnosis. We conclude this section with a Python practice session, where we'll guide you through the implementation of the circuit diagnosis example using the `pgmpy` library.

Bayesian network models

We recall the definition of a Bayesian network model from *Chapter 3*, which we summarize here as follows.

A Bayesian network, with its clear and logical structure, consists of qualitative and quantitative components. The qualitative component is a **directed acyclic graph (DAG)** where nodes represent the network variables, and the directed edges depict the direct probabilistic influence of parent nodes on their respective children. A child node and all its parents form a causal family of the DAG.

The quantitative component specifies the **conditional probability distributions (CPDs)** for each causal family in the DAG. A CPD is the probability distribution of a node conditioned on the value tuple assigned to all its parents.

A Bayesian network provides a compact representation of the joint probability distribution over a set of network variables $\{V_1, \cdots, V_n\}$. This representation is made possible by the factored form of the Conditional Probability Distributions (CPDs). Within this framework, the joint distribution of the network variables is expressed as the product of the CPDs for all the variables.

$$P(V_1, \dots, V_n) = \prod_{i=1}^{n} P\left(V_i \middle| pa(V_i)\right) \quad (7.1)$$

Here, $pa(V_i)$ denotes the node's parents for the variable V_i in the DAG representing the BN.

In this section, we summarized the definition of directed graphical models of Bayesian networks described by directed acyclic graphs.

In the next section, we describe Markov Networks, which are relevant to the join tree clustering algorithm, the main topic of this chapter. Markov networks (MNs) and Bayesian networks (BNs) are both probabilistic graphical models representing dependencies between a set of variables. BNs differ from MNs in their graph structure (directed vs. undirected). Graph-based methods convert between BNs and MNs: moralization (BN to MN) and triangulation (MN to BN) exist to translate between them. Bayesian networks model joint probability as a product of CPDs. Markov networks factor joint probability using potential functions over cliques (fully connected subgraphs).

Markov network models

A Markov network, also known as a **Markov random field (MRF)**, is a probabilistic graphical model representing dependencies between variables. A Markov network, like a Bayesian network, is a factored representation of a joint probability distribution over the set of network variables. An undirected graph, which may contain cycles, represents a Markov network, whereas a directed acyclic graph (DAG) represents a Bayesian network. In a Bayesian network, the directed edges show causal or directional dependencies. In a Markov network, the edges show associations or symmetric dependencies between variables. *Table 7.1* lists the fundamental difference between a BN and an MN in how dependencies are modeled.

Feature	Bayesian Network	Markov Network
Edges	Directed	Undirected
Dependencies	Causal/Directional	Associational/Symmetric
Cycles	Directed Not Allowed	Allowed

Table 7.1—Comparison of the Bayesian networks vs. Markov networks graph representations

Figure 7.1 depicts a Markov network model represented by an undirected graph on eight network variables, $\{A, B, C, D, E, F, G, H\}$. The graph shows six cliques, subgraphs whose nodes are all pairwise connected. Each subgraph represents a subset of variables that are all pairwise associated and probabilistically interdependent. The cliques are $\{A, B, D\}$, $\{A, C, D\}$, $\{C, D, E\}$, $\{D, E, F\}$, $\{E, F, H\}$, and $\{F, G\}$.

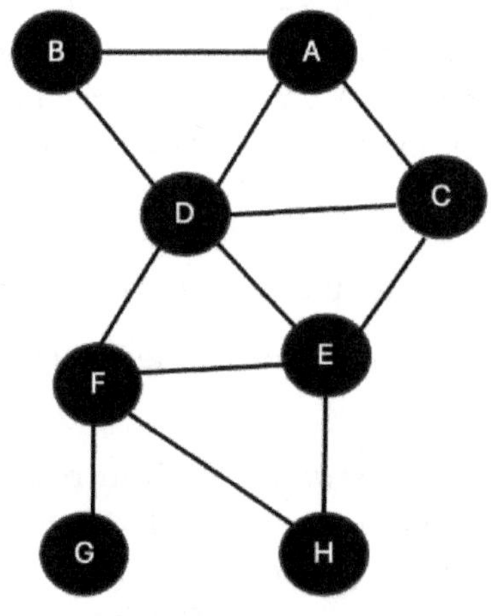

Figure 7.1—An undirected graph of a Markov network example

A Markov network represents a joint probability distribution over a set of network variables, denoted as $V = \{V_1, \cdots, V_n\}$. These variables correspond to the nodes in the undirected graph of the Markov network. The network models this distribution through a factorization that consists of **potential functions** applied to subsets of the network variables. These subsets form the nodes of complete subgraphs known as cliques, where every node is connected to every other node.

Let $\mathcal{C}$ represent the set of cliques in the undirected graph of the Markov network. For clique $c \in \mathcal{C}$, let V_c denote the subset of variables in V associated with that clique Additionally, let $\phi_c(V_c)$ represent the potential function corresponding to clique c. A potential function, also referred to as a **factor**, is defined as a real-valued, non-negative function applied to the variables within the clique.

The factored representation of the joint probability distribution representing a Markov network is the normalized product of the potential functions of all cliques in $\mathcal{C}$, written as:

$$P(V) = \frac{1}{Z} \prod_{c \in \mathcal{C}} \phi_c(V_c) \ (7.2)$$

Here, Z is a normalization constant, called **the partition function**, that ensures the total probability over all possible variable assignments sums to one.

The **Hammersley-Clifford Theorem** states that if a joint probability distribution is strictly positive ($P(V) > 0$), and if all conditional independencies of $P(V)$ are encoded in an undirected graph G, then $P(V)$ can be expressed as a product of potentials over the cliques of G.

For a set of variables $V = \{V_1, \cdots, V_n\}$ and a set of factors ϕ_c over cliques c in the graph, the partition function is defined as:

$$Z = \sum_{v \in \mathcal{V}} \prod_{c \in \mathcal{C}} \phi_c(v_c) \ \mathbf{(7.3)}$$

Here $\mathcal{V}$ is the set of all possible value assignments of V, and v_c denotes the value assignment v of V restricted to the clique $c \in \mathcal{C}$.

Computing the partition function Z of a general Markov network is classified as #P-complete, which implies that it is also NP-hard. The computational intractability of the partition function Z is due to the need to sum over an exponentially large number of configurations of the network variables. This number grows as k^n, where n is the number of variables and k is the size (cardinality) of their value domain, making exact computation infeasible for all but the smallest networks.

To clarify why computing the partition function of *Equation 7.3* is #P-complete, we can represent potential functions as Boolean formulas in conjunctive normal form (CNF). In this representation, satisfying assignments are assigned a weight of 1, while falsifying assignments are given a weight of 0. Consequently, the partition function Z becomes equivalent to counting the number of satisfying assignments of the corresponding CNF formulas. This task is known as the counting version of the Boolean satisfiability problem, referred to as #SAT. The counting SAT problem, #SAT, requires the exact count of satisfying assignments. It is classified as a #P-complete problem, which is fundamentally more challenging than the decision SAT problem. The decision SAT problem determines whether there exists an assignment of truth values (TRUE or FALSE) to the variables of a given Boolean formula such that the entire formula evaluates to TRUE; this problem is known to be NP-complete.

The requirement to compute Z makes exact inference in general probabilistic graphical models (Markov networks) computationally intractable, often requiring approximation methods.

Let's compare the factored representations of a joint probability distribution in Bayesian and Markov networks.

The factors in Bayesian networks consist of conditional probability functions. In contrast, the factors in Markov networks are potential functions, which are real-valued non-negative functions that do not necessarily represent probabilities.

A potential function captures the compatibility or interaction between variables V_c in a specific maximal clique c.

The function assigns weights (non-negative numbers) to various assignments of variables within a clique, which indicate the likelihood of each specific configuration. However, these weights are not probabilities; instead, they represent scores reflecting the relative likelihoods of the different configuration states, based on the value assignments of the clique variables. The higher the weights, the more likely the respective states are.

In the following section, we will describe a graphical representation of Markov networks using factor graphs. These graphs include nodes that reference potential functions and their associated clique variables.

Factor graphs

Unlike in Bayesian networks, where the factorization is explicitly represented by the families of nodes and their parents, in Markov networks, the graph alone does not provide enough information to understand which specific factorization is being assumed. For example, in Markov networks, factors may consist of sub-cliques rather than maximum cliques.

To illustrate that multiple factorizations can exist for the same undirected graph in a Markov network, we can refer to the example presented in *Figure 7.1*. Instead of factoring the joint probability distribution based on potential functions of the six maximum cliques—{A, B, D}, {A, C, D}, {C, D, E}, {D, E, F}, {E, F, H}, and {F, G}—we could consider an alternative factorization based on twelve subcliques. These subcliques represent all the node pairs joined by an edge in the graph: {A, B}, {A, D}, {B, D}, {A, C}, {C, D}, {C, E}, {D, E}, {D, F}, {E, F}, {E, H}, {F, H}, and {F, G}.

The following algorithm presents a simple procedure that takes as input an undirected graph and a defined set of cliques within that graph and produces as output the corresponding factor graph.

Algorithm: Factor Graph

Input: an undirected graph representing the Markov network, which includes the set of cliques $\mathcal{C}$.

Output: the factor graph, $G = (N, E)$

1. Initialize the nodes N of the factor graph G to be the network variables. V and the edges E to be the empty set, $\emptyset$.
2. For each clique $c \in \mathcal{C}$:
    1. Add a node ϕ_c to the set N, which represents the potential function of the clique.
    2. For each variable X in the clique set V_c add an edge (ϕ_c, X) to the set E.
3. Return G.

Let's apply the factor graph algorithm to the Markov network of *Figure 7.1*. The cliques are $\{A, B, D\}, \{A, C, D\}, \{C, D, E\}, \{D, E, F\}, \{E, F, H\}$, and $\{F, G\}$. *Figure 7.2* shows the factor graph having eight nodes for the network variables, six nodes for the potential functions, and fourteen edges connecting potential function nodes to the variables in their associated cliques.

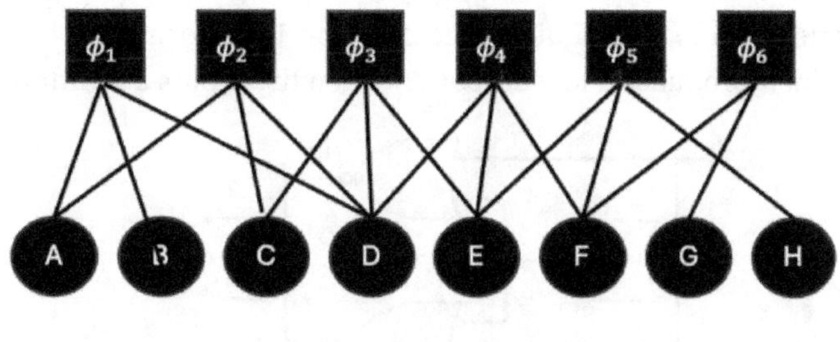

Figure 7.2—Factor graph for the Markov network example

In the following sections, we will examine a use case example in a significant application area that will enhance our understanding of representing and reasoning with probabilistic graphical models.

Using Markov networks to diagnose combinatorial circuits offers a powerful, undirected probabilistic graphical model approach for handling uncertainty, modeling complex dependencies, and performing automated inference in complex systems. In circuit diagnosis, these models represent the system by mapping logic gates to potential functions and variables (inputs, outputs, operational states), which can be converted into a normalized probability distribution to identify faulty components.

Key benefits and insights from using Markov networks for circuit diagnosis include:

- **Effective Representation of Constraints:** Markov networks can represent the functional constraints of combinatorial logic (e.g., AND, OR gates) using non-negative weight potentials (factors), allowing for flexible modeling of both nominal and faulty behavior.
- **Inference of Faulty Components:** By defining a joint distribution over all variables, inference algorithms for computing most probable explanations (MPE) or maximum a posteriori hypotheses (MAP) estimation can compute the most likely state of the circuit, thereby identifying which components function abnormally.
- **Improved Accuracy in Complex Systems:** Markov networks provide a rigorous framework for dealing with noisy data, missing observations, and uncertainty in complex, real-world systems.

While they are powerful for representation and inference in complex systems, Markov networks can be computationally challenging, often requiring techniques that leverage the structural parameters of the graphical model for efficient inference.

Application to circuit diagnosis models

Logic circuits for digital systems contain gates performing information-processing functions expressible in Boolean logic. Fault detection and diagnosis in combinational logic circuits is a challenging engineering application.

Figure 7.3 shows an example of a combinational circuit that outputs the exclusive NOR of its inputs given by the *Boolean logic function* (Mano, 1988):

$$F = XY + \bar{X}\bar{Y} = \overline{(X \oplus Y)} \quad (7.4)$$

The Exclusive NOR (XNOR) digital logic circuit outputs a high signal (1) only when its inputs are equal (both 0 or both 1), acting as a logical equivalence detector. It performs the complement of the XOR (exclusive OR) gate, which produces a low output (0) when the inputs are different.

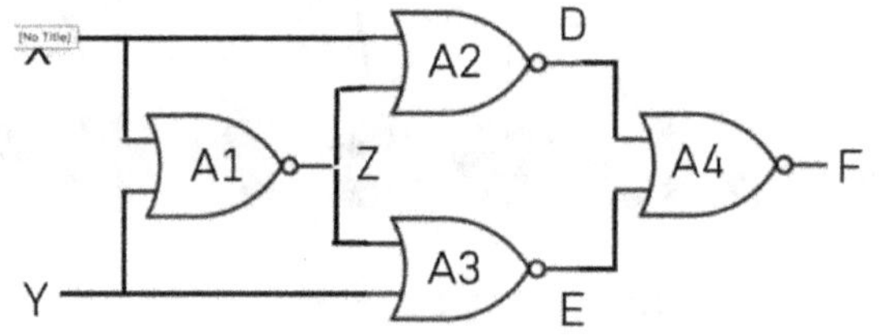

Figure 7.3—Combinatorial digital circuit for the exclusive-NOR function

Figure 7.4 shows the directed acyclic graph for the BN model of the logic circuit. The DAG has 10 nodes and 12 edges. The DAG is sparse and has at most three parents per family.

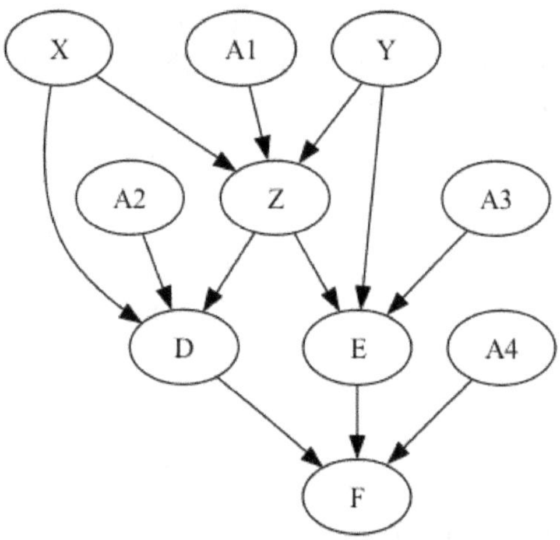

Figure 7.4—Directed acyclic graph for the digital circuit BN model

The circuit BN model is a factored representation of a complete probability distribution over model variables, given by the following product of the conditional probability distributions (CPDs) for all network variables:

$$P(X,Y,Z,D,E,F,A1,A2,A3,A4) = P_{A1}(A1)P_{A2}(A2)P_{A3}(A3)P_{A4}(A4)P_X(X)$$

$$P_Y(Y)P_Z(Z|A1,X,Y)P_D(D|A2,X,Z)P_E(E|A3,Y,Z)P_F(F|A4,D,E) \quad (7.5)$$

Figure 7.5 shows the undirected factor graph representing the structure of the circuit MN model. The MN model represents a complete probability distribution over the set of model variables specified by the set of potential functions and their component variables.

The component variables of the MN's potential functions play a pivotal role in defining the factorization of the MN's probability model. It's important to note that a Bayesian network, represented by a directed acyclic graph (DAG), corresponds to an MN. In this correspondence, the causal families, the variables, and their sets of parents in the DAG are the components of the potential functions of the equivalent MN model. The conditional probabilities of the BN become equivalent to the potential function of the MN. This relationship underscores that a DAG and associated CPDs represent the BN model, and an undirected factor graph and an associated set of potential functions represent the MN model. However, it's crucial to remember that while all Bayesian networks are directly mappable to equivalent Markov networks, the reverse is untrue.

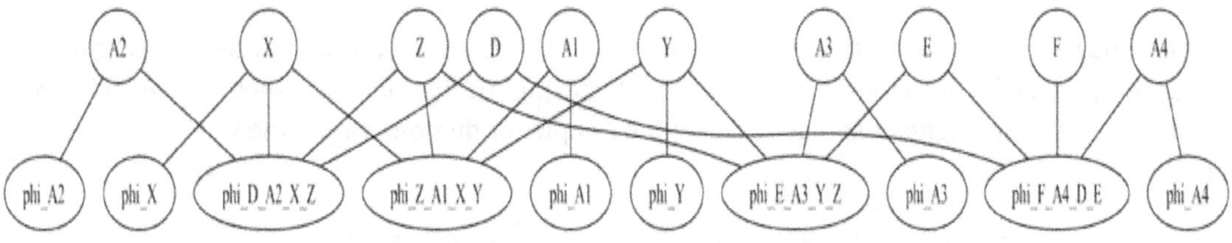

Figure 7.5—Factor graph for the Markov network of the circuit example BN

In our running circuit diagnosis example, we can see the factorization of the complete probability distributions. Here, 'complete probability distributions' refers to the full set of probabilities that describe the behavior of the circuit model. All the potential functions in this factorization are derived from the CPDs of the Bayesian network, and all the components of the potential functions are the causal families of the DAG representing the BN.

$$P(X,Y,Z,D,E,F,A1,A2,A3,A4) = \phi_{A1}(A1)\phi_{A2}(A2)\phi_{A3}(A3)\phi_{A4}(A4)\phi_X(X)$$

$$\phi_Y(Y)\phi_Z(Z,A1,X,Y)\phi_D(D,A2,X,Z)\phi_E(E,A3,Y,Z)\phi_F(F,A4,D,E) \quad (7.6)$$

Since all gates are identical in the circuit example, we only need to show the conditional probability for one gate, which will serve as the template for all four components.

Table 7.2 specifies the conditional probability for Z conditioning on A1, X and Y. A1 denotes the operating condition of the gate; it takes the value 0 to indicate the gate is OK or normal and 1 to indicate the gate is faulty or abnormal. X and Y are the inputs to the NOR gate and are Boolean, which takes the value 0 to indicate False and 1 to indicate True. The top three rows of *Table 7.2* enumerate all possible value assignments of the variables A1, X and Y. The bottom two rows in *Table 7.2* give the conditional probability of the NOR gate's output for the given value tuple of the variables A1, X and Y.

In the four leftmost columns of *Table 7.2*, $A1 = 0$ indicates that the gate is functioning correctly. In the four rightmost columns of *Table 7.2*, $A1 = 1$ signifies that the gate is abnormal or faulty. When $A1 = 0$, the value of Z equals $(\overline{X \oplus Y})$ with a probability of one. However, if $A1 = 1$, the value of Z becomes unpredictable, taking on either 0 or 1 with an equal probability of 0.5.

A1	0	0	0	0	1	1	1	1
X	0	0	1	1	1	1	0	0
Y	0	1	0	1	0	1	0	1
Z(0)	0	1	1	1	0.5	0.5	0.5	0.5
Z(1)	1	0	0	0	0.5	0.5	0.5	0.5

Table 7.2—The conditional probability table for **P(Z|X, Y, A1)**, *which models a NOR gate with output* **Z** *and inputs* **X** *and* **Y**; *here,* **A1** *indicates the operating state of the gate, where* **A1=0** *of signifies a normal state and* **A1=1** *signifies a faulty state*

We will utilize the circuit Bayesian Network (BN) example throughout this chapter to illustrate inference through belief propagation using join tree clustering. In the following section, we will explain why the circuit fault diagnosis serves as a useful example for demonstrating the inference algorithm of join tree clustering.

Why is circuit fault diagnosis a good example?

There are two main reasons that make circuit fault diagnosis an effective case study for demonstrating the inference algorithm of join tree clustering. The following sections will briefly outline each of these reasons.

Circuit diagnosis is a challenging inference task

The directed acyclic graph (DAG) of the circuit's Bayesian Network (BN) model is **multiply connected.** This means that its underlying undirected graph contains cycles (for example, D-F-E-Z-D) or that there are multiple paths between pairs of nodes in the undirected graph (for instance, there are two paths, Z-D-F and Z-E-F, connecting Z and F). Exact inference in multiply connected networks is known to be NP-hard. While efficient exact algorithms, such as belief propagation, are available for **singly connected** networks (also known as **polytrees**), the presence of cycles in multiply connected networks complicates the problem, making it computationally intractable in the worst-case scenario and requiring exponential time. Therefore, it is essential to apply algorithms that transform a multiply connected Bayesian network into a tree structure (a join tree) using the tree clustering method, which enables the use of the belief propagation algorithm for inference.

Circuit models are modular and readily quantifiable

The circuit model features a well-defined modular structure that is particularly suitable for representation through graphical models. Additionally, the functions of the circuit components are articulated using defined Boolean functions. The specifications for conditional probabilities or potential functions are derived from the established Boolean relationships among the variables of the circuit components (gates). Moreover, the failure probabilities of these components are determined based on available reliability data or experiential data drawn from historical records of the components.

The following section presents programming practice using Python to specify the BN and the MN for our circuit diagnosis example.

Practice using Python

This practice is organized in three main sessions. The first session is on graph layouts and drawings. The second session focuses on the circuit representation as a Bayesian network model. The third session focuses on the circuit representation as a Markov network model.

Graph layouts and drawings

In this practice session, we describe the code for two functions, `dagviz` and `gviz`, which are designed to handle the layout and drawing tasks for directed acyclic graphs and undirected graphs, respectively. The code utilizes functions imported from the *Graphviz* library in Python, specifically `pygraphviz`.

Depicting how graphs are structured is an important part of modeling and reasoning in probabilistic graphical models. Graphs provide a powerful and compact means of communicating complex probabilistic models to others, similar to a circuit wiring diagram. This visualization process transforms abstract mathematical concepts related to uncertainty into structured diagrams. For example, making assertions about conditional independence involves checking assertions about separations in the graph. Visualizing the graphs is also important for understanding the reasoning tasks involved in belief propagation within junction trees, which helps in comprehending the reasoning output for problems such as prediction and diagnosis.

The following code block implements the function `dagviz`. The function takes the `nodes` and `edges` for a directed acyclic graph, performs the graph layout, and creates DOT and PNG files for drawing the directed acyclic graph (DAG). It then writes these files using the specified `name`. The function `dagviz` is utilized to visualize the DAG of the circuit's Bayesian network, as illustrated in *Figure 7.4*.

```
import pygraphviz as pgv

def dagviz(nodes, edges, name):
    G = pgv.AGraph(strict=False, directed=True)
    G.add_nodes_from(nodes)
    for edge in edges:
        G.add_edge(edge)
    G.write(name + ".dot")
    # use dot
    G.layout(prog="dot")
    # write previously positioned graph to PNG file
    G.draw(name + ".png")
```

The following code block implements the function `gviz`. This function takes the `nodes` and `edges` for an undirected graph, performs the graph layout, creates DOT and PNG files for drawing the undirected graph, and saves them with the specified `name`. The function `gviz` is used to visualize the factor graph of the circuit's Markov network, as shown in *Figure 7.5*.

```
import pygraphviz as pgv

def gviz(nodes, edges, name):
```

```
    A = pgv.AGraph()
    A.add_nodes_from(nodes)
    for edge in edges:
        A.add_edge(edge)
    A.write(name + ".dot")
    # use dot
    A.layout(prog="dot")
    # write previously positioned graph to PNG file
    A.draw(name + ".png")
```

This completes the first part of the practice. Let's proceed to the second part of the practice, where we will programmatically create the circuit Bayesian network and implement related tasks.

Implementing the circuit Bayesian network model

Implementing the circuit Bayesian network model involves two main steps: structuring the model and quantifying it. We will start with the first step, which is structuring the circuit Bayesian network.

Structuring the circuit Bayesian network

Structuring a Bayesian network (BN) constitutes building its qualitative component, which is defined as the Directed Acyclic Graph (DAG) that maps variables (nodes) and their causal or probabilistic relationships (edges).

The qualitative specification of the circuit Bayesian network is based on the DAG shown in *Figure 7.4*, which consists of 10 nodes and 12 edges.

The nodes in the network consist of the following variables: A1, A2, A3, A4, Y, X, Z, E, D, and F. The variables A1, A2, A3, and A4 represent the assumptions regarding the operational conditions of the four NOR gates in the circuit. The other six variables represent the input and output variables of these gates, as shown in *Figure 7.3*.

All network variables are binary, taking values 0 and 1.

The directed edges of the DAG represent the causal relationships between the inputs, the operational assumptions of each gate, and their corresponding outputs. The directed edges depicted in the DAG shown in *Figure 7.4* are as follows: A1 -> Z, A3 -> E, Y -> Z, Y -> E, A2 -> D, X -> Z, X -> D, Z -> E, Z -> D, E -> F, D -> F, A4 -> F.

The following code block defines a function designed to generate a pgmpy `BayesianNetwork` object for the circuit model. This function incorporates the specifications of the network structure by adding a list of 10 network nodes along with 12 directed edges into the object.

```
from pgmpy.models import BayesianNetwork
def circuit():
    circuit: BayesianNetwork = BayesianNetwork()
    circuit.add_nodes_from([
        "X", "Y", "Z", "D", "E", "F", "A1", "A2", "A3", "A4"])
```

```
    circuit.add_edge("X", "Z")
    circuit.add_edge("Y", "Z")
    circuit.add_edge("A1", "Z")
    circuit.add_edge("X", "D")
    circuit.add_edge("Z", "D")
    circuit.add_edge("A2", "D")
    circuit.add_edge("Z", "E")
    circuit.add_edge("Y", "E")
    circuit.add_edge("A3", "E")
    circuit.add_edge("D", "F")
    circuit.add_edge("A4", "F")
    circuit.add_edge("E", "F")
    return circuit
```

Next, we will create a `pgmpy` `DAG` object, referred to as `G`. This will enable us to query the structure of the circuit's Bayesian network and visualize the directed acyclic graph of the circuit in a later section.

The code below defines a function called `circuit_dag`, which creates a `DAG` object named `G` and adds `nodes` and `edges` to it; these `nodes` and `edges` are properties of the `model`, a `BayesianNetwork` object that is returned by calling the `circuit()` function defined in the previous code block. The code prints a feedback message that confirms the creation of the graph. It also displays the graph `G`, providing information about its nodes and edges.

```
from pgmpy.base import DAG
def circuit_dag():
    model = circuit()
    G = DAG()
    nodes = model.nodes
    edges = model.edges
    G.add_nodes_from(nodes)
    G.add_edges_from(edges)
    return G

print(f"creating the directed acyclic graph for the circuit diagnosis example")
G = circuit_dag()
print(G)
```

Executing the code block above will create the directed acyclic graph `G` for the circuit and display the following feedback message.

```
creating the directed acyclic graph for the circuit diagnosis example
DAG with 10 nodes and 12 edges
```

After creating the DAG object G for the circuit Bayesian network, the next step is to visualize the graph. This involves generating a graph layout that will be output to a text file in DOT format and saved as a PNG file.

The following code utilizes the graph visualization function dagviz, which was defined in a previous step. This code generates both *DOT* and *PNG* files titled "*DAG_Circuit*" for the circuit diagnosis Bayesian Network (BN). The parameters for the dagviz function, specifically the nodes and edges, are derived from the DAG object G that was created in the previous step.

```
dagviz(G.nodes, G.edges, "DAG_circui")
```

Having structured the circuit's Bayesian network, we can now compute the independencies implied by the d-separations within the network's directed acyclic graph.

The following code utilizes the get_independencies() method of the pgmpy DAG object, G, to calculate all the d-separations present in G, which indicate conditional independencies in the circuit Bayesian Network (BN). The results will be written to a text file.

```
from contextlib import redirect_stdout

with open("circuit_dag_d-separations.txt", "w") as f:
    with redirect_stdout(f):
        print(G.get_independencies())
f.close()
print(
    f"Wrote all d-separations for the circuit DAG into file [circuit_dag_d-
separations.txt]"
)
```

Executing the code block above will generate a text file containing all conditional independencies in the circuit Bayesian network (BN) model and will display the following feedback message:

```
Wrote all d-separations for the circuit DAG into file [circuit_dag_d-
separations.txt]
```

This concludes the first part of the practice, where we implemented the qualitative component of the circuit Bayesian network. Now, we will move on to the second part, which focuses on quantifying the network.

Quantifying the circuit Bayesian network

The process of quantifying the circuit Bayesian network begins with creating a BayesianNetwork object. This object is generated using the circuit() function developed in the previous section.

```
model = circuit()
```

The following code block creates pgmpy `TabularCPD` objects that quantify the conditional probability distributions (CPDs) for the assumption variables A1, A2, A3, and A4. These variables represent the operating states of the four NOR gates in the circuit. Additionally, the code generates pgmpy `TabularCPD` objects for the circuit input variables X and Y, as well as the output variable Z of the A1 gate. To conserve space, we do not include the code for creating `TabularCPD` objects for the circuit variables D, E, and F, which serve as the outputs for the other three gates; this code follows the same template as that for the Z variable. It is important to note that the conditional probabilities use infinitesimally small values, epsilon, instead of zeros to ensure strictly positive probability assignments and stable numeric computation.

```
from pgmpy.factors.discrete import TabularCPD

cpd_X = TabularCPD(variable="X", variable_card=2, values=[[0.5], [0.5]])
cpd_Y = TabularCPD(variable="Y", variable_card=2, values=[[0.5], [0.5]])
cpd_A1 = TabularCPD(variable="A1", variable_card=2, values=[[0.95], [0.05]])
cpd_A2 = TabularCPD(variable="A2", variable_card=2, values=[[0.95], [0.05]])
cpd_A3 = TabularCPD(variable="A3", variable_card=2, values=[[0.95], [0.05]])
cpd_A4 = TabularCPD(variable="A4", variable_card=2, values=[[0.95], [0.05]])
cpd_Z = TabularCPD(
    variable="Z",
    variable_card=2,
    values=[
        [1e-9, 1 - 1e-9, 1 - 1e-9, 1 - 1e-9, 0.5, 0.5, 0.5, 0.5],
        [1 - 1e-9, 1e-9, 1e-9, 1e-9, 0.5, 0.5, 0.5, 0.5],
    ],
    evidence=["A1", "X", "Y"],
    evidence_card=[2, 2, 2],
)
```

After creating the CPDs for all network variables, we proceed to quantify the circuit Bayesian network (BN) by adding these CPDs to the model object. We also print a message indicating that the quantification has been completed successfully.

```
model.add_cpds(
    cpd_X, cpd_Y, cpd_Z, cpd_D, cpd_E,
    cpd_F, cpd_A1, cpd_A2, cpd_A3, cpd_A4
)

print(model.check_model())
```

The output displayed below, resulting from the execution of the code above, demonstrates that the quantification of the circuit Bayesian network was successful.

```
True
```

Let's take a moment to review our progress. We have developed a Bayesian network model for the combinatorial circuit, which captures the inputs and outputs of all four gates that constitute the circuit. Additionally, we have included the assumption variables that represent the operational states of these gates. It is important to note that these assumption variables are hidden and unobservable. The circuit Bayesian network model is specifically designed for reasoning about fault diagnosis.

This completes the implementation of the circuit Bayesian network model in Python. In the next section, we proceed to the third and last section of this practice, where we implement the Markov network representation of the circuit.

Implementing the circuit Markov network model

We will start by developing the Markov network model for the circuit and verifying its correctness. After this, we will compute the partition function for the Markov model. Finally, we will create the factor graph and visualize it.

To create the circuit Markov network, we will leverage the work already done to implement the circuit Bayesian network. We will utilize the `to_markov_model()` method from the `BayesianNetwork` object to convert the Bayesian network (BN) model of our circuit diagnosis example into a Markov network representation.

The conversion translates the Conditional Probability Distributions (**CPDs**) from the Bayesian Network into potential functions (**factors**). We utilize the `get_factors()` method for the `MarkovNetwork` object, referred to as `markovmodel`, to retrieve these factors. Subsequently, we print all the factors of the model using a **for** loop.

```
from pgmpy.models import MarkovNetwork
markovmodel = model.to_markov_model()
factors = markovmodel.get_factors()
for factor in factors:
    print(factor)
```

The circuit has ten variables and ten potential functions, as shown in the factor graph represented by the bipartite graph in *Figure 7.5*. The potential functions are `phi_X`, `phi_Y`, `phi_A1`, `phi_A2`, `phi_A3`, `phi_A4`, `phi_Z_A1_X_Y`, `phi_D_A2_X_Z`, `ph_E_A3_Y_Z`, and `phi_F_A4_D_E`.

The following two tables represent the two potential functions for the two input variables X and Y.

```
+------+----------+
| X    |   phi(X) |
+======+==========+
| X(0) |   0.5000 |
+------+----------+
| X(1) |   0.5000 |
+------+----------+
+------+----------+
| Y    |   phi(Y) |
```

```
+======+==========+
| Y(0) |   0.5000 |
+------+----------+
| Y(1) |   0.5000 |
+------+----------+
```

The table below shows the potential functions for the assumption variable A1. The potential functions for the other assumption variables A2, A3, and A4 are identical, as they all share the same probability distribution: 0.95 for state 0 (normal) and 0.05 for state 1 (faulty). Therefore, these functions are not displayed to save space.

```
+-------+-----------+
| A1    |   phi(A1) |
+=======+===========+
| A1(0) |    0.9500 |
+-------+-----------+
| A1(1) |    0.0500 |
+-------+-----------+
```

The table for the potential functions of the four gates—`phi_Z_A1_X_Y`, `phi_D_A2_X_Z`, `phi_E_A3_Y_Z`, and `phi_F_A4_D_E`—demonstrates that all four potential functions model the same NOR logic function; therefore, they are identical. We will present the potential function for one of them, specifically `phi(Z, A1, X, Y), as shown below.`

```
+------+-------+------+------+-----------------+
| Z    | A1    | X    | Y    |   phi(Z,A1,X,Y) |
+======+=======+======+======+=================+
| Z(0) | A1(0) | X(0) | Y(0) |          0.0000 |
+------+-------+------+------+-----------------+
| Z(0) | A1(0) | X(0) | Y(1) |          1.0000 |
+------+-------+------+------+-----------------+
| Z(0) | A1(0) | X(1) | Y(0) |          1.0000 |
+------+-------+------+------+-----------------+
| Z(0) | A1(0) | X(1) | Y(1) |          1.0000 |
+------+-------+------+------+-----------------+
| Z(0) | A1(1) | X(0) | Y(0) |          0.5000 |
+------+-------+------+------+-----------------+
| Z(0) | A1(1) | X(0) | Y(1) |          0.5000 |
+------+-------+------+------+-----------------+
| Z(0) | A1(1) | X(1) | Y(0) |          0.5000 |
+------+-------+------+------+-----------------+
| Z(0) | A1(1) | X(1) | Y(1) |          0.5000 |
+------+-------+------+------+-----------------+
| Z(1) | A1(0) | X(0) | Y(0) |          1.0000 |
```

```
+------+-------+------+------+----------------+
| Z(1) | A1(0) | X(0) | Y(1) |         0.0000 |
+------+-------+------+------+----------------+
| Z(1) | A1(0) | X(1) | Y(0) |         0.0000 |
+------+-------+------+------+----------------+
| Z(1) | A1(0) | X(1) | Y(1) |         0.0000 |
+------+-------+------+------+----------------+
| Z(1) | A1(1) | X(0) | Y(0) |         0.5000 |
+------+-------+------+------+----------------+
| Z(1) | A1(1) | X(0) | Y(1) |         0.5000 |
+------+-------+------+------+----------------+
| Z(1) | A1(1) | X(1) | Y(0) |         0.5000 |
+------+-------+------+------+----------------+
| Z(1) | A1(1) | X(1) | Y(1) |         0.5000 |
+------+-------+------+------+----------------+
```

To verify that the created circuit Markov model, `markovmodel`, is valid, we use the validation function of the `check_model()` method in the `MarkovModel` class.

```
markovmodel.check_model()
```

The output from executing the code above indicates that the circuit Markov model is valid.

```
True
```

Moving on, let's proceed to the task of getting the partition function of the circuit Markov model. Recall, the partition function, Z, defined in *Equation 7.3,* is a normalization constant to ensure the factored form of *Equation 7.2* is a joint probability distribution.

The following code utilizes the `get_partition_function()` method of the `MarkovNetwork` object, `markovmodel`, to obtain the partition function for the Markov model factors. This method returns a value that normalizes the product of all factors, creating a probability distribution when divided by the partition function. A probability distribution must sum to 1 across all network variables and their corresponding domain values. Given that the Markov model factors are a direct mapping from the Bayesian network conditional probability distributions (CPDs), we anticipate that the potential function equals 1. This indicates that the factorization of the Markov model is already normalized.

```
markovmodel.get_partition_function()
```

The output from executing the above code yields a value of 1, which aligns with our expectations. For a general Markov network, the partition function Z is typically not equal to 1 because potential functions (factors) are not required to be normalized probability distributions. Only when the specification of potential functions in a Markov network is derived from a Bayesian network's conditional probability distributions will the partition function equal 1, as in our case.

```
1.0
```

Lastly, we come to the task of creating and visualizing the factor graph for the circuit Markov network.

To obtain the factor graph for the circuit Markov model, we utilize the `to_factor_graph()` method of the `MarkovModel` object `markovmodel` to create a graph object called `factor_graph`.

Afterward, we print the nodes and edges of the `factor_graph` object. Recall that a factor graph is a bipartite undirected graph that includes two distinct subsets of nodes. One subset of the factor graph nodes consists of the model variables, while the other subset comprises the factors representing the Markov model. The edges in the factor graph connect the factor nodes to the variable nodes on which the factors are defined.

```
factor_graph = markovmodel.to_factor_graph()
print(factor_graph.nodes)
print(factor_graph.edges)
```

The output from executing the code produces two lists: one list contains the nodes of the factor graph, while the other list includes the edges of the factor graph.

```
['X', 'Z', 'D', 'Y', 'A1', 'A2', 'E', 'A3', 'F', 'A4', 'phi_X', 'phi_Y',
'phi_Z_A1_X_Y', 'phi_D_A2_X_Z', 'phi_E_A3_Y_Z', 'phi_F_A4_D_E', 'phi_A1', 'phi_
A2', 'phi_A3', 'phi_A4']
[('X', 'phi_X'), ('X', 'phi_Z_A1_X_Y'), ('X', 'phi_D_A2_X_Z'), ('Z', 'phi_Z_
A1_X_Y'), ('Z', 'phi_D_A2_X_Z'), ('Z', 'phi_E_A3_Y_Z'), ('D', 'phi_D_A2_X_Z'),
('D', 'phi_F_A4_D_E'), ('Y', 'phi_Y'), ('Y', 'phi_Z_A1_X_Y'), ('Y', 'phi_E_
A3_Y_Z'), ('A1', 'phi_Z_A1_X_Y'), ('A1', 'phi_A1'), ('A2', 'phi_D_A2_X_Z'),
('A2', 'phi_A2'), ('E', 'phi_E_A3_Y_Z'), ('E', 'phi_F_A4_D_E'), ('A3', 'phi_E_
A3_Y_Z'), ('A3', 'phi_A3'), ('F', 'phi_F_A4_D_E'), ('A4', 'phi_F_A4_D_E'),
('A4', 'phi_A4')]
```

To visualize the factor graph, we utilize the custom `gviz` function, which was defined earlier in this practice, to generate *DOT* and *PNG* files named *"factor_graph"* for the circuit diagnosis Markov model. The nodes and edges parameters for the `gviz` function correspond to the nodes and edges of the graph object named `factor_graph`.

```
gviz(factor_graph.nodes, factor_graph.edges,"factor_graph")
```

The *PNG* file generated by the code output displays the undirected bipartite graph, which is illustrated in *Figure 7.5*.

This concludes all three sessions of this practice section.

Wrapping up this section on probabilistic graphical models, let's summarize the key takeaway points. Bayesian and Markov networks are two categories of probabilistic graphical models. They differ in how they represent a complete probability distribution through factorization. Bayesian networks are represented by directed acyclic graphs, and their factorization is based on conditional probability distributions. Markov networks are represented by undirected graphs and factor graphs, with factorization based on potential functions or factors. All Bayesian networks can be mapped to equivalent

Markov networks. However, a Markov network cannot always be mapped to an equivalent Bayesian network. Circuit fault diagnosis is an application that is well-suited for representation in both Markov networks and Bayesian networks.

Now, let's move on to the next section, where we will formally outline the most common inference tasks in probabilistic graphical models and provide an overview of inference algorithms for reasoning with these models.

Inference tasks and inference algorithms

We start this section by describing the most common inference tasks in probabilistic graphical models, particularly in Bayesian, and Markov networks. Next, we will provide an overview of algorithms for exact inference, with a focus on structure-based algorithms.

Inference tasks

The three primary inference tasks (or queries) defined within probabilistic networks, such as Bayesian and Markov networks, include updating beliefs, calculating the most probable explanation (MPE), and computing the maximum a posteriori (MAP) hypothesis.

The first task, updating beliefs (or computing posterior marginals), serves as the foundational inference task that underpins all other tasks.

The second task, finding the most likely explanation (MPE), is to find the instantiation of all unobserved variables that gives the highest posterior probability based on the evidence from observed variables.

The third task, which involves computing the maximum posterior hypothesis, entails identifying the most likely value assignment for a subset of unobserved variables (referred to as the hypothesis) based on the evidence provided by the observed variables.

In the following section, we provide a formal statement of the first inference task.

Updating beliefs

Updating beliefs is the core of Bayesian inference, which is updating the probability of a hypothesis as evidence becomes available. Bayes' theorem provides the mathematical foundation for the mechanism of belief updating. The mathematical formula helps answer questions like, "What's the probability that a patient has a disease given they tested positive?" when you know the probability of testing positive given they have the disease. For more on the mathematical background of Bayes' formula, please refer to *Chapter 2* in the book.

Let $X = x$ represent a value assignment for a subset of network variables $X \subset V$ that correspond to the evidence. Let Y denote the set of all network variables not included in X, defined by the set difference $V \backslash X$. A hypothesis can be expressed as a value assignment to Y, for example, $Y = y$. According to Bayes' formula, the probability of the hypothesis $Y = y$ given the evidence $X = x$ can be formulated as follows:

$$P(Y = y \mid X = x) = P(X = x|Y = y)P(Y = y)/P(X = x) \quad (7.7)$$

The Bayes formula expresses the posterior probability distribution of Y given the evidence on X. The conditional probability $P(X = x|Y = y)$ and the marginal probabilities $P(Y = y)$ and $P(X = x)$ can be derived from the joint probability $P(X = x, Y = y)$.This joint probability is represented by the factorization formulas found in either *Equation 7.5* for Bayesian networks or *Equation 7.6* for Markov networks.

Next, we state the inference task of calculating the most probable explanation.

Calculating the most probable explanation (MPE)

Let $X = x$ be a value assignment, called evidence, on a subset of network variables $X \subset V$, and let Y be the set difference $V \backslash X$. The task of calculating the MPE is equivalent to finding an assignment $y^* = (y_1^*,...,y_k^*), k = |Y|$, such that:

$$y^* = \underset{y}{\operatorname{argmax}}\, P(Y = y|X = x) \text{ (7.8)}$$

Where $P(Y = y|X = x)$ is the posterior probability of *Equation 7.7.*

The MPE task is about finding the value assignments for all the unobserved variables that align with the evidence on the observed variables and have the highest posterior probability. The MPE task can be seen as performing **abductive reasoning under uncertainty**, where we are provided with evidence on effects, and we aim to find the most likely explanation of all the mechanisms causing the observed effects. This task is particularly useful in fault diagnosis, as it can generate the most probable instantiations for the subset of variables that represent assumptions about the operational status of the system components, thereby assigning blame to the components responsible for the observed evidence.

Let us state the third inference task, namely computing the **Maximum a Posteriori Hypothesis** (MAP).

Computing the maximum a posteriori hypothesis (MAP)

Let M, Y, and X be a partitioning of the network variables V, where X denotes the evidence variables, M the MAP variables called hypotheses, and Y contains the remaining variables in V whose values we are not directly concerned about, and which are the variables we are to marginalize out. The MAP task is to compute the most likely value assignment for the MAP variables conditioning on the evidence $X = x$:

$$m^* = \underset{m}{\operatorname{argmax}} \textstyle\sum_y P(M = m, Y = y|X = x) \text{ (7.9)}$$

where $P(M = m, Y = y|X = x)$ is the posterior probability of *Equation 7.7* where the subset of variables in Y is assigned the value tuple y, and M is assigned the value tuple m, and the summation is over all values of y from the respective domains of variables in Y.

Thus, the MAP task is given the evidence on observed variables; we need to compute a value assignment for a subset of the unobserved variables. For example, in fault diagnosis, the MAP task would generate the most probable instantiations for the subset of variables representing assumptions on the operational status (e.g., normal or abnormal; okay or faulty) of the system components that assign blame to the culprit components consistent with the observed evidence. We will demonstrate an example of the MAP task on our circuit diagnosis example.

Having defined the various inference tasks, let us turn next to an overview of the inference algorithms for performing those inference tasks.

Inference algorithms

Probabilistic inference in Bayesian and Markov networks is governed by the laws of probability, including Bayes' theorem and rules. The emphasis is on developing inference algorithms for accomplishing inference tasks with lesser demands on processing time and space requirements. The algorithms should take advantage of the factorization schemes of the complete probability distribution by leveraging the structural properties of the graphical models of Bayesian and Markov networks.

There are two types of inference algorithms: **Exact** and **Approximate**. The computational complexity of exact inference in Bayesian networks, e.g., finding the posterior probability, is known to be NP-hard, see Cooper (Cooper, 1990). That complexity result follows from the fact that we can map the 3-SAT problem, an NP-complete problem, to a problem of finding posterior probability in Bayesian networks. The 3-SAT problem (Garey, et al., 1979) states that given a collection of clauses $C = \{c_1, \cdots, c_m\}$ on a finite set U of variables such that $|c_i| = 3$ for $1 \leq i \leq m$, determine if there is a truth assignment for U that satisfies all the clauses in C. It has also been proven that even approximate inference in Bayesian networks is also NP-hard, see Dagum and Luby (Dagum, et al., 1993).

The focus of this book is on exact algorithms for inference, and we will describe two categories of algorithms: one is called **tree clustering**, and the other is **variable elimination.** Both algorithms play a crucial role in solving the inference tasks in Bayesian networks. They are part of a class of inferencing algorithms called structure-based algorithms, whose complexity can be quantified using structural parameters (Dechter, et al., 2001). A key parameter is the treewidth, defined in *Chapter 6* on Join Tree clustering. In *Chapter 8*, we will present the variable elimination method. These algorithms, by leveraging the structural properties of the BN's directed acyclic graph, provide practical solutions to the complex inference tasks in Bayesian networks.

The three most common inference tasks in Bayesian and Markov networks are updating beliefs, calculating the MPE, and computing the MAP. Two algorithms for exact inference, the join tree clustering and the variable elimination, are examples of structure-based algorithms whose complexity can be quantified using structural parameters.

In the next section, we will revisit the topic of join tree clustering discussed in *Chapter 6* and explore how to map Markov models into join trees.

From Markov models to join trees

In this section, we focus on join trees and their connection to Markov models. First, we explain why the Join tree structure is important. Next, we describe what constitutes a join tree, a crucial concept in probabilistic graphical models. We follow with a section on the multiplicity of join trees, meaning that numerous equivalent join tree representations exist in a given model, and how the structural parameters of the join tree impact the computational complexity of the inference tasks. Lastly, we demonstrate in a practice section how to transform a Markov model programmatically to a join tree using our running circuit example.

Why the join tree structure is important

Inference in Bayesian and Markov networks is NP-hard, indicating that no polynomial-time algorithm exists for exact probabilistic reasoning in general networks, as established by Cooper in 1990 (Cooper, 1990).

Computing exact posterior probabilities in Bayesian and Markov networks with general graph structure is computationally intractable because it involves summing over an exponential number of possible states.

In singly connected (tree-structured) Bayesian and Markov networks, the belief propagation (BP) algorithm, also referred to as the sum-product algorithm, computes exact inference by updating beliefs in polynomial time.

The join tree structure is important because it represents a singly connected network where we can leverage the BP algorithm for inference tasks.

The algorithm of join tree clustering (also known as the **junction tree algorithm** or **clique tree algorithm**) is a fundamental exact inference method for probabilistic graphical models, such as Bayesian or Markov networks. It transforms a potentially complex, multiply connected network into a tree structure of clusters (a "join tree") to allow efficient belief propagation.

This transformation is necessary because belief propagation is guaranteed to converge on a tree structure (where only one path exists between any two nodes) but can behave uncontrollably on graphs with cycles.

The join tree clustering is designed to perform that mapping from the undirected acyclic graph of the Bayesian network or the factor graph of the Markov network into a tree. Join tree clustering (Lauritzen, et al., 1988) belongs to the class of structure-based algorithms for multiply connected networks, which also includes variable elimination (Dechter, 1999), cutset-conditioning (Pearl, 1988), and hybrid methods (Dechter, et al., 2001). As we will explain in the following sections, structure-based algorithms for inference are practical only if the problem has a sparse graph or small treewidth.

In the next section, we describe what constitutes a join tree.

What is in a join tree?

This section reviews the requirements for a join tree to represent a probabilistic graphical model of a Bayesian or Markov network.

A join tree is derived from the undirected graph of a probabilistic graphical model. In the case of Markov models, the undirected graph is readily available with no preprocessing needed. For a Bayesian network, the undirected graph is computed from the network's directed acyclic graph in a preprocessing step called moralization. Moralization involves two steps: first, connecting all parents of each node in the directed acyclic graph with undirected edges; second, removing the directionality of all edges.

Let us recall from *Chapter 6* that a join tree is computed by first triangulating the undirected graph of the probabilistic graphical model, followed by steps of finding the maximum cliques and connecting the cliques in a clique tree having the **running intersection property** (RIP). Triangulating an undirected graph involves adding edges to make the graph chordal, where every cycle of length greater than 3 has a chord. A chord is an edge that joins two non-consecutive cycle nodes. Recall a clique: a set of variables representing nodes in the triangulated graph that are all pairwise connected. A maximum clique is a clique not contained in any other clique. The RIP means that every pair of nodes in the join tree has their subset of shared variables contained in every edge along the path between the pair in the tree.

The following section illustrates how multiple join tree representations can be used for the same probabilistic graphical model.

Multiplicity of join trees

The join tree representation is not a rigid construct; rather, it serves as a flexible tool. For the same probabilistic graphical model, multiple join trees with varying structures can be derived. This flexibility stems from the ability to triangulate the undirected graph of the probabilistic graphical model in different ways, leading to various clique trees. It is important to note that minimum triangulation, which requires the fewest number of fill-ins and results in the smallest maximum cliques, is unique and is known to be NP-hard (Yannakakis, 1981).

Given a join tree for a probabilistic graphical model, multiple join trees can be derived by merging neighboring clique nodes. This technique of merging cliques results in the creation of secondary join tree models. Such a merger exemplifies a practical approach for trading space for time in the inference tasks related to updating beliefs and computing the Most Probable Explanation (MPE) (El Fattah, et al., 1996).

Two structural parameters for a join tree structure are used to evaluate the inference's time and space complexity with join tree clustering. The two parameters are the induced width, r (the treewidth, which is the maximum clique size minus 1), and the separator size, s, the cardinality of the separator set with the most elements. The treewidth is used to bound the time complexity of the tree-clustering algorithm, and the separator size is used to bound the space complexity of the tree-clustering algorithm. The bounds on the computational complexity of tree clustering are stated in the following theorem.

Theorem 7.1: time-space complexity of join-tree clustering

Given that a belief network has n variables, whose moral graph can be embedded in a clique-tree with induced treewidth r and separator size s, the time complexity for updating beliefs and computing the MPE by a join-tree clustering algorithm is $O(n\ exp(r))$. Its space complexity is $O(n\ exp(s))$.

Theorem 7.1 indicates that the time complexity of inference using tree clustering increases exponentially based on the maximum clique size multiplied by the number of variables. Additionally, the space complexity grows exponentially with the maximum separator width multiplied by the number of variables.

Space complexity often poses a significant bottleneck for inference in graphical models that employ join tree clustering. This issue arises because the algorithm necessitates the storage of large, high-dimensional joint probability tables for the clique separators during the "message passing" phase of belief propagation. To mitigate this space bottleneck in tree clustering inference, it is possible to trade off space for time by utilizing secondary join trees.

When the primary join tree T_0 requires excessive memory, it is possible to generate secondary join trees T_i to reduce space usage, albeit at the expense of increased computation time for inferring the subproblems of the cliques. The process of creating secondary join trees involves taking the primary join tree and merging nodes (cliques) that are separated by large separators. This merging reduces the size of the separators. The separators are organized in descending order by size $s_0, s_1, \cdots, s_n$, and secondary join trees T_i are formed by combining adjacent cliques with separator sizes that exceed a designated threshold, s_i. As the maximum separator size s_i in the secondary tree T_i decreases, the space complexity also decreases. In contrast, the time complexity increases as the size of the maximum clique $r_i + 1$ grows due to the merging process, where r_i denotes the induced tree width of the secondary tree T_i.

Example: time-space tradeoff

Primary Tree T_0: $O(n\, exp(s_0))$ space (maximum memory), $O(n\, exp(r_0))$ time (fastest).

Secondary Tree T_i: $O(n\, exp(s_i))$ space (smaller memory, where $s_i < s_0$), $O(n\, exp(r_i))$ time (slower, where $r_i > r_0$).

Concluding this section, let's summarize the main points. There are multiple join trees representing a Bayesian network or a Markov network. The main structural parameters of a join tree are the tree-width and the separator size, which define bounds on the time and space complexity of the inference tasks of updating beliefs and calculating the MPE.

In the following section, we provide a walkthrough on how to programmatically, using Python and the `pgmpy` library, perform the transformation from the Markov model to a join tree applied to the circuit diagnosis example.

Practice using Python

In this section, we demonstrate a practical hands-on practice that utilizes the powerful Python `pgmpy` library. We design this section to equip you with the necessary skills to implement join tree structures in real-world scenarios. We demonstrate through code how to obtain the join tree representation for the Markov network model of our running circuit diagnosis example. We will use the join tree representation for inference in another practice session in an upcoming section.

The code in this practice is included in the *Jupyter* notebook `'chapter7.ipynb'`. This practice builds upon the code and references objects from the previous practice session.

This practice consists of three tasks. The first task is to transform the circuit Markov model into a join tree. The second task involves listing the maximum cliques, which correspond to the nodes in the join tree. The third task entails plotting the join tree and saving the plot in both DOT and PNG file formats.

We begin with the first task.

Transforming the circuit Markov model into a join tree

The following code gets the junction tree structure built by the pgmpy library using the `to_junction_tree()` method of the Markov model object, `markovmodel`. The code prints the factors for the junction tree model for our circuit diagnosis example.

```
from pgmpy.models import JunctionTree
JT = markovmodel.to_junction_tree()
print(JT.factors)
```

The code generates a representation of the junction tree object. It lists three potential functions, with parameters based on the variables of the corresponding maximum cliques, which serve as the nodes of the tree object. The code output is as follows:

```
[<DiscreteFactor representing phi(A3:2, E:2, Y:2, Z:2) at 0x105581310>,
<DiscreteFactor representing phi(E:2, Z:2, Y:2, A4:2, A1:2, F:2, X:2) at
0x105580e30>, <DiscreteFactor representing phi(A2:2, D:2, X:2, Z:2) at
0x138c75e80>, <DiscreteFactor representing phi(D:2, E:2, Z:2, A4:2, F:2, X:2)
at 0x13fae4710>]
```

Let's proceed to the second task.

Listing the maximum cliques

The following code prints the nodes of the junction tree object, `JT`, computed by the pgmpy library. The result shows three maximum cliques.

```
print(JT.nodes)
```

The output lists three sets of variables which are the maximum cliques in the triangulated graph of the circuit Markov model.

```
[('Z', 'A3', 'E', 'Y'), ('Z', 'X', 'E', 'A4', 'F', 'A1', 'Y'), ('Z', 'X', 'A2',
'D'), ('Z', 'X', 'E', 'A4', 'F', 'D')]
```

We continue with the third and final task of the practice, which involves plotting the join tree and saving it to files.

Visualizing the pgmpy join tree

The following code uses our `gviz()` function to create the *DOT* and *PNG* files named `pgmpy_JT` for the circuit diagnosis Join Tree computed by the pgmpy library. The parameters of the function `gviz()` are the nodes and edges of the *JunctionTree* object `JT`.

```
gviz(JT.nodes, JT.edges, "pgmpy_JT")
```

Figure 7.6 shows the output, which depicts the join tree `JT` computed by the pgmpy library. Note that the join tree JT has three cliques, and the maximum clique has size 6.

```
[('E', 'Z', 'X', 'Y', 'A3', 'A1'), ('E', 'Z', 'X', 'A2', 'F', 'D'), ('E', 'A4',
'F', 'D')]
```

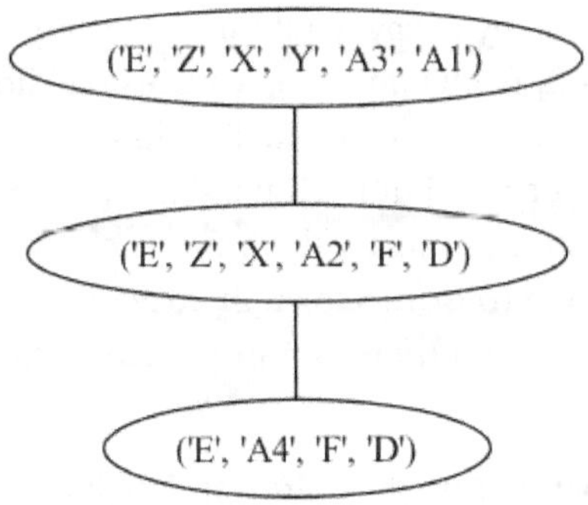

Figure 7.6—Join tree computed by the pgmpy library for the circuit diagnosis BN

Please note the join tree you get computed by pgmpy is more likely than not different from the one shown in *Figure 7.6*. Experimenting with pgmpy confirms the variability of the Join tree, which is computed at different run times. As we demonstrated, there are multiple join tree representations of triangulated graphs and multiple triangulations for the same undirected graph, so this outcome is not unexpected.

In the following section, we compare the join tree in *Figure 7.6* with the join tree from *Chapter 6*.

Comparing the join trees

Figure 7.7 displays a side-by-side comparison of the join trees. On the left, we have the join tree obtained by implementing the tree clustering algorithm, which corresponds to *Figure 6.11* from *Chapter 6*. On the right is the join tree produced automatically by the pgmpy package, derived from the transformation of the Markov model as shown in *Figure 7.6*.

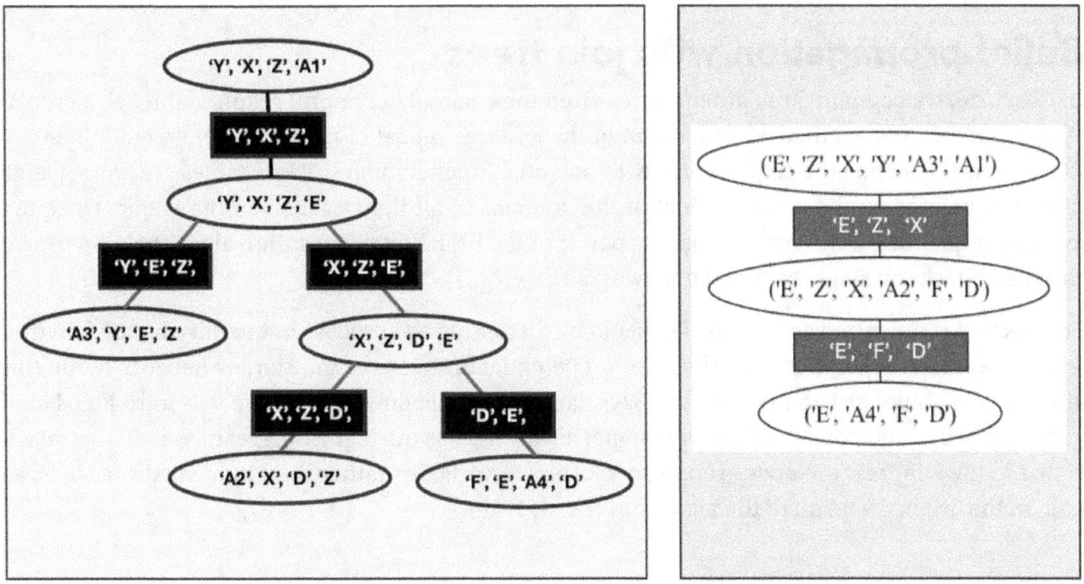

Figure 7.7— Two junction trees for the circuit example; the tree to the left is computed in Chapter 6, while the tree on the right is generated by pgmpy

The join tree on the right, generated by pgmpy, contains three cliques: two of size 6 and one of size 4. In contrast, the join tree on the left, from *Chapter 6*, features six cliques, each of size 4. The join tree computed by pgmpy has a maximum clique size equal to 6 (induced treewidth, $r = 5$) compared to our join tree from *Chapter 6*, which has a maximum clique size equal to 4 (induced width, $r = 3$). Having a smaller induced width can drastically impact the inference's time complexity, as per *Theorem 7.1*. The separator sets (sepsets) for each tree are shown in boxes between neighboring nodes. Both join trees have the same separator size, and therefore inference using either tree has the same space complexity as per *Theorem 7.1*.

The key takeaways from the comparison are the following. The join tree computed in *Chapter 6* through the implementation of the tree clustering algorithm gives you greater control over the representation of the join tree. The use of automated methods, such as those provided by the pgmpy package, is convenient because it handles the triangulation, maximum clique, and join tree construction steps for you. However, this automated process operates in an opaque manner, rendering the individual steps unclear and depriving you of control over the final result.

Concluding this section, let's recap: we reviewed the definition and properties of a join tree representing a probabilistic graphical model. We explained that a graphical model can have multiple join tree representations and cited a theorem that bounds the computation complexity of inference based on structural parameters. In the practice section, we showed how to get the join tree for the probabilistic graphical model of our circuit diagnosis example.

In the following section, we will use both the pgmpy join tree and the join tree we computed step by step by applying the tree clustering algorithm from *Chapter 6* to perform belief propagation for inference tasks within Bayesian networks.

Belief propagation with join trees

In *Chapter 6*, we computed the junction tree from the triangulated moral graph of the BN, a crucial step in building the qualitative component of the join tree model of the BN. We then quantified the join tree by populating the clique nodes with potential function tables. These tables are nonnegative functions defined on the cross product of the domains of all the variables in the clique. Once the join tree is quantified, we can leverage the power of the **BN belief propagation algorithm** to perform various BN tasks on singly connected networks.

This section is organized as follows. We begin by discussing the process of quantifying the join tree, which is a crucial aspect based on the factored potential functions of the Markov network or the conditional probability distributions of the Bayesian network. Following that, we will introduce belief propagation in singly connected networks, specifically trees. A distinguishing feature of this approach is that in the join tree, the nodes represent clusters of variables rather than individual variables, as seen in the original version of the algorithm (Pearl, 1988).

Quantifying the join tree

In this section, we describe the quantification of the join tree representation of graphical models of Bayesian and Markov networks. Quantifying a join tree model amounts to building the clique tables that specify the relations between clique variables for each node in the join tree model.

We begin by describing the process of building the clique tables. Following this, we demonstrate step-by-step how to apply the procedure for building the clique tables on the Bayesian network of our running circuit example.

Building the clique tables

Building the clique tables comprises two steps. The first step involves pairing the factors with buckets designated for each clique. For Bayesian networks, the factors are the conditional probability tables (CPTs), and for Markov networks, the factors are the tables for the potential functions. Following this, the second step is multiplying all the functions for the factors within the bucket to compute the clique's table.

Pairing factors with cliques is done by applying two rules:

- A factor is paired to a bucket whose clique contains all the variables on which the factor's function is defined.
- A factor is paired only to a single bucket.

The clique table for a bucket is calculated as the product of all factors within it. If a bucket is empty—indicating that no factors are assigned to it—the clique table defaults to a unit function, which assigns a value of 1 to all possible instantiations of the variables in the bucket's clique.

The following formally describes, in pseudocode, the procedure for populating the clique tables. We included comments, the lines beginning with '//', aimed at describing what each code block does.

Algorithm: populate clique tables

Input: 1) the clique set: $C_1, \cdots, C_m$; 2) the set of functions of the factors $\Phi = \{\varphi_i(S_i) | i = 1. \cdots, n\}$

Initialize: $f_i(C_i) \leftarrow 1, \forall i = 1. \cdots, m.$

Create empty buckets, one bucket for each clique.

```
// Assign a factor to a clique that contains all the variables in the factor
For each factor φ(S) in Φ:
    Find a bucket for a clique C_i where S ⊆ C_i
    Put φ(S) in the matching bucket.
// Multiply all the factors in every bucket to create the clique tables
    For each C in maxCliques:
        // build the clique table
        f(C) ← ∏_{φ(S)∈bucket(C)} φ(S)
Return {f(C): C ∈ maxCliques}
```

This concludes the description of the procedure for creating the clique tables.

In the next section, we will implement this procedure in the Bayesian network related to our running circuit example.

Quantifying the join tree of the circuit example

In this section, we provide a detailed demonstration of how to implement the construction of clique tables using the join tree from the circuit example. We will utilize the join tree derived from tree clustering in *Chapter 6*, which is illustrated as the tree on the left side of *Figure 7.7*.

Table 7.3 displays the results of the algorithm used to construct the clique tables based on our circuit example. The first two columns show the maximum cliques along with their respective orderings. The third column details the bucket associated with each clique, along with the potential functions (also known as factors), which represent the conditional probability tables for the variables within the clique. The fourth column presents the clique potential function, calculated as the product of all the functions in the clique's bucket. It is important to note that for empty buckets, the function of the cliques equals 1 for all possible value instantiations of the clique variables. This scenario occurs with the second and fourth buckets, which correspond to the second and fourth rows in the table. The upcoming practice section will include a numerical example of the clique tables.

#	Clique	Bucket	Clique Potential
1	$C_1\colon\{Z,X,Y,A1\}$	$\varphi_X = P(X)$ $\varphi_Y = P(Y)$ $\varphi_{A1} = P(A1)$ $\varphi_Z = P(Z\|X,Y,A1)$	$f_1(Z,X,Y,A1) = \varphi_X.\varphi_Y.\varphi_{A1}.\varphi_Z$
2	$C_2\colon\{Z,X,Y,E\}$	$\emptyset$	$f_2(Z,X,Y,E) = 1$
3	$C_3\colon\{Z,A3,Y,E\}$	$\varphi_{A3} = P(A3)$ $\varphi_E = P(E\|Y,Z,A3)$	$f_3(Z,A3,Y,E) = \varphi_{A3}.\varphi_E$
4	$C_4\colon\{Z,X,D,E\}$	$\emptyset$	$f_4(Z,X,D,E) = 1$
5	$C_5\colon\{Z,X,A2,D\}$	$\varphi_{A2} = P(A2)$ $\varphi_D = P(D\|X,Z,A2)$	$f_5(Z,X,A2,D) = \varphi_{A2}.\varphi_D$
6	$C_6\colon\{A4,E,D,F\}$	$\varphi_{A4} = P(A4)$ $\varphi_F = P(f\|E,D,A4)$	$f_6(A4,E,D,F) = \varphi_{A4}.\varphi_F$

Table 7.3— Six buckets corresponding to the six cliques in the join tree of the circuit example; the clique potential is calculated as the product of all factors present in the clique's bucket

Now that we have explained how to construct the clique tables for a join tree and demonstrated the process using our circuit example, we will proceed to the next section. This section will focus on inference with join trees through **belief propagation**.

Inference by propagating beliefs

In this section, we outline the algorithm for inference via belief propagation in the tree-structured representation of the Bayesian Network (BN), as represented by the join tree model and the associated clique tables. We will demonstrate this algorithm step-by-step using a simple BN example illustrated by a join tree, emphasizing its significance and practicality.

The following outlines the belief propagation algorithm using join trees, often referred to as the junction tree algorithm. This algorithm requires as input a join tree, which is represented by cliques arranged in a specific breadth-first order of the tree nodes. In this arrangement, the root node is listed first, with each node preceding its children.

The algorithm requires input that includes the separator sets and the parent-child mappings within the tree. It computes the posterior probability of a subset X from the root clique, given evidence e. This evidence comprises observed values for a subset of the variables Y in the Bayesian network.

Algorithm: inference with join tree clustering

Input: 1) query variables X; 2) evidence e: $(Y = y)$; 3) list of maximum cliques $C_1, \cdots, C_m$ for a breadth-first ordering of the join tree where $C_1 \supseteq X$; the parent $pa(i)$, the children $ch(i)$, and the separator $S_i = C_i \cap C_{pa(i)}$; the clique potentials, $f_i(C_i), i = 1, \cdots, m$.

Output: the belief of X, given e.

For $i = m$ downto 1:

- **Incorporate the evidence**: for every observed variable Y_i member of C_i with value y_i in e, set $Y_i = y_i$ in the clique potential $f_i(C_i)$, removing all tuples with $Y_i \neq y_i$ from the clique table. $f_i(C_i) \leftarrow f_i(C_i|e)$
- **Process received messages:** Multiply the clique potential by the functions on the separator sets, which are the messages received from all the children, $f_i(C_i) \leftarrow f_i(C_i) \times \prod_{k \in ch(i)} f_{k,i}(S_k)$
- **Send messages:** Marginalize the clique potential on variables in C_i, not in the separator set S_i. Marginalizing the function means summing over all values u_i of variables U_i not in the separator set, $U_i = C_i - S_i$. Send the message $f_{i,p(i)} = \sum_{u_i} f_i(C_i)$ to the parent clique $C_{pa(i)}$.

Compute $Bel(X = x)$ by marginalizing $f_1(C_1)$ on all values of variables in $U_1 = C_1 - X$, and dividing by the normalization's constant $Z = \sum_{c_1} f_1(C_1)$.

Return $Bel(x) \leftarrow \frac{1}{Z} \sum_{u_1} f_1(C_1)$.

To illustrate how inference with tree clustering works, we will present an example of a simple join tree. We will apply the steps of the algorithm to compute the beliefs about unobserved variables, based on the evidence provided by one or more observed variables.

Example: inference with tree clustering

Figure 7.8 shows a small Bayesian network we used previously in *Chapter 5*. We use the example to illustrate how the junction tree algorithm works. The figure shows the Bayesian network on the left and the join tree on the right. The figure also shows the conditional probability tables (CPTs) beside the BN's DAG on the left. The join tree on the right shows the cliques, which are the sets, $\{A, B\}$ and $\{B, C\}$, shown as the nodes of the join tree. The singleton set {B} is the separator set.

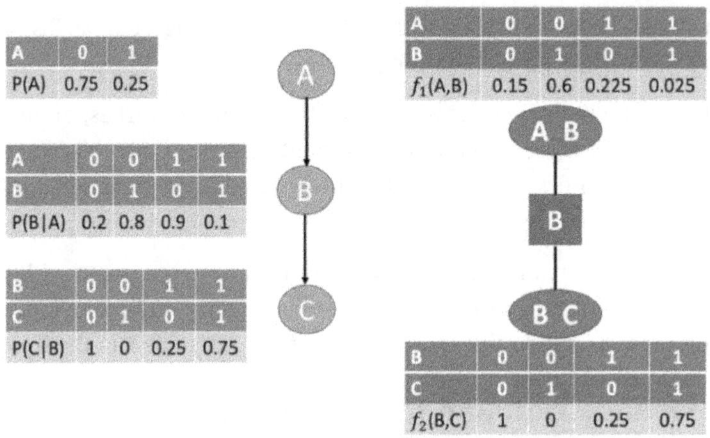

Figure 7.8— A Bayesian network and its CPTs and corresponding junction tree and clique tables

The figure shows the calculated clique tables on the right side, next to the cliques, which are functions over the variables in the clique. For Clique #1, $\{A, B\}$, the clique table is given by the function $f_1(A, B) = P(A)P(B|A)$. For Clique #2, $\{B, C\}$, the clique table is given by the function $f_2(B, C) = P(C|B)$.

Suppose we collected evidence on the variable C having the value 1. Let us apply the belief propagation on the join tree to infer the posterior probability of the variable A.

The first step of the algorithm is to incorporate the evidence. The step implies that all tuples in a clique table with a value of C other than the one given by the evidence are removed. *Figure 7.9* depicts step #1, where the clique table $f_2(B, C)$ has only tuples in which $C = 1$; tuples with $C = 0$ are removed. Step #2 on the right in *Figure 7.9* shows the message $f_{2,1}(B)$ sent to the parent. The message marginalizes $f_2(B, C)$ by summing over all values of C, and since C has only one value, the message will be the projection of the function on the B variable.

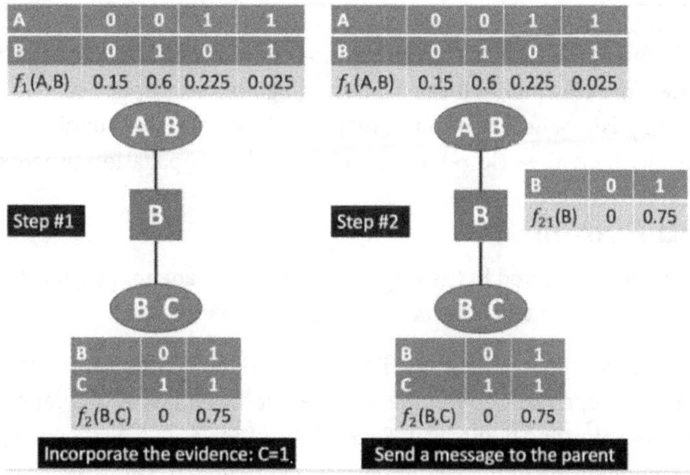

Figure 7.9—Steps to incorporate the evidence and send messages

Figure 7.10 illustrates steps #3 and #4 of belief propagation in the join tree example. In step #3, the root node, which corresponds to the clique $\{A, B\}$, processes the message received from its child node, the clique $\{B, C\}$.This processing involves multiplying the message from the single child clique, represented by the function $f_{2,1}(B)$, with the potential of the root clique, $f_1(A, B)$, to update the clique table of the root node. Step #4, the final step, entails normalizing the clique table. This is achieved by dividing the updated function $f_1(A, B)$ by the partition function Z, which is the sum of $f_1(A, B)$ across all values of A and B, resulting in the probability $P(A, B)$. To obtain the posterior probability for the variable A, we marginalize $P(A, B)$ over the variable B. This yields a probability of 0.96 for $A = 1$ and 0.04 for $A = 0$.

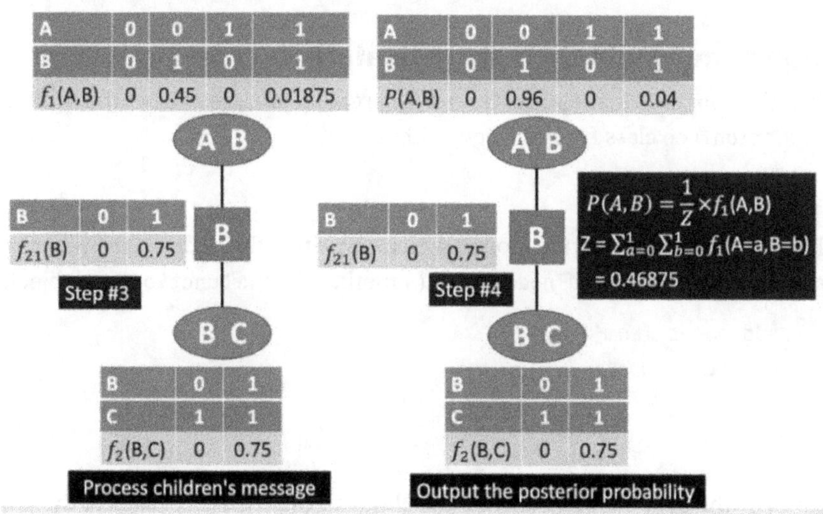

Figure 7.10—Steps to process received messages and compute probability

With the above example, we conclude this section on inference with join-tree clustering. In the following practice section using Python, we will learn how to programmatically populate the clique tables to quantify the computed join tree model for our circuit diagnosis example. This is a crucial step before we perform queries to update beliefs and compute the MAP and MPE for our circuit diagnosis example, which will demonstrate the real-world relevance of the belief propagation process.

Practice using Python

The code for this practice is included in the *Jupyter* notebook, '`chapter7.ipynb`'. This practice session builds upon the code and references objects from the previous session.

This practice will be divided into two sessions. The first session will programmatically implement the qualitative and quantitative specifications of the join tree model of the circuit diagnosis BN, which were detailed in the previous section and depicted in *Figure 7.7* and *Table 7.3*. The second session will be about programmatically implementing multiple inference tasks related to various circuit diagnosis queries.

We begin with the first practice session next.

Implementing the join tree model

In the following section, we will construct a join tree model step by step. This process is based on the join tree structure that we computed using the tree clustering algorithm for the circuit diagnosis Bayesian network discussed in *Chapter 6* and revisited in an earlier section.

This practice session will implement the join tree model in two main steps: first, we will implement the qualitative specification of the model, and second, we will implement the quantitative specification of the model.

Implementing the qualitative specification

First, we utilize the `JunctionTree()` constructor to create a join tree structure, which becomes an object of the `JunctionTree` class from the `pgmpy` library.

```
circuit_JT = JunctionTree()
```

Next, we add the cliques of the join tree model, which constitute the nodes of the junction tree. We implement these steps by using the `add_nodes_from()` method for the `JunctionTree` object, `circuit_JT`.

```
circuit_JT.add_nodes_from([
    ('Z', 'X', 'Y', 'A1'), ('Z', 'X', 'Y', 'E'),
    ('Z', 'A3', 'Y', 'E'), ('Z', 'X', 'D', 'E'),
    ('Z', 'X', 'A2', 'D'), ('A4', 'E', 'D', 'F')])
```

To finish, we specify the connections between the cliques, which are the edges of the join tree model. To do this, we utilize the `add_edges_from()` method for the `JunctionTree` object, `circuit_JT`.

```
circuit_JT.add_edges_from([
    (('Z', 'X', 'Y', 'A1'), ('Z', 'X', 'Y', 'E')),
    (('Z', 'X', 'Y', 'E'), ('Z', 'A3', 'Y', 'E')),
```

```
    (('Z', 'X', 'Y', 'E'),('Z', 'X', 'D', 'E')),
    (('Z', 'X', 'D', 'E'),('Z', 'X', 'A2', 'D')),
    (('Z', 'X', 'D', 'E'),('A4', 'E', 'D', 'F'))
])
```

In order to check that our implementation of the qualitative specification has been successful, we will plot the join tree of our circuit diagnosis Bayesian network (BN).

In order to plot the join tree that we have just programmatically created, we will utilize our `gviz()` function to generate the plot and save it as DOT and PNG files named "circuit_JT."

The following line of code will create the plot, in which the parameters for nodes and edges in the `gviz()` function are taken from the `JunctionTree` object named `circuit_JT`.

```
gviz(circuit_JT.nodes, circuit_JT.edges, "circuit_JT")
```

Figure 7.11 shows the join tree plot saved to the PNG file. This visualization asserts that the qualitative specification of the join tree is implemented correctly.

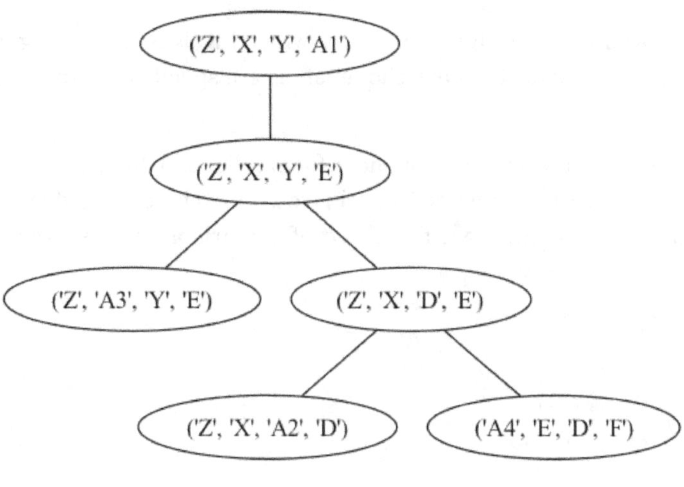

Figure 7.11—Join tree computed by the tree clustering algorithm for the circuit diagnosis BN

At this point, we conclude the first part of implementing the join tree model, where we implement the qualitative specification. Next, we proceed to the second part, in which we programmatically implement the quantitative specification.

Implementing the quantitative specification

We implement the quantitative specification of the join tree model in three main steps. First, we map the conditional probability distributions (CPDs) from the circuit diagnosis Bayesian network (BN) to potential functions. Next, we compute the join tree factors, which serve as the clique tables for the tree nodes. These factors are calculated as the products of the potential functions assigned to the clique buckets, as outlined in *Table 7.3*. Finally, we quantify the join tree by adding the factors computed in the second step to the circuit junction tree object created during the implementation of the qualitative specification.

We begin with the first step implemented by the following code block. The code maps the CPDs of the Bayesian network, which are pgmpy library type TabularCPD objects, to pgmpy DiscreteFactor objects using the to_factor() method applied to the TabularCPD objects. This creates ten potential functions for each variable of the circuit BN model, which we will use to build the clique tables for the join tree in the next steps.

```
phi_A1 = cpd_A1.to_factor()
phi_A2 = cpd_A2.to_factor()
phi_A3 = cpd_A3.to_factor()
phi_A4 = cpd_A4.to_factor()
phi_X = cpd_X.to_factor()
phi_Y = cpd_Y.to_factor()
phi_Z = cpd_Z.to_factor()
phi_D = cpd_D.to_factor()
phi_E = cpd_E.to_factor()
phi_F = cpd_F.to_factor()
```

Moving on, we will now proceed to the second step, which involves calculating the clique tables for the join tree of the circuit diagnosis; these clique tables are specified by the potential functions of the cliques.

The following code calculates the potential functions for the cliques, which are the nodes of the join tree for the circuit diagnosis Bayesian network. The join tree, which is computed by tree clustering, comprises six cliques, each representing specific subsets of the variables in the circuit Bayesian network:

```
C1: ('Z', 'X', 'Y', 'A1'),
C2: ('Z', 'X', 'Y', 'E'),
C3: ('Z', 'A3', 'Y', 'E'),
C4: ('Z', 'X', 'D', 'E'),
C5: ('Z', 'X', 'A2', 'D'),
C6: ('A4', 'E', 'D', 'F')
```

The process for computing the potential function for each clique involves multiplying all potential functions assigned to that clique. If a clique lacks any assigned potential functions, such as the cliques C2 and C4, we designate a potential function for it, which is valued at 1 for all possible instantiations of the clique's variables. We utilize the NumPy library to perform the multiplication of the pgmpy DiscreteFactor objects and to create potential functions that are equal to 1.

```
from pgmpy.factors.discrete import DiscreteFactor
import numpy as np
# clique C1: ('Z', 'X', 'Y', 'A1')
phi_C1 = phi_Z
phi_C1.product(phi_X)
phi_C1.product(phi_Y)
phi_C1.product(phi_A1)
```

```
print(phi_C1)
# clique C2: ('Z', 'X', 'Y', 'E')
phi_C2 = DiscreteFactor(['Z', 'X', 'Y', 'E'], [2, 2, 2, 2], np.ones(16) )
print(phi_C2)
# clique C3: ('Z', 'A3', 'Y', 'E')
phi_C3 = phi_E
phi_C3.product(phi_A3)
print(phi_C3)
# clique C4: ('Z', 'X', 'D', 'E')
phi_C4 = DiscreteFactor(('Z', 'X', 'D', 'E'), [2, 2, 2, 2], np.ones(16))
print(phi_C4)
# clique C5: ('Z', 'X', 'A2', 'D')
phi_C5 = phi_D
phi_C5.product(phi_A2)
print(phi_C5)
# clique C6: ('A4', 'E', 'D', 'F')
phi_C6 = phi_F
phi_C6.product(phi_A4)
print(phi_C6)
```

The code above for computing the potential function for the cliques creates a reference, not a copy, of the potential functions of the Markov networks. For example, when you write `phi_C1 = phi_Z`, it creates a reference instead of a copy. Consequently, when `phi_C1.product(phi_X)` is called, it modifies `phi_Z` in place, as `pgmpy's` `product()` function mutates the object. This behavior is acceptable in this case because `phi_Z` is not reused afterward. The same applies to `phi_C3 = phi_E`, `phi_C5 = phi_D`, and `phi_C6 = phi_F`. If any of the Markov network factors were needed later, they would be corrupted. It is safer to use `phi_C1 = phi_Z.copy()` before calling the `product()` method. For our purposes, the code will function correctly as is.

Executing the above code will print six clique tables, specifying the potential functions for the six respective cliques; each clique corresponds to a node in the join tree. In the following, we only show the first clique table computed for clique C1; the tables for the other five cliques are not shown to save space.

```
+------+------+------+-------+-----------------+
| Y    | Z    | X    | A1    |   phi(Y,Z,X,A1) |
+======+======+======+=======+=================+
| Y(0) | Z(0) | X(0) | A1(0) |          0.0000 |
+------+------+------+-------+-----------------+
| Y(0) | Z(0) | X(0) | A1(1) |          0.0063 |
+------+------+------+-------+-----------------+
| Y(0) | Z(0) | X(1) | A1(0) |          0.2375 |
+------+------+------+-------+-----------------+
| Y(0) | Z(0) | X(1) | A1(1) |          0.0063 |
```

```
+------+------+------+-------+---------------+
| Y(0) | Z(1) | X(0) | A1(0) |        0.2375 |
+------+------+------+-------+---------------+
| Y(0) | Z(1) | X(0) | A1(1) |        0.0063 |
+------+------+------+-------+---------------+
| Y(0) | Z(1) | X(1) | A1(0) |        0.0000 |
+------+------+------+-------+---------------+
| Y(0) | Z(1) | X(1) | A1(1) |        0.0063 |
+------+------+------+-------+---------------+
| Y(1) | Z(0) | X(0) | A1(0) |        0.2375 |
+------+------+------+-------+---------------+
| Y(1) | Z(0) | X(0) | A1(1) |        0.0063 |
+------+------+------+-------+---------------+
| Y(1) | Z(0) | X(1) | A1(0) |        0.2375 |
+------+------+------+-------+---------------+
| Y(1) | Z(0) | X(1) | A1(1) |        0.0063 |
+------+------+------+-------+---------------+
| Y(1) | Z(1) | X(0) | A1(0) |        0.0000 |
+------+------+------+-------+---------------+
| Y(1) | Z(1) | X(0) | A1(1) |        0.0063 |
+------+------+------+-------+---------------+
| Y(1) | Z(1) | X(1) | A1(0) |        0.0000 |
+------+------+------+-------+---------------+
| Y(1) | Z(1) | X(1) | A1(1) |        0.0063 |
+------+------+------+-------+---------------+
```

Let's quickly examine two rows from the output C1 table: the first and the third rows. For the first row, the values are calculated as follows: phi(Y=0, A1=0, Z=0, X=0) = P(Z=0|A1=0, X=0, Y=0) * P(A1=0) * P(X=0) * P(Y=0) = 1e-9 * 0.95 * 0.5 * 0.5, which is approximately 0.0000. This matches the table. For the third row, we have phi(Y=0, A1=0, Z=0, X=1) = P(Z=0|A1=0, X=1, Y=0) * P(A1=0) * P(X=1) * P(Y=0) = (1 - 1e-9) * 0.95 * 0.5 * 0.5, resulting in 0.2375, consistent with the C1 table. Additionally, the sum of the probabilities across all rows equals 1.

Moving along, we proceed to the third and final step for quantifying the join tree model of our circuit diagnosis. The quantification utilizes the potential functions for the cliques that were computed in the second step.

The following code quantifies the join tree by inserting the potential functions of the cliques into the join tree model. These potential functions correspond to the factors of the join tree and are represented as `DiscreteFactor` objects in `pgmpy`. To add the factors to the join tree object, `circuit_JT`, we utilize the `add_factors()` method from the `pgmpy JunctionTree` object.

```
circuit_JT.add_factors(phi_C1, phi_C2, phi_C3, phi_C4, phi_C5, phi_C6)
```

After quantifying the join tree, we conclude the quantitative specification of the circuit's join tree model.

As an additional step, we need to check that the implemented join tree represents a valid junction tree model, specifically that it satisfies the running intersection property. This step is crucial to ensure that the join tree model produces accurate results when querying it for various inference tasks.

We use the `check_model()` method from the `pgmpy JunctionTree` object to verify the validity of the join tree model.

```
circuit_JT.check_model()
```

Executing the above one-line code will output "True," which means the model is valid.

```
True
```

At this point, we conclude the first session of this programming practice, during which we implemented the join tree model for our circuit diagnosis Bayesian network. Next, we will move on to the second session of the practice, where we will programmatically implement multiple diagnostic inference tasks using the join tree model of the circuit Bayesian network.

Implementing multiple inference tasks

In this practice session, you will learn how to implement multiple inference tasks programmatically in Python, specifically focusing on fault diagnosis in combinatorial digital circuits.

The practice session will demonstrate how to frame fault diagnosis as inference tasks within Bayesian networks. It will include formulating diagnostic queries as inference tasks, such as updating beliefs (posterior probabilities), computing the maximum a posteriori hypothesis (MAP), and calculating the most probable explanation (MPE).

Throughout this practice session, we will use **belief propagation** with join tree clustering, as implemented in the `pgmpy` library.

We will perform all inference tasks on each of the two join trees depicted in *Figure 7.7*, specifically the `pgmpy` join tree, `JT`, and the join tree we computed by the tree clustering algorithm in *Chapter 6*, `circuit_JT`.

You will observe that all inference queries conducted on both join trees in the *Jupyter* notebook '`chapter7.ipynb`' produce the same results across all inference tasks. The inference using either model will differ in processing time. The processing time for the `pgmpy` join tree, `JT`, will be greater than that of our tree clustering, `circuit_JT`. The processing time is exponential in the tree width, which is 5 for the `pgmpy` join tree, `JT`, and equals 3 for our join tree, `circuit_JT`.

We made the duplicate effort to validate that performing the inference tasks by querying both join trees produces consistent results. To save space, we will demonstrate all inference tasks with belief propagation based on our join tree model, `circuit_JT`.

We begin with an initialization step before demonstrating the various inference tasks related to our circuit diagnostic queries. The first step involves creating a `pgmpy` `BeliefPropagation` object using the `BeliefPropagation()` constructor, which requires the join tree object as a parameter.

```
from pgmpy.inference.ExactInference import BeliefPropagation

belief_propagation = BeliefPropagation(circuit_JT)
```

We are now ready to begin our multiple inference tasks. Next, we will run two diagnostic queries framed as updating beliefs or calculating the posterior probabilities of the assumption variables of the four circuit components. Recall that an assumption variable is a binary variable indicating whether a circuit component (gate) is ok or faulty.

Task #1: updating beliefs given evidence e1

This task is to compute the posterior probability distribution over all four variables:

```
['A1', 'A2', 'A3', 'A4']
```

conditioned on the evidence e1, consisting of observed input variables 'X' and 'Y' being ON (value 1) and the output variable 'F' being OFF (value 0):

```
{'X': 1, 'Y': 1, 'F': 0}
```

The evidence, e1, indicates that the combinatorial circuit is operating abnormally. The observed inputs are $X = 1$ and $Y = 1$, but the output F is recorded as 0, which contradicts the expected outcome. According to *Equation 7.4*, the normal function of the circuit should produce an output of $F = 1$ rather than $F = 0$:

$$F = \overline{(X \oplus Y)} == \overline{(1 \oplus 1)} = 1$$

Calculating the posterior probabilities of the assumption variables will update the beliefs related to the four gates. This process updates the prior probability of a gate being either functional or faulty to a posterior probability that reflects the evidence provided.

The following code will generate the posterior probability distributions for the four assumption variables. This will provide the probabilities of single and multiple faults within the circuit components.

```
from pgmpy.inference.ExactInference import BeliefPropagation

belief_propagation = BeliefPropagation(circuit_JT)
res1 = belief_propagation.query(
    variables=['A1', 'A2', 'A3', 'A4'],
    evidence={'X': 1, 'Y': 1, 'F': 0})
print(res1)
```

Executing the above code block will produce the following posterior probability:

```
+-------+-------+-------+-------+-------------------+
| A1    | A2    | A3    | A4    |   phi(A1,A2,A3,A4) |
+=======+=======+=======+=======+===================+
| A1(0) | A2(0) | A3(0) | A4(0) |            0.0000 |
+-------+-------+-------+-------+-------------------+
| A1(0) | A2(0) | A3(0) | A4(1) |            0.2981 |
+-------+-------+-------+-------+-------------------+
| A1(0) | A2(0) | A3(1) | A4(0) |            0.2981 |
+-------+-------+-------+-------+-------------------+
| A1(0) | A2(0) | A3(1) | A4(1) |            0.0157 |
+-------+-------+-------+-------+-------------------+
| A1(0) | A2(1) | A3(0) | A4(0) |            0.2981 |
+-------+-------+-------+-------+-------------------+
| A1(0) | A2(1) | A3(0) | A4(1) |            0.0157 |
+-------+-------+-------+-------+-------------------+
| A1(0) | A2(1) | A3(1) | A4(0) |            0.0235 |
+-------+-------+-------+-------+-------------------+
| A1(0) | A2(1) | A3(1) | A4(1) |            0.0008 |
+-------+-------+-------+-------+-------------------+
| A1(1) | A2(0) | A3(0) | A4(0) |            0.0000 |
+-------+-------+-------+-------+-------------------+
| A1(1) | A2(0) | A3(0) | A4(1) |            0.0157 |
+-------+-------+-------+-------+-------------------+
| A1(1) | A2(0) | A3(1) | A4(0) |            0.0157 |
+-------+-------+-------+-------+-------------------+
| A1(1) | A2(0) | A3(1) | A4(1) |            0.0008 |
+-------+-------+-------+-------+-------------------+
| A1(1) | A2(1) | A3(0) | A4(0) |            0.0157 |
+-------+-------+-------+-------+-------------------+
| A1(1) | A2(1) | A3(0) | A4(1) |            0.0008 |
+-------+-------+-------+-------+-------------------+
| A1(1) | A2(1) | A3(1) | A4(0) |            0.0012 |
+-------+-------+-------+-------+-------------------+
| A1(1) | A2(1) | A3(1) | A4(1) |            0.0000 |
+-------+-------+-------+-------+-------------------+
```

The results reveal three potential single-fault scenarios involving A2, A3, or A4, each identified as faulty (value 1) with a probability of 0.2981. *Figure 7.12* illustrates these scenarios, clearly indicating that identifying any of these gates as faulty is equally plausible. The figure emphasizes the single fault using red boxes. Values in parentheses provide predictions for unobserved variables, which help explain the observations within the evidence tuple e1. Each scenario adequately accounts for the observed faulty behavior. Additionally, the probabilities for double and triple faults are found to be minimal. Each fault is diagnosed with an equal probability of P=0.2981. This is because the prior fault probability is consistent across all circuit components, and the observations from the circuit's input and output do not provide sufficient evidence to assign varying levels of responsibility to different components.

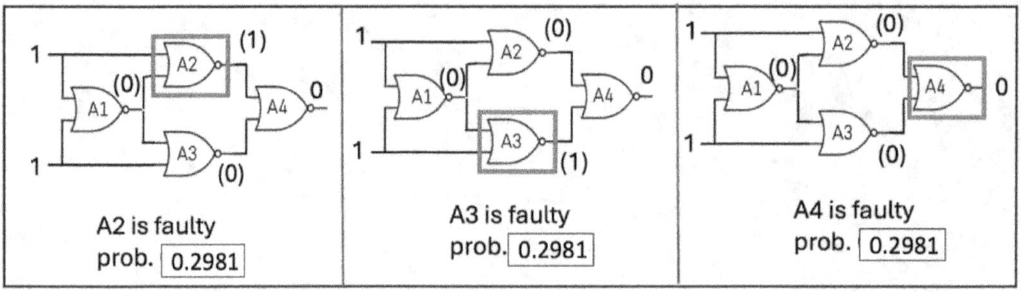

Figure 7.12—Given evidence e1, the highest posterior probability is for the single fault: A2 or A3 or A4

We present another example of an inference task involving the updating of beliefs based on evidence, e2. This evidence expands the observations from e1 by adding one more measurement related to the state of a circuit wiring diagram, as shown in *Figure 7.3*. This measurement pertains to the variable 'D,' which represents the output of the gate 'A2.'

Task #2: updating beliefs given evidence e2

This task is to compute the posterior probability distribution over all four variables:

```
['A1', 'A2', 'A3', 'A4']
```

conditioned on the evidence e2, consisting of observed input variables 'X' and 'Y' being ON (value 1), the output variable 'F' as OFF (value 0), and the output variable 'D' of component 'A2' as ON (value 1):

```
{'X': 1, 'Y': 1, 'F': 0, 'D': 1}
```

This diagnostic query includes an additional observation compared to the previous query. It is anticipated to provide more specific and reliable inferences related to fault diagnosis than the results from the earlier query conditioned on e1.

```
from pgmpy.inference.ExactInference import BeliefPropagation

belief_propagation = BeliefPropagation(circuit_JT)
res2 = belief_propagation.query(
    variables=['A1', 'A2', 'A3', 'A4'],
    evidence={'X': 1, 'Y': 1, 'F': 0, 'D': 1})

print(res2)
```

Executing the above code block will produce the following posterior probability:

```
+-------+-------+-------+-------+------------------+
| A1    | A2    | A3    | A4    |   phi(A1,A2,A3,A4) |
+=======+=======+=======+=======+==================+
| A1(0) | A2(0) | A3(0) | A4(0) |           0.0000 |
+-------+-------+-------+-------+------------------+
| A1(0) | A2(0) | A3(0) | A4(1) |           0.0000 |
+-------+-------+-------+-------+------------------+
| A1(0) | A2(0) | A3(1) | A4(0) |           0.0000 |
+-------+-------+-------+-------+------------------+
| A1(0) | A2(0) | A3(1) | A4(1) |           0.0000 |
+-------+-------+-------+-------+------------------+
| A1(0) | A2(1) | A3(0) | A4(0) |           0.8794 |
+-------+-------+-------+-------+------------------+
| A1(0) | A2(1) | A3(0) | A4(1) |           0.0231 |
+-------+-------+-------+-------+------------------+
| A1(0) | A2(1) | A3(1) | A4(0) |           0.0463 |
+-------+-------+-------+-------+------------------+
| A1(0) | A2(1) | A3(1) | A4(1) |           0.0012 |
+-------+-------+-------+-------+------------------+
| A1(1) | A2(0) | A3(0) | A4(0) |           0.0000 |
+-------+-------+-------+-------+------------------+
| A1(1) | A2(0) | A3(0) | A4(1) |           0.0000 |
+-------+-------+-------+-------+------------------+
| A1(1) | A2(0) | A3(1) | A4(0) |           0.0000 |
+-------+-------+-------+-------+------------------+
| A1(1) | A2(0) | A3(1) | A4(1) |           0.0000 |
+-------+-------+-------+-------+------------------+
| A1(1) | A2(1) | A3(0) | A4(0) |           0.0463 |
+-------+-------+-------+-------+------------------+
| A1(1) | A2(1) | A3(0) | A4(1) |           0.0012 |
+-------+-------+-------+-------+------------------+
| A1(1) | A2(1) | A3(1) | A4(0) |           0.0024 |
+-------+-------+-------+-------+------------------+
| A1(1) | A2(1) | A3(1) | A4(1) |           0.0001 |
+-------+-------+-------+-------+------------------+
```

Figure 7.13 illustrates the result of the diagnostic inference showing only one single fault using a red box around the gate, A2. Values in parentheses provide predictions for unobserved variables, which help explain the observations within the evidence tuple, e2.

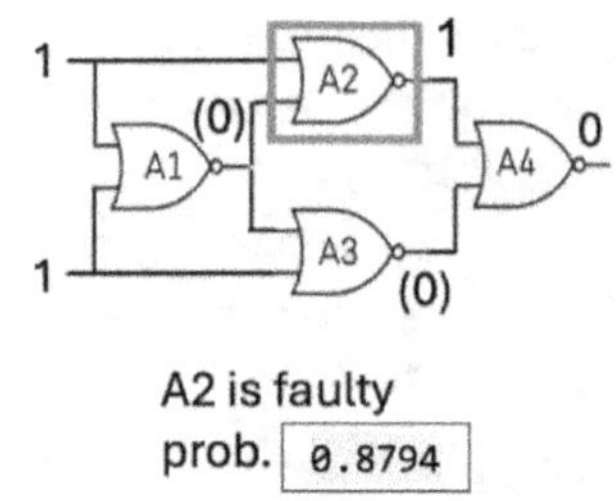

Figure 7.13—Given evidence e2, the highest posterior probability is for the single fault A2

The result indicates that with the additional measurement of the output from gate A2, the probability of A2 being faulty increases nearly threefold, rising from 0.2981 to 0.8794. In contrast, the probabilities of single faults for gates A3 and A4 drop to zero. While there remain non-zero probabilities for double and triple faults, those values are nearly negligible.

Continuing with our circuit diagnosis BN example, let us demonstrate how to programmatically perform the inference task of computing the maximum a posteriori hypothesis.

Query #3: maximum a posteriori hypothesis (MAP) given evidence e1

In this part of the practice, we will utilize the `pgmpy` `map_query` method from `BeliefPropagation` to compute the maximum a posteriori hypothesis (MAP) for the assumption variables:

```
['A1', 'A2', 'A3', 'A4']
```

The MAP query is conditioned on the evidence, e1, given by the observations:

```
{'X': 1, 'Y': 1, 'F': 0}
```

The observations consist of the circuit input variables 'X' and 'Y' being ON (value 1) and the output variable 'F' being OFF (value 0).

The inference task of computing the maximum a posteriori probability involves determining the value assignment for a set of query variables based on evidence obtained from observed variables. This process yields the value assignment to query variables that maximizes the a posteriori probability across all possible configurations of the unobserved variables, as described in *Equation 7.9*.

The following code implements the maximum a posteriori (MAP) inference task, conditioned on the evidence, e1.

```
from pgmpy.inference.ExactInference import BeliefPropagation

belief_propagation = BeliefPropagation(circuit_JT)
res3 = belief_propagation.map_query(
    variables=['A1', 'A2', 'A3', 'A4'],
    evidence={'X': 1, 'Y': 1, 'F': 0})

print(res3)
```

Running the above code will print the following output:

```
{'A1': 0, 'A2': 0, 'A3': 1, 'A4': 0}
```

The result indicates that the maximum a posteriori hypothesis, given the evidence e1, is that there is a single fault: the circuit gate A3 is faulty, as shown in *Figure 7.14*.

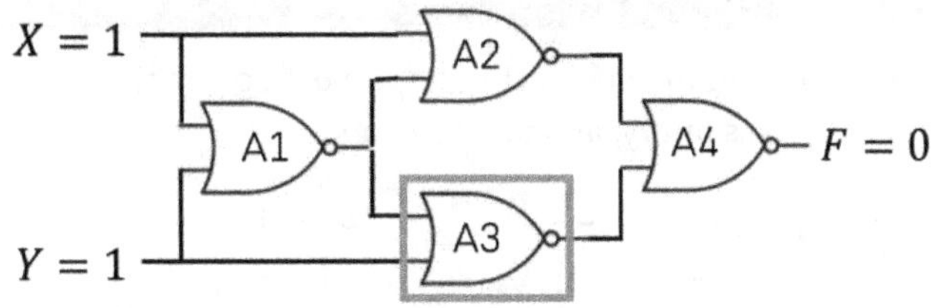

Figure 7.14—Given evidence e1, the single fault A3 is the maximum a posteriori hypothesis

As previously indicated in Task #1, there are three potential single fault diagnoses consistent with the evidence: A2, A3, and A4, all with equal probabilities. The MAP query identifies A3 as the most probable diagnosis; however, A2 and A4 could also be valid answers. Since all options have equal probabilities, the MAP query is arbitrarily breaking the tie.

Next, we will calculate the maximum a posteriori hypothesis based on the evidence, e2. This will allow us to consider both sets of evidence and ensure consistency across all three inference tasks in these practice sessions.

Query #4: maximum a posteriori hypothesis (MAP) given evidence e2

The query is designed to compute the MAP for the four variables:

```
['A1', 'A2', 'A3', 'A4'].
```

The query is conditioned on the evidence, e2, where the input variables 'X' and 'Y' are set to ON (value 1), the output variable 'F' is set to OFF (value 0), and the output variable 'D' of component A2 is set to ON (value 1).

```
{'X': 1, 'Y': 1, 'F': 0, 'D': 1}.
```

We anticipate that the computed MAP hypothesis will be more reliable compared to the MAP hypothesis derived from query #3. This is because we are conditioning on evidence that includes additional information about the 'D' variable, which augments the evidence e1 used in query #3.

The following code implements the maximum a posteriori (MAP) inference task, conditioned on the evidence, e2.

```
from pgmpy.inference.ExactInference import BeliefPropagation

belief_propagation = BeliefPropagation(circuit_JT)
res4 = belief_propagation.map_query(
```

```
    variables=['A1', 'A2', 'A3', 'A4'],
    evidence={'X': 1, 'Y': 1, 'F': 0, 'D': 1})
print(res4)
```

Running the above code will print the following output:

```
{'A1': 0, 'A2': 1, 'A3': 0, 'A4': 0}
```

The result indicates that the maximum a posteriori hypothesis, given the evidence e2, is that there is a single fault: the circuit gate A2 is faulty, as shown in *Figure 7.15*.

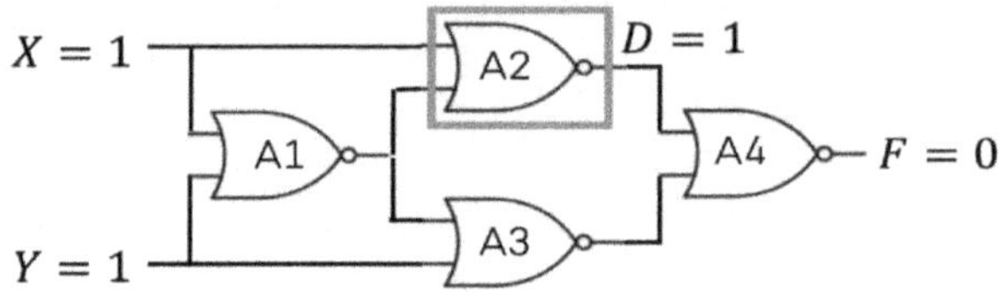

Figure 7.15—Given evidence e2, the single fault A2 is the maximum a posteriori hypothesis

This concludes our practice with the circuit diagnosis Bayesian Network example, which focuses on computing the maximum a posteriori hypothesis using two sets of observed evidence, e1 and e2.

We will now proceed to the final session of this practice, where we will demonstrate diagnostic inferences by calculating the most probable explanation (MPE). Next, we will calculate the MPE based on the evidence, e1.

Query #5: most probable explanation given evidence e1

The inference task of calculating the most probable explanation (MPE) involves determining the value assignment for all unobserved variables based on the evidence obtained from observed variables. This process yields a value assignment to unobserved variables having the highest posterior probability consistent with the evidence gathered from the observed variables.

Calculating the most probable explanation (MPE) is equivalent to computing the maximum a posteriori (MAP) hypothesis for all unobserved variables. These unobserved variables include all network variables except for those that have value assignments in the observed evidence. This relationship defines the MPE and MAP inference tasks, as indicated in *Equations (7.8) and (7.9)*

In this section of the practice, we will use the `map_query` method from the `pgmpy BeliefPropagation` module to compute the most probable explanation (MPE). The query variables will include all network variables except for those that are part of the evidence.

```
['A1', 'A2', 'A3', 'A4', 'Z', 'D', 'E']
```

The query will condition on the evidence, e1, given by the observations:

```
{'X': 1, 'Y': 1, 'F': 0}
```

The observations indicate that the circuit input variables 'X' and 'Y' are both ON, with a value of 1. In contrast, the output variable 'F' is OFF, with a value of 0.

The code provided calculates the MPE inference task, conditioned on the evidence, e1.

```
from pgmpy.inference.ExactInference import BeliefPropagation

belief_propagation = BeliefPropagation(circuit_JT)
res5 = belief_propagation.map_query(
    variables=['A1', 'A2', 'A3', 'A4','Z', 'D','E'],
    evidence={'X': 1, 'Y': 1, 'F': 0})
print(res5)
```

The provided code will produce the following output:

```
{'A1': 0, 'A2': 0, 'A3': 0, 'A4': 1, 'Z': 0, 'D': 0, 'E': 0}
```

The result indicates that the most probable explanation, given the evidence e1, is that there is a single fault: the circuit gate A4 is faulty, as shown in *Figure 7.16*. The figure illustrates the results of the MPE calculation. It depicts the evidence in blue boxes and the corresponding explanation in yellow boxes. The explanation indicates that the fault in gate A4 is the most probable cause of the circuit's abnormal behavior.

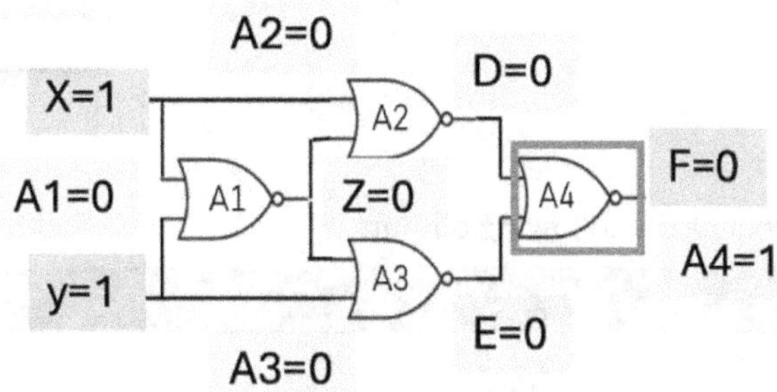

Figure 7.16—Given evidence e1, the single fault A4 is the most probable explanation

Let's discuss why the MPE query produces different diagnoses than the MAP query, even though both are conditioned on the same evidence. The MAP assigns blame to A3, while the MPE assigns blame to A4.

The main reason the diagnoses differ is that Maximum a Posteriori (MAP) identifies the most likely state for a specific subset of unobserved variables, which in this case are the assumption variables regarding the operational state of the circuit's four gates: A1, A2, A3, and A4. This is done by "summing out" or marginalizing other unknown factors, namely the state configurations of the unobserved variables D, E, and Z. In contrast, the Most Probable Explanation (MPE) determines the single most likely combination of all unobserved variables simultaneously, which includes A1, A2, A3, A4, D, E, and Z. For further clarification, refer to *Equations 7.7* and *7.8*, which define the MPE and MAP queries. These equations illustrate that the marginalization of unobserved variables not included in the MAP query is what typically results in different answers for the MAP query compared to the MPE.

Next, we will calculate the most probable explanation based on the evidence, e2. This step will enable us to evaluate both sets of evidence and ensure consistency across all three inference tasks in these practice sessions.

Query #6: most probable explanation given evidence e2

Similar to Query #5, we pose this query to calculate the most probable explanation (MPE) as a query to compute the maximum a posteriori (MAP) hypothesis over all unobserved variables:

```
['A1', 'A2', 'A3', 'A4', 'Z', 'E']
```

The MAP query is conditioned this time on the evidence e2, where the input variables 'X' and 'Y' are ON (value 1), the output variable 'F' is OFF (value 0), and the output variable 'D' of component #2 is ON (value 1):

```
{'X': 1, 'Y': 1, 'F': 0, 'D': 1}
```

The following code block implements the MPE using the `map_query` method from the `pgmpy` `BeliefPropagation` module. The code provided implements the MPE inference task, conditioned on the evidence, e2.

```
from pgmpy.inference.ExactInference import BeliefPropagation

belief_propagation = BeliefPropagation(circuit_JT)
res6 = belief_propagation.map_query(
    variables=['A1', 'A2', 'A3', 'A4', 'Z', 'E'],
    evidence={'X': 1, 'Y': 1, 'F': 0, 'D': 1})
print(res6)
```

The provided code will produce the following output:

```
{'A1': 0, 'A2': 1, 'A3': 0, 'A4': 0, 'Z': 0, 'E': 0}
```

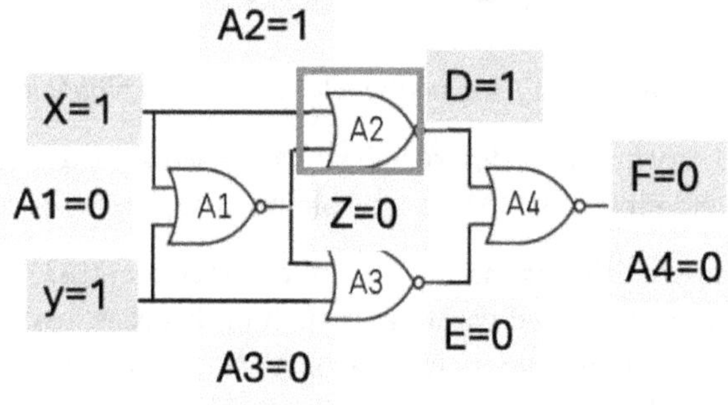

Figure 7.17—Given evidence e2, the single fault A2 is the most probable explanation

The result suggests that the most probable explanation, based on the evidence e2, is that there is a single fault: the circuit gate A2 is faulty. This is illustrated in *Figure 7.17*, which shows the outcomes of the MPE calculation. The figure displays the evidence in blue boxes alongside the corresponding explanation in yellow boxes. This explanation indicates that the fault in gate A2 is the most likely cause of the circuit's abnormal behavior.

The results of task #5 and task #6 demonstrate how inferencing in Bayesian networks performs belief revision. As evidence accumulates, Bayesian inference revises beliefs in hypotheses to leverage the information from additional evidence.

Let's review the six tasks related to fault diagnosis for our XNOR circuit example. *Figure 7.8* lists each task, describing it as an inference task conditioned on evidence representing observed values of circuit variables. The tasks are categorized into three main inference tasks in Bayesian networks: updating beliefs, maximum a posteriori (MAP), and the most probable explanation (MPE). Each task category includes two queries conditioned on two sets of evidence, e1 and e2. The evidence e2 includes an additional observation of gate A2's output (the 'D' variable), along with the observations in evidence e1, which consist of the circuit's inputs 'X' and 'Y' and the circuit's output 'F'.

The purpose of the table in *Figure 7.18* is to illustrate the **nonmonotonic reasoning** aspect of diagnostic inference with Bayesian networks. Evidence e2 serves as an incremental observation relative to e1; thus, the results of the inference queries reflect **belief revision** based on this added evidence.

To update beliefs with the added evidence, we note from the first two rows in the table of *Figure 7.18* that the posterior probability of gate A2 being faulty increases from 0.2981 to 0.8794, while the probabilities of other single faults decrease. For the MAP queries, the third and fourth rows indicate that the inference results are revised based on the new evidence from gate A2's output. This revision retracts the initial diagnosis that A3 is faulty based on evidence e1, updating the conclusion to indicate that gate A2 is now the MAP diagnosis. In the case of the MPE queries, the fifth and sixth rows show that, with the additional evidence, the diagnosis shifts from A4 being faulty under e1 to concluding that gate A2 is the MPE diagnosis instead. The table in *Figure 7.18* shows that all three types of inference queries always come to the same conclusion: gate A2 is the best diagnosis.

Task Type	Task #	Evidence	Descriptions	Diagnosis
Update beliefs (posterior probability)	Task #1-	e1 = {'X': 1, 'Y': 1, 'F': 0}	XNOR inputs X, Y both ON and the output is OFF	P('A1': 0, 'A2': 1, 'A3': 0, 'A4':0) = 0.2981 P('A1': 0, 'A2': 0, 'A3': 1, 'A4':0) = 0.2981 P('A1': 0, 'A2': 0, 'A3': 0, 'A4':1) = 0.2981
	Task #2-	e2 = {'X': 1, 'Y': 1, 'F': 0, 'D': 1}	gate A2's output is ON in addition to e1.	P('A1': 0, 'A2': 1, 'A3': 0, 'A4':0) = 0.8794
Maximum a Posteriori (MAP) hypothesis	Task #3	e1 = {'X': 1, 'Y': 1, 'F': 0}	XNOR inputs X, Y both ON and the output is OFF	'A1': 0, 'A2': 0, 'A3': 1, 'A4': 0
	Task #4	e2 = {'X': 1, 'Y': 1, 'F': 0, 'D': 1}	gate A2's output is ON in addition to e1.	'A1': 0, 'A2': 1, 'A3': 0, 'A4': 0

Task Type	Task #	Evidence	Descriptions	Diagnosis
Most Probable Explanation (MPE)	Task #5	e1 = {'X': 1, 'Y': 1, 'F': 0}	XNOR inputs X, Y both ON and the output is OFF	'A1': 0, 'A2': 0, 'A3': 0, 'A4': 1, 'Z': 0, 'D': 0, 'E': 0
	Task #6	e2 = {'X': 1, 'Y': 1, 'F': 0, 'D': 1}	gate A2's output is ON in addition to e1.	'A1': 0, 'A2': 1, 'A3': 0, 'A4': 0, 'Z': 0, 'E': 0

Figure 7.18—Table summarizing the six inference tasks and the results related to circuit diagnosis; the tasks cover three types of queries: updating beliefs, MAP, and MPE

In the following section, we will discuss the computational cost associated with performing inference tasks, specifically updating beliefs and calculating the Most Probable Explanation (MPE) using join trees.

Evaluation of computational cost of inference using join trees

Belief propagation in a junction tree (clique tree) consists of three main steps: 1) Incorporate evidence. 2) Send messages toward the root in an upward pass. 3) Send messages toward the leaves in a downward pass. Belief propagation passes messages between neighboring cliques to compute exact marginals.

Counting message passing in belief propagation (BP) within a clique tree (or junction tree) involves calculating the number of messages needed to calibrate the tree, which ensures each clique has the correct joint marginal distribution.

In a junction tree with n nodes (cliques), there are $n - 1$ edges. Since message passing requires an upward pass and a downward pass, each edge carries two directed messages, resulting in $(2\,n - 1)$ messages in total.

The table in *Figure 7.19* compares the computational costs of inference using a join tree derived from *Chapter 6* with the join tree generated by `pgmpy`; both trees are illustrated side by side in *Figure 7.7*. According to *Theorem 7.1*, the time complexity for updating beliefs and computing the most probable explanation (MPE) using a join-tree clustering algorithm is $O(n\, exp(r))$. The space complexity is $O(n\, exp(s))$, where n represents the number of variables in the Bayesian network, which is 10 for our circuit model. The variable r denotes the induced treewidth, defined as the size of the maximum clique minus 1, and s represents the maximum size of the separator. The time and space entries in the table do not represent execution run time in seconds or actual memory usage. Instead, they are non-dimensional quantities that help evaluate how the demands for computing time and space vary when performing inference tasks, based on the structural parameters of the Bayesian network's join tree representation.

Join Tree	# cliques	induced treewidth r	separator size s	Time	Space	# Messages
pgmpy (Figure 7.6)	3	5	3	$10 \times 2^5 = 320$	$10 \times 2^3 = 80$	5
Chapter 6 (Figure 7.11)	6	3	3	$10 \times 2^3 = 80$	$10 \times 2^3 = 80$	11

Figure 7.19—Table comparing the time, space, and number of messages for the two join-tree representations of the circuit Bayesian network

With this, we conclude this section on belief propagation with join trees and demonstrate the various inference tasks on a real-life application related to fault diagnosis of digital combinatorial circuits.

Wrapping up, let us emphasize several takeaway points.

Fault diagnosis serves as a key application of reasoning under uncertainty. Inference tasks in Bayesian networks whose graph is multiply connected pose an NP-hard problem for which no polynomial time algorithm is known to exist. The belief propagation message passing algorithm has polynomial time complexity but only on singly connected networks. This is where join tree clustering plays a crucial role, as it converts a generally multiply connected network into a singly connected network. Specifically, it creates a join tree, which consists of nodes that are cliques of subsets of the network's variables. Belief propagation with join trees demonstrated on the circuit fault diagnosis shows a case of non-monotonic reasoning where hypotheses may be retracted or revised upon acquiring new data. Diagnostic reasoning is defeasible; new evidence can lead to the overturning of prior conclusions.

Summary

This chapter focuses on probabilistic inference in Bayesian networks through the use of join tree clustering. It expands on the content covered in *Chapter 6* regarding probabilistic inference in Bayesian networks.

You learned about three of the most common inference tasks in Bayesian and Markov networks: updating beliefs, computing the maximum a posteriori (MAP) hypothesis, and calculating the most probable explanation (MPE).

You also learned about the message-passing belief propagation algorithm for conducting inference with join trees. This was demonstrated using a case study on circuit fault diagnosis, which exemplifies non-monotonic reasoning, where hypotheses can be retracted or revised as new data are obtained.

The knowledge and experience you gained in this chapter equip you with the skills necessary to develop applications for inference tasks across various fields for handling uncertainty and updating beliefs based on new evidence.

The next chapter on probabilistic inference with relational database models will continue with exact inference using algorithms based on variable elimination, an alternative approach to tree clustering. We will utilize the relational database framework to represent probabilistic graphical models, continuing from the point we left off in *Chapter 5*.

References

- Cooper G. F. *The computational complexity of probabilistic inference using Bayesian belief networks* [Journal]. Artificial Intelligence. 1990. Vol. 42. pp. 393-405.
- Dagum P. and Luby M. *Approximating probabilistic inference in Bayesian belief networks is NP-hard* [Journal]. Artificial Intelligence. 1993. Vol. 60. pp. 141–153.
- Dechter J. *Bucket elimination: A unifying framework for reasoning* [Journal]. Artificial Intelligence. 1999. Vol. 113. pp. 41–85.

- Dechter R. and El Fattah Y. *Topological parameters for time-space tradeoff* [Journal]. Artificial Intelligence. 2001. Vol. 125. pp. 93–118.
- El Fattah Y. and Dechter R. *An Evaluation of Structural Parameters for Probabilistic Reasoning: Results on Benchmark Circuits* [Conference]. Proceedings of the Twelfth Conference on Uncertainty in Artificial Intelligence (UAI1996). 1996.
- Garey M.R. and Johnson D.S. *Computers and Tractability: A Guide to the Theory of NP-Completeness* [Book]. New York : W.H. Freeman and Company, 1979.
- Jensen Finn V. and Nielsen Thomas D. *Bayesian Networks and Decision Graphs* [Book]. New York: Springer, 2007.
- Jensen Finn V. *Bayesian Networks and Decision Graphs* [Book]. New York: Springer, 2001.
- Lauritzen S. L. and Spiegelhalter D. J. *Local Computations with Probabilities on Graphical Structures and their Application to Expert Systems* [Journal]. J. R. Statist. Soc. B . 1988. 2: Vol. 50. pp. 157-224.
- Mano M. M. *Computer Engineering: Hardware Design* [Book]. [s.l.]: Prentice Hall, 1988.
- Pearl J. *Probabilistic Reasoning in Intelligent Systems: Networks of Plausible Inference* [Book]. [s.l.]: Morgan Kaufmann, 1988.
- Yannakakis M. *Computing the minimum fill-in is NP-complete* [Journal]. SIAM J. Alg. Disc. Math. 1981. pp. 77-79.

8

Probabilistic Inference with Relational Database Models

This chapter gives an approach to probabilistic inference using relational databases. We use a relational database representation of Bayesian Networks. We define a set of primitive queries extending relational algebra operators to apply to weighted relations. We describe a variable elimination algorithm that compiles a query composition graph that prescribes a structured procedure for compositional queries and solves inference tasks. We apply the algorithm to the inference tasks of posterior probability in BN and maximum expected utility in influence diagrams. Concepts will be reinforced by practice examples with implementation in R. We will explore the following topics:

- Structured Primitive Queries
- Query Composition Graph (QCG) Algorithm
- Computing Posterior Probability
- Computing Maximum Expected Utility

Technical requirements

All the code for this chapter is in R, and you can download it from the GitHub URL: `https://github.com/PacktPublishing/Causal-Inference-with-Bayesian-Networks/tree/main/Chapter_08`

You will be required to install various R libraries to run the R code for the practice and the examples in this chapter, and you can do that from within the **RStudio** IDE. The required packages to install are:

- `dplyr` `https://cran.r-project.org/web/packages/dplyr/readme/README.html`
- `purrr` `https://purrr.tidyverse.org`
- `rlang` `https://cran.r-project.org/web/packages/rlang/index.html`

Structured primitive queries

This section defines three atomic operators that perform inference by executing queries on weighted relations.

First, we revisit the concept of weighted relation, recalling its definition, and illustrating it with examples. Next, we define primitive queries that extend the relational operators to apply to weighted relations and illustrate the defined queries with examples. Then, in a follow-up section, we formally define and implement the three operators in R using R libraries. Let us begin with the following section on weighted relations.

Weighted relations

In *Chapter 5*, we introduce the concept of weighted relations. A weighted relation is represented as a normal relation with one crucial caveat: a weighted relation has a distinguished column, "*Weight*," that must be processed differently than other attributes when performing the relational operations.

The Weight attribute enables using relations to represent uncertainty in graphical models of Bayesian networks. The Weight can also represent a value function in optimization problems, cost in cost-based abduction as in model-based diagnosis, and utility functions in decision-making with influence diagrams. The following defines the notion of weighted relation.

Definition: weighted relation

A weighted relation is a pair: (r, w), where r is a relation and w is a function defined on the scheme of r. For each tuple $t \in r$, a weight is assigned whose value is w(t). A single relation, r', whose scheme can represent a weighted relation (r, w), includes a Weight variable whose value is uniquely determined by the value tuple of all the other variables in the scheme of r'.

Figure 8.1 shows an example of a valid weighted relation given by the table on the left, which is a weighted relation on three variables, $A, B,$ and C, and an added column labeled with the reserved label $Weight$ designating the weight variable.

A	B	C	Weight		A	B	C	Weight
0	1	1	2		0	1	1	2
1	0	1	3		1	0	1	3
0	1	0	1		0	1	1	1
1	0	0	5		1	0	0	5

Valid weighted relation Invalid weighted relation

Figure 8.1— The left table is a valid weighted relation; the right is not

The table satisfies the functional dependency: $A, B, C \rightarrow Weight$; no two rows exist with identical values (a, b, c) for the variables (A, B, C) and different values for the $Weight$. The table on the right is invalid because the first and third rows have the same value tuple (0,1,1) for (A, B, C) but different values in the $Weight$ column.

In summary, a weighted relation is a relation with a particular variable, Weight, that implies a functional dependency for the weighted relation on the Weight attribute.

In the following section, we define three relational operators that extend the familiar relational operators to apply to weighted relations.

Operating on weighted relations

This section describes four operators specifically designed for handling weighted relations. We refer to these operators as `COND`, `PROJECT`, `ELIM`, and `COMB`. They feature customizable functional parameters tailored for particular domain semantics. Relevant application domains include probabilistic databases, Bayesian networks, influence diagrams, optimization problems, and constraint processing. The operators `COND`, `PROJECT`, and `COMB` serve as extensions or generalizations of the three primary relational algebra operators: `SELECT`, `PROJECT`, and `JOIN`. The operator `ELIM` functions as a wrapper around the `PROJECT` operator.

Let's start with a description of the first operator, `COND`, which is the conditioning operator.

The conditioning operator [cond]

The conditioning operator, $cond(r, X, x)$, takes one weighted relation r and a value assignment x for one or more variables X, selects the tuples from the relation having the specific values for each input variable, and returns the new relation formed by the selected tuples. The following formally defines the conditioning operator.

$$cond(r, X, x) = \{t \in r | t[X] = x\} \text{ (8.1)}$$

Here, $t[X]$ denotes the X-value of t, and X is a subset of variables, $X \subset R$, not including the Weight attribute.

The conditioning operator is equivalent to the `SELECT` operator in relational algebra, except that the selection variables do not include one distinct Weight attribute.

Figure 8.2 shows an example of weighted relation and the result of conditioning on a subset of variables. The table on the left shows the input relation containing eight value tuples. The query conditions the weighted relation by selecting tuples t such that $t[X] = 0, t[Y] = 1$. The conditioning operation returns the new relation shown by the table on the left, which contains only two tuples.

X	Y	Z	Weight
0	0	0	1
0	0	1	2
0	1	0	5
0	1	1	3
1	0	-0	7
1	0	1	4
1	1	0	6
1	1	1	6

X	Y	Z	Weight
0	1	0	5
0	1	1	3

Figure 8.2— Applying the conditioning operator **cond(r, [X,Y], [0,1])** *on the weighted relation r given by the left table returns the weighted relation r' shown by the right table*

Next, we will provide a description of the second operator, PROJECT, which functions as the projection operator.

The projection operator [project]

This projection operator $project(r, X, g)$ takes as input a weighted relation r, a subset of variables X, and an optional aggregation function $g(.)$, a summation by default. The projection operator projects r on the "Weight" and the input subset of variables X and aggregates the "Weight" for each unique combination of values of the input variables by applying aggregation function $g(.)$, then, the results are a new weighted relation with the value tuples for the input variables and the associated aggregate Weight.

The projection operator, $project(r(R \cup \{Weight\}), X, g)$, is the weighted relation $r'(X \cup \{Weight\})$ obtained by projecting $r(R \cup \{Weight\})$ on $X \cup \{Weight\}$ and then collapsing all tuples t having the exact value of $t[X]$ and applying the aggregation function $g(.)$ on the values of their Weight attribute. The aggregation function is a summation for probability and a maximization (minimization) for utilities (costs).

If a relation r does not include the distinct "Weight" attribute, then the project operator $project(r, X, g)$ becomes the same as the projection operation $\pi_X(r)$ in relational algebra. We use $\pi_X(r)$ to denote the projection of a relation $r(R)$ onto subset $X \subset R$, $\{t[X] | t \in r\}$ where $t[X]$ denotes the values of X in the tuple t.

The following formally defines the project operator:

$$project(r, X, g) =$$

$$\{< x, w > | t \in \pi_{X \cup \{Weight\}} r, x = t[X], w = g(t[Weight]|_{t[X]=x})\} \quad (8.2)$$

Thus, the project operator $project(r, X, g)$ takes a weighted relation r, a subset of variables X, and aggregation function g and returns a new weighted relation whose tuples are the projection of the input relation r on X after collapsing all tuples with the same value x for X and aggregating the value of their Weight attributes by applying the aggregation function g(.) on them.

The text below demonstrates the projection operator for weighted relations through an example. In this context, a weighted relation denotes a probability distribution, while the aggregation function g(.) corresponds to the summation function Σ(.).

A	B	C	D	Weight
0	0	0	0	0.05
0	0	1	1	0.1
0	1	0	0	0.15
0	1	1	0	0.3
1	1	1	1	0.4

A	Weight
0	0.6
1	0.4

Figure 8.3— Applying the projection operator **project(r, [A]**) *on the weighted relation r given by the left table returns the weighted relation r' shown by the right table*

Figure 8.3 illustrates an example of applying the projection operator to the weighted relation shown in the left table, which corresponds to the joint probability distribution $p(A, B, C, D)$. The projection query, $project(p, [A])$, returns the weighted relation represented in the right table. This relation reflects the marginal distribution p(A), derived from the marginalization of $p(A, B, C, D)$ onto A, expressed as follows:

$$p(A) = \sum_{B,C,D} p(A, B, C, D)$$

Next, we will describe the third operator, `ELIM`, which serves as the elimination operator.

The elimination operator [elim]

The elimination operator is a wrapper of the projection operator. The elimination operator takes a weighted relation $r(R \cup \{Weight\})$ and a subset $X \subset R$ and returns the projection of the relation on the complement set $R \backslash X$. That is

$$elim(r, X, g) = project(r, R \backslash X, g) \quad (8.3)$$

Next, we will describe the fourth and final operator, `COMB`, which represents the combination operator.

The combination operator [comb]

The combination operator $comb(r_1, r_2, f)$ takes two weighted relation tables as inputs along with an optional **combination function** $f(.)$, which defaults to the multiplication function $\prod(.)$. It performs an inner join operation on the two tables. If both tables contain a "Weight" attribute, the operator generates a new "Weight" attribute by applying the combination function $f(.)$ to the original "Weight" attributes from both input relations. Subsequently, it removes the original "Weight" columns and outputs the result as a new weighted relation.

The combination operator, $comb(r_1(R_1 \cup \{Weight\}), r_2(R_2 \cup \{Weight\}), f)$, is the weighted relation $r'(R_1 \cup R_2 \cup \{Weight\})$ obtained by joining the two weighted relations and applying the combination function f(.) on the two Weight attributes, $r_1.Weight$ and $r_2.Weight$, and setting the product to be the Weight attribute of the output relation $r'(R_1 \cup R_2 \cup \{Weight\})$ after dropping the $r_1.Weight$ and $r_2.Weight$ attributes from the join. The following formally defines the elimination operator.

$$comb(r_1(R_1 \cup \{Weight\}), r_2(R_2 \cup \{Weight\}), f) = \{t|\ t[R_1 \cup R_2] \in r_1 \bowtie r_2 \wedge t[Weight] = f(t[r_1.Weight], t[r_2.Weight] \quad]) \quad (8.4)$$

If one or both input relations have no "Weight" attribute, the combination operator is equivalent to a join of the two relations. Recall the join of two relations $r(R)$ and $s(S)$ is defined as follows:

$$r \bowtie s \stackrel{\text{def}}{=} \{t|t[R] \in r \wedge t[S] \in s\} \quad (8.5)$$

To illustrate the `COMB` operator, we will consider an example in which the weighted relations represent probability distributions. In this example, the combination function $f(.)$ defaults to the multiplication function $\prod(.)$ applied to the weights of matched records from the pair of joined weighted relations.

Figure 8.4 shows three weighted relations. The table r1 on the left displays a joint probability distribution over two binary variables, X and Y. The table in the middle, r2, presents a conditional probability distribution for the binary variable Z, given the binary variable Y. The table on the right displays the inner join of the two weighted relations, r1 and r2. It includes two columns: `r1.Weight` and `r2.Weight`, which represent the weight values of the matched records from relations r1 and r2. The first record in the join table corresponds to two matched records: the fourth row from table r1 and the second row from table r2. These records agree on the value of the shared variable Y, which is Y=1 in both cases.

r_1(X, Y, Weight)		
X	Y	Weight
0	0	0.15
0	1	0.6
1	0	0.225
1	1	0.025

r_2(Y, Z, Weight)		
Y	Z	Weight
0	0	1
1	0	0.25
1	1	0.75

r_1(X, Y, Weight) r_2(Y, Z, Weight)				
X	Y	r_1.Weight	Z	r_2.Weight
1	1	0.025	0	0.25
0	1	0.6	0	0.25
1	1	0.025	1	0.75
0	1	0.6	1	0.75
0	0	0.15	0	1
1	0	0.225	0	1

Figure 8.4— Two weighted relations and their natural join

The combination operator, using a multiplication function, results in a new weighted relation, as shown in *Figure 8.5*. This table is created by multiplying the `r1.Weight` and `r2.Weight` from the join table, which is displayed on the right side of *Figure 8.4*. The resulting product forms the Weight column in the combination table presented in *Figure 8.5*, specifically representing the output of the combination operation.

X	Y	Z	Weight
1	1	0	0.00625
0	1	0	0.15000
1	1	1	0.01875
0	1	1	0.45000
0	0	0	0.15000
1	0	0	0.22500

Figure 8.5— The output of the combination operator on the two relations

$r_1(X, Y, Weight)$ and $r_2(Y, Z, Weight)$ in *Figure 8.4* with multiplication as combination

Having defined the four fundamental operators of weighted relations—`COND`, `ELIM`, `PROJECT`, and `COMP`—in *Equations 8.1* through *8.4*, we will now proceed to the practice section. In this section, we will demonstrate how to implement these operators using R and the *relations* library. Additionally, we will practice applying the operators to examples of weighted relations.

Practice using R

The following section discusses the implementation of R functions that perform four weighted relation operations: projection, conditioning, elimination, and combination. These operators are defined as functions with parameters that can be adjusted to suit different applications, accommodating the semantics of various domains. This includes reasoning tasks related to graphical models, such as Markov and Bayesian networks, influence diagrams, cost-based abduction, and optimization problems.

Before we proceed, we describe in the following section the required libraries you need to install and load to run the R code.

Required libraries

The code requires the following libraries:

```
library(dplyr)
library(purrr)
library(rlang)
```

The dplyr R library is one of the main packages in the *tidyverse*, and it provides a variety of functions that allow you to manipulate dataframes in R. Notable functions include `filter()` for subsetting rows, `select()` for choosing columns, `mutate()` for adding new variables, `summarize()` for aggregating data, and a set of joining functions like `left_join()`, `inner_join()`, etc.

The *purrr* R library is a core part of the *Tidyverse* ecosystem, designed to provide a complete and consistent toolkit for **functional programming** for vectors and lists. It is primarily used to replace complex for loops with more readable and robust iteration methods. It stands out for its ability to map functions across these data structures with enhanced consistency and safety compared to the base R equivalents, such as `lapply()`.

The *rlang* R library, a foundational R package, provides a unified framework for working with core R language features and the *Tidyverse.* It is primarily a "tool for tool-builders," designed to help developers create robust functions and packages. It allows you to program language objects, such as formulas or expressions, and create tidy evaluation functions. Its most notable feature is the `parse_expr()` function, which converts a string into an R expression that can be evaluated later. Additionally, *rlang* provides the "`bang-bang`" (!!) operator, which is used to unquote the expression in the `filter()` function, demonstrating its role in enabling dynamic R code evaluation.

With the above overview of the R libraries, let's describe the code for implementing the four operators for querying weighted relations, which is the topic of the next section.

Implementing R functions for querying weighted relations

This section will outline the R code for implementing four relational queries associated with the weighted relation operators: PROJECT, ELIM, COND, and COMB. These operators were described in the previous section. All the code provided here is contained in the R script file *'functions.R'*. You can run the R code in *RStudio* for hands-on practice.

Let us begin by describing the code for the R function that implements the `project` operator used to perform projection queries on weighted relations.

Projection query

The R function `project` implements the projection operator that projects the relation on the "Weight" and given `vars` and then aggregates the result by applying a function `FUN` to the "Weight" for each unique combination of vars values. The `FUN` argument equals the `sum` by default. If you change the `FUN`, the function will adjust the projection and aggregation accordingly. Note that the input data frame should have a "Weight" column.

```
project <- function(df, vars, FUN = sum){
  # Create a unique grouping key across the specified vars
  group_key <- do.call(paste, c(df[vars], sep = "_"))
  # Add this key to the dataframe
  df$group_key <- group_key
  # Split the dataframe by the key and apply the aggregation function to each
group's weight
  df_grouped <- split(df, df$group_key)
  res <- do.call(rbind, lapply(df_grouped, function(group_df) {
    aggr_weight <- FUN(group_df$Weight)
    # Return a dataframe with the unique variable combinations
    # and aggregated weight
    unique_rows <- unique(group_df[vars])
    unique_rows$Weight <- aggr_weight
    return(unique_rows)
  }))
  # Remove temporary group_key from the original dataframe
  df$group_key <- NULL
  return(res)
}
```

Next, we will describe the R function code that implements the `elim` operator for performing elimination queries on weighted relations.

Elimination query

The R function `elim` implements the elimination query defined earlier. It is basically a wrapper of the function `project`. The `elim` function takes a `relation` and a variable `var` to be eliminated and calls the `project` function on the variables `vars`, which are the complement of `var` in the set of variables in the scheme of the `relation`. The `elim` function also takes an optional aggregation function `FUN` that it passes on to the `project` function.

```
elim <- function(relation, var, FUN = sum){
  # vars are all the variables in the relation scheme other than var and
"Weight"
```

```
  vars <- setdiff(names(relation), c(var, "Weight"))
  # Return the projected relation using the vars
  return(project(relation, vars, FUN))
}
```

We will now describe the R function code that implements the cond operator, which is used for performing conditioning queries on weighted relations.

Conditioning query

The following function cond implements the conditioning operator for weighted relations. The cond function takes three input parameters: `relation`- a weighted relation; `vars`- a vector representing a subset of variables; and vector `values` representing value assignment for the respective variables in `vars`. Both `vars` and `vals` must have the same length, and the relation should be defined on a superset of `vars`. The function selects only tuples from `relation` satisfying `vars = vals`.

```
cond <- function(relation, vars, vals){
  if (length(vars) != length(vals)) {
    stop("Length of variables and values do not match.")
  }
  # Create the expressions
  filter_expr <- map2_chr(vars, vals, ~ paste0(.x, "==", .y)) %>%
    paste(collapse = " & ") %>%
    parse_expr()
  # Apply the expressions to relation
  relation <- relation %>%
    filter(!!filter_expr)
  return(relation)
}
```

Next, we will describe the fourth and final R function code that implements the comb operator, which is used to perform combination queries on weighted relations.

Combination query

The combination query is a comb function that takes two weighted relations and an optional parameter FUN, which allows the combination of the two relations by multiplying or summing their 'Weight' attributes. The 'Weight' of two relations is multiplied by default in the given code.

This comb function takes two data frames, `df1` and df2, as inputs, performs a join operation on them, and if both of them have a "Weight" attribute, creates a new "Weight" attribute by applying the combining function FUN to the original "Weight" attribute of both data frames. It then drops the original "Weight" columns and converts the resulting data frame to a new weighted relation.

This function's usage should be straightforward: you input two data frames representing the two relations you want to process and the combination function `FUN`. The `comb` function will return a relation object as the result. After running the function, you can check the result by invoking `relations::relation_table(result)`.

```
comb <- function(df1, df2, FUN = "*"){
  common_names <- intersect(names(df1), names(df2))
  common_names <- common_names[common_names != "Weight"]
  join_df <- merge(x = df1, y = df2, by = common_names,
                   all = TRUE, suffixes = c('_df1', '_df2'))
  if (FUN == "*") {
    # Multiply the weights
    join_df <- join_df %>% mutate(Weight = case_when(
      !is.na(Weight_df1) & !is.na(Weight_df2) ~ Weight_df1 * Weight_df2,
      !is.na(Weight_df1) & is.na(Weight_df2) ~ Weight_df1,
      is.na(Weight_df1) & !is.na(Weight_df2) ~ Weight_df2
    ))
  } else if (FUN == "+") {
    # Sum the weights
    join_df <- join_df %>% mutate(Weight = case_when(
      !is.na(Weight_df1) & !is.na(Weight_df2) ~ Weight_df1 + Weight_df2,
      !is.na(Weight_df1) & is.na(Weight_df2) ~ Weight_df1,
      is.na(Weight_df1) & !is.na(Weight_df2) ~ Weight_df2
    ))
  } else {
    stop("Invalid operation. Only + (sum) and * (multiply) are allowed.")
  }
  # Drop the original weight columns.
  join_df <- join_df %>% select(-Weight_df1, -Weight_df2)
  return(join_df)
}
```

Having described the implementation of the primitive queries as functions using R, we illustrate in the next section how to use those functions to perform various queries on weighted relations.

All the examples are included in a quarto document named `'primitive_queries.qmd'`. The code uses the primitive queries R functions defined in the R script file `'functions. R'`.

Applying the primitive queries on weighted relations examples

In the following, we give various illustrative examples of using the projection, conditioning, and combination queries on multiple weighted relations examples.

Example: projection query

The input data frame represents the table specifying a weighted relation on three variables: X, Y, and Z. The projection query projects the input relation on the input variables X and Y. It's important to note that the query uses summation as the aggregation function, which plays a vital role in the calculation process.

```
r1_X_Y_Z <- data.frame(X=c(0,0,0,1,1),
                       Y=c(0,0,1,0,0),
                       Z=c(0,1,1,0,1),
                       Weight = c(0.3,0.3,0.2,0.1,0.1),
                       stringsAsFactors = FALSE)
result <- project(r1_X_Y_Z, c("Y","Z"))
print(result)

# Output:
    Y Z Weight
0_0 0 0    0.4
0_1 0 1    0.4
1_1 1 1    0.2
```

Example: projection query

The input data frame is a table of a weighted relation on X, Y, and Z. The following projects the relation on X and Y. We explicitly specify the `FUN` parameter as `max`, which results in applying the `max` of the Weights to obtain the maximum utility for X and Y over all values of Z.

```
r2_X_Y_Z <-
        data.frame(X = c(0,0,0,0,1,1,1,1),
                   Y = c(0,0,1,1,0,0,1,1),
                   Z = c(0,1,0,1,0,1,0,1),
                   Weight = c(1,2,5,3,7,4,6,5),
                   stringsAsFactors = FALSE)

result1 <- project(r2_X_Y_Z, c("X","Y"),FUN=max)
print(result1)

# Output:
    X Y Weight
0_0 0 0      2
0_1 0 1      5
1_0 1 0      7
1_1 1 1      6
```

Example: conditioning query

The following uses the `*cond*` operator on the weighted relation `r2_X_Y_Z` to select only tuples with value assignment `X=0` and `Y=1`.

```
r2_X_Y_Z <-
        data.frame(X = c(0,0,0,0,1,1,1,1),
                   Y = c(0,0,1,1,0,0,1,1),
                   Z = c(0,1,0,1,0,1,0,1),
                   Weight = c(1,2,5,3,7,4,6,5),
                   stringsAsFactors = FALSE)

result <- cond(r2_X_Y_Z, c("X","Y"), c(0, 1))
print(result)

# Output:
  X Y Z Weight
1 0 1 0      5
2 0 1 1      3
```

Example: combination query

The example combines two potential functions: 1) `phi_X_Y` on the variables X and Y, and 2) `phi_Y_Z` on Y and Z. The `comb` function joins `phi_X_Y` and `phi_Y_Z`, multiplying the "Weight" attribute of joined tuples, tuples matching on the value of the shared Y attribute. The result is a joint probability on the combined variables X, Y and Z.

```
phi_X_Y <- data.frame(X = c(0,0,1,1),
                      Y = c(0,1,0,1),
                      Weight = c(0.15, 0.6, 0.225, 0.025),
                      stringsAsFactors = FALSE)

phi_Y_Z <- data.frame(Y = c(0,1,1),
                      Z = c(0,0,1),
                      Weight = c(1, 0.25, 0.75),
                      stringsAsFactors = FALSE)


# Run the function
rel <- comb(phi_X_Y, phi_Y_Z, FUN="*")

# Print the resulting relation.
print(rel)

# Output:
```

```
  Y X Z  Weight
1 0 0 0 0.15000
2 0 1 0 0.22500
3 1 0 0 0.15000
4 1 0 1 0.45000
5 1 1 0 0.00625
6 1 1 1 0.01875
```

Example: combination query

Consider two utility functions, f1(X, Y) and f2(Y, Z), represented by the two weighted relations r1 and r2 in the code below. We combine the two relations to get the sum of the two functions, a function of the three variables f(X, Y, Z), defined by the result of the combination query. Note that we specify the combining function as "+" for summing the "Weights" of the relations r1 and r2 joined tuples instead of multiplying them as is the case by default.

```
r1 <- data.frame(X = c(0,0,1,1),
                 Y = c(0,1,0,1),
            Weight = c(4,8,0,5),
            stringsAsFactors = FALSE)

r2 <- data.frame(Y = c(0,1,1),
                 Z = c(0,0,1),
            Weight = c(0,5,6),
            stringsAsFactors = FALSE)
# Run the function
rel <- comb(r1, r2, FUN="+")

# Print the resulting relation.
print(rel)

# Output:
  Y X Z Weight
1 0 0 0      4
2 0 1 0      0
3 1 0 0     13
4 1 0 1     14
5 1 1 0     10
6 1 1 1     11
```

In concluding this section on structured primitive queries, let us summarize what is learned in this section. We began with the definition of weighted relation, which is a relation that includes an additional attribute labeled with a reserved name called "Weight" to be treated in non-standard ways when operating on those relations. We defined four operators, `COND`, `ELIM`, `PROJECT`, and `COMB`, to perform conditioning, elimination, projection, and combination queries. The conditioning query `COND` corresponds to the `SELECT` query in relational algebra on non-weighted relations. The `PROJECT` query differs from the `PROJECT` query in relational algebra by treating the distinct "Weight" attribute differently when performing the projection to conform with domain-specific semantics specified by an aggregation function parameter. The `ELIM` query is a wrapper of the `PROJECT` query, which takes a variable to be eliminated from a relation and projects the relation on all attributes of the relation scheme other than the Weight attribute and the specified elimination variable. The query `COMB` is equivalent to the `INNER JOIN` query on non-weighted relations. Still, it significantly differs on weighted relations as it processes the distinct weight attributes on the pair of input relations by combining them according to some input combination function to apply domain-specific semantics.

In the next section, we describe an algorithm called query composition graph, a variant of the variable elimination approach for exact inference in Bayesian networks. The algorithm represents the Markov Bayesian network model using weighted relations and formulates the inference as a composition of primitive queries on weighted relations.

Query composition graph (QCG) algorithm

This section presents a query composition algorithm based on the weighted relation approach to performing reasoning tasks in deterministic and probabilistic networks. The algorithm belongs to variable elimination algorithms for exact inference with graphical models.

This section is organized as follows. We begin with a review of the factored representation of the graphical models of Bayesian networks. Next, we give a motivational introduction to illustrate the basic idea of variable elimination in the context of inference in Bayesian networks. Following this, we formally describe the QCG algorithm for inference based on compositional queries on weighted relations. Finally, in a practice section, we describe the code of various functions for implementing the QCG algorithm and give an example demonstrating the application of one of those functions. Let's dive into it.

Factored representation of Markov models

In this section, we summarize the factored representation of Markov graphical models and define the concept of factors. We consider a Bayesian network example, which we will use in the next section.

Let us recall the factored representation of the Markov model of a Bayesian network given by a set of functions that represent the factors. The joint probability distribution over the network variables is the product of all the factors up to a normalizing constant called the partition function.

A Markov network represents a complete probability distribution on a set of variables $V = \{V_1, \cdots, V_n\}$ using an undirected graph $G = (V, E)$ where V are the nodes, and E is the edges, sets of distinct pairs of nodes in V.6

The Markov model is a factorization of a complete joint probability distribution over V given as the product of a set of non-negative functions called potential functions or simply factors, defined on subsets of variables in V; every subset contains variables that are all pairwise connected in the moral graph of the Bayesian network, or a causal family in the directed acyclic graph. We write the factored representation as follows:

$$P(V_1, \cdots, V_n) \propto \prod_{i=1}^{m} f_i(S_i)|_{S_i \subset V} \quad (8.6)$$

A factor is a more general concept than a probability distribution. A joint or conditional probability represents factors, but the converse is invalid, as a factor need not be a probability or sum out to 1. In *Chapter 6*, we saw how we define factors on the cliques of a join tree by combining functions propagated between the join tree nodes. Those intermediary functions are not probabilities but messages that enable the computation of probabilities.

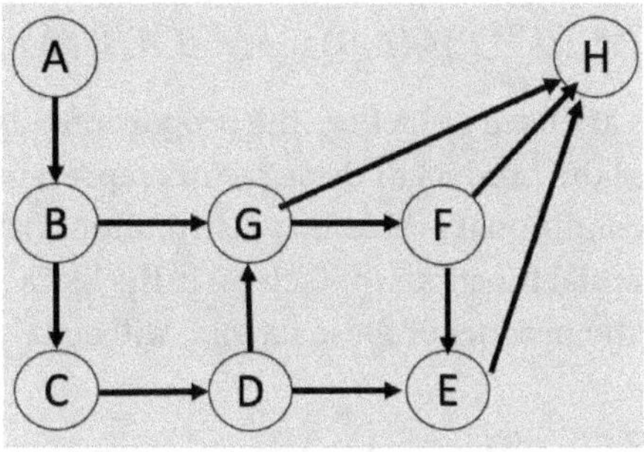

Figure 8.6— Directed acyclic graph of the example Bayesian network

Figure 8.6 shows the directed acyclic graph of a Bayesian network example from (Dechter and El Fattah 2001) representing the following factored probability model,

$$P(A, B, C, D, E, F, G, H) = P(A)\, P(B|A)\, P(C|B)\, P(D|C)$$

$$P(G|B, D)\, P(E|D, F)\, P(F|G)\, P(H|G, F, E) \quad (8.7)$$

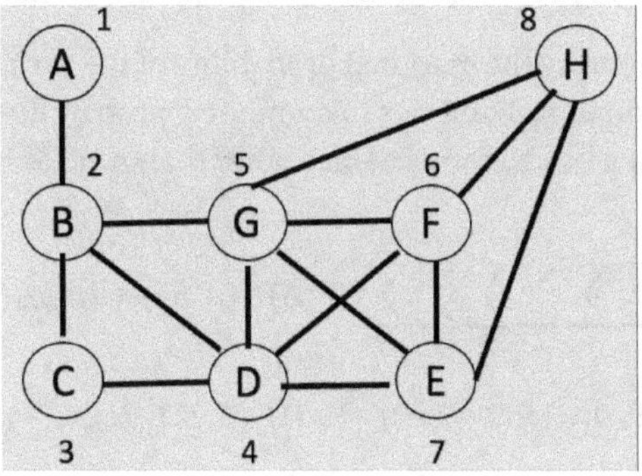

Figure 8.7—Moral graph of the example Bayesian network

Figure 8.7 shows the moral graph for the Bayesian network. The moral graph has all the families in the directed acyclic graph forming cliques where all nodes in a clique are pairwise connected. The causal family of the G node has an undirected edge connecting nodes B and D so that $\{B, D, G\}$ forms a clique. We also have the parents of the E node connected by an edge $D - F$. *Figure 8.7* shows the moral graph for the Bayesian network. The moral graph has all the families in the directed acyclic graph forming cliques where all nodes in a clique are pairwise connected. The causal family of the G node has an undirected edge connecting nodes BB and DD so that $\{B, D, G\}\{B, D, G\}$ forms a clique. We also have the parents of the E

$D - F$ and the parents of the H node getting connected by an edge $E - G$. The conditional probabilities can be directly used as the factors on all the factored families, and we write,

$$P(A, B, C, D, E, F, G, H) \propto f_1(A)\, f_2(A, B) f_3(B, C) f_4(C, D) f_5(B, D, G)$$

$$f_6(D, E, F) f_7(F, G) f_8(E, F, G, H) \quad (8.8)$$

When conditional probabilities are used as factors, the proportionality sign in *Equation 8.8* becomes an equality. This occurs because the product of these factors represents a joint probability that sums to 1. In other words, when the conditional probability distributions from the factored representation of Bayesian networks act as potential functions (or factors) in the Markov network model, the partition function, Z, which normalizes the product of these factors, will equal 1. Please refer to *Equations 7.2* and *7.3* in *Chapter 7*.

In the upcoming section, we will utilize the previously described Bayesian network example to provide a step-by-step explanation of the variable elimination approach to inference in Bayesian networks.

Introducing variable elimination

This section gives a motivational introduction to the main ideas behind the variable elimination (VE) algorithms. We use the Markov model example from the previous section to illustrate the main ideas, considering the problem of determining the marginal probability of a query variable in a Bayesian network represented by a factored representation of a complete probability distribution.

Consider the problem of computing the marginal probability of the variable A. A brute force approach is multiplying all the conditional probabilities to compute a probability table of the joint probability distribution and then summing out all the variables other than A. That is, we compute the sum of the products of the factors,

$$\sum_B \sum_C \sum_D \sum_G \sum_F \sum_E \sum_H (f_1\,(A) f_2(A, B) f_3(B, C) f_4(C, D)$$

$$f_5(B, D, G) f_6(D, E, F) f_7(F, G) f_8(E, F, G, H)) \qquad (8.9)$$

The brute force approach is computationally inefficient and becomes prohibitively expensive as the number of network variables becomes large since the computational complexity grows exponentially with the number of variables.

Variable elimination exploits the factored representation of the Markov network model. It begins with selecting a total ordering of the network variables. In the following section, we discuss the importance of variable ordering and describe various variable ordering methods based on various heuristics.

Variable ordering for variable elimination

The ordering of network variables in variable elimination is critical, as it determines the computational complexity of exact inference by defining the size of induced factors on variable cliques. A good ordering minimizes the size of these factors, leading to fast inference, while a poor ordering can create large cliques, causing an exponential increase in computation time. As previously pointed out in *Chapter 7*, finding the optimal variable ordering that minimizes the maximum induced factor size is NP-hard, requiring the use of heuristics for finding a good but not necessarily optimal ordering.

The following four methods are based on different ordering heuristics (Dechter and El Fattah 2001): **maximum-cardinality ordering, minimum-width ordering, minimum-degree ordering**, and **causal ordering.** It is important to note that the causal ordering is applicable only to directed acyclic graphs. The maximum-cardinality ordering is created by starting with an arbitrarily chosen node and then repeatedly selecting the unordered node that is adjacent to the maximum number of already ordered nodes. The minimum-width ordering is computed by selecting nodes from last to first, choosing the node with the fewest neighbors, removing that node along with its incident edges from the graph, and continuing this process until the graph is empty. The min-degree ordering (Bertelè and Brioschi 1972) is similar to the min-width ordering, but with the distinction that it connects the neighbors of the selected nodes. The causal ordering is just a topological sort of a directed acyclic graph.

Figure 8.7 shows the moral graph of our Bayesian network example, with the variables being ordered according to maximum cardinality. Since the moral graph of the example is also chordal, the maximum cardinality ordering is also a **perfect elimination ordering** (PEO) as explained in *Chapter 7*.

Describing variable elimination by example

The idea behind all the variable elimination algorithms is to decompose a problem into smaller subproblems and assemble the problem's solution from the subproblems' solutions.

Let us demonstrate the process of variable elimination using the Bayesian network example. We will consider the variable ordering based on maximum cardinality, as illustrated in *Figure 8.7*. The variable elimination will proceed in reverse order, so we begin by eliminating H.

Eliminating H

Eliminating H means rearranging the summation and multiplication in *Equation 8.7*.

$$\sum_B \sum_C \sum_D \sum_G \sum_F \sum_E f_1(A) f_2(A,B) f_3(B,C) f_4(C,D)$$

$$f_5(B,D,G) f_6(D,E,F) f_7(F,G) \textstyle\sum_H f_8(E,F,G,H)) \quad (8.10)$$

The rightmost summation is a function on the variables G, F and E, representing a new factor that replaces all the factors inside the summation. In this instance, there is only one term, $f_8(E, F, G, H)$, which is replaced by an induced factor.

$$h_H(G, F, E) = \sum_H f_8(E, F, G, H) \quad (8.11)$$

The next variable to eliminate in reverse order is E.

Eliminating erosion

To eliminate E, we move all the terms involving the E variable in *Equation 8.10* to the rightmost side of the equation, and so do we move the summation over E. We write,

$$\sum_B \sum_C \sum_D \sum_G \sum_F \sum_E f_1(A) f_2(A, B) f_3(B, C) f_4(C, D)$$

$$f_5(B, D, G) f_7(F, G) \sum_E f_6(D, E, F) h_H(G, F, E)) \ (8.12)$$

Eliminating E involves multiplying the two factors $f_6(D, E, F)$ and $h_H(G, F, E)$ and then summing out E to produce the following induced function overG, F, D,

$$h_E(D, F, G) = \sum_E f_6(D, E, F) h_H(G, F, E) \ (8.13)$$

We replace the sum of the product of the two factors $f_6(D, E, F)$ and $h_H(G, F, E)$ in *Equation 8.12*by thenew induced factor $h_E(D, F, G)$. The summation multiplication formula of *Equation 8.12* now becomes:

$$\sum_B \sum_C \sum_D \sum_G \sum_F f_1(A) f_2(A, B) f_3(B, C) f_4(C, D) f_5(B, D, G) f_7(F, G) h_E(D, F, G) \quad (8.14)$$

The next variable to eliminate in reverse order is F.

Eliminating F

To eliminate F, we move all factors on variables that include F in equation 8.14 to the rightmost side of the equation and move the summation over F most inward. The summation multiplication formula becomes

$$\sum_B \sum_C \sum_D \sum_G f_1(A) f_2(A, B) f_3(B, C) f_4(C, D) f_5(B, D, G) \sum_F f_7(F, G) h_E(D, F, G). \ (8.15)$$

Eliminating F produces a new induced factor on D and G.

$$h_F(D, G) = \sum_F f_7(F, G) h_E(D, F, G) \quad (8.16)$$

The next variable to eliminate is G.

Eliminating G

To eliminate G, we substitute the induced factor $h_F(D,G)$ in the summation multiplication formula of *Equation 8.15* and move the summation over G most inward. The summation multiplication formula becomes

$$\sum_B \sum_C \sum_D f_1(A)\, f_2(A,B) f_3(B,C) f_4(C,D) \sum_G h_F(D,G) f_5(B,D,G) \quad (8.17)$$

Performing the multiplication and summing out G eliminates G and produces the new induced factor.

$$h_G(B,D) = \sum_G h_F(D,G) f_5(B,D,G) \ (8.18)$$

We continue in the same way to eliminate the next variable, D.

Eliminating duplicates

To eliminate D, we substitute the induced factor $h_G(B,D)$ in the summation multiplication formula of *Equation 8.17* and move the summation over D most inward. The summation multiplication formula becomes

$$\sum_B \sum_C f_1(A)\, f_2(A,B) f_3(B,C) \sum_D f_4(C,D) h_G(B,D) (8.19)$$

Performing the multiplication and summing out D eliminates D and produces the new induced factor.

$$h_D(B,C) = \sum_D f_4(C,D) h_G(B,D) \quad (8.20)$$

Moving on, the next step is to eliminate the variable C.

Eliminating C

To eliminate C, we substitute the induced factor $h_D(B,C)$ in the summation multiplication formula of *Equation 8.19* and move the summation over C most inward. The summation multiplication formula becomes

$$\sum_B f_1(A)\, f_2(A,B) \sum_C f_3(B,C)\, h_D(B,C) \quad (8.21)$$

Performing the multiplication and summing out C eliminates C and produces the new induced factor.

$$h_C(B) = \sum_C f_3(B,C)\, h_D(B,C) \quad (8.22)$$

The next step is eliminating B.

Eliminating B

To eliminate B, we substitute the induced factor $h_C(B)$ in the summation multiplication formula of *Equation 8.21* and move the summation over B most inward. The summation multiplication formula becomes

$$f_1(A) \sum_B f_2(A,B) h_C(B) \quad (8.23)$$

Performing the multiplication and summing out B eliminates B and produces the new induced factor.

$$h_B(A) = \sum_B f_2(A,B) h_C(B) \quad (8.24)$$

We now reach the conclusion of the variable elimination process and are ready to compute the marginal probability of the variable A, which we will do next.

Computing the marginal probability of the variable A

At the end of the elimination process, we only have a multiplication of two functions on A, obtained by substituting the induced function $h_B(A)$ into *Equation 8.23*, which yields,

$$f_1(A)h_B(A) \quad (8.25)$$

The marginal probability of A, $\pi(A)$, is obtained by normalizing the multiplication output to sum it to 1. That is,

$$\pi(A) = \frac{f_1(A)h_B(A)}{\sum_A f_1(A)h_B(A)} \quad (8.26)$$

Comparing variable elimination and tree clustering

Variable elimination (VE) and join tree clustering (also known as the junction tree algorithm) are fundamentally equivalent exact inference methods that differ primarily in their organization and intended query efficiency. Both rely on the distributive law to push summations (marginalization) inside products (factors), and both are constrained by the treewidth of the graph, which dictates their exponential time and space complexity.

The process of eliminating a variable in variable elimination (VE) generates an induced factor, which is the product of all present factors that include that variable, followed by marginalizing it out. Each induced factor corresponds to a clique in the chordal graph. This clique includes the eliminated variable as well as the other variables that define the function of the induced factor. The maximal cliques found in a join tree, also known as a junction tree, represent the largest cliques produced during the VE process.

The communication between nodes in a join tree mimics the flow of information in variable elimination. A separator set is the intersection of two adjacent cliques in the join tree. In the Join Tree algorithm, a message sent from one clique to another across a separator set is precisely the induced factor generated in VE when variables in the first clique (but not the second) are eliminated. The scope of an induced factor in VE is identical to the set of variables in the corresponding separator set.

The variable elimination steps on our example produce five maximal cliques, which are $\{A, B\}$, $\{D, B, C\}$, $\{B, D, G\}$, $\{E, D, F, G\}$, and $\{H, G, F, E\}$. The size of the maximum clique is 4, and the induced width of the ordering is 3.

The maximal cliques can be represented in a tree-structured representation known as a join tree (clique tree, or junction tree). *Figure 8.8* shows the junction tree equivalent to the variable elimination for computing the marginal probability $p(A)$. The junction tree shown in the figure has the maximal clique $\{A, B\}$ as the root node because it contains the variable A for which the marginal probability is to be evaluated. The junction tree clearly satisfies the running intersection property (RIP) as discussed in *Chapter 7*.

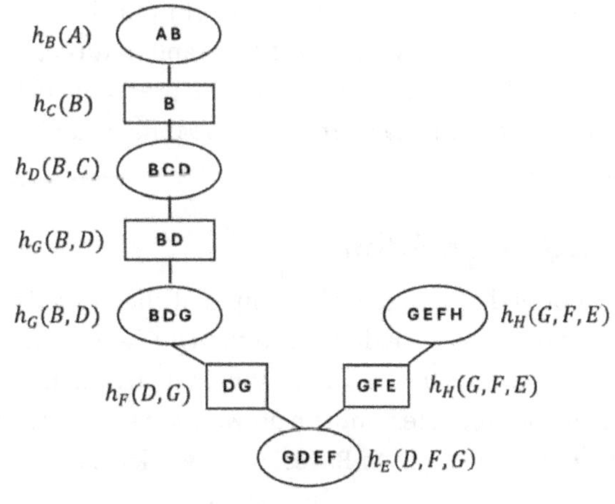

Figure 8.8—Join tree equivalent to variable elimination for computing marginal probability p(A); the induced factors are shown besides the nodes and edges

Figure 8.8 shows the induced factors resulting from the ordered variable elimination steps beside the associated maximal cliques, which are the nodes of the tree, and the associated separator sets, which are the edges of the tree. We note that the induced factor beside each separator is a function of the variables, which are all the elements of the separator set. We note that the induced factors beside each clique are functions of the eliminated variable and the variables that define the function of the induced factor. For example, the induced factor $h_F(D,G)$ is defined on the variables D and G, which are the elements of the separator set $\{D,G\}$. The induced factor $h_G(B,D)$ beside the clique $\{B,D,G\}$ indicates that the clique is formed from the eliminated variable G and the set $\{B,D\}$, which are the variables on which the factor is defined. Note also that the induced factors have the same ordering as that of the variable elimination, namely in the following order: (H,E,F,G,D,C,B,A).

Going through the variable elimination steps on the example points to the underlying connection between the maximal cliques computed in the join tree approach to inference and the inference approach of variable elimination.

Variable Elimination

When we eliminate a variable, we are decomposing the inference problem into smaller-size subproblems. Each variable elimination step reduces the factored representation of a complete probability distribution on n variables to a factored representation of a complete probability distribution on `n-1` variables, with the eliminated variable being the only one missing from the new representation.

Let us recap the takeaway point of this section. The variable elimination process is a systematic and structured approach that follows a clear and logical sequence. It starts with ordering the variables and then eliminates them, proceeding from the highest-order variable to the lowest-order variable. Each elimination step creates an induced factor on lower-order component variables connected with the eliminated variable in the triangulated moral graph. The components of the induced factors are cliques of variables that are all pairwise connected.

In the following section, we will delve into the QCG algorithm for inference based on variable elimination. This algorithm constructs a query composition graph, a directed acyclic graph with nodes representing primitive queries and edges indicating execution precedence. The QCG algorithm provides a computational framework for reasoning tasks in both probabilistic and deterministic networks, and it's a powerful tool for optimization tasks.

Describing the QCG algorithm

In the previous section, we provided a practical example of the variable elimination approach for inference in a Bayesian network. Now, we delve into the QCG algorithm, a crucial computational framework that automates this approach. This algorithm is instrumental in performing various reasoning tasks in both probabilistic and deterministic networks, as well as accomplishing optimization tasks, making it a valuable tool in the fields of Bayesian networks and R programming.

A query composition graph is a structured representation of compound queries composed of primitive queries.

Figure 8.9 shows the generic structure of a query composition graph. The graph has a modular structure, with modules representing elimination steps. The composition graph is a directed acyclic graph having three types of nodes: `elim`, `comp`, and `assoc`. The `elim` and `comp` type nodes correspond to the *elim* and *comp* queries. The `assoc` node is a **data dictionary** that dynamically tracks the execution stages. The directed edges represent the messaging between the nodes.

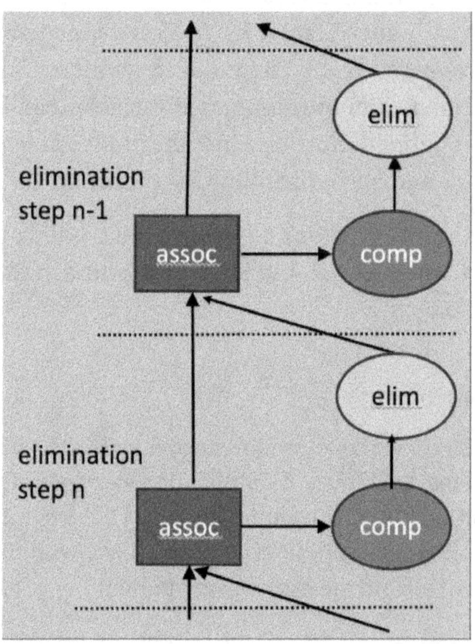

Figure 8.9—Query composition graph of consecutive elimination steps

The query composition graph is a compilation of the variable elimination process. Once the graph is compiled for a reasoning task, it can be reused to perform the task on new data.

The QCG is not just a one-time solution. It's a tool that can be built once for a specific task and then reused for multiple task input data instances. For instance, consider a QCG used to compute the posterior probability on a query variable, conditional on value assignment for a subset of conditioning variables. We can apply the QCG algorithm to compile a composition graph, which can be reused on different value assignments of conditioning variables. This reusability feature adds practicality and value to the QCG algorithm, making it a valuable asset in your Bayesian network and R programming toolkit.

Let's delve into the formal description of the QCG algorithm. This comprehensive algorithm takes a database as a set of weighted relations defined on a set of variables V with a subset $C \subset V$ designated as conditioning variables whose values are assigned and an ordering of all non-conditioning variables. It then performs variable elimination on the non-conditioning variables and compiles a network of primitive queries represented by a query composition graph. This thorough approach ensures that no aspect of the reasoning task is overlooked, giving you confidence in the QCG algorithm's capabilities.

The algorithm first conditions all weighted relations by removing all tuples in a relation with value assignment on the conditioning variables other than the conditioning values and then removing the conditioning variables from the relation scheme. Next, the algorithm eliminates the non-conditioning variables in an elimination ordering that goes backward in the variable ordering. In each eliminating a variable X_i step, the algorithm fetches from the dictionary `assoc` the subset of relations Δ_i that associates with X_i; it combines them with the `comb` query q_{comb} then eliminates X_i from q_{comb} by an `elim` query q_{elim}. Next, the algorithm removes the eliminated variable from the `assoc` dictionary and adds the new relation q_{elim} to `assoc`, associating q_{elim} with the highest-ordered variable in the q_{elim} scheme. The algorithm compiles the query graph nodes and edges by tracking the query composition steps.

Algorithm query composition graph (QCG)

Input: 1) A database of weighted relations $\Delta = \{r(R \cup \{Weight\}) | R \subset V\}$; 2) A set of conditioning variables, $C \subset V$; 3); An ordering $d = (V_1, \cdots, V_n)$.

Output: A query composition graph $G = (N, E)$, the nodes N are primitive queries and the directed edges represent the dependencies between queries.

Initialization: $N \leftarrow \emptyset, E \leftarrow \emptyset$.

Description:

```
// conditioning
For each r(R) ∈ Δ:
S ← R ∩ C

If S ≠ ∅:
q_cond ← cond(r(R), S, s)
Δ ← Δ ∪ {q_cond} \ r(R)


// create variable-relations dictionary
assoc_n ← assoc(Δ, d)
For i = n down to 1:
```

// combination step

$q_{comb} \leftarrow comb\_list(assoc_i\$V_i)$

// elimination step

$q_{elim} \leftarrow elim(q_{comb}, V_i)$

// update the assoc dictionary

$assoc_{i-1} \leftarrow add\_relation\_remove\_var(q_{elim}, assoc_i, X_i)$

// update the graph

$N \leftarrow N \cup \{q_{comb}, q_{elim}, assoc_i, assoc_{i-1}\}$

$E \leftarrow E \cup \{q_{comb} \rightarrow q_{elim}, assoc_i \rightarrow q_{comb}, q_{elim} \rightarrow assoc_{i-1}, assoc_i \rightarrow assoc_{i-1}\}$

Output $G = (N, E)$

All the elements of the QCG algorithm are implemented using R. In an earlier practice section, we already demonstrated the R code for the *cond*, *comb*, and *elim* queries and practiced those operators on some examples.

In the following section, we provide a detailed explanation of the R code for three key functions in the QCG algorithm: `assoc`, `add_relation_remove_var`, `comb_list`. These functions play a crucial role in the implementation of the QCG algorithm. They are included in the R script of the '`functions.R`' file, which you can refer to for a comprehensive understanding of their implementation.

Practice using R

In this section, we describe the implementation of the functions mentioned in the QCG algorithm, and we give an example of the usage of the function that combines multiple weighted relations on an example to compute the joint probability distribution of a Bayesian network.

Combining multiple weighted relations

The function `comb_list` combines a list of weighted relations. It takes a list of two or more relations and combines them by calling `` `comb.` `` The function takes a list, `rels`, of data frames representing weighted relations. Each data frame should have two or more columns: A weighted relation should have one "Weight" column providing the weight of the relationship. Each row specifies a relationship instance. The function also takes a parameter, `FUN`, a combination function that defaults to "*" meaning multiplication. The function `comb_list` returns a data frame representing the combined weighted relationships of the relations list.

```
comb_list <- function(rels, FUN = "*"){
  if(length(rels) == 1){
    return(rels[[1]])
  } else {
    combined_relation <- comb(rels[[1]], rels[[2]], FUN)

    if(length(rels) > 2){
      for(i in 3:length(rels)){
```

```
        combined_relation <- comb(combined_relation, rels[[i]], FUN)
      }
    }
    return(combined_relation)
  }
}
```

Computing dictionary of variable-relation associations

The assoc function takes a list of weighted relations (rels) and a vector of variables (vars). It returns a dictionary whose keys are the variable names, and values are those relations from rels where the scheme includes the variable and no variable succeeding it in the vars vector. The weighted relations (rels) list is a list of data frames, each representing a weighted relation. Each data frame should have two or more columns, with one distinct column named "Weight". The vector of variables (vars) is a character vector of variable names.

```
assoc <- function(rels, vars) {
  # Initialize an empty list to store the relations
  dict <- setNames(vector("list", length(vars)), vars)
  # Create a named vector that maps variable names to their order
  var_order <- setNames(seq_along(vars), vars)
  # Loop through each variable
  for (var in vars) {
    # Loop through each relation
    for (rel in rels) {
      # Extract variables in relation directly from column names
      rel_vars <- colnames(rel)[colnames(rel) != "Weight"]
      # If the relation contains the variable and all variables in the relation
come before or at the same position in the order
      if (var %in% rel_vars && all(var_order[rel_vars] <= var_order[var])) {
        # Add it to the dictionary
        dict[[var]] <- c(dict[[var]], list(rel))
      }
    }
  }
  return(dict)
}
```

Modifying the variable-relation associations

The add_relation_remove_var function takes a dictionary (assoc) that pairs variables with their associated list of weighted relations, a new weighted relation (relation), and a variable (var). The function first removes the entry for var from assoc. Then, it updates the modified dictionary by appending the new relation to the list of relations belonging to the highest-ordered variable present in the new relation.

```
add_relation_remove_var <- function(assoc, relation, var) {
  # remove var from assoc dictionary
  assoc1 <- remove_var(assoc, var)
  # add relation to assoc1 dictionary
  assoc2 <- add_relation(assoc1, relation)
  return (assoc2)
}
```

Variable removal function

The remove_var function takes a dictionary (assoc) that pairs variables with lists of weighted relations and a variable (var). The function returns a new dictionary where the entry corresponding to var is removed.

```
remove_var <- function(assoc, var) {
  # Create a new dictionary excluding the entry for var
  new_assoc <- assoc[!names(assoc) %in% var]
  return(new_assoc)
}
```

Relation addition function

The add_relation function takes a dictionary (assoc) that pairs variables with their associated list of weighted relations and a new weighted relation (relation). The function updates the dictionary by appending the new relation to the list of relations belonging to the highest-ordered variable present in the new relation.

```
add_relation <- function(assoc, relation) {
  # Extract variables from the relation
  rel_vars <- colnames(relation)[colnames(relation) != "Weight"]
  # Create a map from variable name to their order
  var_order <- setNames(seq_along(names(assoc)), names(assoc))
  # Identify the variable in the relation with the highest order
  highest_var <- rel_vars[which.max(var_order[rel_vars])]
  # Add relation only to the variable with the highest order
  if (highest_var %in% names(assoc)) {
```

```
      assoc[[highest_var]] <- c(assoc[[highest_var]], list(relation))
    }
    return(assoc)
}
```

The following examples are included in the practice R Markdown document '*queries.qmd*'.

Example: combining a list of relations

We consider an example of a Bayesian network (BN) with conditional probabilities as weighted relations. The directed acyclic graph of the BN is a chain having three nodes $\{X, Y, Z\}$ and two directed edges $\{X \to Y, Y \to Z\}$. The quantification of the network is given by the probabilities $P(X)$, $P(Y|X)$, and $P(Z|Y)$ represented as weighted relations specified by the dataframes `P_X`, `P_Y_X`, and `P_Z_Y` in the code below. We use the `comb_list` function to combine the conditional probabilities for the three-node network to compute the joint probability distribution $P(X,Y,Z) = P(X)P(Y|X)P(Z|Y)$.

```
P_X <- data.frame(X = c(0,1),
                  Weight = c(0.75, 0.25),
                  stringsAsFactors = FALSE)
P_Y_X <- data.frame(X = c(0,0,1,1),
                    Y = c(0,1,0,1),
                    Weight = c(0.2,0.8,0.9,0.1),
                    stringsAsFactors = FALSE)
P_Z_Y <- data.frame(Y = c(0,0,1,1),
                    Z = c(0,1,0,1),
                    Weight = c(1,0,0.25,0.75),
                    stringsAsFactors = FALSE)
JPD <- comb_list(list(P_X,P_Y_X,P_Z_Y))
print(JPD)

# Output:
  Y X Z  Weight
1 0 0 0 0.15000
2 0 0 1 0.00000
3 0 1 0 0.22500
4 0 1 1 0.00000
5 1 0 0 0.15000
6 1 0 1 0.45000
7 1 1 0 0.00625
8 1 1 1 0.01875
```

Example: add relation remove variable

The following example shows how to use the `assoc` function to create the variable-relations association dictionary and how to update the dictionary throughout the execution of the QCG algorithm by using the '`add_relation_remove_var`' function to add the induced relations from the elimination step to the dictionary and also remove the eliminated variable from the dictionary.

```
f_X <- data.frame(X = c(0,0,1), Weight = c(3,5,7), stringsAsFactors = FALSE)
f_Y <- data.frame(Y = c(1,0,1), Weight = c(2,10,6), stringsAsFactors = FALSE)
f_Z <- data.frame(Z = c(1,1,0), Weight = c(4,3,1), stringsAsFactors = FALSE)

dict = assoc(list(f_X,f_Y,f_Z),c("X","Y","Z"))
f_X_Y <- data.frame(X = c(0,0,1,1), Y = c(0,1,0,1), Weight = c(9,3,4,7),
stringsAsFactors = FALSE)

dict2 <- add_relation_remove_var(dict, f_X_Y, "Z")
print(dict2)

# Output:
$X
$X[[1]]
  X Weight
1 0      3
2 0      5
3 1      7

$Y
$Y[[1]]
  Y Weight
1 1      2
2 0     10
3 1      6

$Y[[2]]
  X Y Weight
1 0 0      9
2 0 1      3
3 1 0      4
4 1 1      7
```

In conclusion, this section on the QCG algorithm highlighted several key points. First, a Bayesian network's factored representation of a joint probability distribution naturally aligns with our framework for weighted relations. Second, the QCG algorithm translates the variable elimination process for inference on graphical models into a query composition problem involving weighted relations. Finally, in the practice section, we demonstrated how to implement the compositional process of the QCG algorithm using functions in R, providing illustrative examples of these functions in action.

In the next section, we will apply the QCG algorithm to compile compositional queries to solve the inference task of computing the posterior probability on network variables based on given evidence, which is a value assignment to conditioning variables.

Computing posterior probability

This section is not just about theory but about practical application. We introduce our query composition approach to variable elimination, a powerful tool for inference in Bayesian networks. We start by presenting the problem of computing the posterior probability and its mapping to a composed query on weighted relations. Then, we dive into the QCG algorithm, a proper tool for this task, and provide a detailed, step-by-step demonstration of its application on a Bayesian network example. Finally, we provide a hands-on practice section using R, showing you how to apply the QCG algorithm to compute the posterior probability on the same example using the Bayesian network representation as a database of weighted relations.

Problem statement

First, we define the problem of computing the posterior probability in Bayesian networks. Next, we illustrate how this problem is equivalent to a compositional query on weighted relations representing the Bayesian network.

Defining the posterior probability task

Given a Bayesian network as a factorization of a complete probability distribution on the set of network variables V, where the probability distribution on the set V is equivalent to the product of a set of factors specified as non-negative functions on component variables formed from subsets $S_i \subset V$ as given by *Equation 8.6*, which we rewrite,

$$P(V) \propto \prod_{i=1}^{m} f_i(S_i) \quad (8.27)$$

The task of computing the posterior probability is as follows: Given a value assignment, $\boldsymbol{e}$, called "evidence" as$C = c$, to a subset of network variables $C \subset V$, called "conditioning variables," compute the probability of a non-conditioning variable given the evidence. Let X be the subset of non-conditioning variables $X = V \backslash C$, and $X_1 \in X$ be the non-conditioning variable for which we want to compute its posterior probability. The task of computing the posterior probability $P(X_1|\boldsymbol{e})$ is equivalent to the following sum-multiplication formula, where we sum the product on the right-hand side of *Equation 8.27* over all the $(n-1)$ other non-conditioning variables, where n equals |X|, the cardinality of X. We write,

$$P(X_1|\boldsymbol{e}) \propto \sum_{X_2} \cdots \sum_{X_n} \prod_{i=1}^{m} f_i(S_i\,|\boldsymbol{e}) \quad (8.28)$$

In the following section, we formulate the problem of computing the posterior probability in Bayesian networks as a problem of composing queries on weighted relations.

Applying the QCG algorithm to the posterior probability task

In the previous section, we saw that the problem of computing the posterior probability reduces to performing multiplication of probabilities and summing out the product over marginalized variables.

In our weighted relation framework, we represent the non-negative functions, $f_i(S_i)$, specifying the Bayesian network conditional probability tables, or more generally, the factors of the Markov model, as weighted relations, $r_i(S_i \cup \{Weight\})$, where the tuples in $t \in r_i$, have the value of the $Weight$ attribute set as follows:

$$t(Weight) = f_i(t(S_i)) \quad (8.29)$$

The set of weighted relations forms the database input to the QCG algorithm. To compute the posterior probability using the QCG algorithm, we need to provide as input not only the database of weighted relations Δ but also the set of conditioning variables $C \subset V$ and an ordering d of the database variables V. When the variable elimination process is complete, the posterior probability for any query variable can be determined by normalizing the combination query associated with that variable. Normalization involves dividing the Weight attribute of the combination query's resulting weighted relation by the total sum of the Weight attributes for all elements in the relation. This process ensures that the total sum equals 1, thereby creating a proper probability distribution.

Demonstrating the working of the QCG algorithm on a worked example

In this section, we illustrate with a worked example how the QCG algorithm computes the posterior probability in Bayesian networks by composing queries on a database of weighted relations.

The QCG algorithm eliminates the Bayesian network variables one by one in a given elimination ordering. Each elimination step first executes a combination query using multiplication as a combination function and then eliminates the variable from the result of the combination query by projecting the weighted relation resulting from the combination query using summation as an aggregation function on the Weight attribute of the relation. Thus, each elimination step involves summing out the eliminated variable from the product of the weighted relations associated with the eliminated variable. The input of an elimination step comes from the dictionary of variable-relations associations, `assoc`, from the previous elimination step. The output of an elimination step is an update of the `assoc` dictionary by removing the eliminated variable and adding the new relation resulting from the elimination query, which is a relation on variables of lower order than the eliminated variable.

Let us consider a Bayesian network on three variables, X, Y, and Z, having the following factorization of the joint probability distribution,

$$P(X,Y,Z) = P(X)P(Y|X)P(Z|Y) \quad (8.30)$$

Consider the inference task of computing the posterior probability of the variable *X* given the evidence,*Z* = *1*.

P_X			P_Y_X					P_Z_Y				
X	0	1	X	0	0	1	1	Y	0	0	1	1
Weight	0.75	0.25	Y	0	1	0	1	Z	0	1	0	1
			Weight	0.2	0.8	0.9	0.1	Weight	1	0	0.25	0.75

Figure 8.10—Condition probability tables of a Bayesian network example

Figure 8.10 quantifies the Bayesian network example, specified by the weighted relations P_X, P_Y_X, and P_Z_Y, representing the conditional probabilities P(X), P(Y|X), and P(Z|Y), respectively, in the factored representation of *Equation 8.30*. The "Weight" attribute in the tables indicates the conditional probabilities. *Figure 8.11* illustrates the directed acyclic graph of the Bayesian network, including the specified variable ordering d = (X,Y,Z).

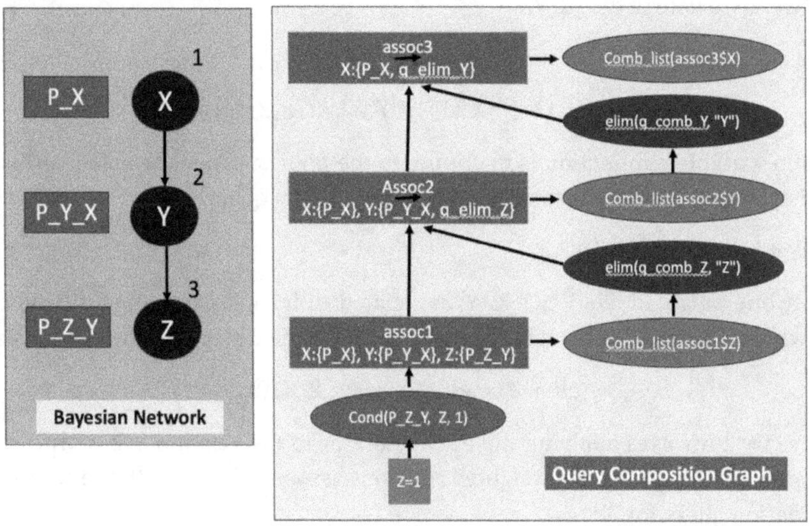

Figure 8.11—A Bayesian network and its query composition graph for computing the posterior probability P(X|Z = 1)

Figure 8.11 illustrates the Bayesian network on the left and the directed acyclic graph representing the query composition graph (QCG) on the right. The QCG algorithm is used to compute the posterior probability *P(X|Z = 1)*.

We will describe how the QCG algorithm operates in this example and illustrate its workflow step by step.

The first step of the QCG algorithm involves conditioning the weighted relations based on the provided evidence, which corresponds to the query `cond(P_Z_Y, Z, 1)`. This process is illustrated by the graph on the right in *Figure 8.11*. The conditioning query eliminates all tuples in the weighted relation `P_Z_Y` where Z is not equal to 1, resulting in a table that contains only two tuples instead of the original four, as shown in *Figure 8.12*.

P_Z_Y		
Y	0	1
Z	1	1
Weight	0	0.75

Figure 8.12—Conditioning the probability table P_Z_Y on the evidence Z=1

At each variable elimination step, the QCG algorithm computes a dictionary associating network variables with factors, represented by weighted relations. The dictionary associates network variables with relations such that (1) a relation associates with only one variable and (2) a variable associates only with a relation defined on the variable and possibly other lower-ordered variables but no variable of higher order.

The initial dictionary associates the network variables (X, Y, Z) in the given order with the initial factors, which are the weighted relations `P_X`, `P_Y_X`, and `P_Z_Y`. X will associate with `P_X`, Y will associate with `P_Y_X`, and Z will associate with `P_Z_Y`.

The following equation defines the initial dictionary, `assoc1`.

$$assoc1 = [X:\{P\_X\}, Y:\{P\_Y\_X\}, Z:\{P\_Z\_Y\}] \ (8.31)$$

The first step in variable elimination is to eliminate the highest-order variable, Z. This elimination process consists of two queries, starting with the combination query.

$q\_comb\_Z = comb\_list(assoc1\$Z)$ (8.32)

As there is only one weighted relation, `P_Z_Y`, associated with Z in *assoc*1, the function `comb_list` returns `P_Z_Y`, which is detailed in the table of *Figure 8.12*. The next step involves the elimination query.

$$q\_elim\_Z \ = elim(q\_comb\_Z, Z) \ (8.33)$$

The query `q_elim_Z` involves applying the `elim` operator to the relation `P_Z_Y` with respect to the variable Z. *Figure 8.13* illustrates the weighted relations associated with both the combination query and the elimination query for Z.

q_comb_Z			q_elim_Z		
Y	0	1	Y	0	1
Z	1	1	Weight	0	0.75
Weight	0	0.75			

Figure 8.13—the weighted relation of the combination and elimination queries for Z

Upon eliminating Z, the dictionary *assoc*2 is generated from *assoc*1 by removing the entry for Z and adding the weighted relation `q_elim_Z`, which will be linked to the Y variable. The dictionary *assoc*2 is defined as follows:

$$assoc2 = [X:\{P\_X\}, Y:\{P\_Y\_X, q\_elim\_Z\}] \ (8.34)$$

The next step in variable elimination is eliminating the variable Y. The computation for that step involves performing two queries. The first query is the combination query `q_comb_Y`, which combines the relations associated with Y in the current dictionary, *assoc*2, given by $assoc2\$Y$, which are {`P_Y_X`, `q_elim_Z`}.

$$q\_comb\_Y \; = comb\_list(assoc2\$Y) \; (8.35)$$

The second query is the elimination query, `q_elim_Y`, which eliminates Y from the weighted relation, `q_comb_Y`, written as

$$q\_elim\_Y \; = \; elim(q\_comb\_Y, Y) \; (8.36)$$

The results of the two queries for the BN example are shown in the two weighted relations in *Figure 8.14*.

q_comb_Y					q_elim_Y		
Y	0	0	1	1	X	0	1
X	0	1	0	1	Weight	0.6	0.075
Weight	0	0	0.6	0.075			

Figure 8.14—the weighted relation of the combination and elimination queries for Y

Upon eliminating Y, the dictionary `assoc3` is generated from `assoc2` by removing the entry for Y and adding the weighted relation `q_elim_Y`, which will be linked to the X variable. The dictionary *assoc*3 is defined as follows:

$$assoc3 \; = \; [X\colon \{P\_X, q\_elim\_Y\}] \; (8.37)$$

Eliminating X is the final step. Since X is the last variable, there is no elimination query; rather, there is only one query, the combination query, `q_comb_X`, which combines the relations associated with X in the current dictionary, *assoc*3, given by $assoc3\$X$, which are $\{P\_X, q\_elim\_Y\}$.

$$q\_comb\_X \; = comb\_list(assoc3\$X) \; (8.38)$$

The computed result of the combination query for X is displayed in the table on the left in *Figure 8.15*. However, the weighted relation does not represent a valid probability distribution for X, as the sum of the weights is not equal to 1. To convert the weighted relation into a probability, we must normalize it by dividing each tuple's Weight attribute value by the total sum of the Weight attribute across all tuples. After normalizing the weighted relation from the combination query, we obtain the weighted relation that represents the probability $P(X \mid Z = 1)$, which is shown on the right row of *Figure 8.15*.

q_comb_X			Prob(X|Z = 1)		
X	0	1	X	0	1
Weight	0.45	0.01875	Weight	0.96	0.04

Figure 8.15— The combination query for X is presented in the table on the left, while the posterior probability for X is shown in the table on the right

This concludes the walkthrough of the QCG algorithm, which is applied to computing the posterior probability in a Bayesian network example. Although the example network is relatively simple, the purpose of this detailed illustration is to clarify how variable elimination works for exact inference. Additionally, it demonstrates how the QCG algorithm compiles the variable elimination process into a directed acyclic graph (DAG), allowing for reuse when answering queries with different input data. The computational effort invested in the compiled query composition graph is amortized over the process of answering a query for multiple instances of the problem.

In the following section, we practice using R to apply the QCG algorithm to compute the posterior probability and verify the results of the Bayesian network example.

Practice using R

The entire code for this practice is included in the R Markdown file `'posterior_probability.qmd'` and rendered files with the same name in HTML and Docx format.

The purpose of the practice is to demonstrate the workflow of the QCG algorithm step by step and validate the results in the previous section on the Bayesian network example.

First we code the Bayesian network factors of the example as weighted relations given by the following data frames.

```
P_X <- data.frame(X = c(0,1),
                  Weight = c(0.75, 0.25),
                  stringsAsFactors = FALSE)
P_Y_X <- data.frame(X = c(0,0,1,1),
                    Y = c(0,1,0,1),
                    Weight = c(0.2,0.8,0.9,0.1),
                    stringsAsFactors = FALSE)
P_Z_Y <- data.frame(Y = c(0,0,1,1),
                    Z = c(0,1,0,1),
                    Weight = c(1,0,0.25,0.75),
                    stringsAsFactors = FALSE)
```

Initialization

1. The first step is to incorporate the evidence Z=1 in the relations that include the Z variable; there is only one such relation P_Z_Y, so we apply the `cond` function with the parameters P_Z_Y, Z, and 1 (the value of the evidence on Z)

```
P_Z_Y <- cond(P_Z_Y, c("Z"), c(1))
print(P_Z_Y)

# Output:
Y Z Weight
```

```
0 1 0.00
1 1 0.75
```

2. The second step is to create the initial dictionary data structure assoc1 of key-value pairs where the keys are the network variables and the value is the list of relations associated with the variable. Recall that relations are associated with the highest order variable of their schemes so that a relation is associated with only one variable based on the variable ordering.

```
assoc1 <- assoc(list(P_X, P_Y_X, P_Z_Y), c("X", "Y", "Z"))
print(assoc1)

# Output:
$X
$X[[1]]
  X Weight
1 0   0.75
2 1   0.25

$Y
$Y[[1]]
  X Y Weight
1 0 0    0.2
2 0 1    0.8
3 1 0    0.9
4 1 1    0.1

$Z
$Z[[1]]
  Y Z Weight
  0 1 0.00
  1 1 0.75
```

We begin the variable elimination process after initialization using the QCG algorithm. We eliminate the variables in backward order. The example has three variables, X, Y, and Z, ordered in increasing order from 1 to 3. The variable elimination goes in backward order, so the first step in executing the QCG algorithm is to eliminate Z, followed by step 2 to eliminate Y, and then step 3, eliminate X. We show the code below for each elimination step.

Take note of the systematic nature of each elimination step in the code below. Recall the systematic approach of each elimination step in the QCG algorithm, which involves three substeps. In each substep, we execute a systematic query on relevant weighted relations.

a. *combination query*, where we call the function `comb_list` to combine the list of weighted relations associated with the variable from the current *assoc* dictionary.

b. *elimination query*, where we call the function `elim` to eliminate the variable from the weighted relation, which is the result returned by the combination query.
c. *updating the assoc dictionary*, where we call the function `add_relation_remove_var`, which adds the induced weighted relation, which is the result returned by the elimination query to the dictionary associating the induced relation with the highest order variable while removing the key corresponding to the eliminated variable from the dictionary.

Step 1: eliminate Z

Step 1-A: combination query for Z

The following code runs the `comb_list` function on the set of relations `assoc1$Z` associated with Z in the `assoc1` dictionary. We print the result `q_comb_Z` returned from executing the combination query on the Z variable.

```
q_comb_Z = comb_list(assoc1$Z)
print(q_comb_Z)

# Output:
Y Z Weight
 0 1 0.00
 1 1 0.75
```

Step 1-B: elimination query for Z

The following code runs the `elim` function to eliminate the variable Z from the weighted relation `q_comb_Z`, which we computed from running the combination query for Z.

```
q_elim_Z = elim(q_comb_Z, "Z")
print(q_elim_Z)

# Output:
  Y Weight
1 0   0.00
2 1   0.75
```

Posterior probability for Y

The computation of the posterior probability for Y is a straightforward process. The weighted relation `q_elim_Z`, obtained from the elimination query for Z, is a factor function on Y. By simply normalizing this weighted relation, ensuring the sum of the Weight column equals 1, we can compute the posterior probability for Y. This normalization step, executed in the second line of the code below, is all it takes. The result, the posterior probability for Y, clearly indicates that Y=1 with probability 1.

```
prob_Y <- q_elim_Z
prob_Y$Weight <- prob_Y$Weight/sum(prob_Y$Weight)
print(prob_Y)
```

```
# Output:
  Y Weight
1 0      0
2 1      1
```

Step 1-C: updating assoc dictionary for Z

The following code uses the `add_relation_remove_var` function to update the `assoc` dictionary for Z. The function takes three arguments: the current dictionary `assoc1`, the weighted relation `q_elim_Z`, which resulted from eliminating Z, and the variable we eliminated in this step, Z. We print the new dictionary `assoc2`, resulting from updating `assoc1`. Notice that `assoc2` removed the key for the eliminated variable and updated the relations associated with the variable Y, where the induced relation from eliminating Z is now added to its list of associated relations.

```
assoc2 <- add_relation_remove_var(assoc1, q_elim_Z, "Z")
print(assoc2)

# Output:
$X
$X[[1]]
  X Weight
1 0   0.75
2 1   0.25

$Y
$Y[[1]]
  X Y Weight
1 0 0    0.2
2 0 1    0.8
3 1 0    0.9
4 1 1    0.1

$Y[[2]]
  Y Weight
1 0   0.00
2 1   0.75
```

Step 2: eliminate Y

Step 2-A: combination query for Y

The following code runs the `comb_list` function on the set of relations `assoc2$Y` associated with Y in the current *assoc* dictionary, `assoc2`. We print the result `q_comb_Y` returned from executing the combination query on the Y variable.

```
q_comb_Y <- comb_list(assoc2$Y)
print(q_comb_Y)

# Output:
  Y X Weight
1 0 0  0.000
2 0 1  0.000
3 1 0  0.600
4 1 1  0.075
```

Step 2-B: elimination query for Y

The following code runs the `elim` function to eliminate the variable Y from the weighted relation `q_comb_Y`, which we computed by running the combination query for Y.

```
q_elim_Y <- elim(q_comb_Y, "Y")
print(q_elim_Y)

# Output:
  X Weight
1 0  0.600
2 1  0.075
```

Step 2-C: updating assoc dictionary for Y

The following code uses the `add_relation_remove_var` function to update the `assoc` dictionary for Y. The function takes three arguments: the current dictionary `assoc2`, the weighted relation `q_elim_Y`, which resulted from eliminating Y, and the variable we eliminated in this step, Y. We print the new dictionary `assoc3`, resulting from updating `assoc2`. Notice that `assoc3` removed the key for the eliminated variable and updated the relations associated with the variable X, where the induced relation from eliminating Y is now added to its list of associated relations.

```
assoc3 <- add_relation_remove_var(assoc2, q_elim_Y, "Y")
print(assoc3)

# Output:
$X
$X[[1]]
```

```
  X Weight
1 0    0.75
2 1    0.25

$X[[2]]
  X Weight
1 0  0.600
2 1  0.075
```

Step 3: eliminate X

Step 3-A: (combination query for X)

The following code runs the `comb_list` function on the set of relations `assoc3$X` associated with X in the current *assoc* dictionary, `assoc3`. We print the result `q_comb_X` returned from executing the combination query on the X variable.

```
q_comb_X <- comb_list(assoc3$X)
print(q_comb_X)

# Output:
  X  Weight
1 0 0.45000
2 1 0.01875
```

X is the last variable in the elimination order, so we need not run an elimination query. The combination query for X is the weighted relation representing a factor function of X only, so we compute the posterior probability for X.

Posterior probability of X

The posterior probability for X is derived from the normalization of the weighted relation factor `q_comb_X`, which is the result of the combination query for X. The normalization, performed in the second line of the code below, is crucial. It involves dividing the values of the Weight column in the weighted relation by the sum of the weights. This ensures that the factor function is a probability.

```
prob_X <- q_comb_X
prob_X$Weight <- prob_X$Weight/sum(prob_X$Weight)
print(prob_X)

# Output:
  X Weight
1 0   0.96
2 1   0.04
```

Let us review what we covered. First, we defined the inference task of computing the posterior probability in Bayesian networks. We explained how we map the problem to an equivalent problem of composing queries on weighted relation databases. Using an example Bayesian network, we demonstrated how the QCG algorithm works and how the variable elimination is carried out by composing queries from primitive queries on the weighted relation representation of the Bayesian network model. We then practiced all the presented concepts with hands-on practice, implementing the steps of the QCG algorithm to compute the posterior probability for the Bayesian network example.

In the next section, we will apply the QCG algorithm and the weighted relation queries to the problem of computing the maximum expected utility and computing the decision policy for the maximum expected utility in decision networks and influence diagrams.

Computing maximum expected utility

This section introduces the crucial QCG compilation procedure, a practical and efficient method for finding the MEU and computing the MEU policy. It's specifically tailored for decision networks represented by influence diagrams, particularly those that depict a single decision-maker who retains the information. The no-forgetting assumption underscores that the direct predecessor set for one decision must be a subset of the direct predecessor set of any subsequent decision.

First, we delve into the powerful and versatile variable elimination approach and its application in solving the MEU and deriving the MEU policy. We then demonstrate how the QCG algorithm harnesses this approach to compute the **maximum expected utility** (**MEU**) policy of a decision network example. To further enhance your understanding, we provide a practice section using R, where we compute the MEU decision policy on the quantified decision network model of the example implemented in our weighted relations representation framework with code.

Solving the MEU problem by variable elimination

In *Chapter 5*, in the section on *Influence Diagrams*, we formally defined decision networks and their graphical representation in influence diagrams. We also defined the MEU problem and the associated MEU decision policy in that section.

This section delves into the practical application of variable elimination, a powerful tool, for computing the MEU decision policy for decision networks represented by influence diagrams.

Explaining variable elimination on a decision network example

To understand the basic ideas on how variable elimination applies to the MEU problem in decision networks, we will use a commonly used example called the weather decision problem (Gao, Lecture 17 Decision Networks 2021).

The weather decision problem involves deciding whether to take an umbrella, weighing the uncertainty of future weather against the costs of getting wet or carrying the umbrella unnecessarily. This decision-making process combines probability, as represented by weather forecasts, with preferences related to utility in order to maximize expected utility for rational choices.

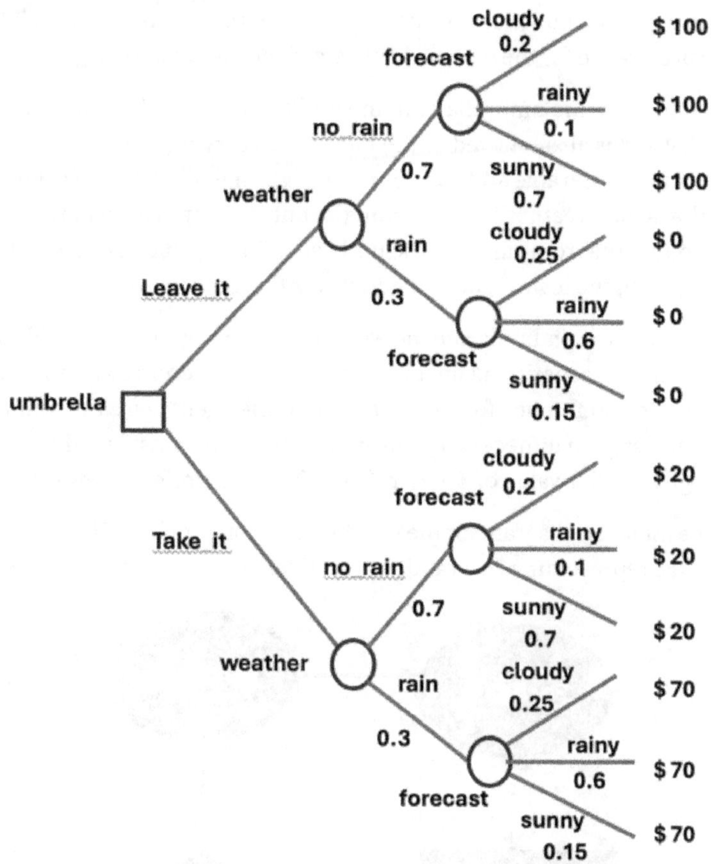

Figure 8.16—Decision tree diagram for the umbrella decision problem

Figure 8.16 depicts a decision tree representing a concrete example of the weather decision problem. The numerical values shown in the decision tree correspond to the example from Gao (Gao, Lecture 17 Decision Networks 2021) so that you can compare the variable elimination approach using the QCG algorithm with other techniques and verify the results. We will mention the distinction between the QCG algorithm and the other approaches for answering queries on finding the maximum expected utility with influence diagrams at the end of this section.

The root of the decision tree is the decision node (umbrella), which has two branches: `'take_it'` and `'leave_it'`. Each branch leads to the chance variable (weather), which has two branches: 'no rain' and 'rain.' Following each branch of the weather node is another chance variable (forecast) with three branches: 'cloudy,' 'rainy,' and 'sunny.' The decision tree contains a total of 12 branches, and the leaves correspond to the outcomes of each decision scenario, quantified by a dollar amount representing the utility of that scenario. The numbers next to the two branches stemming from the chance node (weather) represent the prior probabilities of each outcome. The numbers next to the three branches from the chance node (forecast) show the conditional probabilities of each forecast instance based on the weather. Note the symmetry of the decision tree in this problem.

The utility function assigns a numerical value (e.g., from 0 to 100) to each possible scenario. It represents the agent's preference for the outcomes of each decision scenario (a path in the decision tree).

If the agent decides to leave the umbrella at home and the weather does not turn to rain, the agent will enjoy staying dry and is also relieved of the burden of carrying the umbrella. The agent regards this scenario as the most preferable and assigns it the highest dollar value of $100. If the agent leaves the umbrella and the actual weather turns to rain, the outcome of that scenario is that the agent will be soaked wet, although free from carrying the umbrella. The agent considers this scenario the least preferable and assigns $0, the lowest dollar amount, for this outcome.

If the agent decides to take the umbrella and the weather turns rainy, he or she will stay dry by carrying it. The agent considers this scenario the next best in order of preference and assigns a dollar value of $70 to quantify his or her preference for that outcome. If the agent takes the umbrella and it doesn't rain, they stay dry but carry an unnecessary umbrella. The agent considers this scenario moderately preferable and assigns a dollar value of $20 to reflect his or her preference for that outcome.

Figure 8.17 shows the influence diagram for the weather decision problem. The diagram has two **chance nodes** shown by ovals, representing two random variables: '`weather`' and '`forecast`'.

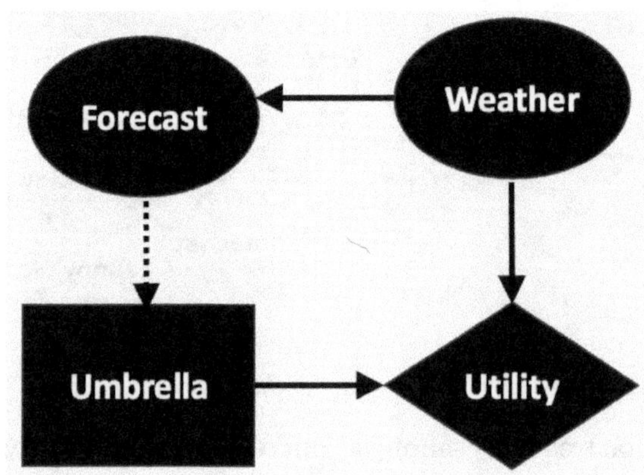

Figure 8.17—Influence diagram for the umbrella decision problem

The weather node represents a random variable whose value domain is binary: '`rain`,' '`no_rain`,' meaning whether it does rain on any given day. The forecast node represents a random variable that depends on the weather node and is regarded as a noisy weather sensor. The forecast node can take one of three values: sunny, cloudy, or rainy.

There is one **decision node** represented by a box called '`umbrella`.' The decision node can take one of two values: '`take_it`' or '`leave_it`,' meaning whether to take the umbrella when going out or leave it at home.

It's important to note the dotted line that connects the forecast node to the decision node. This line signifies that the information provided by the forecast value is available to the decision-maker when making the umbrella decision. In other words, the forecast influences the decision to take or leave the umbrella.

The Influence diagram has a **utility node** represented by the diamond shape, the value of which is a function of the value tuple for the weather variable and the umbrella decision.

Let W, F, D denote the weather, forecast, and umbrella decision variables, respectively. Let $u(W, D)$ be the utility function of the weather random variable and the umbrella decision. Let $P(W)$ denote the probability distribution of the weather variable W and $P(F|W)$ the conditional probability of the forecast variable F conditional on the weather variable W.

The expected utility is a function parameterized by the decision variable D. We can write the expected utility as a function of D,

$$E[U|D] = \sum_F \sum_W u(W, D)P(F|W)P(W) \quad (8.39)$$

If there is no forecast information, then the optimal decision is

$$D^* = \underset{D}{\operatorname{argmax}} \sum_W u(W, D)P(W) \quad (8.40)$$

In the weather decision problem, we seek the optimal policy for making the umbrella decision based on the forecast information. A decision policy is a functional mapping from the information domain, which is the value of the forecast variable, to the domain of the decision variable, which is the umbrella decision.

The umbrella decision is binary, meaning there are two choices: to take the umbrella or to leave it. If we have, for example, three possible weather forecasts, this results in a total of eight possible decision policies. Among these, two policies involve always taking the umbrella or always leaving it. The other six policies consist of scenarios where we take (or leave) the umbrella for one forecast while leaving (or taking) it for the remaining two forecasts.

If the cardinality of the information domain is n and the cardinality of the decision domain is k, then the size of the space of possible policies is k^n.

For the weather decision problem, the policy is a functional mapping $\delta(F, D)$, where for each forecast F, there corresponds a unique decision D.

The MEU and the MEU policy $\delta^*(F, D)$ are as follows:

$$\text{MEU} = \sum_F \max_D \sum_W u(W, D)P(F|W)P(W) \quad (8.41)$$

$$\delta^*(F, D) = \{(F, D): \psi(F, D) = \psi^*(F)\} \quad (8.42)$$

$$\psi(F, D) = \sum_w u(w, d)P(f|w)P(w) \quad (8.43)$$

$$\psi^*(F) = \max_d \sum_w u(w, d)P(f|w)P(w) \quad (8.44)$$

Equation 8.41 implies that the MEU is obtained by the following variable elimination steps:

1. Eliminating W by summing it out of the product of the probabilities $P(F|W), P(W)$, and the utility function $u(W, D)$, which results in the function $\psi(F, D)$ of *Equation 8.43*.
2. Eliminating the decision D by maximizing the function $\psi(F, D)$ over the variable D results in the function $\psi^*(F)$ on F.
3. Eliminating F by summing $\psi^*(F)$, which equals the MEU.

Notice that the elimination order is constrained so that all predecessors of a decision variable in the directed acyclic graph that influence the decision variable must be eliminated after the decision variable is eliminated.

The following section outlines the computation of the MEU policy using our QCG algorithm along with the weighted relation representation framework.

Now, let's examine how the QCG algorithm effectively addresses the MEU problem.

Solving the MEU problem with the QCG algorithm

To solve the MEU problem outlined in the previous section using the QCG algorithm, we first need to express it in terms of weighted relations. This task can be accomplished in the following ways.

To address the MEU problem described in the previous section using the QCG algorithm, we must first represent it in terms of weighted relations. This representation can be accomplished in the following manner.

- We create weighted relations for each conditional probability table or factor in the factorized Bayesian network model in the influence diagram. For the example influence diagram in *Figure 8.17,* we have two chance nodes, 'weather' and 'forecast,' and we can represent them by the weighted relations *weather(weather,Weight)* and *forecast(weather, forecast, Weight)* respectively. *Figure 8.18* quantifies the two weighted relations based on the data of the decision tree representation in *Figure 8.16.*
- We create weighted relations for each utility node representing a utility factor in the influence diagram. The utility function gives the relation. For the example influence diagram in *Figure 8.17,* we have one utility function for which we create a weighted relation *utility(weather, umbrella, Weight),* given by the utility function of its child utility node. *Figure 8.18* encodes the utility function as per the values in the decision tree representation in *Figure 8.16.*

forecast

weather	forecast	Weight
no_rain	sunny	0.7
no_rain	cloudy	0.2
no_rain	rainy	0.1
rain	sunny	0.15
rain	cloudy	0.25
rain	rainy	0.6

weather

weather	Weight
no_rain	0.7
rain	0.3

utility

weather	umbrella	Weight
no_rain	take_it	20
no_rain	leave_it	100
rain	take_it	70
rain	leave_it	0

Figure 8.18—Weighted relations for the weather decision problem

The QCG algorithm orders the variables using the variable ordering *(forecast, umbrella, weather)*. It creates a dictionary where the weighted relations are paired with the highest-ordered variable in their schemes. The QCG algorithm eliminates the variables beginning with the highest-ordered variable and continues in the reverse variable ordering.

The first step is to eliminate *weather* from the result of the combination query on the set of relations associated with 'weather' in the dictionary, which are the three weighted relations. The result of the combination and elimination queries from that step is shown in *Figure 8.19*.

q_comb_weather

weather	forecast	umbrella	Weight
no_rain	sunny	take_it	9.8
no_rain	sunny	leave_it	49
no_rain	cloudy	take_it	2.8
no_rain	cloudy	leave_it	14
no_rain	rainy	take_it	1.4
no_rain	rainy	leave_it	7
rain	sunny	take_it	3.15
rain	sunny	leave_it	0
rain	cloudy	take_it	5.25
rain	cloudy	leave_it	0
rain	rainy	take_it	12.6
rain	rainy	leave_it	0

q_elim_weather

forecast	umbrella	Weight
cloudy	leave_it	14
cloudy	take_it	8.05
rainy	leave_it	7
rainy	take_it	14
sunny	leave_it	49
sunny	take_it	12.95

Figure 8.19—results of combination and elimination queries for the chance variable 'weather'

The variable-relations pairs are updated by adding the '`q_elim_weather`' to and removing the variable `'weather'` from the dictionary. The second step is eliminating `'umbrella'` from the relations associated with `'umbrella'` in the dictionary. There is only one relation, the `'q_elim_weather`,' so there is no need for a combination query, and we proceed with the elimination query `'q_elim_umbrella`.' Eliminating the decision variable '`umbrella'` uses `FUN='max'` for the aggregation function as the relation `'q_elim_weather'` has the '`Weight'` attribute representing expected utility. The result of the elimination query is shown in *Figure 8.20*.

q_elim_umbrella'	
forecast	Weight
cloudy	14
rainy	14
sunny	49

Figure 8.20—results of elimination query for the decision variable 'umbrella'

The MEU is the sum of the Weight column, which equals 77. The MEU policy is computed by performing an inner join of the two relations, `q_elim_umbrella` and `q_elim_weather`, which yields the table shown in *Figure 8.20*.

MEU policy		
forecast	Weight	umbrella
cloudy	14	leave_it
rainy	14	take_it
sunny	49	leave_it

Figure 8.21—MEU policy and expected utility in the Weight column

Figure 8.22 illustrates the results as a decision tree, with the root representing the forecast and branches indicating all possible states. At each branch, a decision is made regarding whether to take or leave the umbrella. This tree comprises six distinct paths representing the policy space. The values at the end of each path in the tree reflect the expected utility for each policy. Policies that maximize expected utility are highlighted with bold branches, and their corresponding expected utility values are shown in boxes.

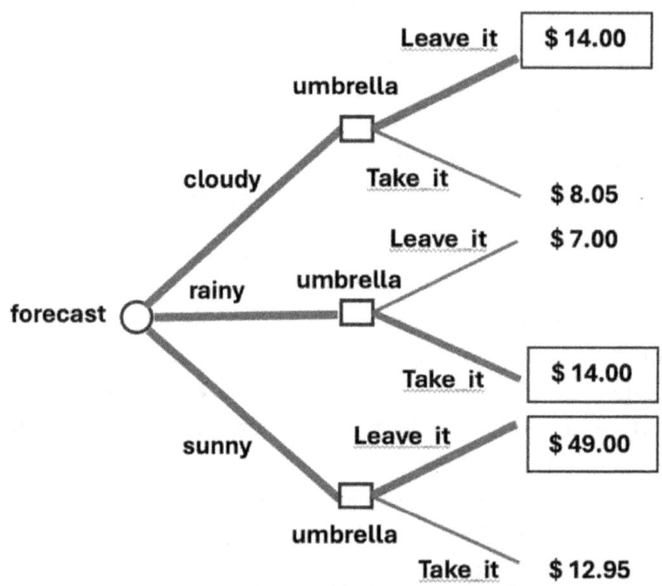

Figure 8.22— Decision tree with bold branches illustrating the MEU (Maximum Expected Utility) policy

In the next section, we show how to compute the MEU policy using our QCG algorithm implemented in R and run on the weather decision problem.

Practice using R

The code for this section is included in the Quarto markdown document '*decision.qmd.*' The following describes the step-by-step variable elimination for the weather decision problem using the QCG algorithm, as explained in the previous section.

First, we define the weighted relations for the chance nodes and their utility function. The weighted relations are data frames with a distinct column named `Weight`. The data frames, `weather` and `forecast`, represent the probability factor, while the data frame for `utility` represents the utility factor.

```
weather <- data.frame(
  weather = c('no_rain',"rain"),
  Weight = c(0.7, 0.3),
  stringsAsFactors = FALSE)

forecast <- data.frame(
  weather=c('no_rain','no_rain','no_rain','rain','rain','rain'),
  forecast=c('sunny','cloudy','rainy','sunny','cloudy','rainy'),
  Weight=c(0.7,0.2,0.1,0.15,0.25,0.6),
  stringsAsFactors = FALSE
)
utility <- data.frame(
  weather = c('no_rain','no_rain','rain','rain'),
  umbrella = c('take_it','leave_it','take_it','leave_it'),
  Weight = c(20, 100, 70, 0),
  stringsAsFactors = FALSE
)
```

Having specified the weighted relational tables that quantify the influence diagram for our weather decision problem, we are now ready to apply the QCG algorithm. This algorithm implements the variable elimination (VE) method for exact inference for decision-making, assisting us in identifying the maximum expected utility (MEU) and developing the corresponding MEU policy.

The first stage of applying the algorithm is performing the initialization steps.

Initialization

The initialization steps are the conditioning step and the *assoc* dictionary creating step. Here, we have no evidence for conditioning as in the posterior probability use case, so we are left with the necessary step of initializing the *assoc* dictionary that pairs the variables with weighted relations respecting the ordering of the variables so that relations are associated with the variable in their scheme that ranks highest in the variable ordering. Recall the variables that are the influence diagram's nodes are ordered as follows: 1) forecast, 2) umbrella, and 3) weather.

Create variable-relations association dictionary: assoc1

To create the initial variable-relations association dictionary `assoc1`, we use the function `assoc`, which takes the list of the variables in their respective order and the set of all weighted relations for the decision problem. The variables properly ordered is the `assoc` function's first argument `list(forecast, utility, weather)`, and the set of weighted relations is the `assoc` function's second argument `c("forecast," "umbrella," "weather")`. The code below prints the dictionary `assoc1` computed by the function `assoc`. Notice that all the weighted relations are associated with the variable `weather` because all relations have the variable `weather` in their scheme, and relations are associated with the highest-order variable.

```
assoc1 <- assoc(
  list(forecast, utility, weather),
  c("forecast", "umbrella", "weather")
)
print(assoc1)

# Output:
$forecast
NULL

$umbrella
NULL

$weather
$weather[[1]]
  weather forecast Weight
1 no_rain    sunny   0.70
2 no_rain   cloudy   0.20
3 no_rain    rainy   0.10
4    rain    sunny   0.15
5    rain   cloudy   0.25
6    rain    rainy   0.60

$weather[[2]]
  weather umbrella Weight
1 no_rain  take_it     20
2 no_rain leave_it    100
3    rain  take_it     70
4    rain leave_it      0

$weather[[3]]
```

```
  weather Weight
1 no_rain    0.7
2    rain    0.3
```

After the initialization, we are ready to write the code for writing the queries to implement each variable elimination step. Recall that elimination ordering reverses the variable ordering, so we begin eliminating the highest-ordered variable and continue backward until we reach the variable with the lowest rank in the ordering.

The following code consists of three blocks. Each block corresponds to eliminating each of the three variables of the weather decision problem. As weather is the variable with the highest rank in the ordering, we begin by eliminating weather.

Step 1: eliminate chance node (weather)

Eliminating the chance node of the variable `weather` requires three substeps:

a. To combine the list of relations associated with the variable `weather`.
b. To eliminate the variable `weather` from the weighted relation resulting from the previous combination step.
c. To update variable-relations *assoc* dictionary.

Now, let's put theory into practice. The following code demonstrates how to implement the three substeps we just discussed.

Step 1-A: combination query (q_comb_weather)

Step 1-A involves a crucial Combination query: `q_comb_weather`. This query is designed to combine all the weighted relations associated with the `weather` variable, a key step in our process of eliminating the chance node.

Here's how the code works: It uses the `comb_list` function to combine all weighted relations associated with the weather variable. These relations are obtained from the `assoc1` dictionary, specifically `assoc1$weather`, which serves as the input parameter for the `comb_list` function. The result, `q_comb_weather`, is then written to a Comma-Separated Values (CSV) file that you can easily view in *Excel*.

The table on the left in *Figure 8.19* shows the results from the combination query.

```
q_comb_weather = comb_list(assoc1$weather)
write.csv(q_comb_weather,file = "q_comb_weather.csv")
```

Step 1-B: elimination query (q_elim_weather)

The 'elimination query' is a crucial step in our process. It computes an induced weighted relation by eliminating the variable `weather` from the result of the above combination query. This step is necessary to isolate the impact of other variables on our data analysis.

The code below uses the function `elim` with the two input parameters: 1) the weighted relation from the combination query `q_comb_weather`, and 2) the variable's name to eliminate, which is "weather." The code then writes the result from the `elim` query `q_elim_weather` to a CSV file. The result you can open in *Excel* is the table shown on the right in *Figure 8.19*.

```
q_elim_weather <- elim(q_comb_weather,"weather")
write.csv(q_elim_weather,file = "q_elim_weather.csv")
```

Step 1-C: create variable-relations association dictionary (assoc2)

After running the elimination query and creating the new induced relation on variables of lower rank in order than weather, we must update the *assoc* dictionary. We must mutate the current dictionary `assoc1` to a new dictionary `assoc2` that removes the eliminated variable "weather" and adds the induced weighted relation `q_elim_weather`. The following code does the *assoc* updating by calling the function `add_relation_remove_var` with the three input parameters:

1. The current dictionary: assoc1
2. The induced weighted relation `q_elim_weather`
3. The variable's name to remove from the dictionary: the eliminated variable `weather`

The outcome of this step is a transformation of the current dictionary `assoc1` into the new dictionary `assoc2`. In this updated dictionary, the induced relation of the query `q_elim_weather` is added, and the eliminated variable `weather` is removed. We then print the `assoc2` dictionary, which clearly demonstrates the success of our update: only the decision node variable `umbrella` is now associated with the new induced relation; the list of relations associated with the variable for the chance node `forecast` remains NULL.

```
assoc2 <- add_relation_remove_var(assoc1, q_elim_weather, "weather")
print(assoc2)

# Output
$forecast
NULL

$umbrella
$umbrella[[1]]
                 forecast umbrella Weight
cloudy_leave_it    cloudy leave_it  14.00
cloudy_take_it     cloudy  take_it   8.05
rainy_leave_it      rainy leave_it   7.00
rainy_take_it       rainy  take_it  14.00
sunny_leave_it      sunny leave_it  49.00
sunny_take_it       sunny  take_it  12.95
```

Step 2: eliminate decision node (umbrella)

Just as we did with the `weather` chance node, we now turn our attention to the `umbrella` decision node. Its elimination is a crucial step in our probabilistic graphical model, and it involves three substeps:

1. To combine the list of relations associated with the variable `umbrella`;
2. To eliminate the variable `umbrella` from the weighted relation resulting from the previous combination step;
3. To update variable-relations *assoc* dictionary.

Since there is only one weighted relation associated with the decision node `umbrella` in the dictionary `assoc2`, we can bypass the combination query substep and proceed directly to the elimination substep:

Step 2-A: elimination query (q_elim_umbrella)

It's important to recognize the distinction in the process of eliminating a decision node compared to a chance node. When eliminating a chance node, the aggregation function FUN defaults to `sum`. However, when we eliminate a decision node, such as the variable "umbrella," we need to change the default value of the aggregation function FUN to `max`. This adjustment is necessary because the Weight attribute in the weighted relation is interpreted as utility, rather than probability, as it is when eliminating a chance node.

The code below uses the `elim` function to compute the induced weighted relation `q_elim_umbrella`. The input parameters to the `elim` function are:

1. The weighted relation associated with umbrella from the `assoc2` dictionary, `assoc2$umbrella`, which equals `q_elim_weather`.
2. The name of the variable to eliminate is "umbrella"
3. The `FUN` parameter, set to `max`, plays a significant role in our elimination process. It specifies the aggregation function on the Weight attribute, which represents a utility that we aim to maximize in our model

The following code writes the result of the elimination query to a CSV file. The result you can open in *Excel* is the table in *Figure 8.20*.

```
q_elim_umbrella <- elim(q_elim_weather, "umbrella", FUN = max)
write.csv(q_elim_umbrella,file = "q_elim_umbrella.csv")
```

Step 2-B: create variable-relations association dictionary (assoc3)

The function `add_relation_remove_var` is used to update the current dictionary. The input parameters include the current dictionary, `assoc2`, the relation to be added, `q_elim_umbrella`, and the name of the variable to be removed, "umbrella." This function mutates the current dictionary, `assoc2`, into a new dictionary, `assoc3`, where the induced relation from the query `q_elim_umbrella` is added and the entry for the variable "umbrella" is removed.

As a result, the assoc3 dictionary now contains only one entry for the remaining variable: the chance node variable "forecast," which is linked to a single relation—the induced weighted relation from the elimination query q_elim_umbrella.

```
assoc3 <- add_relation_remove_var(assoc2, q_elim_umbrella, "umbrella")
print(assoc3)

# Output:
$forecast
$forecast[[1]]
       forecast Weight
cloudy   cloudy     14
rainy     rainy     14
sunny     sunny     49
```

Step 3: computing the MEU and the MEU policy

The 'forecast' node, being the last in the sequence, plays a crucial role in our computation of the maximum expected utility and the associated decision policy.

Only one relation, r_forecast, is linked to the last node, the 'forecast' node in the 'assoc3' dictionary. In a manner similar to the previous elimination step, there is also a single relation, r_umbrella, associated with the elimination variable 'umbrella' in the 'assoc2' dictionary.

It is crucial to note that:

- The 'r_umbrella' relation is essential for providing the expected utility in the Weight column for all situation-decision pairs, which are the value tuples of the (forecast, umbrella) pair. This is a key step in our MEU computation.
- The relation r_forecast gives us the maximum expected utility for each situation, the value of the forecast column.

To compute the MEU, we need to sum the values in the Weight column of the r_umbrella table. Additionally, to identify the MEU policy, we must perform an inner join between the 'r_forecast' and 'r_umbrella' relations.

We divide the calculation of the MEU policy and the MEU into three substeps.

Step 3-A: finding the weighted relations for consecutive situation decision nodes

The following code extracts the single weighted relations r_umbrella and r_forecast associated with the consecutive eliminated nodes umbrella and forecast, respectively. The relation r_umbrella is associated with the variable' umbrella' in the *assoc* dictionary assoc2, and the relation r_forecast is associated with the variable 'forecast' in the assoc3 dictionary. The code below demonstrates how to extract and print the two weighted relations.

```
r_forecast <- assoc3$forecast[[1]]
r_umbrella <- assoc2$umbrella[[1]]
```

```
print(r_forecast)

# Output:
        forecast Weight
cloudy    cloudy     14
rainy      rainy     14
sunny      sunny     49
print(r_umbrella)

# Output:
                  forecast umbrella Weight
cloudy_leave_it    cloudy leave_it  14.00
cloudy_take_it     cloudy  take_it   8.05
rainy_leave_it      rainy leave_it   7.00
rainy_take_it       rainy  take_it  14.00
sunny_leave_it      sunny leave_it  49.00
sunny_take_it       sunny  take_it  12.95
```

Now, we can readily compute the MEU policy using the inner join of the above two relations, which will be done in the next step.

Step 3-B: computing the MEU policy

To compute the MEU policy, we need to identify all forecast-umbrella pairs with the maximum expected utility, which is the value of the Weight column in the 'r_forecast' table. We achieve this by performing an inner join of the two relations' r_forecast' and 'r_umbrella' based on their forecast and Weight attributes. The following code demonstrates this process using the inner-join operator imported from the 'dplry' package.

```
library(dplyr)
# Perform inner join
policy_EU <- dplyr::inner_join(r_forecast, r_umbrella, by =
c("forecast","Weight"))
print(policy_EU)

# Output:
  forecast Weight umbrella
1   cloudy     14 leave_it
2    rainy     14  take_it
3    sunny     49 leave_it
```

Step 3-C: computing the MEU

The MEU, a key metric in decision theory, is obtained by summing the Weight column in the policy_EU relation of the above *dataframe*.

Computing the MEU

The MEU is given by summing the expected utility given by the `Weight` column in the `policy_EU` relation of the above dataframe.

```
MEU <- sum(policy_EU$Weight)
print(MEU)

# Output:
[1] 77
```

In the following section, we discuss the main points of distinction of the QCG algorithm.

Juxtaposing the QCG algorithm with other approaches

In this section, we emphasize the differences between the QCG algorithm for calculating the MEU policy and other computational methods that rely on variable elimination or dynamic programming.

The QCG algorithm employs a relational database framework for representing the elements of a decision network (influence diagram), comprising utility functions, chance variables, decision variables, and their interdependencies.

The QCG algorithm compiles the relational database schema representing a decision network into a directed acyclic graph, called a query DAG. The relational queries are composed of well-defined primitive relational operators: ELIM, COMB, and COND. The compositional queries represent steps of a variable elimination process.

The main distinguishing feature of the QCG algorithm is that once the compiled query DAG is generated, it can be reused to solve the MEU problem across different problem instances. The composed queries, which form the nodes in the query DAG, represent views in a relational database. A relational database view is essentially a virtual table that is populated dynamically from the query that defines the relational view.

The computational cost of the QCG algorithm is distributed over the duration of solving the MEU policy across various problem instances.

Other methods for determining the MEU policy in influence diagrams incur the same computational cost on every individual problem instance, which can lead to inefficiencies when scaling more complex decision networks where policies need to be determined under different datasets based on network parameters (conditional probability distributions, prior probabilities, and utility functions), particularly because they do not take advantage of the computational cost distribution amortized across multiple instances, resulting in longer processing times and increased resource consumption compared to the QCG algorithm.

The QCG algorithm effectively utilizes a relational database framework to transform the variable elimination process into a query DAG that can be executed across multiple problem instances.

The QCG algorithm is demonstrated in this chapter using R; however, it can also be implemented using structured query language (SQL), with influence diagrams represented as relational tables in a SQL database. *Chapter 5* of the book already illustrated the use of a relational database, specifically PostgreSQL, to represent an influence diagram.

Implementing the relational queries of the QCG algorithm in a SQL database ensures efficient querying and fast performance. Modern database management systems (DBMS) feature query optimizers and indexing capabilities that automatically determine the most efficient execution plan for any SQL statement. Additionally, specialized SQL engines can quickly process millions of records and handle complex joins across multiple tables, returning results in mere seconds.

In conclusion, we will summarize the key points. The Maximum Expected Utility (MEU) problem involves inferring the decision policy that maximizes expected utility in decision problems represented by probabilistic graphical models known as influence diagrams. We presented an example influence diagram related to a weather decision problem to demonstrate how the MEU policy can be computed using variable elimination. Additionally, we described how our QCG algorithm is capable of calculating the MEU policy for decision problems as represented by influence diagrams. In the practice section, we demonstrated how the QCG algorithm is implemented in R to infer the MEU policy for the weather decision problem using our weighted relation representation.

Summary

The chapter presents a structured approach to exact inference in Bayesian networks and influence diagrams by representing probabilistic and utility models as weighted relations in a relational database framework. It introduces four core primitive queries—COND (conditioning), PROJECT (projection with aggregation), ELIM (variable elimination), and COMB (factor combination via joins)—and shows how they generalize relational algebra to support probabilistic and decision-theoretic reasoning. Building on these primitives, the chapter describes the Query Composition Graph (QCG) algorithm, a compiled, reusable representation of variable elimination (VE) that supports tasks such as computing posterior probabilities and maximum expected utility (MEU). Extensive R implementations and worked examples demonstrate how inference and decision policies can be computed efficiently and reused across multiple problem instances, highlighting both the theoretical foundations and practical benefits of the compositional query approach.

The topics in this chapter are important because they provide a practical and rigorous framework for exact inference and decision making in complex probabilistic models using familiar relational concepts. By showing how Bayesian networks and influence diagrams can be expressed as weighted relations and manipulated through structured queries, the chapter bridges theory and implementation, enabling scalable inference, systematic reuse of computation, and integration of probabilistic reasoning into datacentric workflows. These ideas are especially valuable for readers in AI, data science, and decision analytics because they clarify why variable elimination works, how it can be compiled into reusable query plans (QCG), and how the same machinery supports both probabilistic inference and optimal decision policies—making advanced reasoning techniques more declarative, reusable, and tightly integrated with data.

We will transition to a new topic next in *Chapter 9*, where we describe Pearl's do-calculus for reasoning about causal effects, enabling the transformation of interventional distributions into observational ones.

References

- Bertelè, Umberto, and Francesco Brioschi. 1972. *Nonserial Dynamic Programming*. New York: Academic Press.
- Dechter, J. 1999. *Bucket elimination: A unifying framework for reasoning.* Artificial Intelligence 113: 41–85.
- Dechter, Rina, and Yousri El Fattah. 2001. *Topological parameters for time-space tradeoff.* Artificial Intelligence 93-118.
- El Fattah, Y. 1998. *An Elimination Algorithm for Model-based Diagnosis.* Ninth International Workshop on Principles of Diagnosis (Dx98). Cape Cod, Massachusetts, USA.
- El Fattah, Yousri, and Mark A. Peot. 2000. *A compositional structured query approach to automated inference.* KR'00: Proceedings of the Seventh International Conference on Principles of Knowledge Representation and Reasoning. San Francisco, CA, United States: Morgan Kaufmann Publishers Inc. 213–224.
- El Fattah, Yousri, and Rina Dechter. 1995. *Diagnosing tree-decomposable circuits.* IJCAI'95: Proceedings of the 14th international joint conference on Artificial intelligence. Montreal Quebec Canada.
- Gao, A. 2021. *Lecture 13 Variable Elimination Algorithm.* November 2. `https://cs.uwaterloo.ca/~a23gao/cs486686_f21/lecture_notes/Lecture_13_on_Variable_Elimination_Algorithm.pdf`.
- 2021. *Lecture 17 Decision Networks.* November 2. `https://cs.uwaterloo.ca/~a23gao/cs486686_f21/lecture_notes/Lecture_17_on_Decision_Network_2.pdf`.
- Howard, R. A., and J. E. Matheson. 1984. *Influence diagrams.* In The Principles and Applications of Decision Analysis, Vol. 2. Menlo Park: Strategic Decision Group.
- Huang, C., and A. Darwiche. 1994. *Inference in belief net- works: A procedural guide.* International Journal of Approximate Reasoning 11.
- Jensen, F., F. V. Jensen, and S. L. Dittmer. 1994. *From influence diagrams to junction trees.* Proceedings of the Tenth Conference on Uncertainty in Artificial Intelligence (UAI-94).
- Wong, S.K.M., C.J. Butz, and D. Wu. 1999. *On the Implication Problem for Probabilistic Conditional Independency.* Regina,Saskatchewan, Canda: University of Regina.
- Wong, S.K.M., C.J. Butz, and D. Wu. 1999. *On the Implication Problem for Probabilistic Conditional Independency.* Regina, Saskatchewan, Canada: University of Regina.

Join our community on Discord

Join our community's Discord space for discussions with the author and other readers: `https://packt.link/hYCWN`

Join our community on Discord

9

Causal Inference with Structural Causal Models

We present Pearl's do-calculus, which consists of three inference rules that allow interventional and observational distributions to be rewritten when certain conditions hold in a causal diagram. To establish the identifiability of a causal effect query, we repeatedly apply the rules of do-calculus until the resulting expression no longer contains a do-operator; the query is then estimable from non-experimental data, and the final do-free expression can be used as an estimator. When do-calculus can be used to compute a causal effect, we say the effect is **`identifiable`**; otherwise, it is **`unidentifiable`**. We also review the back-door and front-door criteria and outline the steps for the symbolic identification of causal effect queries. The practice sections reinforce these concepts with examples using the `dagitty` and `dosearch` packages in R.

In this chapter, we are going to cover the following main topics:

- Causal effect identifiability
- Back-door criterion
- Front-door criterion
- Pearl's do-calculus
- Symbolic derivation of causal effects

Technical requirements

All the code for this chapter is in R, and you can download it from the GitHub URL: `https://github.com/PacktPublishing/Causal-Inference-with-Bayesian-Networks/tree/main/Chapter_09`

You will be required to install various R libraries to run the R code for the practice and the examples in this chapter, and you can do that from within the **RStudio** IDE. The required packages to install are:

- `dagitty` `https://cran.r-project.org/web/packages/dagitty/index.html`
- `ggdag` `https://cran.r-project.org/web/packages/ggdag/index.html`

- `ggm` `https://cran.r-project.org/web/packages/ggm/index.html`
- `dosearch` `https://cran.r-project.org/web/packages/dosearch/index.html`
- `ggplot2` `https://cran.r-project.org/web/packages/ggplot2/index.html`
- `mvtnorm` `https://cran.r-project.org/web/packages/mvtnorm/index.html`
- `DiagrammeR` `https://cran.r-project.org/web/packages/DiagrammeR/index.html`

Causal effect identifiability

This section defines the causal inference task and the identifiability problem.

It is organized as follows: First, we describe the representational framework of structural causal models. Next, we define the causal inference task. Following that, we state the identifiability problem. Lastly, we define adjustment sets and their connection to identifiability.

Structural causal models

Structural causal models are represented by directed acyclic graphs $G = (V, E)$ representing the factored representation of a complete probability distribution $P(V)$, which gives the probability of the random variables V, which are the nodes of G, taking the value v from the variables' domains. The graph must be acyclic since cycles create contradictions. For example, a cycle can represent X causes Y causes X which is logically impossible.

The DAG G represents the factorization of the probability distribution $P(V)$ expressed as the product of the conditional probabilities of each variable conditioned on the value tuple assigned to all its parents as follows:

$$P(V) = \prod_i P(V_i \,|pa_i) \text{ (9.1)}$$

Here pa_i is the value tuple of the parents of the variable $V_i \in V$ in the DAG G. This factorization holds if and only if the Markov condition is satisfied, meaning that each variable is conditionally independent of its non-descendants given its parents.

For the probabilistic model to admit causal interpretation, the probability distribution $P(V)$ and its factored representation should be derived from structured equation model M in which each variable $V_i \in V$ is determined by a functional mechanism in a **recursive structured equation model** M, under the **modularity assumption,** formally stated in the following definition.

A structural causal model is a generative model that defines the causal mechanisms between a set of **observable variables** $V = \{V_1, \cdots, V_n\}$, depending on a set of **unobserved (latent) variables** $U = \{U_1, \cdots, U_n\}$.

The variables V represent the **endogenous variables,** and U is the **exogenous** variable that depends on mechanisms outside the model's scope. The exogenous variables have no parents in DAG $\underline{G}$.

The structural causal model consists of n functions $F = \{f_1, \cdots, f_n\}$. Each function in $f_i \in F$ defines a **structural equation** for a variable V_i given U_i and V_i's parents, $pa(V_i)$, in a directed acyclic graph G, called the **causal graph.** The causal graph encodes all the causal assumptions of the structural equation model.

Each structural equation given by a function $f_i \in F$ describes a **causal mechanism** for a **causal family** consisting of a node $V_i \in V$ and the node's parents pa_i in the causal graph.

$$V_i = f_i(pa_i, U_i), i = 1, \cdots, n \quad (9.2)$$

In the structural causal model, the set of variables, U, is considered random and given by a probability distribution $P(U)$. The distribution $P(U)$ and the recursive structural equations uniquely define the probability distribution on the union set of observed and unobserved variables

The following is the formal definition of a structural causal model, based on *Pearl, Causality: Models, Reasoning, and Inference, 2009*:

A **structural causal model (SCM)** is a quadruple $M = \langle U, V, F, P(U) \rangle$, where:

- U is a set of latent unobserved (**exogenous**) variables determined by factors outside the model.
- V is a set $\{V_1, \cdots, V_n\}$ of observed (**endogenous**) variables.
- F is a set of functions $\{f_1, \cdots, f_n\}$ such that a function $f_i \in F$ determines each endogenous variable $V_i \in V$.
- $P(U)$ is a joint probability distribution of the variables in the set U.

A causal graph G of the SCM has a set of nodes $V \cup U$, and directed edges are drawn such that for every $f_i \in F$, a directed edge is drawn from every argument in f_i to the node V_i. We omit the exogenous nodes to enhance clarity and add a bidirected dashed edge between two endogenous nodes if they share an exogenous parent.

A key concept in structural causal models is the modularity assumption, which we will describe next.

Describing the modularity assumption

The modularity assumption in structural causal models (SCMs)—also known as autonomy or invariance—is the foundational assumption that causal mechanisms are independent of one another.

The modularity assumption ensures intervention invariance, which means that the relationships between variables remain stable, even when the system is manipulated. When an intervention, represented by Pearl's do operator, $do(V_i = v_i)$, occurs, the structural equation f_i for V_i is modified, while all other structural equations for different variables remain unchanged.

For a variable V_i in a structural causal model (SCM), its causal mechanism is represented by its conditional distribution $P(V_i|pa_i)$. The modularity assumption implies the independence of causal mechanisms; altering one distribution $P(V_i|pa_i)$ will not impact the other mechanisms $P(V_j|pa_j)$ for all $(j \neq i)$.

Formally, let $P^M(V_i|pa_i)$ be the conditional probability of each variable V_i as a function of the value of its parents in model M. The modularity assumption means that if an intervention model $\tilde{M}$ is derived from M by replacing the mechanism for a variable V_k by overriding its mechanism and replacing it with an equation that sets the value of V_k to a specific value from its domain, then only the conditional probability for the variable V_k will be changed to:

$$P^{\tilde{M}}(V_k|pa_k) = \delta_{v_k} \quad (9.3)$$

Here, δ_{v_k} denotes the Kronecker delta function, which assigns the value 1 to the value v_k222222and zero for all other values in the domain of the variable V_k. Under the modularity assumption, all the conditional probabilities for all other variables will remain unchanged in the intervention model $\widetilde{M}$ as they are in the model M. Hence,

$$P^M(V_i|pa_i) = P^{\widetilde{M}}(V_i|pa_i), \forall i \neq k \text{ (9.4)}$$

To illustrate the modularity assumption, consider the causal model represented by the directed acyclic graph (DAG) shown on the left side of *Figure 9.1*.

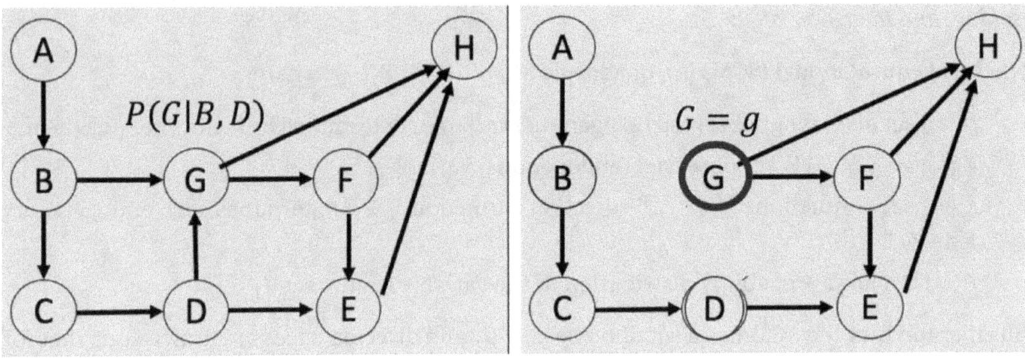

Figure 9.1—Explanation of structural causal model modularity; a unit intervention on G does not influence the causal mechanisms of other variables

The causal mechanism for the variable G is defined by its conditional probability distribution $P(G|D, B)$, which describes its probability based on the value assignments of its parent variables B and D. When an intervention is made using the do operator, do(G=g), which fixes the value of G, the effect will be to eliminate the influence of its parent variables by removing the inward-directed edges from B and D. Under the modularity assumption, the remaining causal mechanisms will remain intact, as depicted in the mutilated DAG on the right side of *Figure 9.1*.

An important category of structural causal models is the Markovian causal models, which we will describe next.

Markovian causal models

The following definition specifies restrictions on the probabilistic causal model to ensure that the model is equivalent to a Markov Bayesian network; that is, the model satisfies the local and global Markov conditions stated in *Chapter 3* on Bayesian networks. Recall the **local Markov condition:** a variable is independent of all its non-descendants in the directed acyclic graph conditioning on all its parents. The **global Markov property** states that for any disjoint node sets X, Y, and Z in a DAG, if Z **d-separates** X and Y, then X and Y are conditionally independent given Z in the associated probability model. A set Z d-separates X and Y if it blocks every undirected path between them—by conditioning on a **non-collider** on the path, or by leaving a **collider** and all of its descendants unconditioned.

We call a structural causal model **Markovian** if the exogenous variables are mutually independent, that is:

$$U_i \perp U_j \ \forall i,j = 1,\cdots,n, i \neq j \ (9.5)$$

The causal graph of a Markovian SCM is a directed acyclic graph. If the probability distribution P of the SCM admits the factorization of *Equation 9.1* according to a DAG, G, then G and P are said to be **compatible.**

Under the Markovian assumption, the exogenous disturbances U are mutually independent (see *Equation 9.5*), so there are no latent common causes inducing dependence among the endogenous variables. Consequently, the causal graph is acyclic and the observational distribution $P(V)$ factorizes according to the DAG structure, exactly as in a Bayesian network.

Markovian structural causal models (SCMs) ensure that causal effects, represented as $P(y|do(x))$, can be identified from observational data based on the assumption of no unmeasured confounding. This means that all exogenous disturbances or error terms U are independent, and the relationships among variables are acyclic. Identification is achieved through graphical criteria—specifically, the **back-door criterion** and the **front-door criterion,** which will be explained in later sections of this chapter. These graphical criteria enable the symbolic evaluation of causal effects in terms of observational conditioning probabilities, eliminating the need for experiments when using observational data.

The next section focuses on semi-Markovian structural causal models (SMCMs), which extend Markovian SCMs by allowing unobserved (latent) confounding among observed variables.

Semi-Markovian causal models

The following definition specifies the class of semi-Markovian probabilistic causal models.

We call the causal model **semi-Markovian** if the exogenous variables are dependent and each $U_i \in U$ appears in at most two structural equation functions in F. The causal graph of a semi-Markovian SCM is an **acyclic directed mixed graph (ADMG),** in which latent common causes are represented by **bidirected edges** between pairs of observed variables.

ADMGs have directed edges (→), bidirected edges (↔), and no directed cycles. A semi-Markovian causal model example with a latent unobserved confounder and its acyclic directed mixed graph is shown in *Figure 9.2.*

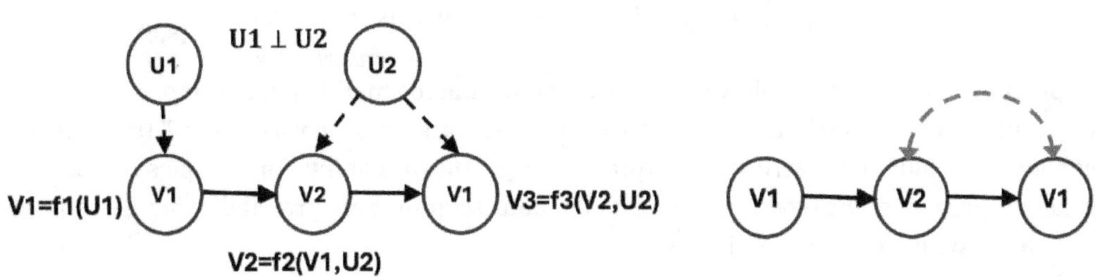

Figure 9.2— Example semi-Markovian structural causal model and its acyclic directed mixed graph

As mentioned previously in *Chapter 4*, the semi-Markovian causal models imply the presence of an **unobserved confounder**, that is, an exogenous variable that influences the causal mechanisms of two endogenous variables. The presence of an unobserved confounder is graphically represented by a bidirectional edge between a pair of endogenous variables while not explicitly depicting the exogenous variable.

Semi-Markovian structural causal models (SCMs) are models that allow for latent confounding (unobserved common causes) represented by bidirected edges, making them not guaranteed to be identifiable. While Markovian models (no unobserved confounding) are always identifiable, the presence of latent variables in semi-Markovian models can make it impossible to uniquely determine the causal effect from observational data alone.

Analysis of causal effect identifiability in semi-Markovian causal models often requires Pearl's do-calculus, a set of three rules that can determine if a causal query, $P(Y|do(X))$, can be reduced to a formula involving only observed probabilities, $P(V)$.

You will learn about the three rules of do-calculus and how they serve as the machinery for analyzing identifiability and non-identifiability in the later sections of this chapter.

Having formally defined the structural causal models, let's describe the causal inference task in the following section.

The causal inference task

The causal inference task involves determining the probability distribution of a select subset of variable Y, known as **outcome** conditioning, on intervention on another subset of variable T, referred to as **treatment** or **exposure**. We use both terms, treatment or exposure, to reference intervention or action variables. T and Y are two disjoint subsets of model variables V.

Let M be a structural causal model that induces the probability distribution $P(V)$, where V is the set of model variables, observed and unobserved variables.

Let (T, Y) be an ordered pair of disjoint subsets of variables in V. The problem of determining the causal effect of a treatment T on outcome$Y$ is the problem of computing $P\big(Y\big|do(T = t)\big)$, which gives the probability distribution of the outcome Y as a function of observed variables in V for every intervention value assignment $T = t$. The post-intervention distribution is given by:

$$P(V = v) = \begin{cases} \prod_{\{i|V_i \notin T\}} P(v_i|pa_i) & v \text{ consistent with } t. \\ 0 & v\ inconsistent\ with\ t \end{cases} \quad (9.6)$$

The expression in *Equation 9.6* follows from *Equation 9.1* and the modularity assumption. It is obtained by truncating all conditional probability factors for every intervention variable in T from the product in *Equation 9.1*, evaluating the truncated expression to get the probability for all tuples compatible with the value assignment to the intervention variables, and assigning zero probability for all incompatible value tuples assigned to model variables.

The causal effect of T on Y can be computed from *Equation 9.6* by marginalizing all variables in V, not in Y, as follows:

$$P(Y = y|do(t)) = \sum_{v_{V \setminus Y}} P(V = v|do(t)) |_{v_Y = y} \quad (9.7)$$

Here, v_Y denotes the restriction of the value tuple v on the set of variables V to the subset $Y \subset V$.

In the next section, we proceed to formally state the identifiability problem.

The identifiability problem

The following is a formal definition of identifiability of causal effect in a probabilistic causal model.

Given a structural causal model M represented by a causal graph G and probability distribution $P(V)$ on observed variables V, we say the causal effect of an action $do(T = t)$ on a set of variables Y is identifiable from P in G if $P(Y|do(T))$ is uniquely computable from any positive probability of the observed variables $P(V)$ in any model that induces G. That is, for every pair of models M_1 and M_2 inducing G, $P^{M_1}(Y|do(t)) = P^{M_2}(Y|do(t))$ whenever $P^{M_1}(\mathrm{V}) = P^{M_2}(\mathrm{V}) > 0$ and $G(M_1) = G(M_2) = G$.

For Markovian models whose causal graph is a directed acyclic graph not containing bidirected edges (i.e., confounders) and whose variables are all observed variables, identifiability is well established. Specifically, in the Markovian model when all variables are observed, the causal effect is computed by *Equation 9.7*.

For semi-Markovian models, the presence of unobserved confounders introduces a layer of complexity. The causal graph becomes a mixed graph containing bidirected edges, and identifiability is only assured under certain criteria, which we will explore in the following sections of this chapter.

The following two examples illustrate two causal graphs: one in which the causal effect is identifiable and the other unidentifiable. We begin with the identifiable case.

Example 9.1: identifiable causal effect

The causal graph in *Figure 9.3* shows the structure of a causal model consisting of three variables X, Y, and Z. We are adopting the conventions and terminologies explained in the "Categorizing Causal Influence Structures" section from *Chapter 4*.

Here X represents an exposure, alternatively a treatment, Y represents the outcome of the exposure, and Z is a third variable that causally impacts both the treatment and its outcome. There are two causal mechanisms implied by the model. One is a direct causation $X \to Y$, and the other is a confounder mechanism $X \leftarrow Z \to Y$. If the confounder variable is observable then the causal effect of X on Y becomes identifiable and the probability of Y conditioning on an intervention on X is as follows:

$$P(y|do(x)) = \sum_z (P(z)P(y|x,z)) \quad (9.8)$$

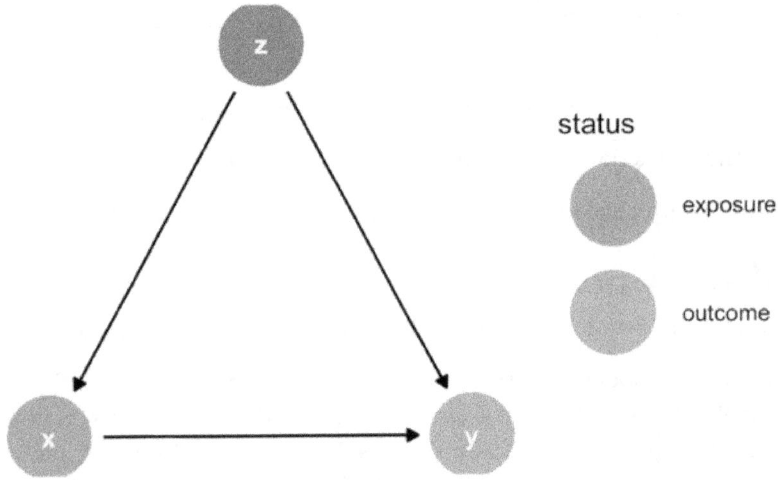

Figure 9.3—Causal model with observable confounder Z- causal effect of X on Y is identifiable

Please note that this equation is valid only when there is no confounding (no common cause of X and Y). Here, the *do* operator cuts the arrow from Z into X, so it is not a confounder anymore. *Equation 9.8* for the intervention probability follows from *Equation 9.7*. To explain, we depict the intervention graph corresponding to the truncated factorization formula for the conditional probability under intervention. See *Figure 9.4*. The intervention graph has the directed edge from Z to X removed because the intervention on X obliterates the causal mechanism implied by the directed edges directed to the intervention variable from all its parents. The probability of Y conditioning on intervention on X will equal the conditional probability of Y conditioning on X in the intervention graph, precisely the expression given by *Equation 9.7*.

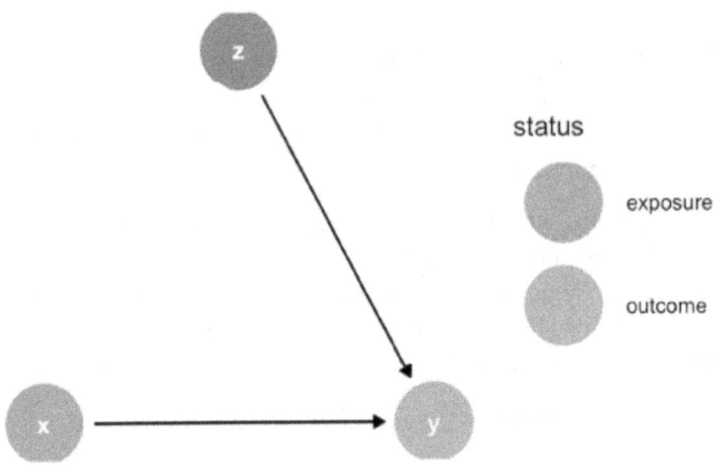

Figure 9.4—Intervention graph for do(X) for the measured confounder model

Next we give an example of the unidentifiable case.

Example 9.2: unidentifiable causal effect

Figure 9.5 shows the same causal structure as the one in *Example 9.1*, with the distinction that the confounder variable Z is unobservable, which is to say Z is a latent variable or exogenous variable determined by mechanisms outside the model. In this case, Z opens a path between X and Y, which is the path given by $X \leftarrow Z \rightarrow Y$. Not knowing "Z" opens a biasing path between "X" and "Y," and knowing "Z" blocks that path. Unlike the model in *Example 9.1*, the causal model in the case where Z is unobservable, the causal effect of X on Y becomes unidentifiable.

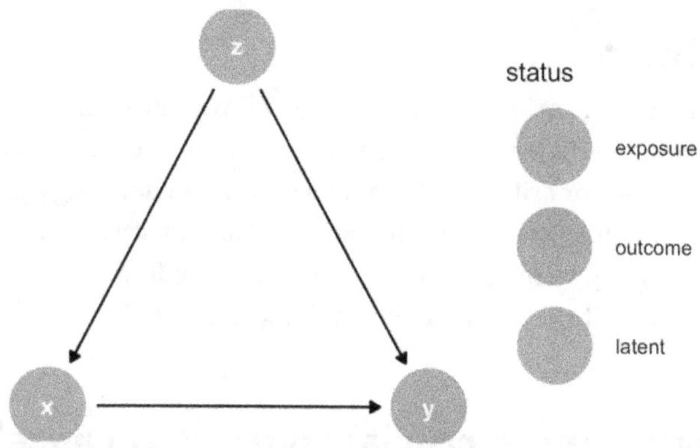

Figure 9.5— Causal model with unobservable confounder Z- causal effect of X on Y is unidentifiable

The following section defines the notion of adjustment sets as they relate to the identifiability of causal effects.

Identifiability and adjustment sets

Let $M = \langle U, V, F, p(u) \rangle$ be a structural causal model represented by causal graph G. Let X and Y be two disjoint subsets of V. Let Z be a third subset of observable covariates in V disjoint with X and Y. If we can express $P(y|\ do(x))$ in terms of the distribution on the observed variables Z, X, and Y only, then we say Z is an **adjustment set** and the causal effect of X on Y is identifiable.

The causal effect of X on Y in the confounder model of *Example 9.1* is identifiable, and Z represents an adjustment variable. This example is an instance of identifiable causal models that satisfy the crucial back-door criterion, making Z a back-door adjustment. In the next section, we will delve into the back-door criterion and back-door adjustment, two critical concepts in our study of causal inference.

The following is the general definition of a valid adjustment set.

Let $M = \langle U, X, F, p(u) \rangle$ be a structural causal model; we say a subset $Z \subset X \backslash \{X, Y\}$ is a valid adjustment set relative to the ordered pair of variables(X, Y) if:

$$P(y|do(x)) = \Sigma_z(P(z)P(y|x, \mathbf{z})) \ (9.9)$$

In conclusion, the main points covered in this section on causal effect identifiability are as follows. A structural causal model enables a causal interpretation of a Bayesian network. The causal interpretation sets the foundation for evaluating the causal effect of assigning a treatment represented by variable X on an outcome represented by variable Y. The causal inference task is to compute the probability distribution on the outcome conditioning of the intervention where we set the treatment value, represented by the `do()` operator. If the interventional conditioning can be computed uniquely from the pre-intervention probability on the observed variables P(V), then the causal effect query is identifiable. A sufficient condition for the causal effect to be identifiable is the existence of valid adjustment sets.

Let us proceed to the next section, where we describe the back-door criterion and the back-door adjustment, which provide one element in our causality toolbox for identifying patterns of identifiable causal effects based on the non-parametric representation of a causal model described by its causal diagram.

Back-door criterion

This section on the back-door criterion is organized as follows: First, in a motivational section, we describe the basic ideas and the underlying intuition behind the back-door criterion. We follow with a section describing the back-door criterion. Next, in a section on defining the back-door adjustment sets, we state the back-door adjustment theorem and illustrate it with an example causal model with multiple back-door paths. Lastly, in the section on practice using R, we implement the example causal model using the `dagitty` R library and run several queries to enhance the understanding of the back-door criterion.

The back-door criterion: a crucial concept in causal inference

In this section, we revisit the two examples given in the previous section as a motivational introduction to build an understanding of the intuition behind the back-door criterion. We will explain why the causal effect is identifiable in *Example 9.1* and non-identifiable in *Example 9.2*. The two examples present a case of confounder bias due to the back-door path between the exposure and the outcome. A back-door path is a non-causal, biasing path connecting the exposure and the outcome, with an edge incident to the exposure node having the arrow pointing in.

The previous section presented an example of a non-identifiable causal effect (*Example 9.2, Figure 9.5*). This example, which involves an unobserved confounder, illustrates the difficulty in identifying the effect of X on Y in the causal graph from the observed joint distribution $P(x, y)$ alone. To demonstrate the non-identifiability, we build a counterexample based on the identifiability *Definition 9.1* by constructing two models, M1 and M2, conforming with the causal graph G of *Figure 9.5*, and we can achieve an identical $P(x, y)$ for both models by selecting probability distributions for $P(z)$, $P(x \mid z)$, and $P(y \mid z)$ appropriately. We can construct$P(y \mid x, z)$ to be independent of the value of x in one model and not in the other. The quantification of the conditional probabilities for the two models will ensure that the intervention probabilities are guaranteed to be different while $P(x, y)$ is the same in both models.

Example 9.1 in the previous section is a practical instance of back-door adjustment. It demonstrates how we can use the observed confounder variable Z to make an adjustment and recover the causal effect from the observational data, which could not be determined by observing x and y alone, as $P\big(y \big| do(x)\big)$ is not the same as $P(y|x)$.

In the next section, we will describe the back-door criterion.

Defining the back-door criterion

The back door criterion, a practical guide in the complex landscape of observational studies, provides a sufficient condition for identifiability. It guides us to the adjustment variables Z that, when controlled, allow us to estimate, using the observation data described by the probability distribution the causal effect of treatment X on outcome Y.

Now, let's delve into Pearl's formal definition of the back-door criterion (Pearl, 2009). This definition is a graphical test on the topology of the causal graph to determine whether a subset Z of the graph node V is sufficient to identify $P(y \mid do(x))$ from the pre-intervention distribution $P(v)$.

A set Z satisfies the back-door criterion relative to an ordered pair of variables (X, Y) in a causal diagram G if:

1. No node in Z is a descendant of X, and,
2. Z blocks every path between X and Y, which contains an arrow that goes into X.

The back-door criterion identifies a set of variables Z to control for, ensuring that only causal paths between treatment or exposure X and outcome Y remain open. This criterion blocks all biasing, spurious, and confounding paths without introducing new ones. A set Z is considered valid if it effectively blocks all paths between X and Y that include an arrow pointing into X (known as back-door paths) and does not include any descendants of X. To illustrate the back-door criterion, consider the three causal graphs in *Figure 9.6*.

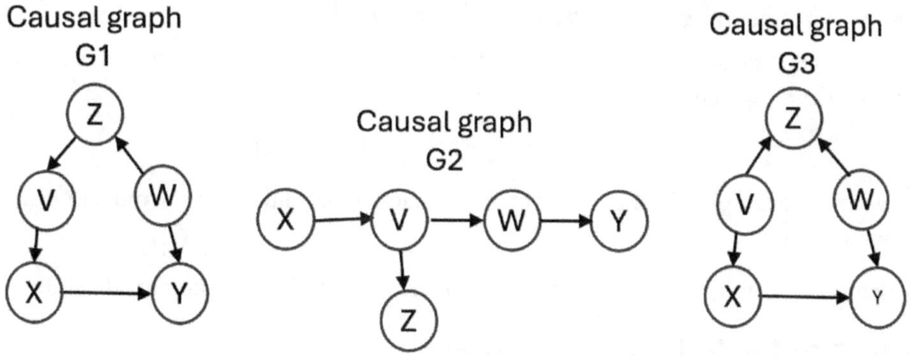

Figure 9.6—Z satisfies the back-door criterion in graph G1, whereas in graphs G2 and G3, it does not

In graph G1, the node Z satisfies both conditions of the criterion by blocking the biasing path between X and Y. This path is represented as $\leftarrow V \leftarrow Z \leftarrow W \rightarrow Y$ and includes an arrow pointing into X. There are no spurious paths in G1. However, for the other two graphs, G2 and G3, Z fails to meet one of the two conditions of the back-door criterion. In graph G2, Z is a descendant of X, violating the first condition. To illustrate the significance of this condition, assume we control for Z. Controlling Z would open a path between X and the exogenous variable U_v (not shown), represented by the path $X \rightarrow V \leftarrow U_V$. The spurious path arises because V is a collider, and we are conditioning on its descendant Z. In graph G3, Z violates the second condition of the back-door criterion, as Z is a collider that opens the biasing path $X \leftarrow V \rightarrow Z \leftarrow W \rightarrow Y$.

Pearl's back-door criterion, a pivotal tool in our research arsenal, plays a significant role in identifying a set of covariates Z that are sufficient for determining treatment X's **average causal effect** (ACE) on outcome Y. Its importance cannot be overstated in our quest for accurate causal inference.

The total causal effect is the sum of all direct and indirect effects from causal and biasing paths. A causal path is one in which all edges are directed from cause to effect. In contrast, a biasing path is an information path that connects cause to effect regardless of edge directionality. A biasing path is bidirectional, indicating correlation or association between end variables, while a causal path is unidirectional, pointing from a cause to an effect.

The covariates Z that satisfies the back-door criterion are instances of what Pearl (Cinelli, et al., 2024) calls 'good control.' These variables will significantly reduce bias when conditioned on and included in a regression model for predicting the ACE. The back-door criterion thus guides us in identifying these robust controls for conditioning, effectively eliminating or countering the effect of confounding bias due to confounding variables.

In *Example 9.1*, we showed how the measured confounder is a back-door adjustment, which enabled us to compute the intervention probability in *Equation 9.8*. In the next section we describe Pearl's back-door adjustment.

Describing the back-door adjustment

The following is Pearl's back-door adjustment theorem (Pearl, 2009).

Theorem 9.1: Pearl's back-door adjustment

If a set of variables Z satisfies the back-door criterion relative to (X, Y), then the causal effect of X on Y is identifiable and is given by the formula.

$$P(y|do(x)) = \sum_z P(y|x,z)P(z) \quad (9.10)$$

We already saw an example of the application of the back-door adjustment formula in *Example 9.1* in the previous section.

Let us work on an example to illustrate the back-door criterion and the back-door adjustments in detail.

Example 9.3: back-door adjustments

Figure 9.7 unveils a sophisticated causal diagram for a causal model involving five observed variables: A, V, Z, E, and D. The variable E is the intervention variable denoting an exposure or treatment, while D represents the outcome of the exposure. The causal diagram reveals that all three variables, A, B, and Z, are confounding variables for the causal effect of E on D and are elements of biasing paths for the direct causal link from E to D. There are four biasing paths, adding to the complexity of the diagram. The paths are:

```
"E <- A -> Z -> D"
"E <- A -> Z <- B -> D"
"E <- Z -> D"
"E <- Z <- B -> D"
```

Verifying that both sets {A, Z} and {B, Z} satisfy the back-door criterion is reassuring. For the first set {A, Z}, it's clear that it meets the first condition in the criterion: no element in the set is a descendant of the exposure variable E. The second condition of the back-door criterion is also met, as the two nodes A and Z block all biasing paths listed above. It's important to note that both sets require minimal adjustments, as either A, B, or Z alone cannot block all biasing paths. The singleton subset {Z} will block only one path, "E <- Z -> D," leaving the three other paths open (unblocked). The singleton subset {A} will block two paths, "E <- A -> Z -> D" and "E <- A -> Z <- B -> D" while leaving two others unblocked. Similarly, the singleton subset {B} will block two paths, "E <- Z <- B -> D" and "E <- A -> Z <- B -> D" while leaving two others unblocked.

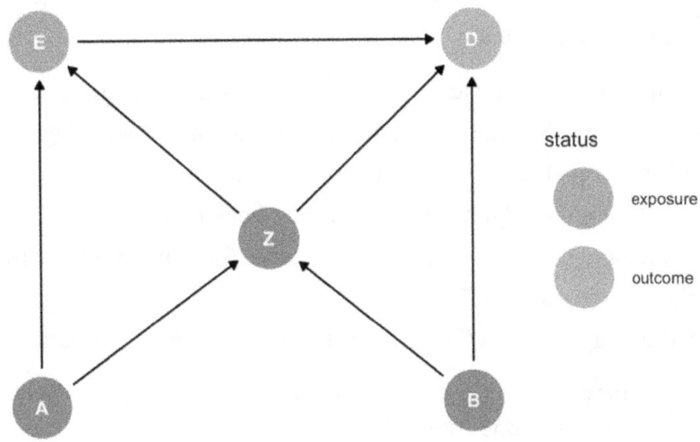

Figure 9.7—Causal graph with multiple back-door paths

We can generate simulation data for the example; assuming the true model for observational data is given by the following set of linear Gaussian equations representing the quantification of the underlying SCM. The equations are as follows:

$$A = U_A$$

$$B = U_B$$

$$Z = 0.3\,A + 0.7\,B + U_Z$$

$$E = 0.5\,Z + 0.35\,A + U_E$$

$$D = 1.2\,E + 0.9\,B + 0.65\,Z + U_D$$

The parameters in the linear equations are **path coefficients** on a **path diagram** representation, which annotates the directed edges of the causal graph of *Figure 9.7*. In the linear SCM model, the adjustment sets are the variables to regress using linear regression. Suppose the causal effect of E on outcome D is identifiable using the back-door adjustment as the case here, supposing the model is accurate. In that case, we can recover the path coefficient from D to E, equal to 1.2, as per the equations above.

The path coefficient will provide the **direct causal effect** of E on D. In the following practice, using the R section, we will generate simulation data and perform linear regression on the observed variables regressing on the back-door adjustment sets.

In the following section, we illustrate how to programmatically implement *Example 9.3* and verify the results demonstrated in the example.

Practice using R

In this practice, we implement the tasks in R to demonstrate the back-door criterion and the back-door adjustment for the SCM in *Example 9.3*.

The tasks for this practice are as follows:

1. Getting the confounding model and displaying the causal diagram.
2. Listing and visualizing all paths between exposure and outcome.
3. Listing and visualizing all back-door adjustment sets.
4. Verifying every back-door adjustment satisfies the back-door criterion.
5. Generating Simulated observational data for a linear Gaussian model of the back-door SCM.
6. Estimating the average causal effect.
7. Visualizing the average causal effect on a scatter plot of simulation data.

Let us delve into the code for practice. The entire code is made available on the book website contained in the *Rmarkdown* document `back-door.Rmd`.

First we load the required libraries:

```
library(dagitty)
library(ggdag)
```

Getting the confounding model and displaying the causal diagram

The SCM "confounding" of *Example 9.3* is already included in the `dagitty` library. We can load the model using the `dagitty::getExample` function, as in the code below. The function returns the causal graph in text format. We use the `dagitty::dagitty` function in the code below to construct a `dagitty` graph object from the textual description. Next, we use `ggdag` to plot the `dagitty` graph object. The result is depicted in *Figure 9.7*.

```
library(dagitty)
library(ggdag)
g <- dagitty::getExample("confounding")
g2 <- dagitty::dagitty(g)
ggdag::ggdag_status(g2) + ggdag::theme_dag()
```

We proceed to the second task next.

Listing and visualizing all paths between exposure and outcome

We use the function `dagitty::paths` to generate all paths between the exposure "E" and the outcome "D". The results are five paths: one causal path, "E -> D," and the other four are biasing back-door paths: `"E <- A -> Z -> DE <- A -> Z <- B -> D"`, `"E <- Z -> D"`, and `"E <- Z <- B -> D"`.

```
dagitty::paths(g, "E", "D", directed= FALSE)$paths
## [1] "E -> D"      "E <- A -> Z -> D"      "E <- A -> Z <- B -> D"
## [4] "E <- Z -> D"      "E <- Z <- B -> D"
```

We can visualize the five paths between "E" and "D" using the function `ggdag::ggdag_paths`.

```
ggdag::ggdag_paths(g2)
```

Figure 9.8 depicts the output plots of all open paths between "E" and "D".

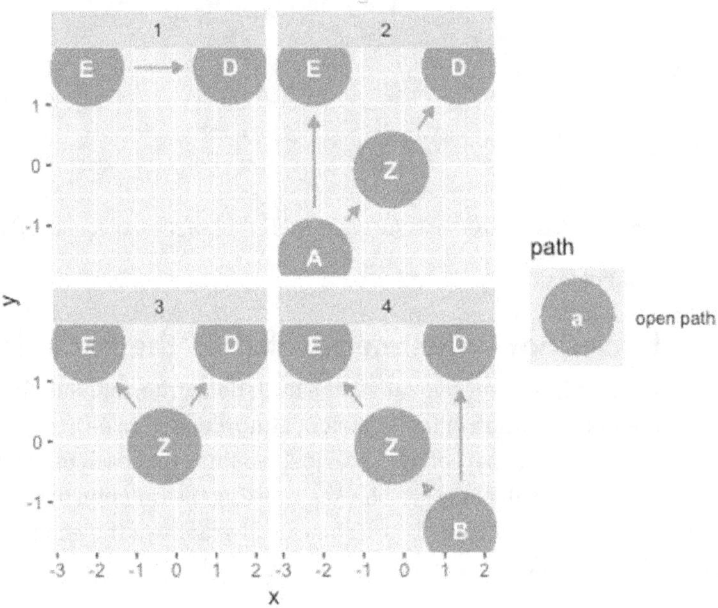

Figure 9.8—Visualization of all paths between the node "E" and the node "D"

Figure 9.8 illustrates the four open paths between nodes "E" and "D." It omits the closed path `"E <- A -> Z <- B -> D"` because of the presence of the collider node Z.

Let us proceed to the third task next.

Listing and visualizing all back-door adjustment sets

We can use the function `dagitty::adjustmentSets` to get the back-door adjustment sets.

```
dagitty::adjustmentSets(g2)
## { B, Z }
## { A, Z }
```

We can visualize the adjustment sets by calling the function `ggdag::ggdag_adjustment_set` on the `dagitty` graph object `g2`, which returns two sets $\{A, Z\}$, and $\{B, Z\}$ depicted in *Figure 9.9*.

```
ggdag::ggdag_adjustment_set(g2)
```

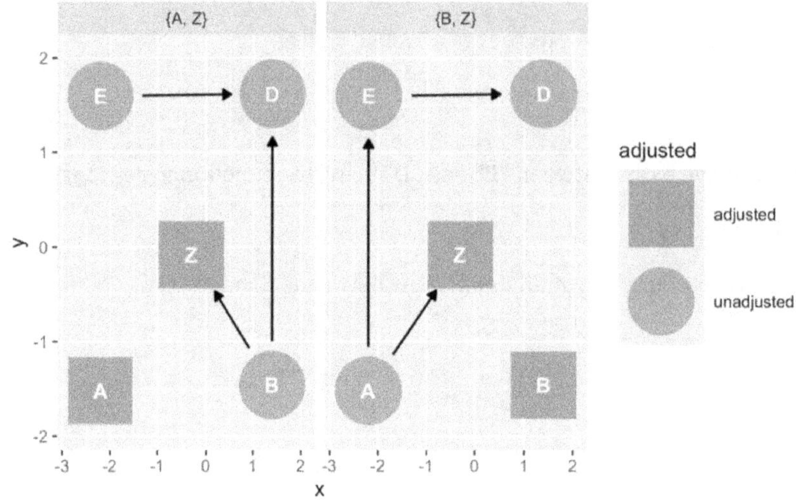

Figure 9.9—Two back-door adjustments for the causal graph of the "confounding" model

Next we proceed to the fourth task.

Verifying every back-door adjustment satisfies the back-door criterion

In the following, we verify the adjustment sets satisfy the back-door criterion. It is already clear that the adjustment sets satisfy the first condition of the criterion, as no element in either set is a descendant of the exposure "E." To verify the second part of the criterion, we show that each of the two sets blocks all four back-door paths. We show that all paths are blocked by using the function `dagitty::dseparation` to check the d-separation between "E" and "D" on the causal graphs of the example after excluding the causal link "E -> D." The result confirms the d-separation on both sets.

```
dagitty::dseparated( "dag {A -> E ; A -> Z ; B -> D ; B -> Z ;  Z -> D ; Z ->
E}","E","D",list("A","Z"))
## [1] TRUE
dagitty::dseparated( "dag {A -> E ; A -> Z ; B -> D ; B -> Z ;  Z -> D ; Z ->
E}","E","D",list("B","Z"))
## [1] TRUE
```

Next to the fifth task.

Generating simulated observational data for a linear Gaussian model of the back-door SCM

We generate data to simulate observational data assuming a linear Gaussian structural equation model with specified parameters. The exogenous variables, which are assumed to be independent and in standard normal form, play a crucial role in this model. The model data is stored in a data frame with n rows and five columns, with each column corresponding to a distinct observed variable.

```
# simulate data (linear model)
n <- 1e4
A <- rnorm(n)
B <- rnorm(n)
Z <- 0.3 * A + 0.7 * B + rnorm(n)
E <- 0.5 * Z + 0.35 * A + rnorm(n)
D <- 1.2 * E + 0.9 * B + 0.65 * Z + rnorm(n)
model.data <- data.frame (
  A <- A,
  B <- B,
  Z <- Z,
  E <- E,
  D <- D
)
```

Moving on to the sixth task.

Estimating the average causal effect

Estimating the path coefficient using linear regression on the simulated observational data determines the average causal effect. We can adjust using either one of the two back-door adjustments. The results show that the estimated path coefficient returned from linear regression in both cases is almost identical and is fairly accurate relative to the real value 1.2.

Adjusting for {"A", "Z"}:

```
coef(lm(D ~ E + A + Z, model.data))
## (Intercept)            E           A           Z
## -0.01794663  1.21495120 -0.13039998  1.07486571
```

Adjusting for {"B", "Z"}:

```
coef(lm(D ~ E + B + Z, model.data))
## (Intercept)            E           B           Z
## -0.02063401  1.21142586  0.90659135  0.64322075
```

Finally, we proceed to the last task.

Visualizing the average causal effect on a scatterplot of simulation data

In this script, the scatter plot is generated using `geom_point()`, and the linear regression line is added with `geom_smooth()` where `method = "lm"` specifies a linear model. The caption includes the estimated slope from the linear regression.

```
# Load required library
library(ggplot2)

# Recreate model and extract coefficients
model <- lm(D ~ E + A + Z, model.data)
model.coefs <- coef(model)

# Create scatter plot of D versus E
ggplot(model.data, aes(x = E, y = D)) +
  geom_point(alpha = 0.5, size = 1) +  # Draw points
  geom_smooth(method = "lm", col = "red") +  # Add linear regression line
  labs(
    title = "Scatter plot of D versus E with linear regression line",
    x = "E",
    y = "D",
    caption = paste0("Slope: ", round(model.coefs["E"],2))
  ) +
  theme_minimal()
```

Figure 9.10 shows the output plot.

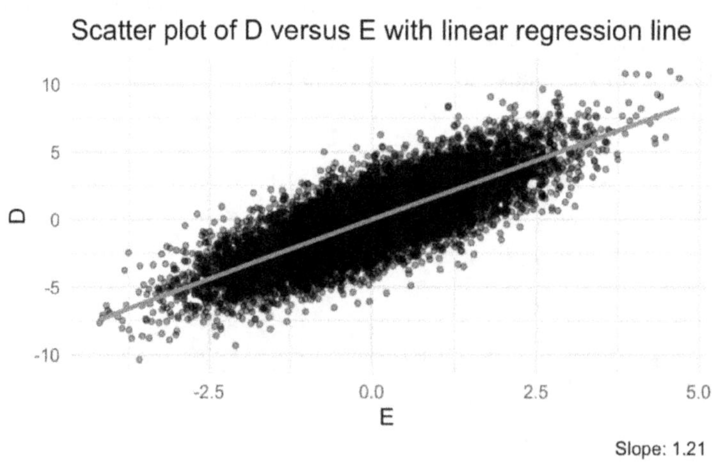

Figure 9.10—visualizing the estimated average causal effect vs the scatter plot of the simulated observational data

In conclusion, the main points covered in this section on the back-door criterion are as follows.

The back-door criterion, a key concept in causal inference, provides a sufficient condition to determine whether a causal effect is identifiable based only on the topology of the DAG, a crucial element in representing a causal model. The back-door criterion sets two conditions; when met by a set of nodes in the DAG, the causal effect can be computed using the back-door adjustment formula. The two conditions of the back-door criterion relative to an ordered pair of variables (X, Y) regarding a treatment and outcome, respectively, are as follows. One, no node in the adjustment set Z is a descendant of the intervention variable X. Two, Z blocks every path between X and Y that contains an arrow into X. Using R, the dagitty library provides a set of functions for specifying the DAG, testing the conditions of the back-door criterion, and for finding the adjustment sets.

In the next section, we consider another vital result related to identifiability based on another criterion called the front-door.

Front-door criterion

This section is organized as follows: First, we define the front-door criterion, which gives conditions on the causal DAG for the identifiability of a causal effect between an ordered pair of nodes (X, Y) in the causal DAG. Next, we describe the front-door adjustment, a formula stated as a theorem that computes $P(Y|do(x))$ from the preintervention distribution on the causal model variables. Following this, in the practice section, we illustrate coding examples to reinforce an understanding of the front-door criterion and use related R libraries.

Defining the front-door criterion

In the previous section on the back-door criterion, we learned how to calculate the causal effect in the presence of an observed confounder. The back-door criterion specifies conditions for back-door adjustment sets that can be used to calculate the causal effect from the observational data according to the back-door adjustment theorem and *Equation 9.10*.

We cannot apply the back-door adjustment in the presence of an unmeasured confounder. We saw in *Example 9.2* an unmeasured confounder in which the causal effect is unidentifiable.

The front-door criterion addresses the identifiability of the causal effect for causal models with unmeasured confounders. In essence, it specifies sufficient conditions for identifiability under unobserved confounders in models where forward paths exist between the exposure and the outcome variables intercepted by mediator variables, as in *Figure 9.11*.

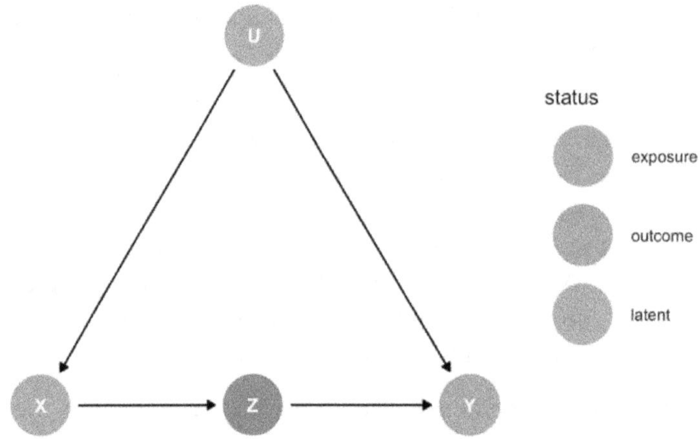

Figure 9.11—causal graph containing an unobserved confounder U - the causal effect of X on Y is mediated by Z

The front-door criterion specifies sufficient conditions to render the effect of X on Y identifiable through the mediator variable Z. The following is the formal definition of Pearl's front-door criterion (Pearl, 2009).

Definition 9.2: Pearl's front-door criterion

A set of variables Z satisfies the front-door criterion relative to an ordered pair of variables (X, Y) in a DAG G if:

1. Z intercepts all directed paths from X to Y,
2. There is no unblocked back-door path from X to Z; and,
3. All back-door paths from Z to Y are blocked by X.

The three conditions of Pearl's front-door theorem are essential because they facilitate the identification of a causal effect, $P(Y|do(X))$, even in cases where the confounder U (which exists between X and Y) is not measured.

Condition 1 is important because it stipulates that Z intercepts all directed paths from X to Y. This ensures that there are no "hidden" directed paths from X to Y that are not mediated by Z. As a result, it guarantees that Z functions as a full mediator, meaning that all causal effects must flow through it. This condition allows for the decomposition of the causal effect into two components: the effect of X on Z and the effect of Z on Y.

Condition 2 is crucial because it claims that there is no unblocked back-door path from X to Z. This assertion ensures that the causal effect of X on Z is identifiable, as there are no unobserved confounders influencing both X and Z. Consequently, the correlation between X and Z is attributed solely to the causal effect of X on Z.

Condition 3 is significant because it states that all back-door paths from Z to Y are blocked by X. This blocking guarantees that the effect of Z on Y can be identified when controlling for X, which serves as the treatment.

The three conditions of the front-door criterion make it possible to estimate the average treatment effect (ATE) of X on Y.

In the next section, we describe the front-door adjustment theorem, which gives a general expression of the probability distribution of outcome Y conditioned on intervention variable treatment X.

Describing the front-door adjustment

The front-door criterion enables us to compute the average treatment effect of X on Y in three steps. In the first step, we compute the probability of Z conditioning on intervention on X, $P(Z|\ do(X))$. In the second step, we compute the probability of Y conditioning on Z and intervention on X, $P(Y|Z, do(X))$. In the third step, we multiply the intervention probabilities from the two steps and sum up Z to get the probability of Y conditioning on intervention on X as follows:swswsw

$$P(Y|do(X)) = \sum_Z P(Y|Z, do(X))P(Z|do(X)) (9.11)$$

By condition 2 of the front-door criterion, there can be no open back-door paths from X to Z. Therefore:

$$P(Z|do(X)) = P(Z|X) \ (9.12)$$

As Z intercepts all directed paths from X to Y, it follows that:

$$P(Y|Z, do(X)) = P(Y|do(Z)) \ (9.13)$$

Condition 3 of the criterion, X, is a back-door adjustment for the causal effect of Z on Y. It follows from the back-door adjustment theorem that we can write:

$$P(Y|do(Z)) = \sum_X (p(X)p(Y|X, Z)) \ (9.14)$$

Equations (9.11) through *(9.14)* yield the front-door adjustment formula in the following theorem due to Pearl (Pearl, 2009).

Theorem 9.2: front-door adjustment

If Z satisfies the front-door criterion relative to (X, Y) and if $P(X, Z) > 0$ then the causal effect of X on Y is identifiable and is given by the formula:

$$P(Y|\text{do(X)}) = \sum_Z \left(p(Z|X) \sum_X (p(X)p(Y|X, Z)) \right) (9.15)$$

In the following practice using R, we will programmatically demonstrate the front-door adjustment of the causal model depicted in *Figure 9.7*.

Practice using R

In this practice, we will implement several tasks in code using R to demonstrate the front-door criterion.

The tasks for this practice are as follows:

- Constructing and displaying the causal diagram
- Simulating observational data set

- Performing linear regression
- Plotting scatter diagram

Let us delve into the code for practice. The entire code is available on the book website, contained in the *Rmarkdown* document `front-door.Rmd`.

First, we load the required libraries:

```
library(dagitty)
library(ggdag)
library(mvtnorm)
library(ggplot2)
```

Constructing and displaying the causal diagram

We use the `dagitty::dagitty` function to define the DAG of the front-door model and visualize it using the ggdag functions. The output is the graph in *Figure 9.7*.

```
g <- dagitty("dag{
  X <- U -> Y
  X -> Z -> Y
  X [exposure]
  Y [outcome]
  U [unobserved]
}")
# set coordinates for graph layout
coords <- list(X=c(X=1, Z=2, Y=3, U=2),
               Y=c(X=1, Z=1, Y=1, U=2))
coords_df<-coords2df(coords)
coordinates(g)<-coords2list(coords_df)
#Plot causal graph
ggdag::ggdag_status(g) + ggdag::theme_dag()
```

We proceed to the second task next.

Simulating observational dataset

The following code generates a dataframe containing simulated observational data on the observed model variables X, Z, and Y. We specify a linear Gaussian model for the front-door SCM. The influence of the unobserved confounder is simulated by generating two correlated random variables, Ux and Uy. The observed data record is stored in the dataframe model data.

```
# simulate data (linear Gaussian model)
n <- 1e5
# we will generate correlated, normal random variables with the program
rmmvnorm in the mvtnorm library
out <- rmvnorm(n, mean = c(0, 0),
```

```
                    sigma = matrix(c(1,.6, .6, 1), 2, 2))
Ux <- out[,1]
Uy <- out[,2]
Uz <- rnorm(n)
X <- 0.1 * Ux
Z <- 1.2 * X + 0.3 * Uz
Y <- 2.0 * Z + 0.2 * Uy
model.data <- data.frame (
  X <- X,
  Z <- Z,
  Y <- Y
)
```

Next, we get to the third task.

Performing linear regression

The following code performs three linear regressions. First, it regresses the mediator Z on treatment X to estimate the path coefficient for X → Z. Second, it regresses the output Y on the mediator Z to estimate the path coefficient for Z→ Y. Third, it regresses the output Y on treatment X.

Estimating path coefficient X → Z

```
model_1 <- lm(Z ~ X, model.data)
model_1.coefs <- coef(model_1)
print(model_1.coefs)
## (Intercept)           X
## 0.001449566 1.200510056
```

Estimating path coefficient Z → Y

```
model_2 <- lm(Y ~ Z, model.data)
model_2.coefs <- coef(model_2)
print(model_2.coefs)
##   (Intercept)            Z
## -8.950966e-05  2.139136e+00
```

Regressing Y on X

```
model_3 <- lm(Y ~ X, model.data)
model_3.coefs <- coef(model_3)
print(model_3.coefs)
## (Intercept)           X
## 0.003187794 3.609510185
```

Next, to the fourth and final task.

Plotting scatter diagram

The estimated ATE for the front-door model is the product of the two path coefficients of `model_1` and `model_2`. The following code plots the scatter diagram for the `model.data` and two lines: one solid red line displaying the estimated ATE, and the other dotted blue line shows the regression line Y~X. The estimated ATE is close to the true value given by the product of the actual path coefficients in the front-door linear SCM.

```
# Plot data points
p <- ggplot(model.data, aes(x = X, y = Y)) +
  geom_point(alpha = 0.1, size = 0.1)
# Creating the lines
line1_data <- data.frame(
  X = range(model.data$X),
  Y = model_1.coefs["X"] * model_2.coefs["Z"] * range(model.data$X)
)
line2_data <- data.frame(
  X = range(model.data$X),
  Y = model_3.coefs["X"] * range(model.data$X)
)
# Add informative text box
textbox <- paste0(
  "Average Treatment Effect (ATE)\n",
  "True value = 2.4\n",
  "Estime value (Red regression slope) = ",
  round(model_1.coefs["X"] * model_2.coefs["Z"],4), "\n",
  "Estimate value (Blue regression slope) = ", round(model_3.coefs["X"],4)
)
# Plotting the lines
p +
  geom_line(data = line1_data, aes(X, Y), colour = "red", size = 2) +
  geom_line(data = line2_data, aes(X, Y),
            linetype = "dashed", colour = "blue", size = 2) +
  annotate(
    "text",
    x = min(model.data$X),
    y = max(model.data$Y),
    label = textbox,
    hjust = 0,
    vjust = 1,
    size = 4,
    colour = "black"
  )
```

Figure 9.12 shows the output plots.

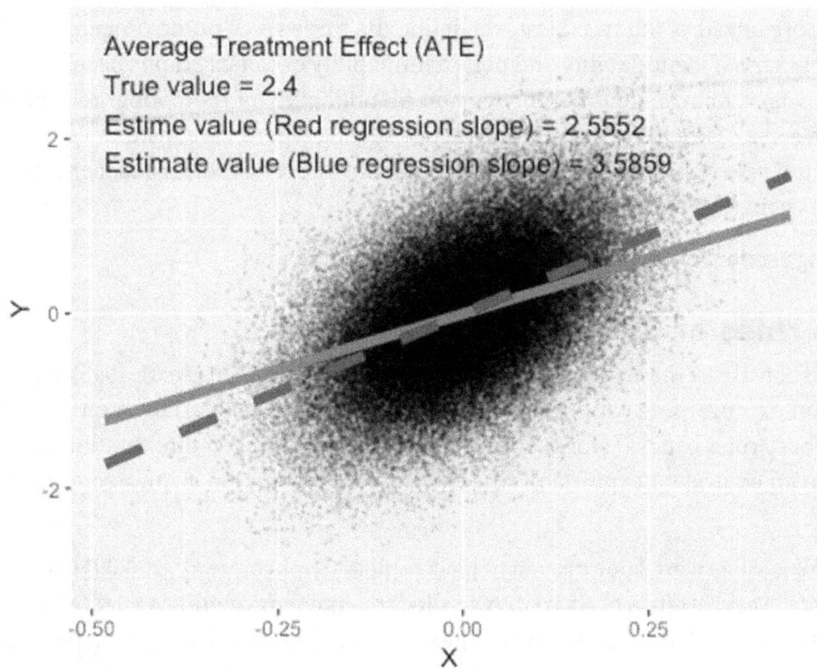

Figure 9.12—Scatter diagram along with the ATE line and the Y~X regression line

As we wrap up this section on the front-door criterion, let us summarize the main takeaway points.

The front-door criterion comes into play when variables Z intercept all directed paths from treatment X to outcome Y, and all back-door paths from X to Z and Z to Y are blocked.

The conditions in the front-door criterion allow computing the causal effect P(Y|do(X)) by the front-door adjustment formula.

The intuition behind the front-door criterion is that the conditions of the criterion guarantee the identifiability of the effect of X on Y, as it can be estimated by computing the effect of X on Z and then the effect of Z on Y, which are both identifiable by their blocked back-door paths, and then combining the two causal effects.

By applying the front-door criterion in a practical example using R, we've demonstrated how to compute the ATE by linear regression. This process involves estimating the path coefficients for the direct paths X → Z and Z → Y and multiplying them, providing a hands-on understanding of the concept.

Next, we'll delve into the do-calculus rules, which provide a rewritten map between observational and interventional conditioning. It's important to note that the do-calculus subsumes both the front-door and the back-door criteria, making it a vital tool in the machinery of identifiability and non-identifiability, and showing the interconnectedness of these concepts.

Pearl's do-calculus

This section is organized as follows: First, we define the three rules of the do-calculus. Next, we describe the machinery of identifiability and non-identifiability of causal graphs based on the existence of adjustment sets or known identifiability or non-identifiability of their subgraphs or supergraphs. We follow with a use case application in the practice section where we consider a causal graph from sports biomedical research and establish identifiability of direct and total causal effects from finding adjustment sets using the *dagitty* library in R.

In the following section, we describe the three rules of do-calculus.

The three rules of do-calculus

The do-calculus consists of three rules that Pearl presented with proof in (Pearl, 1995). Each rule applies in a specific context regarding conditional independencies between variables on perturbed graphs. For instance, these rules can transform an expression involving mixed intervention and observation conditioning to an equivalent expression where observation or action is ignored or exchanged with an observation.

Pearl's do-calculus rules were later proven to be complete (Shpitser, et al., 2006) (Huang, et al., 2006). This means that if a causal effect of a variable X called treatment on another variable Y called outcome, given by the interventional probability distribution $\mathrm{P}\big(\mathrm{Y}\big|\mathrm{do}(\mathrm{X})\big)$, is identifiable, then the distribution can be transformed into an expression function only of the pre-intervention distribution, the distribution on the observed variables, P(V). This transformation ensures that the expression is free of the do( ) operator, enabling the estimation of the average treatment effect using observational data.

The rules of the do-calculus adopt the following notations. Let G be a directed graph and X, Y, Z, and W be disjoint subsets of nodes of G.

- The perturbed graph $G_{\overline{X}}$ is obtained from G after removing all the incident edges directed toward X
- The perturbed graph $G_{\underline{X}}$ is obtained from G after removing all the incident edges whose arrows point away from X
- We write Z(W) to denote the set of nodes in Z that are not ancestors of W ($Z\backslash an(W)$), where an(W) denotes the ancestors of W

The following theorem formally defines Pearl's do-calculus (Galles, et al., 1995).

Theorem 9.3: Pearl's do-calculus

Let M be a structural causal model represented by the directed acyclic graph G and the probability distribution $P(.)$. For any disjoint subsets of variables X, Y, Z, and W, we have:

Rule 1: Insertion and deletion of observations:

$$P(y|do(x), z, w) = P(y|do(x), w) \text{ when} (Y \perp Z|X, W)_{G_{\overline{X}}}$$

Rule 2. Exchanging actions and observations:

$$P(y|do(x), do(z), w) = P(y|do(x), z, w) \text{ when } (Y \perp Z|X, W)_{G_{\overline{X}\underline{Z}}}$$

Rule 3: Insertion and deletion of actions:

$$P(y|do(x), do(z), w) = P(y|do(x), w) \text{ when } (Y \perp Z|X, W)_{G_{\overline{X}\,\overline{Z(W)}}}$$

Rule 1 states that the insertion or deletion of the observation of Z into the conditioning in the expression $(y|do(x), w)$ or $P(y|do(x), z, w)$ does not change the valuation of either expression if Y is d-separated from Z given X and W in the perturbed graph $G_{\overline{X}}$. Rule 1 is the familiar rule of conditional independence applied to the model after intervention on X, represented by the perturbed graph $G_{\overline{X}}$, where all incident edges pointing to the nodes in X are removed.

Rule 2 delineates the conditions for exchanging actions and observations in the context of the perturbed graph $G_{\overline{X}\underline{Z}}$, where directed edges into X and out of Z are removed. It states that if Y and Z are d-separated given X and W in the perturbed graph $G_{\overline{X}\underline{Z}}$), then the causal effect on outcome Y from acting on X and Z, conditioning on observing W, has the same probability distribution as if we are to act only on X and observe Z in addition to W.

Rule 3 specifies the conditions for inserting and deleting actions in the context of the perturbed graph $G_{\overline{X}\,\overline{Z(W)}}$, where all edges into X are removed, and the edges into Z excluding the subset of Z whose elements are the ancestors of W are also removed. The rule states that if Y and Z are d-separated given X and W in the perturbed graph $G_{\overline{X}\,\overline{Z(W)}}$, then we can insert the intervention conditioning on Z into the conditioning in the expression $P\ (y|do(x), w)$ or delete it from the conditioning in the expression $P(y|do(x), do(z), w)$,which will not affect the valuation of the probability distribution on the outcome Y.

In a later section, we will see how the do-calculus rules are applied to derive symbolic expressions of the causal effect.

In the next section, we will delve into the machinery for identifiability and non-identifiability based on the non-parametric representation of SCM given by the causal graph structure.

Machinery for identifiability and non-identifiability

This section is organized as follows: First, in a section on machinery for identifiability, we review results in the context of do-calculus. We follow with a section on machinery for non-identifiability. We conclude with a practice section where we consider a detailed example of running various queries on a use case model from sports biomedical research.

Let us begin with the machinery for identifiability.

Machinery for identifiability

The do calculus provides a powerful machinery for investigating whether a causal effect query $P(y|do(x))$ is identifiable. The strength of the do-calculus stems from the fact that it has been proven that the rules of the do-calculus are complete (Huang, et al., 2006). The completeness of the do-calculus rules means that if a causal effect is identifiable in a causal model given by a causal graph G and distribution on observed variables $P(v)$, then there exists a sequence of the three rules of the do-calculus to derive an expression for the causal effect of an intervention X on outcome Y free of intervention conditioning, i.e. an expression to compute $P(y|do(x))$ as a function only of $P(v)$ free of the do operator.

A causal effect is identifiable if a valid adjustment set exists. Specifically, suppose X is the intervention variable, Y the outcome, and Z is a valid adjustment set for the ordered pair (X, Y) as in *Definition 9.2*. In that case, the causal effect is identifiable, and the intervention distribution $P(y|do(x))$ can be computed using *Equation 9.9*.

For example, if X, Y, and the parents of X, pa(X), are all observed variables in the causal model, then the parents of X are valid adjustment variables.

Figure 9.13 gives samples of causal graph G where the causal query of the effect of X on Y is identifiable. Let us go through each graph in the sample and explain why $P(y|do(x))$ is identifiable.

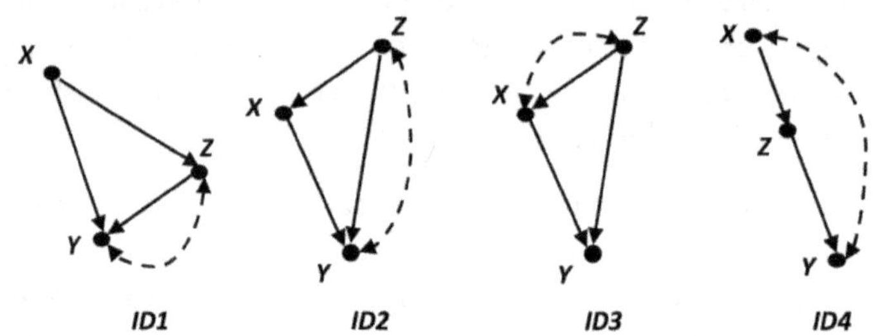

Figure 9.13—Samples of causal graphs where the effect of X on Y is identifiable

For graph ID1, the causal effect requires no adjustment. It is given by the conditional probability $P(y|x)$ obtained by marginalizing out z from $P(x, y, z)$ and then conditioning on x. From Rule 2, it follows that we can exchange the action X with the observation of X, and thus, we get $P(y|do(x)) = P(y|x)$.

For the graph, ID2 has Z as a valid back-door adjustment. To derive $P(y|do(x))$ we can write from probability axioms:

$$P(y|do(x)) = \sum_{z} P(y|do(x), z)P(z|do(x))$$

It follows from rule 3 that $P(z|do(x)) = P(z)$. It follows from rule 2 that $P(y|do(x), z) = P(y|x, z)$. We can compute $P(z)$ by marginalization of x and y in $P(x, y, z)$ and compute $P(y|x, z)$ by conditioning on x and z in $P(x, y, z)$. Hence, the expression $P(y|do(x))$ can be computed from the joint distribution $P(x, y, z)$ and given by:

$$P(y|do(x)) = \sum_{z} P(y|x, z)P(z)$$

Graph ID3 is another example of back-door adjustment, just like ID2, where $\{Z\}$ is a valid adjustment set. The causal effect is identifiable and evaluated using the same expression as in graph ID2.

The graph ID4 is identifiable as it is a model of front-door adjustment. The expression $P(y|do(x))$ is given by *Equation 9.15;* in the following section, we will go over the derivation steps using the rules of the do-calculus to derive the formula for the front-door adjustment.

The lemma by Tian and Pearl (Tian, et al., 2003) asserts that if a causal query $Q = P(Y|do(X))$ is identifiable in a graph G, Q is still identifiable in subgraphs of G resulting from removing a directed or bidirected edge from G. This result has a practical application, enabling us to infer the identifiability of a causal effect in a graph G from knowing the causal effect is identifiable in a supergraph of G. This application is particularly relevant to real-world problems in causal inference.

Let us delve into the machinery for non-identifiability next.

Machinery for non-identifiability

Figure 9.14 depicts a sample of causal graph G where the causal query $Q = P(Y|do(X))$ is non-identifiable.

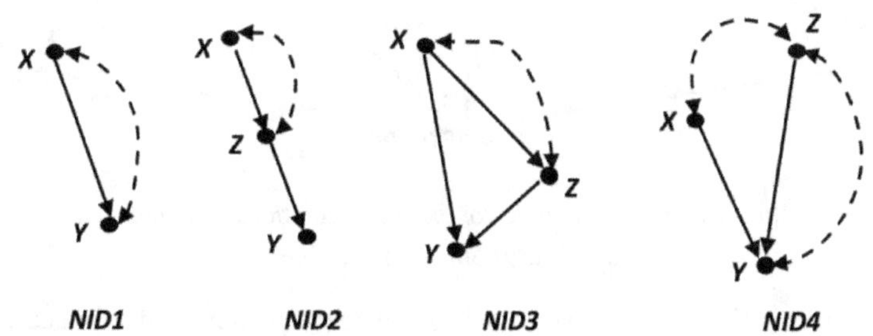

Figure 9.14— Samples of causal graphs where the effect of X on Y is non-identifiable

The lemma by Tian and Pearl (Tian, et al., 2003) states that if a causal query $Q = P(Y|do(X))$is non-identifiable in a graph G, it remains non-identifiable in all graphs that result from adding a directed or bi-directed edge to G. For instance, if we introduce the directed edge $Z \to X$ to the graph NID4 in *Figure 9.14,* the query Q continues to be non-identifiable in the resulting graph. This result has a practical application, enabling us to infer the non-identifiability of a causal effect in a graph G from knowing the causal effect is non-identifiable in a subgraph of G. This application is particularly relevant to real-world problems in causal inference.

Let us proceed to a practice section next.

Practice using R

In this section, we apply the identifiability machinery to a causal graph from an application in sports biomedical research. The Shrier model, depicted in *Figure 9.15,* illustrates the causal relationship between warming up before exercise and the injury outcome, expressed as the causal effect query.

$$Q = P(Injury \mid do(WarmUpExercises)).$$

Before conducting observation studies, epidemiologists seek to select proper measurements to achieve unbiased causal effect estimates. Estimation bias can result from conditioning on covariates containing confounder or selection variables, inducing a biasing path: a non-causal path from exposure to outcome.

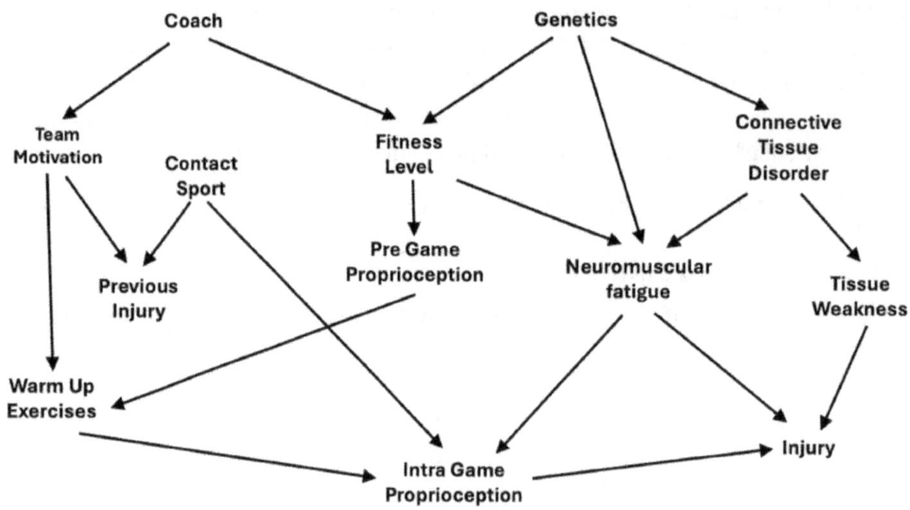

Figure 9.15— Causal graph of the Shrier model for the causal relationship between warm-up exercises and injury (Shrier, et al., 2008)

In this practice, we will use the *dagitty* R package to analyze the Shrier model and compute adjustment sets to identify both direct and total causal effects of $WarmUpExercises$ on $Injury$.

The benefit of listing alternative adjustments is that we can test the validity of the causal model by comparing the causal effect estimation resulting from the different adjustment sets.

The tasks for this practice are as follows:

1. Specify the DAG and display the graph.
2. Find the testable implications.
3. List the exogenous variables.
4. Find minimal adjustment sets for estimating the direct causal effect.
5. Find a canonical adjustment set to estimate the total causal effect.
6. Find all adjustment sets for estimating the total causal effect.
7. Find minimal adjustment sets for estimating the total causal effect.
8. Find conditional instrumental variables to estimate the total causal effect.

Let us delve into the code for practice. The entire code is made available on the book website contained in the *Rmarkdown* document `usecase.Rmd`.

First we load the required libraries.

```
library(ggm)
library(dagitty)
```

We begin with the first task.

Specify the DAG and display the graph

The first line of the code chunk below uses the ggm::DAG function to create the directed acyclic graph shrier_dag, representing the Shrier causal model. The second line of code employs the ggm::plotGraph function to visualize this graph, producing a DAG with the same layout as shown in *Figure 9.15*.

```
shrier_dag <- ggm::DAG(FitnessLevel ~ Coach,
  TeamMotivation ~ Coach,
  NeuromuscularFatigue ~ ConnectiveTissueDisorder,
  TissueWeakness ~ ConnectiveTissueDisorder,
  IntraGameProprioception ~ ContactSport,
  PreviousInjury ~ ContactSport,
  NeuromuscularFatigue ~ FitnessLevel,
  PreGameProprioception ~ FitnessLevel,
  ConnectiveTissueDisorder ~ Genetics,
  FitnessLevel ~ Genetics,
  NeuromuscularFatigue ~ Genetics,
  Injury ~ IntraGameProprioception,
  Injury ~ NeuromuscularFatigue,
  IntraGameProprioception ~ NeuromuscularFatigue,
  WarmUpExercises ~ PreGameProprioception,
  PreviousInjury ~ TeamMotivation,
  WarmUpExercises ~ TeamMotivation,
  Injury ~ TissueWeakness,
  IntraGameProprioception ~ WarmUpExercises)
ggm::plotGraph(shrier_dag)
```

Let us proceed to the second task.

Find the testable implications

The testable implications are the conditional independencies the probability model must satisfy, which are implied by the list of all the d-separations in the Shrier causal graph.

The following line of code uses the dagitty::getExample function to get the Shrier DAG from the dagitty library and assign the DAG to the variable **gs** used in the causal analysis next.

```
gs <- dagitty::getExample("Shrier")
```

The code below uses the `dagitty::impliedConditionalIndependencies` function to get all the d-separations in the causal graph **gs**. There are 197 conditional independencies, which is too many to list all of them. The output shows only the first ten. The list of implied conditional independencies uses an abbreviated version of the variables' long names.

```
x <- dagitty::impliedConditionalIndependencies( gs )
print(length(x))
## [1] 197

print( x )
## Coch _||_ CnTD
## Coch _||_ CntS
## Coch _||_ Gntc
## Coch _||_ Injr | InGP, NrmF, TssW
## Coch _||_ Injr | NrmF, TssW, WrUE
## Coch _||_ Injr | NrmF, PrGP, TmMt, TssW
## Coch _||_ Injr | FtnL, NrmF, TmMt, TssW
## Coch _||_ Injr | CnTD, InGP, NrmF
## Coch _||_ Injr | CnTD, NrmF, WrUE
## Coch _||_ Injr | CnTD, NrmF, PrGP, TmMt
```

Let us get to the third task.

List of exogenous variables

The following uses the `dagitty::exogenousVariables` to get the exogenous variables for the sports DAG **gs**. The output shows three variables, which are the three root nodes of the DAG **gs**.

```
dagitty::exogenousVariables(gs)
## [1] "Coach"         "ContactSport" "Genetics"
```

Let us proceed to the fourth task.

Find minimal adjustment sets for estimating the direct causal effect

The following line of code uses the `dagitty::adjustmentSets` with set parameters to list all minimal sufficient adjustment sets for identifying the average causal effect of "`WarmUpExercises`" on "`Injury`." Direct effects are the path coefficients in a structural equation (Gaussian linear model). The result that the code returns consists of seven sets. *Figure 9.16* depicts the first adjustment set on the causal graph, with the three elements shown on the graph nodes in red boxes. We note that one element, "`IntraGameProprioception`," is a mediator, appearing in all seven adjustment sets. The other two variables in the adjustment set depicted in *Figure 9.16* effectively block all back-door paths between the exposure "`WarmUpExercises`" and the outcome "`Injury`."

```
dagitty::adjustmentSets( gs, "WarmUpExercises", "Injury", effect="direct", type
= "minimal" )
```

```
## { IntraGameProprioception, NeuromuscularFatigue, TissueWeakness }
## { ConnectiveTissueDisorder, IntraGameProprioception,
##   NeuromuscularFatigue }
## { FitnessLevel, Genetics, IntraGameProprioception, NeuromuscularFatigue
##   }
## { Coach, FitnessLevel, IntraGameProprioception, NeuromuscularFatigue }
## { FitnessLevel, IntraGameProprioception, NeuromuscularFatigue,
##   TeamMotivation }
## { Coach, IntraGameProprioception, NeuromuscularFatigue,
##   PreGameProprioception }
## { IntraGameProprioception, NeuromuscularFatigue, PreGameProprioception,
##   TeamMotivation }
```

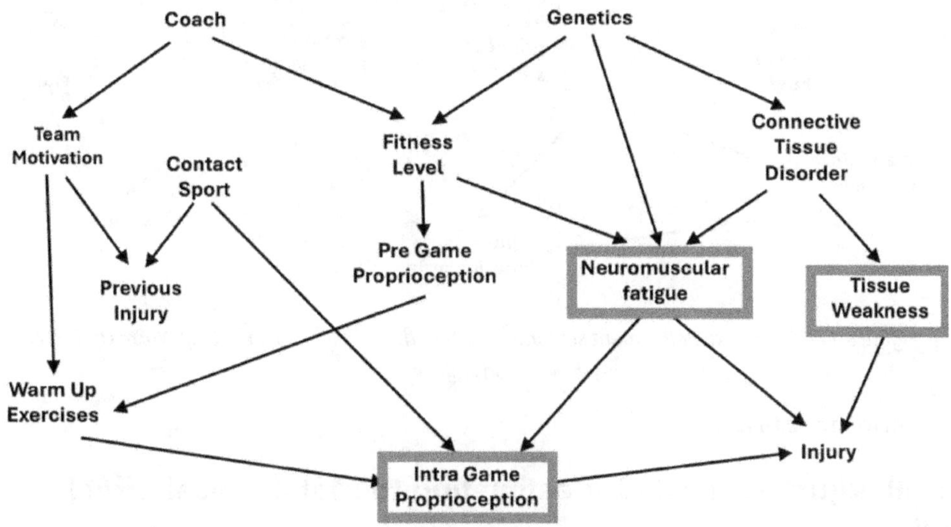

Figure 9.16—Minimal adjustment set for estimating the direct causal effect of "Warm Up Exercises" on "Injury"

Next we proceed to the fifth task.

Find a canonical adjustment set to estimate the total causal effect

The following code uses the `dagitty::adjustmentSets` to get a canonical adjustment set for estimating the total causal effect of "`WarmUpExercises`" on "`Injury`" for the DAG **gs**. A **canonical adjustment set** is a single set that consists of all (possible) ancestors of exposures and outcomes minus (possible) descendants of nodes on proper causal paths. This canonical adjustment set is always valid if any valid set exists. Notice the parameters `effect = "total"` and `type = "canonical"` in the call of `dagitty::adjustmentSets` in order to get a canonical adjustment set. The code returns a canonical set comprising nine elements.

```
dagitty::adjustmentSets(gs, "WarmUpExercises", "Injury", effect="total", type =
"canonical" )
```

```
## { Coach, ConnectiveTissueDisorder, ContactSport, FitnessLevel,
##   Genetics, NeuromuscularFatigue, PreGameProprioception,
##   TeamMotivation, TissueWeakness }
```

Figure 9.17 shows the elements of the canonical adjustment set marked by red boxes on the causal diagram.

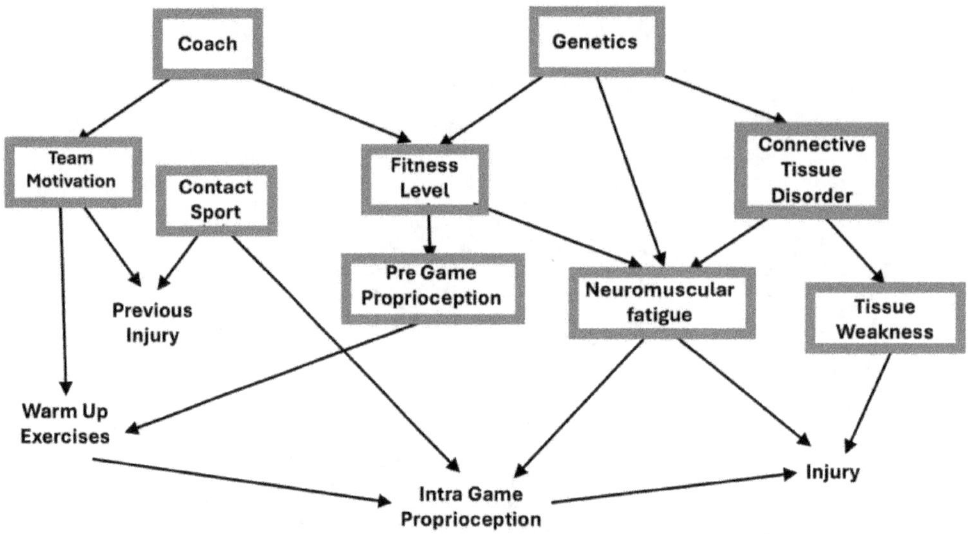

Figure 9.17—Canonical adjustment set for estimating the total causal effect of "Warm Up Exercises" on "Injury"

Moving on to the sixth task.

Find all adjustment sets for estimating the total causal effect

The following code uses the `dagitty::adjustmentSets` to get all adjustment sets to estimate the total causal effect of "`WarmUpExercises`" on "`Injury`" for the DAG **gs**. Notice the parameters `effect = "total"` and `type = "all"` in the call of `dagitty::adjustmentSets` in order to get all adjustment sets for the total causal effect. The code returns a list of 696 adjustment sets, of which we list the first ten.

```
list <- dagitty::adjustmentSets(gs, "WarmUpExercises", "Injury",
effect="total", type = "all" )
length(list)
## [1] 696
print(list)
## { Coach, FitnessLevel }
## { Coach, ConnectiveTissueDisorder, FitnessLevel }
## { Coach, ContactSport, FitnessLevel }
## { Coach, ConnectiveTissueDisorder, ContactSport, FitnessLevel }
## { FitnessLevel, Genetics }
```

```
## { Coach, FitnessLevel, Genetics }
## { ConnectiveTissueDisorder, FitnessLevel, Genetics }
## { Coach, ConnectiveTissueDisorder, FitnessLevel, Genetics }
## { ContactSport, FitnessLevel, Genetics }
## { Coach, ContactSport, FitnessLevel, Genetics }
```

Figure 9.18 shows the members of the first adjustment set for estimating the total causal effect of "WarmUpExercises" on "Injury," marked by red boxes on the causal diagram.

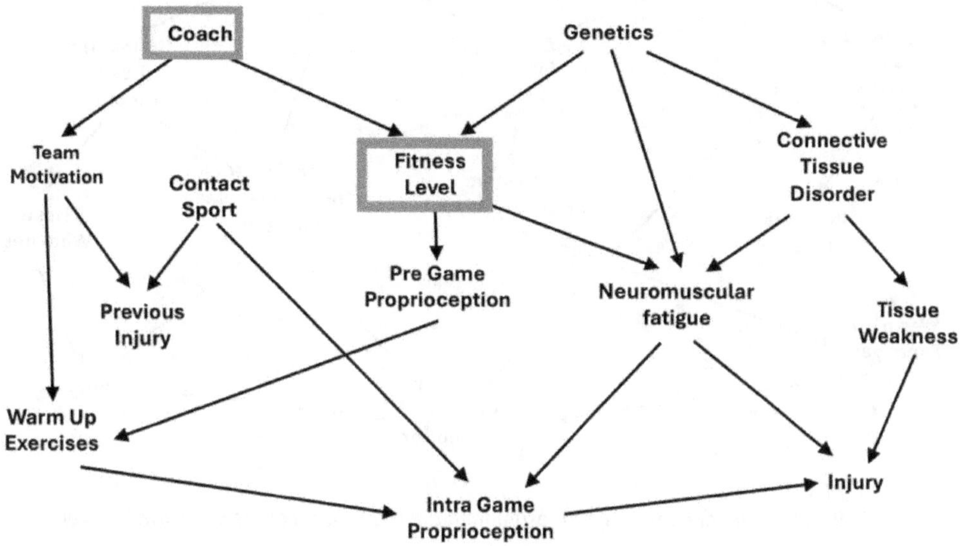

Figure 9.18—An adjustment set for estimating the total causal effect of "Warm Up Exercises" on "Injury"

Next we proceed to the seventh task.

Find minimal adjustment sets for estimating the total causal effect

The following code uses the `dagitty::adjustmentSets` to get the list of minimal adjustment sets to estimate the total causal effect of "`WarmUpExercises`" on "`Injury`" for the DAG **gs**. The parameters `effect = "total"` and `type = "minimal"` are used in the call to `dagitty::adjustmentSets` to obtain the list of minimal adjustment sets for estimating the total causal effect. The code returns seven adjustment sets, each providing an option of a minimal adjustment set for epidemiologists to consider when adjusting for confounding bias.

```
dagitty::adjustmentSets(gs, "WarmUpExercises", "Injury", effect="total", type =
"minimal" )
## { NeuromuscularFatigue, TissueWeakness }
## { ConnectiveTissueDisorder, NeuromuscularFatigue }
## { FitnessLevel, Genetics }
## { Coach, FitnessLevel }
## { FitnessLevel, TeamMotivation }
```

```
## { Coach, PreGameProprioception }
## { PreGameProprioception, TeamMotivation }
```

Figure 9.19 shows the fifth adjustment set from the returned list where the two nodes marked by red boxes on the causal diagram block all back-door paths, making the causal effect identifiable by the back-door adjustment.

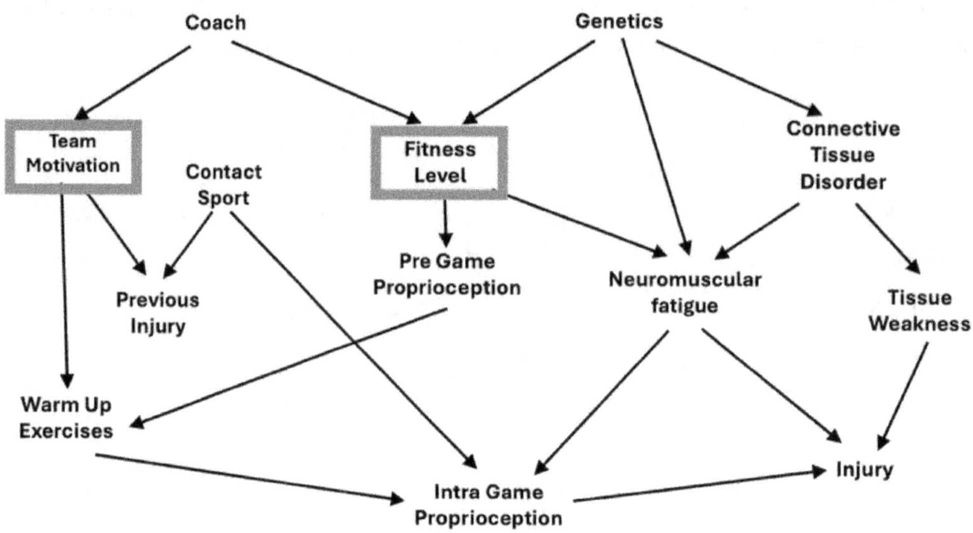

Figure 9.19— Minimal adjustment set for estimating the total causal effect of "Warm Up Exercises" on "Injury"

Finally, we get to the eighth and final task.

Find instrumental variables to estimate the total causal effect

The following code uses the `dagitty::instrumentalVariables` to generate a list of instrumental variables that can be used to infer the total effect of "`WarmUpExercises`" on "`Injury`." The function returns a list of instrumental variables Z or conditional instrumental variables Z | W, as in our use case per the code output below. Instrumental variables (IVs) are exogenous variables used to identify causal effects in the presence of latent confounding, under the specific linearity assumptions for the structural causal model.

```
dagitty::instrumentalVariables( gs,"WarmUpExercises", "Injury" )
##  Coach |  NeuromuscularFatigue, TissueWeakness
##  ConnectiveTissueDisorder |  NeuromuscularFatigue, TissueWeakness
##  FitnessLevel |  NeuromuscularFatigue, TissueWeakness
##  Genetics |  NeuromuscularFatigue, TissueWeakness
##  PreGameProprioception |  NeuromuscularFatigue, TissueWeakness
##  PreviousInjury |  ContactSport, NeuromuscularFatigue, TissueWeakness
##  TeamMotivation |  NeuromuscularFatigue, TissueWeakness
```

As we conclude this section on the rules of do-calculus, let us summarize the takeaway points. The do-calculus consists of three rules. Each rule is applicable when specific conditional independencies in a perturbed graph are met. The rules enable rewriting a probability distribution expression containing mixed interventional and observational conditioning. The first rule enables inserting (removing) an observation; the second enables exchanging an observation with an intervention; the third enables inserting (removing) an intervention. A perturbed graph is the DAG of the causal graph with directed edges incident to specific nodes, being removed depending on their arrows, whether pointed inward or outward relative to the nodes. Identifiability and non-identifiability of a causal effect can be established for specific causal graph structures with samples depicted in Figures 9.7 and 9.8. The practice section illustrates how to programmatically specify a causal graph, list testable implications, and determine adjustment sets for identifiable direct and total causal effects in a real-life use case.

The following section will explain how to apply the do-calculus rules to determine whether a causal model is identifiable. We will repeatedly apply the do-calculus rules to either succeed or fail to derive a symbolic expression for the interventional probability as a function only of the observed variable distribution.

Symbolic derivation of causal effects

This section is organized as follows: First, we describe the task of symbolic derivation of causal effects. Next, we illustrate a worked example, carefully guiding you through each step of the symbolic derivation task. Next, we review selected identification algorithms and automated tools. We top off with a practice section demonstrating how to use R to derive the symbolic expression of the causal effect and visualize the derivation digraph.

Describing the symbolic derivation task

In the previous section, we introduced the three rules of the do-calculus. While theoretical concepts, these rules are practical tools of utmost relevance when dealing with an identifiable causal effect. They allow you to express the intervention distribution as a function of the observed variables' probability distribution P(V), guiding you through causal inference in a real-world context.

We can define the task of symbolic derivation as follows:

- Input:
    - Directed acyclic graph $G = (N, E)$
    - Set of observed variables V⊆N having a joint probability distribution $P(v)$.
    - Causal query$Q = P(Y|do(x))$ where X and Y are two disjoint subsets of V.
- Output:
    - Identifiable: a symbolic expression for Q as a function only of the distribution on the observed variables P(v) is found.
    - Non-identifiable: No symbolic expression for Q free of the $do(\ )$operator can be found.

The derivation of a symbolic expression of a causal effect query that is a function of only the distribution of observed variables if the causal effect is identifiable, or the conclusion that the causal effect is non-identifiable, amounts to a search problem where in each step of the search, we have a probability expression and a set of methods for mapping the probability expression to another expression. The rules for mapping conditional probability expressions include the three rules of do-calculus and the rules of probability axioms. The search space can be represented as a directed graph (digraph) whose nodes are conditional probability expressions, and the edges are annotated by the rules applied to transform a node expression to a child node expression. The leaves of the digraph are conditional probability expressions. If a leaf node is reached which is free of the *do* operator, then the causal query is identifiable, and if none is found, then the query is non-identifiable.

In the next section, we will use a worked example to illustrate step by step how to carry out the symbolic derivation task for a causal effect query and visualize the process of the derivation on a digraph.

Worked example on derivation with do-calculus

Let us consider the causal model with an unobserved confounder shown in *Figure 9.20*. We will illustrate how to derive the result of the front-door adjustment that determines a symbolic representation of $P(Y|do(X))$ as a function of the distribution over the observed variables given by $P(X,Y,Z)$.

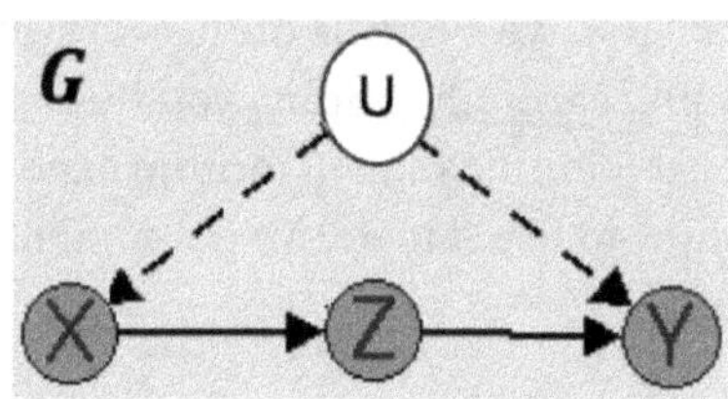

*Figure 9.20—Causal graph **G** with latent variable: X Exposure; M mediator; Y output; U unobserved*

Using the laws of probability, we can write:

$$P(Y|do(X)) = \Sigma_Z P(Y,Z|do(X)) = \Sigma_Z P(Y|Z,do(X))\, P(Z|do(X)) \quad (9.16)$$

In the following, we will repeatedly apply the do-calculus rules to rewrite the expression on the right-hand side of *Equation 9.16* such that it is free of the *do* operator. The goal is to prove the front-door adjustment formula of *Theorem 9.2, Equation 9.15*.

Step 1: applying rule 2 to rewrite P(Z|do(X))

We can apply Rule 2 to exchange the intervention conditioning on X with observation conditioning on X in the expression $P(Z|do(X))$. Rule 2 allows us to exchange action with observation when Z is conditionally independent of X in the perturbed graph $G_{\underline{X}}$, specifically when $(Z \perp X)_{G_{\underline{X}}}$. *Figure 9.21* illustrates the perturbed graph $G_{\underline{X}}$, where Z is d-separated from X. Therefore, we can express this relationship as:

$$P(Z|do(X)). = P(Z|X) \quad (9.17)$$

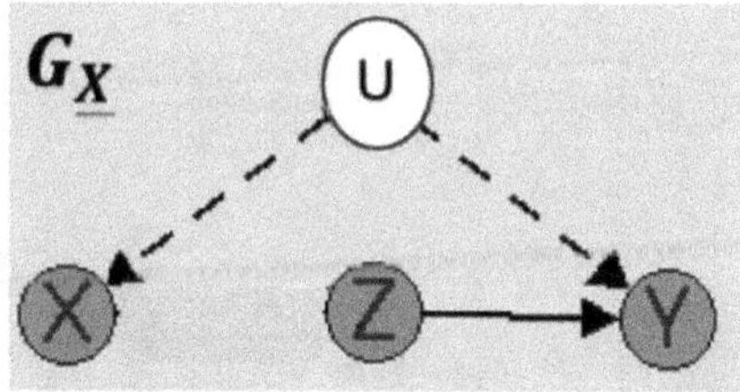

Figure 9.21—Perturbed graph $G_{\underline{X}}$*, derived from graph G in figure 9.20, is created by all incident edges pointing away from X*

Substituting $P(Z|X)$ for $P(Z|do(X))$ in *Equation 9.16* yields:

$$P(Y|do(X)) = \Sigma_Z P(Y|Z, do(X))\, P(Z|X) \quad (9.18)$$

Step 2: applying rule 2 to rewrite P(Y|Z, do(X))

We can apply *Rule 2* to exchange the observation conditioning on Z with intervention conditioning on Z in the expression $P(Y|Z, do(X))$. Rule 2 allows us to exchange action $do(Z)$ with observation Z when Y is conditionally independent of Z given X in the perturbed graph $G_{\overline{X}\underline{Z}}$, specifically when$(Y \perp Z|X)_{G_{\overline{X}\underline{Z}}}$. *Figure 9.22* depicts the perturbed graph $G_{\overline{X}\underline{Z}}$, where Y is d-separated from Z given X. Therefore, we can express this relationship as:

$$P(Y|do(Z,X)) = P(Y|Z, do(X)) \quad (9.19)$$

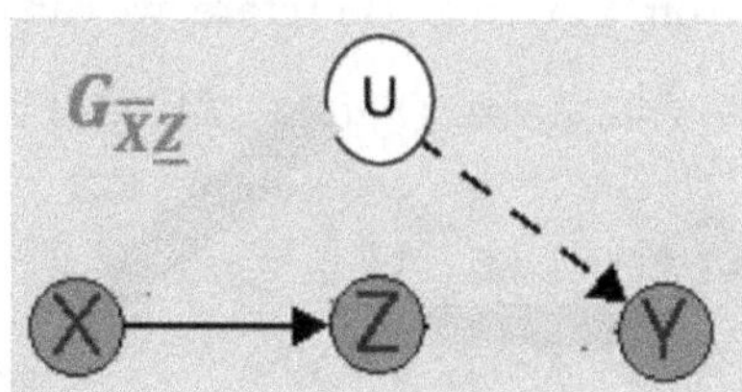

Figure 9.22—Perturbed graph $G_{\overline{X}\underline{Z}}$*, derived from graph G in Figure 9.20, is created by removing all incident edges directed toward X and all incident edges pointing away from Z*

Substituting $P(Y|do(Z,X))$ for $P(Y|Z, do(X))$ in *Equation 9.18* yields

$$P(Y|do(X)) = \Sigma_Z P(Y|do(Z,X))\, P(Z|X) \quad (9.20)$$

Step 3: applying rule 3 to rewrite P(Y|do(Z,X))

We can apply *Rule 3* to drop the intervention conditioning on X from $P(Y|do(Z,X))$. Rule 3 allows us to drop the intervention conditioning on X from $P(Y|do(Z,X))$ when Y and X are conditionally independent given Z in the perturbed graph $G_{\overline{X}\,\overline{Z}}$, specifically when $(Y \perp X|Z)_{G_{\overline{X}\,\overline{Z}}}$ *Figure 9.23* depicts the perturbed graph $G_{\overline{X}\,\overline{Z}}$, where Y is d-separated from X given Z. Therefore, we can express this relationship as:

$$P(Y|do(Z)) = P(Y|do(Z,X)) \quad (9.21)$$

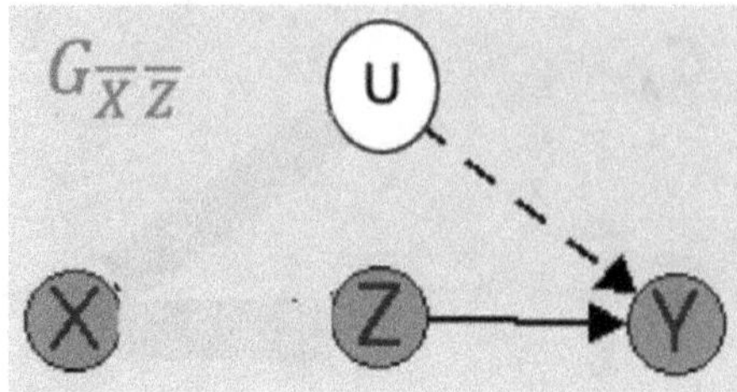

Figure 9.23—Perturbed graph $G_{\overline{X}\,\overline{Z}}$, derived from graph G in figure 9.20, is created by removing all incident edges directed toward X and all incident edges directed toward Z

Substituting $P(Y|do(Z))$ for $P(Y|do(Z,X))$ in *Equation 9.20* yields:

$$P(Y|do(X)) = \Sigma_Z P(Y|do(Z))\, P(Z|X) \quad (9.22)$$

Step 4: applying rule 2 to rewrite P(Y|do(Z))

Using probability axioms we can write:

$$P(Y|do(Z)) = \Sigma_X P(Y,X|do(Z)) = \Sigma_X P(Y|do(Z),X)P(X|do(Z)) \quad (9.23)$$

We can apply *Rule 2* to exchange intervention on Z with observation of Z in the expression $P(Y|do(Z),X)$ in *Equation 9.23*. *Rule 2* allows us to exchange action with observation when Y and Z are d-separated given X in the perturbed graph $G_{\underline{Z}}$, specifically when $(Y \perp Z|X)_{G_{\underline{Z}}}$. *Figure 9.24* depicts the perturbed graph $G_{\underline{Z}}$, where Z is d-separated from Y given X. Therefore, we can express this relationship as:

$$P(Y|Z,X) = P(Y|do(Z),X) \quad (9.24)$$

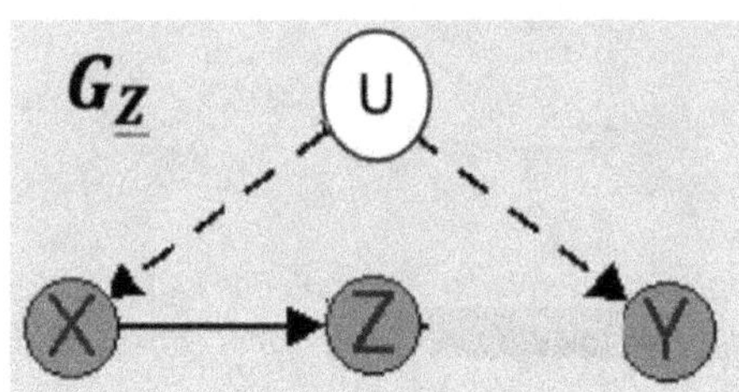

Figure 9.24—Perturbed graph $G_{\underline{Z}}$, derived from graph G in figure 9.20, is created by removing all incident edges pointing away from Z

Substituting $P(Y|Z,X)$ for $P(Y|do(Z),X)$ in *Equation 9.23* yields:

$$P(Y|do(Z)) = \Sigma_X P(Y|Z,X)P(X|do(Z)) \quad (9.25)$$

Step 5: applying rule 3 to rewrite P(X|do(Z))

We can apply *Rule 3* to eliminate the intervention conditioning on Z in the expression $P(X|do(Z))$. *Rule 3* states that when X and Z are d-separated in the perturbed graph $G_{\overline{Z}}$, specifically when $(X \perp Z)_{G_{\overline{Z}}}$. *Figure 9.25* depicts the perturbed graph $G_{\overline{Z}}$, where X and Z are d-separated. Therefore, we can express this relationship as:

$$P(X) = P(X|do(Z)) \quad (9.26)$$

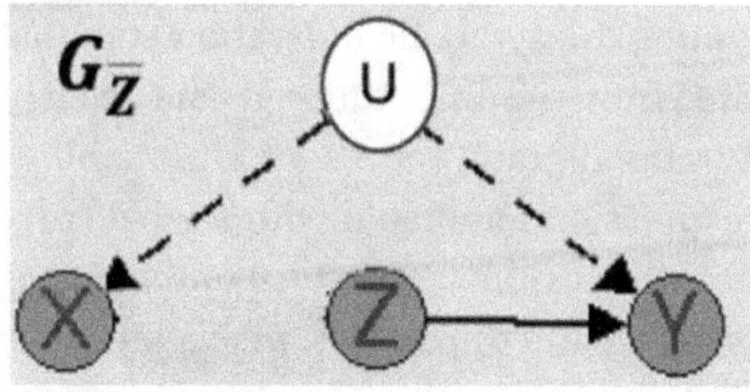

Figure 9.25—Perturbed graph $G_{\overline{Z}}$, derived from graph G in Figure 9.20, is created by removing all incident edges directed toward Z

Substituting $P(X)$ for $P(X|do(Z))$ in *Equation 9.25*, we get:

$$P(Y|do(Z)) = \Sigma_X P(Y|Z,X)\, P(X) \quad (9.27)$$

Substituting $P(Y|do(Z))$ from *Equation 9.27* into *Equation 9.22*, we get the symbolic formula for calculating $P(Y|do(X))$ from the distribution of observed variables, which proves the causal effect of X on Y is identifiable and given by:

$$P(Y|do(X)) = \Sigma_Z(P(Z|X)\, \Sigma_X(P(X)P(Y|Z,X))) \quad (9.28)$$

Equation 9.28 gives the same formula as the one stated in *Theorem 9.2* on front-door adjustment.

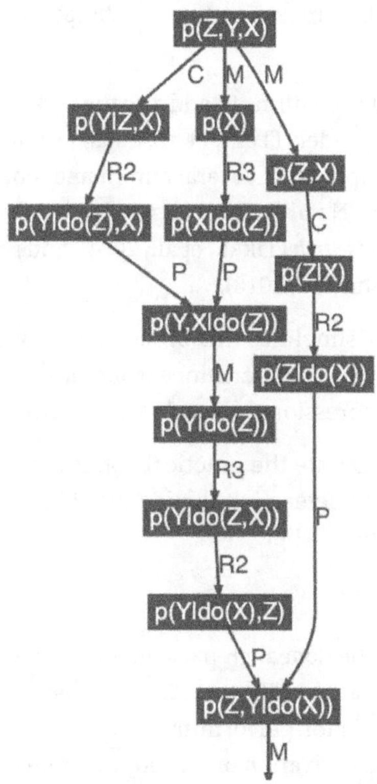

Figure 9.26—Directed graph representing the symbolic derivation steps for proving identifiability of the causal effect $\mathbf{P(Y|do(X))}$ for the front door model

Figure 9.26 is a digraph representing the symbolic derivation step that shows how the causal effect query $P(Y|do(X))$ can be evaluated from the observational data given by the distribution on observed variables $P(X, Y, Z)$. The nodes in the digraph represent a probability expression and the directed edges explain a rewrite of the expression annotated with a label corresponding to the applied rule of the do-calculus or of the probability axioms. The rules of the do-calculus are labeled R1, R2, and R3, corresponding to the first, second, and third rules, respectively. The rules of probability axioms are C, M, and P. The C label indicates conditioning. The label M means marginalization. The label P indicates the product of the expressions of the parent nodes.

In the next section, we overview algorithms and automated tools for determining the identifiability of a causal effect query and computing the formula for an identifiable causal effect.

Identifiability algorithms and automated tools

In the previous section, we saw an example of applying the do-calculus rules to determine whether a causal effect is identifiable and, as a by-product, compute an expression for the interventional distribution that uses only the pre-intervention distribution on observed variables. The take-away point of the exercise is that the process involves a sequence of steps with decisions made at every step on what rewrite rule to use from the probability axioms and the rules of the do-calculus. Applying a do-calculus rule at a step requires conditional independency conditions to be verified on a perturbed graph associated with the rule. The process can be cumbersome to perform manually for a causal model with many variables.

Many researchers have developed algorithms to address the task of determining the identifiability of causal effects in structural causal models (Tian, et al., 2003) (Shpitser, et al., 2006) (Tikka, 2022). The algorithms have spurred the development of several automated tools implemented as software library packages written in R and Python. Notable R packages for determining causal effect identifiability include *causaleffect* (Tikka, 2022), *dosearch* (Tikka, et al., 2024), and *pcalg* (Kalisch , et al., 2024). Notable Python packages include `dowhy` (Sharma, 2018),

Significantly, the `dosearch` library simplifies this process. It provides a function that takes a causal effect query and, using heuristic search, determines whether the causal effect is identifiable. If it is, the function returns a symbolic expression that evaluates in terms of observed variables.

In the next section, we will demonstrate the practical application of the `dosearch` package. We will use it on the front-door example to determine the identifiability of the causal effect and verify the symbolic derivation steps illustrated earlier.

Practice using R

In this practice, we utilize R and the `dosearch` package to conduct causal inference tasks based on the front-door causal model illustrated in *Figure 9.20*. The objective of this exercise is to gain practical experience with automated tools for performing causal inference tasks. Additionally, we aim to reinforce our understanding of the derivation of symbolic formulas for causal effects by applying the rules of do-calculus that we previously explored step-by-step in the same example.

The tasks for this practice are as follows:

1. Check identifiability of a causal effect query.
2. Get the symbolic formula for the identifiable causal effect.
3. Get the derivation digraph of the search for the symbolic formula.
4. Plot the derivation digraph.

Let us delve into the code for practice. The entire code is made available on the book website contained in the *Rmarkdown* document `symbolic.Rmd`.

First we load the required libraries:

```
library(dosearch)
library(DiagrammeR)
```

Check identifiability

To check the identifiability of a causal query, we use the `dosearch::is_identifiable` function, which takes input from the result returned by the primary `dosearch::dosearch` function. The `dosearch:dosearch` function identifies a causal query from available data in a semi-Markovian causal model described by a directed acyclic graph (DAG). In a semi-Markovian causal model, each unobserved variable has exactly two children, and bidirected edges denote them. The `dosearch::dosearch` function has three main arguments: 1) data: a character string describing the available distributions, 2) query: a character string describing the target distribution, 3) graph: a character string describing a DAG using the `dagitty` package. See the documentation for more details on the `dosearch::dosearch` function (Tikka, et al., 2024).

In the following code chunk, we find the identifiability of the causal query `"P(Y|do(X))"`. The data parameter is the distribution of the observed variables `"P(X, Y, Z)"`. We specify the DAG for the semi-Markovian front-door model as the object `g_dagitty` using the `dagitty` syntax. Running the code chunk returns TRUE, meaning the causal query is identifiable.

```
## Graph input with 'dagitty'
if (requireNamespace("dagitty", quietly = TRUE)) {
  data <- "P(X,Y,Z)"
  query <- "P(Y|do(X))"
  g_dagitty <- dagitty::dagitty("dag{
  X <- U -> Y
  X -> Z -> Y
  X [exposure]
  Y [outcome]
  U [unobserved]
}")
res <-  dosearch::dosearch(data, query, g_dagitty)
```

```
}
dosearch::is_identifiable(res)
## [1] TRUE
```

Let us proceed to the second task.

Get the symbolic formula for the identifiable causal effect

To get the symbolic formula, we use the dosearch::get_formula function, which returns the identifying formula that describes a causal query of an object of class dosearch. If no formula is available, it returns NULL. In the code below, the dosearch::get_formula function takes input from the result, **res**, returned by the primary dosearch::dosearch function. We concatenate the output from the dosearch::get_formula function with the "$" at the beginning and the end of the formula to render it using LaTeX; the outputs from running the code chunk and the LaTeX output are as follows:

```
# convert to latex equation
cat(paste("$", dosearch::get_formula(res), "$"))
## $ \sum_{Z}\left(p(Z|X)\sum_{X}\left(p(X)p(Y|X,Z)\right)\right)
```

$$\sum_{Z}\left(p(Z|X)\sum_{X}\left(p(X)p(Y|X,Z)\right)\right)$$

The formula we obtain from running the code is the same as the formula obtained earlier, *Equation 9.28*, by applying the rules of do-calculus that we previously explored step by step for this front-door causal model.

Let us get to the third task, next.

Derivation digraph of the search for the symbolic formula

Throughout the process, the dosearch::get_derivation function is continuously involved. It is the key to obtaining the derivation digraph, returning the derivation of a causal query of an object of class dosearch. If no derivation is available, it returns NULL. In the following code chunk, we use the dosearch::get_derivation function with the three arguments: the result, **res**, returned by the primary dosearch::dosearch function, the run_again option set to TRUE, and the draw_all set to TRUE. Setting the run_again option to TRUE is a flag for repeating the search. Setting the draw_all option to TRUE is a flag for the derivation that contains every step the search takes. The function returns the derivation trace as a string specifying a digraph named "InferenceTree". The syntax of the digraph follows the graphviz language syntax. Due to the lack of space, we do not include the text output here.

```
digraph <- dosearch::get_derivation(res, run_again = TRUE, draw_all = TRUE)
print(digraph)
```

Let us proceed to the fourth and final task.

Plot the derivation digraph

The derivation digraph obtained in the previous task is not just a depiction but a visual representation of the search for the symbolic formula that describes the identifiable causal query of the front-door causal model. It provides a clear and concise view of the process. To plot the derivation digraph, we use the `DiagrammeR::grViz` function to render the graph. The following code chunk plots the derivation graph, as shown in *Figure 9.27*. The digraph shows the derivation path from the root node, which is the probability distribution on observed data "P(X, Y, Z)" and the leaf node labeled by the causal query "P(Y|do(X))" proving the causal query is identifiable.

```
# Render the graph
grViz(digraph)
```

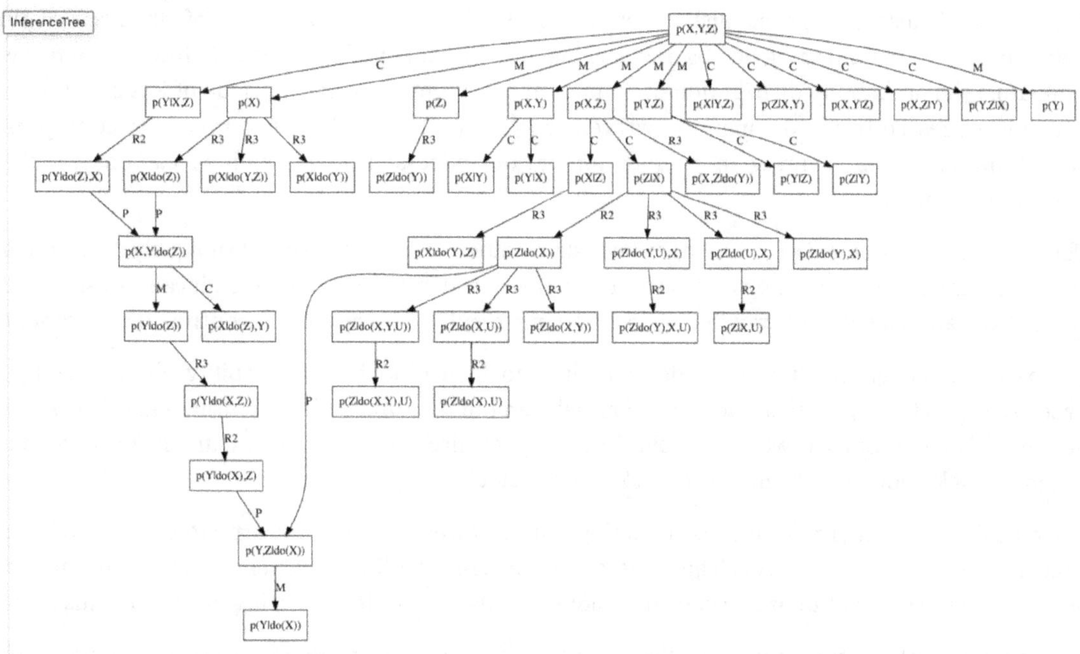

Figure 9.27– The derivation digraph depicts the search for the symbolic formula that describes the identifiable causal query of the front-door causal model

Figure 9.27 uses a different search strategy than *Figure 9.26* to derive the same symbolic expression for the causal effect in the front-door model. Unlike *Figure 9.26*, it contains multiple leaf nodes besides the target $P(Y \mid do(X))$; these leaves represent dead ends in the search and therefore mark points where the derivation must backtrack.

Let's now review the key points we've learned in the section on the symbolic derivation of causal effects. A causal query Q is an interventional conditional probability expression $P(Y|do(x))$ of an output variable Y, conditioned on an intervention $do(x)$ on treatment variables X. A causal query is identifiable if we can find an equivalent expression that is a function only of the distribution on observed variables $P(V)$. The process of finding this symbolic formula is a search problem in mixed interventional and observational conditional probability expressions. We can map these conditional probability expressions from one to another equivalent expression by applying the rules from the do-calculus and the probability axioms. This search can be represented as a digraph, with its nodes being probability expressions and the directed edges representing the rule used for mapping from one expression to another equivalent expression. We've delved into an example where we derived the front-door adjustment formula using a sequence of mapping steps, employing the rules of the do-calculus and the probability axioms. This example was not just theoretical—we've practiced hands-on, using R code and available R libraries, to depict the digraph and derive the symbolic formula.

Let us wrap up the chapter with an overall summary next.

Summary

Chapters 7 and 8 discussed probabilistic inference in Bayesian networks, where beliefs about a variable Y are updated by conditioning on observed evidence X. As explained in *Chapter 4*, the causal effect is evaluating the probability distribution of Y under an intervention on X; the value of X is set or fixed, which is expressed by the do operator, written as $P(Y|do(X))$. The *do* operator is a key concept in causal inference, representing an intervention or controlled setting in which we can observe the effect of a variable.

The identifiability problem in causal inference asks whether a causal effect can be uniquely determined from the joint distribution of observed variables. If a causal effect is identifiable, it can be estimated from observational data in a way that yields the same result as a randomized controlled experiment.

The back-door criterion, offers a practical condition to determine the identifiability of a causal effect based solely on the topology of the DAG, a crucial element in representing a causal model. The front-door model comes into play when variables Z intercept all directed paths from treatment X to outcome Y, and all back-door paths from X to Z and Z to Y are blocked.

The do-calculus, a set of rules for manipulating causal expressions, is a powerful tool in causal inference. It consists of three rules, each applicable when specific conditional dependencies in a perturbed graph are met. The front-door and back-door adjustments, can be derived using the do-calculus rules.

A causal query Q is a mixed interventional and observational conditional probability expression $P(Y|do(x), W)$ of an output variable Y, conditioned on an intervention on treatment variables X and possible observations W. A causal query is identifiable if we can find an equivalent expression that is a function only of the distribution on observed variables $P(V)$.

Using R packages, we can programmatically specify a causal graph, list testable implications, determine adjustment sets for identifiable direct and total causal effects, compute whether a causal query is identifiable, and compute a symbolic expression for an identifiable causal query.

Chapter 10 introduces statistical approaches for causal inference with observational data, focusing on the potential outcomes framework based on the Neyman-Rubin causal model and using counterfactual outcomes to formulate causal inference.

References

- Cheng Debo [et al.] *Conditional Instrumental Variable Regression with Representation Learning for Causal Inference* [Online]. 2024. `https://openreview.net/pdf?id=qDhq1icp08`.
- Shrier I. and Platt R. *Reducing bias through directed acyclic graphs* [Article]. BMC Medical Research Methods. 2008.
- Cinelli Carlos, Forney Andrew and Pearl Judea. *A Crash Course in Good and Bad Controls* [Journal. Journal Sociological Methods and Research. 2022.
- Galles David and Pearl Judea. *Testing Identifiability of Causal Effects* [Conference]. Proceedings of the Eleventh Conference on Uncertainty in Artificial Intelligence (UAI1995). 1995.
- Huang Y, and Valtorta M Pearl. *Calculus of Intervention is Complete* [Conference]. Proceedings of the Twenty-Second Conference on Uncertainty in Artificial Intelligence. 2006.
- Kalisch Markus, Hauser Alain and Maechler Martin. Package 'pcalg' [Online]. 12 February 2024. `https://cran.r-project.org/web/packages/pcalg/pcalg.pdf`.
- Kyono T.M. Commentator: *A Front-End User-Interface Module for Graphical and Structural Equation Modeling* [Report]. Technical Report R-364. Los Angeles: Cognitive Systems Laboratory Department of Computer Science, UCLA, 2010.
- Lemberger Pirmin and Oblin Denis Reconciling Causality and Statistics [Online]. 25 September 2020. `https://arxiv.org/pdf/2007.03940`.
- Marc Bellemare, Bloem Jeffrey and Wexler Noah. *The Paper of How: Estimating Treatment Effects Using the Front-Door Criterion* [Online]. 12 September 2020. `http://marcfbellemare.com/wordpress/wp-content/uploads/2020/09/BellemareBloemWexlerFDCSeptember2020.pdf`.
- Pearl J. *Causal Diagrams for Empirical Research* [Journal]. Biometrika. 1995. Vol. 82. pp. 669–688.
- Pearl J. *Causality: Models, Reasoning and Inference* [Book]. New York, NY, USA : Cambridge University Press, 2009.
- Sharma Amit. *DoWhy : A library for causal inference* [Online]. Microsoft Research Blog. 21 August 2018. 24 July 2024. `https://www.microsoft.com/en-us/research/blog/dowhy-a-library-for-causal-inference/`.
- Shpitser I and Pearl J. *Identification of Joint Interventional Distributions in Recursive Semi-Markovian Causal Models* [Conference]. Proceedings of the 21st National Conference on Artificial Intelligence. 2006.
- *Smith Louisa Selection Mechanisms and Their Consequences: Understanding and Addressing Selection Bias* [Article]. Current Epidemiology Reports. 2020. pp. 179–189.
- Tian Jin and Pearl Judea. *A General Identification Condition for Causal Effects* [Conference]. AAAI. 2002.
- Tian Jin and Pearl Judea. *On the Identification of Causal Effects* [Online]. 2003. `https://ftp.cs.ucla.edu/pub/stat_ser/R290-L.pdf`.
- Tian Jin and Pearl Judea. *On the Identification of Causal Effects* [Online]. 2003. `https://ftp.cs.ucla.edu/pub/stat_ser/R290-L.pdf`.
- Tikka S., Hyttinen A. and Karvanen J. *Causal Effect Identification from Multiple Incomplete Data Sources: A General Search-Based Approach.* [Journal]. Journal of Statistical Software. 2021. pp. 1–40.

- Tikka Santtu. Package 'causaleffect' [Online]. 12 October 2022. `https://cran.r-project.org/web/packages/causaleffect/causaleffect.pdf`.
- Tikka Santtu, Hyttinen Antti and Karvanen Juha. Package 'dosearch' [Online]. 9 July 2024. `https://cran.r-project.org/web/packages/dosearch/dosearch.pdf`.

Get this book's PDF version and more

10

Causal Inference with Observational Data

This chapter presents the potential outcomes framework (Rubin or Neyman-Rubin causal model), which uses mathematical notation to describe counterfactual outcomes. The approach can be used to describe the causal effect of an exposure or treatment on an outcome in statistical terms. The causal effect is the difference in the potential outcomes for a particular population given different counterfactual scenarios. Causal effects are not directly observable since they involve comparing unobserved counterfactual outcomes that would have happened under different circumstances. We review the assumptions under which a causal effect is identifiable, i.e., it can be estimated using observable data. We review the methods for estimating causal effects (causal estimands). We give demonstrations with examples using real-life datasets and practice using the `CausalModels` package in R.

In this chapter, we are going to cover the following main topics:

- The Potential Outcomes framework
- Causal diagrams (directed acyclic graphs)
- Defining causal effects
- Identifying causal effects
- Estimating causal effects

Technical requirements

All the code for this chapter is in R, and you can download it from the GitHub URL: `https://github.com/PacktPublishing/Causal-Inference-with-Bayesian-Networks/tree/main/Chapter_10`

You will be required to install various R libraries to run the R code for the practice and the examples in this chapter, and you can do that from within the **RStudio** IDE. The required packages to install are:

- *haven* `https://cran.r-project.org/web/packages/haven/index.html`
- *MatchIt* `https://cran.r-project.org/web/packages/MatchIt/index.html`

- *dplyr* `https://cran.r-project.org/web/packages/dplyr/readme/README.html`
- *boot* `https://cran.r-project.org/web/packages/boot/index.html`
- *ggplot2* `https://cran.r-project.org/web/packages/ggplot2/index.html`
- *tidyverse* `https://cran.r-project.org/web/packages/tidyverse/index.html`
- *CausalModels* `https://cran.r-project.org/web/packages/CausalModels/index.html`
- *coefplot* `https://cran.r-project.org/web/packages/coefplot/index.html`
- *broom* `https://cran.r-project.org/web/packages/broom/index.html`
- *survey* `https://cran.r-project.org/web/packages/survey/index.html`

The potential outcomes framework

We begin this section by describing the potential outcomes framework (POF) . Next, we explain the missing data problem related to experimental trials and observational studies. Following this, we describe a counterfactual experiment to illustrate the potential outcomes model and the distinction between causation and association. After this, we explain the role of confounding in causal inference. Next, we describe observational studies and randomized trials and the role of the propensity score. Finally, we outline the three phases of the causal inference workflow.

What is POF?

The **potential outcomes framework** (POF), the **Neyman-Rubin** model, is based on statistical methods (Neyman 1923) (Rubin 1974). The POF formulates causal inference as a statistical analysis problem with missing data or counterfactual outcomes.

The POF requires the definition of three variables: treatment, outcome, and covariates. The definition of the treatment and outcome variables is inherent to the causal question that we want to investigate with the observational study. The treatment, also called exposure, represents an exogenous variable that, when exercised on or by a statistical unit in a population relevant to the causal question, results in an effect manifested by the outcome variable, an endogenous variable of the unit. The exercised treatment must precede the observed outcome. The covariates are other endogenous variables, called confounders, that are supposed to influence the treatment and the outcome. The confounders and the treatment must be the only variables affecting the outcome.

The missing data problem

The missing data stems from the fact that we can only observe one potential outcome but cannot observe the counterfactual outcomes, which is what Holland called the fundamental problem of causal inference (Holland 1986). Suppose we have two random variables, T and Y, where T is a dummy variable indicating **exposure** or **treatment** status in a population. T takes value one if a population's unit U_i receives treatment and 0 if not. Define Y_i (1) as the potential outcome if we give treatment for unit U_i and $Y_i(0)$ if we do not. See *Figure 10.1*.

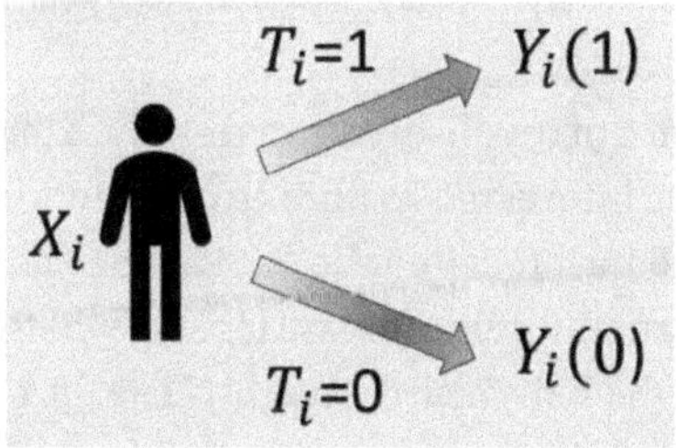

Figure 10.1— Potential outcomes of binary treatment

Each unit U_i has a set of **covariates**, which are features that can influence treatment decisions and outcomes. Let X be a k-dimensional vector of covariates with $k \geq 2$ representing a unit's features. X represents a set of characteristics such as sex, age, income, etc., referred to as covariates, pretreatment variables, or exogenous variables.

We can observe only one of the potential outcomes for any unit U_i, which is the realized outcome corresponding to the treatment assignment T_i.

$$Y^{obs} = T.Y(1) + (1 - T) \cdot Y(0) \quad (10.1)$$

We do not observe the complement PO, constituting the missing outcome.

$$Y^{mis} = (1 - T).Y(1) + T \cdot Y(0) \quad (10.2)$$

Figure 10.2 shows that for a binary treatment T_i on a unit U_i, there are two potential outcomes, $Y_i(0)$ and $Y_i(1)$, corresponding to the value assignments $T_i = 0$ and $T_i = 1$.

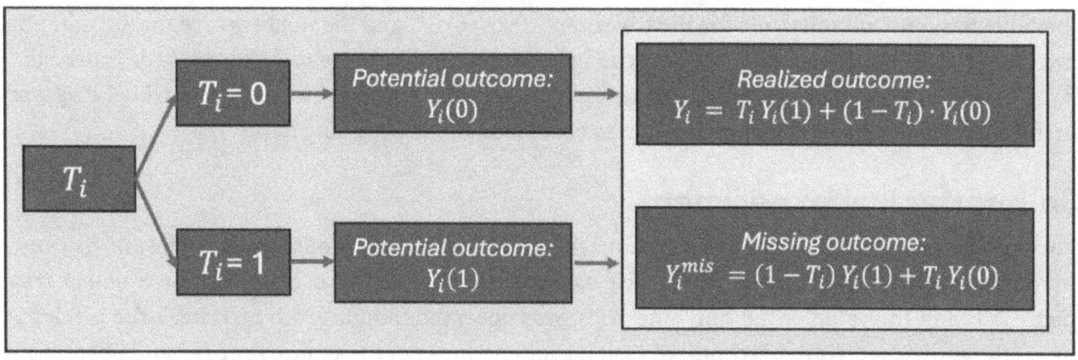

Figure 10.2—Potential outcomes and realized and missing outcomes for binary treatment $\boldsymbol{T_i}$ on unit $\boldsymbol{U_i}$

The observational data include only the realized outcome Y_i, a function of T_i, and map to one of the potential outcomes. The missing data Y_i^{mis} is the other potential outcome not observed for the unit U_i. We can only observe one of the potential outcomes, not both, for any given unit, depending on whether it receives treatment, $T_i = 1$, or it receives control, $T_i = 0$. Hence, we have a missing data problem. The data are the potential outcomes for the imaginary situation in which we reverse the actual treatment or control decision for all units. Because the missing data are for a condition that did not happen, we call the case the **counterfactual outcome**.

Use case example

To illustrate, let's consider a use case of an observational study, a use case that we will use throughout this chapter. In the use case, we are interested in finding out how smoking during pregnancy affects a mother's baby's birth weight. The dataset, called the `Cattabeo2` dataset, is found on the `Stata` dataset website (Stata 2014). In that observational study, the variable of whether the mother smoked, `'mbsmoke'`, corresponds to the treatment, and the baby's birth weight, `'bweight'`, is the outcome. *Table 10.1* shows a few rows of the dataset for this use case.

mbsmoke	bweight(1)	bweight (0)	bweight
1	3345	?	3345
0	?	3300	3300
0	?	4678	4678
0	?	3487	3487
0	?	2211	2211
1	2438	?	2438
1	2579	?	2579

Table 10.1: A partial sample of observational data from the Cattaneo2 dataset; the treatment is mbmoke (binary one if the mother smokes; 0 if not); bweight is the observed baby birthweight in grams; bweight (1) and bweight(0) are the potential outcomes for each treatment value- question marks are the missing data the counterfactuals

The table has four columns: one for the treatment, `'mbsmoke'`, and the next two are for the potential outcomes, `'bweight(1)'`, `'bweight(0)'`, for the baby's birth weight if the mother smoked, `'mbsmoke = 1'`, and if the mother did not smoke, `'mbsmoke'` = 0. The fourth column is the observed baby's grams birthweight (`'bweight'`). The question marks correspond to the missing data.

Counterfactual experiment

The hypothetical counterfactual experiment can explain the potential outcomes model and its causal inference approach. In that experiment, we assigned every member of a population a binary treatment, T=1, and the value of the outcome, Y(1), was observed. Suppose we turn back the clock and repeat the experiment, but this time, we assign every member of the population T=0 and observe the outcome Y(0). Because we can never turn back the clock, an unobservable counterfactual is one of the two experimental outcomes for every individual explaining the source of the missing data problem.

Association is not causation

We all know the idiom "association is not causation." This section will use the counterfactual experiment to emphasize the distinction between association and causation.

We say that an exposure or treatment T is a **causal determinant** of the outcome Y if the mean value of the outcome $Y(1)$ in a counterfactual experiment differs from the mean value of $Y(0)$. That is:

$$E[Y(1)] \neq E[Y(0)] \quad (10.3)$$

In our counterfactual experiment, all individual characteristics and experimental conditions that influence the outcome are held constant, except exposure/treatment, to observe each individual's outcome under both conditions. As a result, the population means $E[Y|T = 1]$conditioned on exposure $(T = 1)$ and exposure $(T = 0)$ show an average over the same distribution of unique characteristics. Thus, for the counterfactual experiment.

$$E[Y(1) - Y(0)] = E[Y(1)] - E[Y(0)] = E[Y|T = 1] - E[Y|T = 0] \quad (10.4)$$

We note that the quantity on the left-hand side of *Equation 10.4* computes the mean value of the outcome Y over the entire population being exposed minus the mean value of the outcome Y over the entire population being unexposed, quantifying the **causation** effect of T on Y. The quantity on the right-hand side of *Equation 10.4* computes the average outcome Y conditioning on the exposed population minus the average outcome Y conditioning on the unexposed population, which quantifies the **association** effect of T on Y. In the context of the counterfactual experiment, the two quantities on both sides of *Equation 10.4* are equivalent because the exposed and unexposed population groups conditioned upon are identical to the entire population, which means that all other causal determinants of outcome values are perfectly balanced in the exposed and unexposed populations.

In the next section, we explain the role of confounding in the difference between association and causation when making inferences about the effect of treatment. In association, we compute the effect based on observing T, and in causation, based on intervention for setting the value of T.

The role of confounding in causal inference

Of course, causal effects cannot be estimated in counterfactual experiments. The outcome is generally observable for each individual under only one of the two conditions. In place of a counterfactual experiment, we usually have to compare mean values of the outcome in two distinct populations, one composed of exposed individuals represented by a vector of covariates X having a probability density function $F_X(X = x\ |T = 1)$ and the other of unexposed represented by covariates X having probability distribution $F_X(X = x\ |T = 0)$. Therefore, if the probability distributions of the covariates in the exposed $(T = 1)$ and unexposed $(T = 0)$ subpopulations are not equal, we cannot guarantee that the difference in mean values of the outcome between the two subpopulations defined by a potentially causal variable of interest, T, is not equal to its causal effect on that outcome as per *Equation 10.4*. We say **confounding** is present and that X are **confounders**. The presence of confounding from the differences in the distributions of the covariates in the exposed and unexposed groups indicates an unbalanced data experiment. *Equation 10.4* is valid only if there is no confounding, in which case association is causation. In the presence of confounding, the quantities on both sides of *Equation 10.4* will differ, leading to the inequation:

$$Cofounding \Leftrightarrow E[Y(1) - Y(0)] \neq E[Y|T = 1] - E[Y|T = 0] \quad (10.5)$$

In the next section, we describe the experimental approaches for collecting data and making statistical inferences to compute statistical estimands of the causal effect of treatment. The two main approaches are observational studies and randomized trials.

Observational studies and randomized trials

In the previous section, we learned the importance of mitigating the confounding effect to make causation equivalent to association so we can compute causal effects using statistical measures from experimental data.

An important metric of confounding is the propensity score, which maps multivalued features of the units in a population to a single number between zero and one. Therefore, we begin this section by defining the propensity score. Next, we will describe randomized controlled trials and observational studies and compare their main characteristics.

Propensity score

The **propensity score** (**PS**) is essential for causal inference in experimental and clinical trials (Cleophas, et al. 2009). The propensity score (PS) is the conditional probability of assignment to a specific treatment given the value assignment x to the vector of observed covariates X.

$$\pi(x) = Pr(T = 1|X = x) = E[T|X = x] \quad (10.6)$$

The propensity score is an important parameter that maps any given stratum $x = \{x_1, \cdots, x_k\}$ of baseline covariates $X = \{X_1, \cdots, X_k\}$ to a single number between 0 and 1. The propensity score plays a vital role in computational methods for the estimation of treatment effects in experimental trials and observational studies (Rosenbaum and Rubin 1983).

Randomized trials

Randomized trials, or **randomized controlled trials** (RCTs) (Clinical Research - randomized-control-trial n.d.), are called the gold standard when investigating the causal effect of a treatment on a particular outcome. RCTs are often used to investigate the causal effect of a treatment on a particular outcome. RCTs are often used in the medical field when investigating the effectiveness of new drugs for treating a specific illness, e.g., testing a new cancer drug on a patient's life extension. In an RCT, participants are randomly assigned to one of two groups. One group, called treatment, is exposed ($T = 1$), e.g., given the drug, and the other, called control, is unexposed ($T = 0$), e.g., given a placebo instead. The RCT aims to mitigate the confounding to produce an unbiased estimate of the treatment effect by randomly assigning participants to treatment and control, producing two balanced populations.

The propensity score in an RCT is a known function because a predetermined random mechanism determines treatment assignment to a unit. Consequently, in an RCT, there is only one recognized specification for $\pi(X)$ for the PS.

Observational studies

An observational study is an alternative to RCTs for investigating the effects of a treatment on an outcome. Conducting RCTs is not always possible or practical due to various reasons, such as the difficulty of recruiting participants or ethical or economic reasons. For example, suppose we want to investigate the potential causal influence of maternal smoking during pregnancy on the baby's birth weight. In that case, we cannot conduct an RCT, as it would be unethical to ask pregnant women to smoke.

Observational studies conducted on a sample of n units from a given population generate datasets $\{(T_i, X_i, Y_i)\ i = 1 \dots, n\}$ consisting of tuples; each tuple contains data for a single unit U_i containing the value for the treatment T_i, the covariates X_i, and the observed outcome Y_i. The values (T_i, X_i, Y_i) in the dataset are n random samples from a generative joint probability distribution $P(T, X, Y)$ induced by the random sampling from the superpopulation.

Unlike RCTs, in **observational studies**, there is not one accepted specification for propensity score, $\pi(x)$, which needs to be estimated from observed data using, for example, a logistic regression model. In this model, treatment (exposure) status is regressed on observed characteristics, the covariates X. The estimated propensity score is the predicted probability of the fitted regression model (Moisset, et al. 2017). After estimating the propensity score, there are methods of using this score to control covariates for estimating treatment effects, such as **standardization, inverse probability of treatment weighting** (**IPTW**) and **propensity score matching** (**PSM**).

In the following section, we will cover an overview of the causal inference workflow.

Causal inference workflow

Figure 10.3 shows a schematic of the causal inference workflow. The schematic shows a three-phase process for investigating causal research questions that conform with a consensus on the workflow process of causal inference (Heckman and Pinto 2022) (Imbens 2004) (J. Pearl 2009).

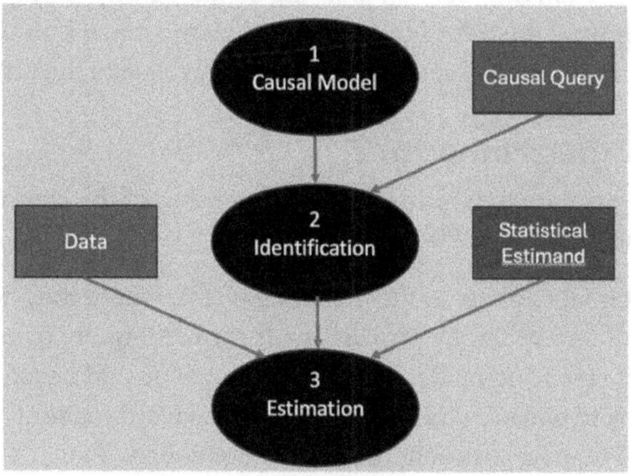

Figure 10.3—Causal inference workflow

The workflow begins with a causal model, such as the Neyman-Rubin or structural causal models. The next step in the workflow is the identification phase, where the researcher investigates whether the causal query can be identified. The last phase is the causal effect estimation phase, which takes place only if the identifiability of the causal question is positive. If a causal effect is identifiable, we can then compute the causal effect based on the probability distribution of the observed variables induced by the observational data. The causal estimation phase takes as input: a) the data based on experiments or observations; b) the statistical estimand mapping the causal query to a target estimand, e.g., **average treatment effect** (**ATE**). The final output from the estimation phase quantifies the estimate of the estimand based on the model and the data.

To conclude, let us summarize the takeaway points of this section on the potential outcomes framework. The framework is based on the concept of potential outcomes in counterfactual experiments on all individuals (units) in a population when assigned treatment (T=1) and when not assigned treatment (T=0). Under this counterfactual experiment, the causal effect can be computed by comparing the potential outcomes. In reality, we have a missing data problem: the counterfactual outcomes, the unrealized outcomes, because we can observe only the outcome for one of the assignments: treatment or non-treatment for each unit. Because of the missing data problem, we can only rely on estimating the counterfactual outcomes from the observed data using statistical methods. The accuracy of estimating the potential outcomes and, hence, the causal effect are greatly affected by the presence of confounding. One of the main objectives in observational studies and randomized trials is to reduce the impact of confounding variables by implementing different techniques, with a focus on those involving propensity scores. Causal inference involves three stages: modeling, identification, and estimation.

The following section is on causal diagrams, studied in previous chapters. We compare them with the potential outcomes model as both models are alternative tools serving the first phase in the causal inference workflow.

Causal diagrams: directed acyclic graphs

We begin this section by explaining why we use causal diagrams. Next, we describe what is in causal diagrams and what information they convey, focusing on diagrams describing the building blocks of causal mechanisms. Following this, we highlight an overview comparison of the potential outcomes and the causal diagrams as alternative causal models as they relate to the process of causal inference.

What are causal diagrams for?

Causal diagrams are the primary tool for the initial two phases of the causal inference workflow described in the previous section.

Causal diagrams are essential in phase 1, the modeling phase, where they help articulate our causal assumptions and make them explicit by articulating the directionality of causal influence between the nodes of the DAG representing model variables, endogenous and exogenous. When we draw a directed edge from node X to node Y, we assume a direct causal influence of X on Y. More importantly, the absence of a directed edge implies conditional independence.

Causal diagrams are instrumental in phase 2, the identification phase, where we can decide whether the causal effect sought by a given causal query, the subject of a causal investigation study, is identifiable. Determining the identifiability of a causal effect question is necessary before conducting phase 3, estimating causal effects through statistical techniques.

What is in causal diagrams, and what do they tell us?

Causal diagrams are directed acyclic graphs consisting of nodes shown as ovals and directed edges connecting pairs of nodes. Arrows in the directed edges point in the direction of causal influence. The absence of edges between nodes indicates that there is no direct causal influence between those nodes. Below, we illustrate causal diagrams of fundamental causal mechanisms and explain the information conveyed by each diagram.

- **Causation.** *Figure 10.4* displays a causal diagram illustrating how variable T, representing a treatment, directly impacts variable Y, referred to as outcome. The diagram suggests that manipulating the value of T will directly impact the probability distribution of Y.

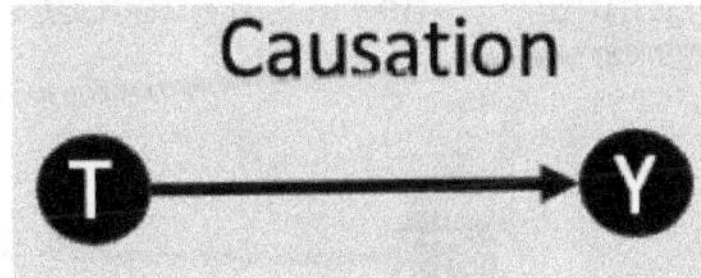

Figure 10.4—Causation mechanism where treatment T has direct causal effect on outcome Y

- **Mediation:** In the mediation mechanism, *Figure 10.5*, treatment T has a causal effect on outcome Y through an intermediary variable M called a mediator. The directed path from T to Y gives the causal path, where all edges point in the same direction from the cause to the effect.

Figure 10.5—Mediation mechanism where M mediates the effect of treatment T on outcome Y

- **Confounder:** *Figure 10.6* shows a causal diagram in which a covariate X confounds the causal effect of T on Y via the biasing path$T \leftarrow X \rightarrow Y$, creating an association between T and Y that results in a confounding bias when$X$ is not controlled or adjusted for.

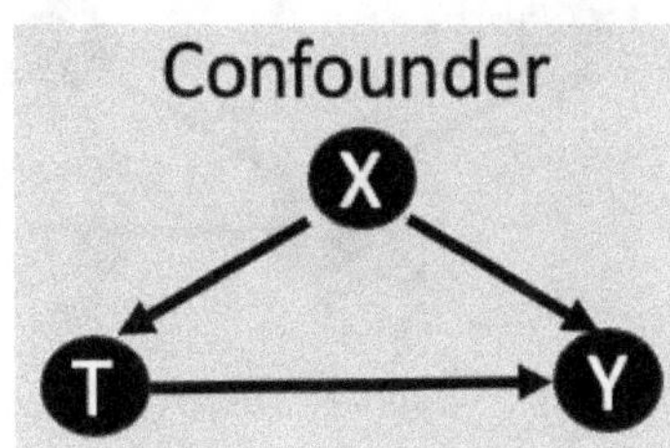

Figure 10.6—Covariate X confounds the causal effect of treatment T on outcome Y

- **Chain:** *Figure 10.7* shows a causal mechanism where covariate X is on a causal path from T to Y, a chain $T \rightarrow X \rightarrow Y$. The chain mechanism has to be accounted for when computing the causal effect of T on Y by controlling for X or adjusting for X.

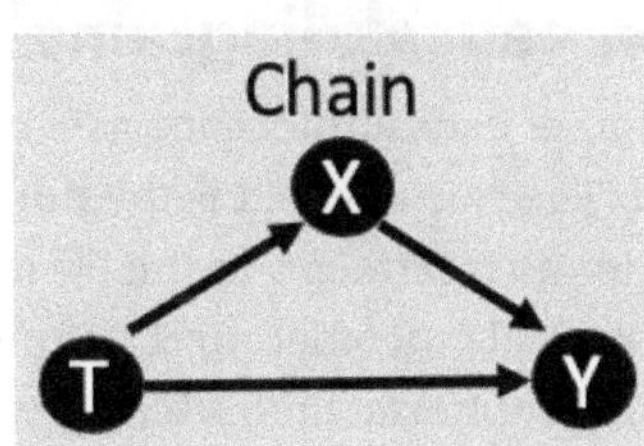

Figure 10.7—Covariate X is on a causal path from treatment T to outcome Y, forming a chain

- **Collider**: A collider mechanism, *Figure 10.8*, is a biasing path on which two consecutive edges have opposing arrows both pointing to a node X. Controlling or adjusting for X or any of its descendants in the DAG creates an association between treatment T and outcome Y that results in a collider bias when computing the causal effect of T on Y.

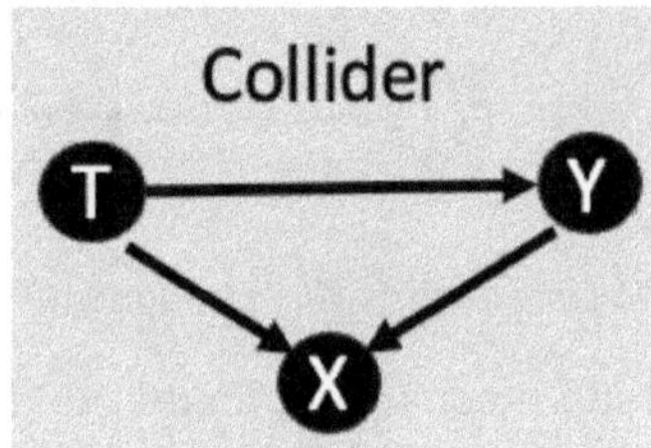

Figure 10.8— Covariate X is a collider on the biasing path from treatment T to outcome Y, creating bias when controlling for or adjusting

- **Instrument**: The instrument mechanism serves the situations where confounder X is unobserved, making it impossible to estimate the effect of treatment T on outcome Y reliably. The mechanism uses an instrument variable I that must meet three criteria: a) I must have a causal effect on T, b) All the causal paths from I to Y have to go through T, i.e., the causal effect of I on Y must be fully mediated by T, and c) There can be no backdoor paths from I to Y, i.e., a biasing path connecting I to Y starting with an edge with an arrow pointing to I.

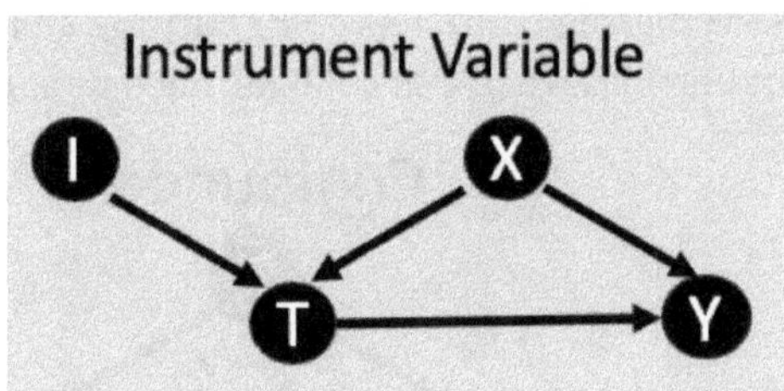

Figure 10.9— The instrument variable I is used as a predictor of T and to mitigate the T and Y association due to unobserved confounder X

In the next section, we relate causal diagrams to the potential outcomes model and discuss how both approaches, though using different languages, are viable approaches to causal inference with mathematically equivalent results.

Potential outcomes vs. causal diagrams

The PO model frames the problem of causal inference as a counterfactual experiment in which all potential outcomes are observed. The causal effect is the difference between the potential outcome under treatment and the one under no treatment. In the PO model, the fundamental problem is that it's impossible to observe both potential outcomes directly, only the one that occurs. The non-observable potential outcome is called the counterfactual, which is what has not happened or is not the case.

The causal diagrams are graphical depictions of underlying structured equations in structural causal models as described in previous chapters. The SCM approach specifies modularity conditions on the structured equations to enable interventions on the model endogenous variables without affecting the structural equations for the other nonintervention variables. Each structural equation is viewed as an autonomous causal mechanism that determines the value of an endogenous variable V as a deterministic function of the parents of V in the DAG of the causal diagram, $pa(V)$, and an exogenous variable, u_V, with the exogenous variables being jointly independent. The notion of intervention on a variable V is represented by the do-operator, with the following operational semantics: $do(V = v)$ means replacing the structural equation for V by the equality $V = v$. Following the notations and formulation due to Pearl (J. Pearl 2010), if we carry an intervention setting $V = v$ in model M, then we denote the post-intervention model as M_v and the probability distribution of variable Y as $P_M(y|do(v))$ or equivalently $P_{M_v}(y)$. In the SCM framework, we can represent the average treatment effect as

$$E|Y|do(T = 1)] - E|Y|do(T = 0)] \quad (10.7)$$

Here, the expected value of Y under the intervention $do(T = 1)$ uses the probability distribution $P(y|do(T = 1))$. Similarly, the expected value of$Y$ under the intervention do(T=0) uses the probability distribution $P(y|do(T = 0))$.

While the PO and SCM models use different approaches, they lead to a mathematically equivalent formulation of the causal effect. The PO model uses a counterfactual approach, while the SCM and causal diagrams use structural equations and the notion of intervention of the do-operator, which engenders the do calculus enabling the derivation of causal effect from observational data if the causal effect is identifiable. The causal diagrams have the advantage of formally deciding the identifiability of causal effect. On the other hand, the PO model sets conditions for identifiability without explicitly committing to a specific structure of causal assumptions between model variables, as in the SCM causal diagrams.

To conclude, let us highlight the main points of this section on causal diagrams. Causal diagrams are directed acyclic graphs representing causal assumptions about model variables' relationships—the causal diagram structure conveys information about causal interactions and underlying causal mechanisms. Causal families of nodes and their parents correspond to autonomous causal mechanisms given by structured equations. The causal effect of treatment variable T on outcomes Y in the causal graph is obtained by computing the probability distribution for the perturbed graph where the causal mechanism for T is replaced by $T = t$. There is a precise mathematical equivalence between expressing treatment effect based on the counterfactual approach of potential outcomes and the one based on the intervention of the causal diagrams.

In the next section, we define the causal effects of treatment in the framework of potential outcomes.

Defining causal effects

In this section, we define several estimands for quantifying treatment effects relevant to observational and experimental studies. The term estimand refers to the quantity that needs to be estimated in an experimental study. An estimator is the statistical algorithm used to compute the estimand, and an estimate is the numerical value the estimator computes for the quantity specified by the estimand.

The term **treatment effect** refers to the causal effect of a binary (0–1) variable representing action, decision, or some intervention on an outcome variable representing implications for real-world scenarios. The term 'treatment effect' has its roots in medical applications, but the term is now used much more generally. Economics and healthcare examples include: the effects of smoking mothers on low birth weight (LBW) of infants (Cattaneo 2010); the impact of LBW on costs to society (Almond, Chay and Lee 2005); the effect of subsidized employment and training programs on the earnings of disadvantaged workers and jobseekers (LaLonde 1986).

In the following section, we utilize this example from Garrido et al. (Garrido, et al. 2016) to exemplify the different treatment effect estimands.

Example (Care Coordination Program): Consider the scenario of scientists who want to utilize observational data to calculate the impact of a hospital-based care coordination program on the probability of being readmitted to the hospital within 30 days. The output variable is binary: 1 if readmitted to the hospital within 30 days and 0 if not. The treatment variable is binary (1=care coordination intervention, 0=usual care).

Individual treatment effect

The **individual treatment effect** (ITE), also known as the **individual causal effect** (ICE), for a unit U_i, is the difference between the two potential outcomes $Y_i(0)$ and $Y_i(1)$ for the individual, expressed as follows:

$$\delta_i = Y_i(1) - Y_i(0) \quad (10.8)$$

Since we can only observe one of the potential outcomes for any given unit, the ITE can only be quantified from the data if we make assumptions for estimating the unobserved counterfactuals and the assignment to treatment.

Example: The ITE assesses how the care coordination intervention impacts the likelihood of hospital readmission within 30 days for a particular patient i. That is, the ITE represents the discrepancy in the likelihood of $Y_i(1)$ being one under the care coordination program and $Y_i(0)$ being 1 in the absence of the program.

Average treatment effect on the treated

The **average treatment effect on the treated** (ATT) is a treatment effect estimand for the subpopulation at which the treatment was targeted. It is defined as follows:

$$ATT = E[Y(1) - Y(0) | T = 1] \quad (10.9)$$

The ATT compares the effect of the treatment in the treated subpopulation with what would have happened in this subpopulation if they had not received the treatment. For example, when deciding whether a treatment cures an illness, the ATT helps us understand the expected difference in the treated subpopulation, the proportion of individuals who were cured versus the proportion that would have been cured if they had not received the treatment. Similarly, in the case of deciding whether to raise the minimum wage, the ATT gives us insight into the expected difference in the unemployment rate in states that enacted a minimum wage increase compared with a situation where these states did not raise the minimum wage.

Example: Calculate the average difference in the probability of being readmitted within 30 days due to the care coordination intervention, but only for those who received the intervention. ATT equals the probability of $Y(1)$ being one given T is one minus $Y(0)$ being one given T is 1.

Average treatment effect on the untreated (ATU)

The **average treatment effect on the untreated** (ATU) is a treatment effect estimand for the untreated subpopulation. It is defined as follows:

$$ATU = E[Y(1) - Y(0)|T = 0] \quad (10.10)$$

The ATU compares the effect of the treatment in the untreated subpopulation with what would have happened in this subpopulation if they had received the treatment. For example, when deciding whether a treatment cures an illness, the ATU helps us understand the expected difference in the untreated subpopulation: the proportion of individuals who would have been cured had they received the treatment versus those who were cured despite the non-treatment. Similarly, in the case of deciding whether to raise the minimum wage, the ATU gives us insight into the expected difference in the unemployment rate in states that did not enact a minimum wage increase if they had enacted the minimum wage increase.

Example: Calculate the average difference in the probability of being readmitted within 30 days due to the care coordination intervention, specifically for those who did not receive the intervention. The likelihood of $Y(1)$ being one when T is zero minus the likelihood of $Y(0)$ being one when T is zero is equivalent to ATU.

Average treatment effect

The **average treatment effect** (ATE), also known as the **average causal effect** (ACE), measures the expected average effect of treatment at the population level. It is defined as the expected value of the difference in potential outcomes when units received treatment versus when units did not receive any treatment:

$$ATE = E[Y(1) - Y(0)] \quad (10.11)$$

The ATE represents a weighted average of the ATT and the average treatment effect on the untreated ATU. Estimating the ATE can help policymakers make decisions by predicting the effect of treatment before enacting the policy. When deciding on whether a treatment cures an illness, the ATE provides insight into the expected difference in the proportion of individuals who were cured of an illness when everyone received the treatment versus when no one received the treatment. When policymakers decide whether to raise the minimum wage, the ATE can provide insight into the expected change in the unemployment rate when the minimum wage is raised for everyone versus when the minimum wage stays the same.

Example: Calculate the average variation in the probability of being readmitted within 30 days between patients who received the care coordination intervention and those who did not. ATE estimates $P[Y(1)]$ minus $P[Y(0)]$.

Conditional average treatment effect

The **conditional average treatment effect** (CATE) is the treatment effect estimand focusing on subpopulations in which all units have the same value x assigned to a covariate vector X. The CATE is a function $\tau(x)$ defined as the expected difference between the two POs conditioned on the covariates $X = x$:

$$\tau(x) = E[Y(1) - Y(0) | X = x] \quad (10.12)$$

The CATE provides insight into the effect of treatment cases when the effect differs between subgroups, also known as the **heterogeneous treatment effect** (Athey and Imbens 2016), (Kunzel, et al. 2019), (Nie and Wager 2021).

Example: Calculate the average difference in 30-day readmission likelihood for patients 65 years and older who received the intervention compared to those who did not. CATE calculates the difference in risk within the population group aged over 65 years as $P[Y(1)| \, age > 65] - P[Y(0)| \, age > 65]$.

In summary, this section on defining causal effects describes various estimands relevant to quantifying the effect of treatment, which include the ITE, the ATE, the ATT, the ATU, and the CATE. The following section describes the three main assumptions in the potential outcomes framework to ensure the causal effects are identifiable.

Identifying causal effects

In this section, we present the conditions the potential outcomes model requires to ensure that the treatment effect is identifiable. Identifiability implies that the causal effect estimand can be estimated from the observational dataset. The observational dataset contains n-value tuples $\{(T_i, X_i, Y_i), i = 1, \cdots, n\}$ for the treatment T, the confounders X, and the outcome Y, respectively. The dataset represents an n-**independent identically distributed (iid)** sample from an underlying distribution $P(T, X, Y)$.

This section is organized as follows. First, we describe the potential outcome framework (POF) assumptions to identify the ATE estimand. Next, we demonstrate that we can derive the ATE based on the probability distribution of the observational data under the conditions of the POF assumptions.

POF assumptions to identify the ATE

The POF must make three assumptions to ensure the causal effect is identifiable and consistently estimate the ATE from observational data. These assumptions are the **Exchangeability** assumption, the **Positivity** assumption, and the **Stable Unit Treatment Value Assumption** (**SUTVA**).

Exchangeability

The exchangeability assumption is alternatively named the **unconfoundedness assumption**, the **conditional-independence assumption**, or the **ignorability** assumption. It states that conditional on the covariates X, the treatment T is independent of the potential outcomes $((Y(0), Y(1))$. That is:

$$T \perp (Y(0), Y(1)) \mid X \quad (10.13)$$

exchangeability assumption requires that the potential outcomes for any two units U_i, U_j within strata defined by covariates X, the probability of their potential outcomes should be independent of the treatment assignment.

$$P(Y(t)|T = t, X) = P(Y(t)|T = t', X), \text{for all}, t.t' \in \{0,1\} \quad (10.14)$$

The exchangeability assumption permits the counterfactual (unobservable) outcomes in an exposed group (T=1) to be proxied by the observed results in an unexposed group (T=0) (Igelström, et al. 2022).

Positivity

The positivity assumption, also called the **overlap assumption,** is as follows. Conditional on the covariates X_i, every unit U_i has some non-zero probability of receiving treatment ($T = 1$) or control ($T = 0$). In other words, the treatment assignment should be stochastic, which means the propensity score must be a positive number and must be less than 1:

$$0 < \pi(X) < 1 \quad (10.15)$$

Unlike the case of the exchangeability assumption, which is not possible to verify from the data alone, the positivity assumption can sometimes be verified from the data (Hernán and Robins, Causal Inference: What If 2020, Lee and Lee 2022).

Stable unit treatment value assumption (SUTVA)

The SUTVA is composed of two assumptions: **Consistency** and **No interference.** The consistency assumption says that if treatment $T = t$ is applied, the observed outcome is the potential outcome for treatment t, that is:

$$(T = t) \Rightarrow (Y = Y(t)) \quad (10.16)$$

The consistency assumption states that for each unit U_i under every treatment scenario, $T_i = t_i, t_i \in \{0.1\}$, the potential outcomes $\{Y_i(0), Y_i(1)\}$ are well-defined and have a single value. This assumption will be broken if multiple treatment versions are in place, as might be the case with surgery, for example, if different surgeons perform the surgery and if these various treatment versions result in different possible outcomes (VanderWeele and Hernán 2013). The consistency assumption is also called the **no-multiple-versions-of-treatment** assumption. Cole et al. (Cole and Frangakis 2009) pointed out that consistency is a challenge in observational studies with treatments that are hard to imagine manipulating as in cases where the treatment is a biological feature like body weight; there are numerous competing methods to (potentially) give a person a body mass index of 25 kg/m2, and each method may have a distinct causal effect on the result. Pearl (J. Pearl 2010) has argued that the consistency assumption should be framed as a "theorem in the logic of counterfactuals."

The no interference assumption is that no interference or spillover effect creates interdependencies between the unit's potential outcomes. When there is no interference, then the potential outcomes for a unit U_i are functions of the treatment assignments $T_i = t_i, t_i \in \{0.1\}$ to the unit U_i regardless of the treatment $T_j = t_j, t_j \in \{0.1\}$ assignments to all the other units $U_j, j \neq i$. That is:

$$Y_i(t_1, \cdots, t_{i-1}, t_i, t_{i+1}, \cdots, t_n) = Y_i(t_i) \quad (10.17)$$

Gerber and Green (Gerber and Green 2012) describe many situations where the no interference assumption may be violated. For instance, in the case of vaccination treatment, it is conceivable that the potential outcomes of contracting a disease in response to treatment of a unit are likely to be influenced by the treatment of other units.

Identification under the POF assumptions

The following demonstrates that the average causal effect is identifiable under the POF assumptions. That means we can estimate the ATE using the observational data.

The average treatment effect that we want to estimate is expressed as follows:

$$\tau = E[Y(1) - Y(0)] \quad (10.18)$$

To prove identifiability, we need to show that the ATE can be expressed using only observable quantities, the covariates X, which are the observed confounders.

Using the law of total expectation, we write:

$$\tau = E_X\left[E[(Y(1) - Y(0))|X]\right] \quad (10.19)$$

Furthermore:

$$\tau = E_X\left[E\left[(Y(1) - Y(0))|X\right]\right] \quad (10.20)$$

Based on the positivity assumption, *Equation 10.9*, we can rewrite *Equation 10.15* as:

$$\tau = E_X[E[Y(1)|X, T = 1]] - E_X\left[E\left[Y(0)|X, T = 0\right]\right] \quad (10.21)$$

It follows from the exchangeability assumption, $T \perp (Y(0), Y(1))\ |X$, that the potential outcomes Y(1) and Y(0) are independent of the treatment T conditional on the covariates X (*Equation 10.9*). Therefore, we can rewrite *Equation 10.16* as follows:

$$\tau = E_X[E[T\,Y(1) + (1 - T)\,Y(0) \mid X, T = 1]]$$

$$-E_X\left[E\left[TY(1) + (1 - T)Y(0)|\, X, T = 0\right]\right] \quad (10.22)$$

By *Equation 10.1*, which is valid under the SUTVA assumption, we have:

$$\tau = E_X[E[Y^{obs}|X, T = 1]] - E_X\left[E\left[Y^{obs}|X, T = 0\right]\right] \quad (10.23)$$

which is a function of the observed data, and we can turn our attention to estimating this quantity.

The following sums up the main points in this section on identifying causal effects. The potential outcomes framework makes three assumptions: exchangeability, positivity, and SUTVA. Exchangeability sets the condition that the potential outcomes are independent of the treatment conditioned on knowing the confounders, the covariates X. Positivity requires that the propensity score be positive and less than 1. The SUTVA is composed of two assumptions: consistency and non-interference. Consistency requires that if t is applied to a unit, the potential outcomes Y(t) for the unit be well-defined, e.g., no multiple treatment versions. Non-interference means that the potential outcomes of a unit are influenced only by the treatment of the unit and not impacted by the treatment of other units.

Having demonstrated the identifiability under the POF assumptions, we are ready to proceed to the main task of estimating the ATE, which is the topic of the following section.

Estimating causal effects

We organize this section as follows. First, we describe a use case of an observational study related to the effect of maternal smoking on a baby's birth weight. Following this, we will present several methods for estimating the average treatment effect and illustrate their application on the use case in practice sessions using R. We will use the use case throughout this section, where we present three estimation methods. The estimation methods will be described in the following order: we begin with the standardization method. Next, we present the inverse probability weighting method. Finally, we present the propensity score matching method.

Estimation under the POF assumptions

An observational study is an alternative to random controlled trials (RCTs) for investigating the effects of a treatment on an outcome. The problem that arises in observational studies is the presence of confounders, that is, covariates that influence the treatment assignment, which results in a selection bias that confounds the estimate of the treatment effect.

To formally represent the estimation problem, consider an observational study in which participants are randomly assigned to a binary treatment T. Let $X = \{X_1, \cdots, X_k\}$ be a vector of categorical baseline covariates and let Y be the result. We are interested in knowing how baseline characteristics affect the variation of T's effect on Y. Within any given stratum $x = \{x_1, \cdots, x_k\}$; the treatment effect can be defined as follows.

Without incorporating the confounders in the causal model for the treatment and outcome, the causal effect estimation will be based on association rather than causation. For example, in our use case dataset for the observational study on the effect of a mother smoking on the baby's birthweight, we would not be able to compute the actual effect of a mother's smoking on her baby's birthweight without controlling for covariates such as marital status ('`mmarried`'), the mother's age ('`mage`'), whether the mother had a prenatal doctor's visit in the baby's first trimester ('`prenatal1`'), and whether this baby is the mother's first child ('`fbaby`').

In the following, we describe three ATE estimators under the POF assumptions. First, we derive statistical formulas for the regression adjustment estimator. Next, we do the same for the inverse propensity score estimator. Lastly, we describe a hybrid estimator called the doubly robust estimator.

Regression adjustment estimator

Under the POF assumptions, we showed earlier in the identifying causal effect section that the average treatment effect estimand can be expressed using two quantities.

$$\mu_0(x) = E_X[E[Y^{obs}|X, T = 1]] \quad (10.24)$$

$$\mu_1(x) = E_X[E[Y^{obs}|X, T = 0]] \quad (19.25)$$

The two quantities of *Equations 10.24* and *10.25* are functions of the covariates X. One is the average outcome of the treatment group as a function of the covariates, and the other is the average outcome of the control group as a function of the covariates.

Estimating a conditional expectation is a problem known as regression, which is perhaps the most common problem in statistics. Let us consider some regression methods to establish estimators $\widehat{\mu_0}$ and $\widehat{\mu_1}$ of the two functions of *Equations 10.24* and *10.25*.

Then, we can compute an estimation of the average treatment effect from *Equation 10.24* by plugging in the regression estimated functions:

$$\hat{\tau} = E_X[\widehat{\mu_1(X)} - \widehat{\mu_0(X)}] \quad (10.26)$$

We can estimate the quantity in *Equation 10.26* by taking the sample average:

$$\hat{\tau} = \frac{1}{n} \sum_{i=1}^{n} \widehat{\mu_1(x_i)} - \widehat{\mu_0(x_i)}] \quad (10.27)$$

The average treatment effect estimate, *Equation 10.27*, is just the average of the difference between two regression functions. The estimate based on the regression adjustment, as given by the formula in *Equation 10.27*, can be interpreted as a regression method to impute the missing data by substituting estimates for the missing potential outcomes, the counterfactual, for each observational unit.

Inverse propensity score estimator

This estimator version makes use of the propensity score. Recall the propensity score, *Equation 10.6*, represents the probability that a unit with covariates x receives treatment.

$$E[T \mid X = x] = \pi(x) \quad (10.28)$$

$$E[1 - T \mid X = x] = 1 - \pi(x) \quad (10.29)$$

Let us go back to the ATE definition of *Equation 10.18*:

$$\tau = E[Y(1) - Y(0)]$$

Let us rewrite it by first taking the expected difference between the potential outcomes conditioning on the covariates X and then determining the expected value of the expected value conditioning on X:

$$\tau = E_X[E[(Y(1) - Y(0))| X]]$$

$$= E_X[E[Y(1)| X]] - E_X[E[Y(0)| X]] \quad (10.30)$$

From the POF exchangeability assumption, *Equation 10.13* or *10.14*, which says the potential outcomes $Y(1)$ and $Y(0)$ are conditionally independent of the treatment given X, it follows that:

$$E[Y(1).T|X] = E[Y(1)|X]\,E[T|X] = \pi(X)\,E[Y(1)|X] \quad (10.31)$$

From the POF positivity assumption, *Equation 10.15*, that says the propensity score $\pi(X)$ takes values between 0 and 1, it follows that we can rewrite *Equation 10.31* as follows:

$$E[Y(1)|X] = E\left[\frac{Y(1)\,T}{\pi(x)} |X = x\right] \quad (10.32)$$

Similarly, we can write:

$$E[Y(0)|X] = E\left[\frac{Y(0)\,(1-T)}{1-\pi(x)}|X = x\right] \quad (10.33)$$

Substituting *Equations 10.32* and *10.33* into *Equation 10.30* yields:

$$\tau = E_X\left(E_T\left[E\left[\frac{Y(1)\,T}{\pi(x)}|X = x\right]\right] - E_T\left[E\left[\frac{Y(0)\,(1-T)}{1-\pi(x)}|X = x\right]\right]\right) \quad (10.34)$$

It follows from the exchangeability assumption, $T \perp (Y(0), Y(1))\,|X$, that the potential outcomes $Y(1)$ and $Y(0)$ are independent of the treatment T conditional on the covariates X, (*Equation 10.14*). We can rewrite *Equation 10.34* in terms of only observed quantities:

$$\tau = E_X\left(E_T\left[E\left[\frac{Y^{obs}\,T}{\pi(x)}|X = x\right]\right] - E_T\left[E\left[\frac{Y^{obs}\,(1-T)}{1-\pi(x)}|X = x\right]\right]\right) \quad (10.35)$$

$$= E\left[\frac{Y^{obs}\,T}{\pi(x)}\right] - E\left[\frac{Y^{obs}\,(1-T)}{1-\pi(x)}\right] \quad (10.36)$$

Since the observational data are independent and identically distributed (iid) samples from the induced probability distribution on T, Y, X, we can estimate the quantities in *Equation 10.36* as the mathematical average over the n samples. Hence, the estimate is:

$$\hat{\tau} = \frac{1}{n}\sum_{i=1}^{n}\left[\frac{Y_i^{obs}T_i}{\pi(X_i)} - \frac{Y_i^{obs}(1-T_i)}{1-\pi(X_i)}\right] \quad (10.37)$$

This inverse propensity score estimator is called the **Horvitz-Thompson estimator** or the inverse propensity score estimator (Horvitz and Thompson 1952). In practice, we do not know the propensity score, so we estimate it from the data: this is again a problem of regression, except the outcome is binary. To estimate the propensity score, $\pi(x) = E[T|x]$, we use the **logistic regression** model (Hosmer and Lemeshow 2000),

$$\pi(x) = \frac{e^{\beta_0 + \sum_{j=1}^{k}\beta_j x_j}}{1 + e^{\beta_0 + \sum_{j=1}^{k}\beta_j x_j}} \quad (10.38)$$

A transformation of $\pi(x)$ used in logistic regression is the **logit transformation**, defined as:

$$g(x) = logit(\pi(x)) = ln\left[\frac{\pi(x)}{1-\pi(x)}\right] = \beta_0 + \sum_{j=1}^{k}\beta_j\,x_j \quad (10.39)$$

In summary, the central intuition in the inverse propensity score estimator is to balance the observational data by weighing the observations for the treated and untreated groups by the inverse of their propensity scores, and then calculating the difference between the average of the weighted outcomes of the treated group and the average of the weighted outcomes of the untreated group.

Doubly robust estimator

Hybrid estimators have been proposed that combine the two approaches of regression adjustment and inverse probability weighting using the propensity score. The hybrid estimators named **doubly robust** have the property that they provide unbiased estimates for the treatment effect if either the regression model or the propensity score inverse weighting model is specified well, even if one of the models is misspecified. For an extensive review of methods for estimating causal effect from observational data, see Shi, Zivich and Chu (2024).

At this point, we conclude this section on estimation under the POF assumptions, where we derived the equations for two estimator methods: the regression adjustment estimator and the inverse propensity score estimator. We also mentioned in passing a third estimator called doubly robust, a hybrid estimator that combines the two derived estimators.

The following section describes algorithmic statistical methods for implementing estimator methods for estimating the ATE with observational data.

Algorithmic statistical methods

In this section, we describe three algorithmic statistical methods used for estimating the causal effect, e.g., the ATE. The statistical methods are **standardization, inverse probability of treatment weighting,** and **propensity score matching.** All three methods rely on regression models, which we describe under the standardization method, but the description carries over to the other two methods.

Standardization

The standardization method uses G-computing and GLMs (generalized linear models) to perform parametric standardization by adjusting bias by including all confounders as covariates.

A GLM is a flexible generalization of ordinary linear regression that allows the response variable to have different distributions (e.g., normal, binomial, Poisson). GLMs are regression models that model the relationship between a response variable and one or more explanatory variables. GLMs consist of three components: random component, systematic component, and link function. The random component specifies the distribution of the response variable (e.g., Gaussian, binomial, Poisson). The systematic component specifies a parameterized predictor, a linear combination of the explanatory variables (e.g., $\beta_0 + \beta_1 X_1 + \beta_2 X_2$). The link function links the mean of the response variable to the linear predictor (e.g., identity link for Gaussian, logit link for binomial, log link for Poisson).

G-computation (or the G-formula) is a versatile method used in causal inference to estimate causal effects from observational data. It can be applied to various scenarios, simulating the potential outcomes under different treatment scenarios. The main steps of G-computation are as follows. First, the fitted regression model is used to predict the potential outcomes for each unit in the dataset in the treated and untreated groups. Next, the average of the predicted potential outcomes of all units in each group is computed. The last step is estimating the ATE by taking the difference between the average predicted potential outcomes for the treated and the average predicted potential outcomes for the untreated.

Standardization uses bootstrapping to measure the uncertainty of the calculated ATE. Bootstrapping proceeds in three main steps: 1) repeatedly sampling the initial dataset with replacement, 2) fitting the outcome model to each sampled dataset, and 3) calculating the ATE for each bootstrap sample.

The following is a mock example to illustrate G-computation and GLM.

Example: Consider a study estimating the causal effect of smoking (`mbsmoke`) on birth weight (`bweight`), accounting for covariates like marital status (`mmarried`).

1. Fit a GLM to model the outcome:

```
outcomemodel <- glm(bweight ~ mbsmoke + mmarried + othercovariates, data
= data)
```

2. Predict potential outcomes under treatment (smoking) and control (non-smoking):

```
data$treatment <- 1
predtreatment <- predict(outcomemodel, newdata = data, type = "response")
data$treatment <- 0
predcontrol <- predict(outcomemodel, newdata = data, type = "response")
```

3. Estimate the ATE:

```
ate <- mean(predtreatment - predcontrol)
```

In the next section, we describe the following algorithmic statistical method.

Inverse probability of treatment weighting (IPTW)

The IPTW method adjusts for measured confounding in observational studies by assigning weights based on the estimated inverse of the propensity score. The weights are used to reshape the probability distributions of the covariates (the observed confounders), which amounts to synthetically reshaping the populations of the treatment and control groups so that they have similar characteristics. In other words, the IPTW method uses the weights to adjust for confounding by balancing the probability distributions of the confounders across the treatment and control groups.

The last algorithmic statistical method considered here is described next.

Propensity score matching (PSM)

Propensity score matching aims to process the control and treatment groups so that their propensity scores have similar distributions. In PS matching, a unit in the treatment group is selected randomly and matched with a unit in the control group whose propensity score is close to that of another unit in the treatment group; matching without replacement matches on a one-to-one basis, a unit from the treatment group with a unit in the control group. Matching with replacement allows pairing up a unit from the treatment group with multiple units in the control group, which is useful where there are fewer units in the control group than those in the treatment group. For an overview of PS matching methods used in causal inference, see (Stuart 2010).

In summary, there are three algorithmic statistical methods for estimating the ATE: standardization, IPTW, and PSM. All three methods use regression models to estimate the outcome and propensity score models to estimate the causal effect of treatment, e.g., the ATE.

The following section mentions available software for computing causal effects with observational data.

Software for estimating treatment effects

There is commercial software for estimating treatment effects, one by STATA (StataCorp. 2023) and the other by SAS (SAS Institute Inc. 2016). For a review of the STATA software and a demonstration with examples, see (Drukker 2016). For the use of causal effect estimators in the SAS software's **CAUSALTRT** procedure, see (Lamm and Yung 2017).

We use the R software and library packages to implement and demonstrate three approaches to estimating causal effects. The first is based on standardization, the second on inverse probability weighting, and the third on propensity score matching. R software is a cost-effective choice. Being open-source software, it is free, and it allows fine-grained access to the minute details of the implementation and performance, as well as control for producing custom results.

In the next section, we will describe the observational study use case, which we will use in our next R practice.

Use case: the Cattaneo2 dataset of smoking and birth weight

Research shows that birth weight is the primary measure of a baby's health, and there is a correlation between low birth weight (LBW), weight less than 2500 grams, and severe health problems such as high blood pressure, cerebral palsy, deafness, blindness, asthma, and lung disease. Almond et al. (Almond, Chay and Lee 2005) provide evidence on the impact of maternal smoking on birth weight (and LBW incidence) and analyze the impact of LBW-preventing intervention policies on health and cost consequences.

Our specific research goal is to estimate the average treatment effect of a mother's smoking on the birth weight of her newborn.

We use an observational dataset, based on (Cattaneo 2010), available from Stata (Stata 2014). In the example at hand, the treatment is: '`mbsmoke`' (cigarette smoked during pregnancy), which has the value 0 if the mother did not smoke and 1 if she did. The output is '`bweight`' (infant birthweight in grams).

The dataset includes several covariates that may influence the treatment (smoking) and the baby's birth weight. The covariates include variables such as whether the mother had a prenatal exam in the first trimester, whether the mother is married, the mother's age, whether this is the first birth and the education level of the father, the mother's marital status, the education level of the mother, her age, whether she had a prenatal exam in the first trimester, whether this baby is her first baby, etc.

The following practice section uses the described dataset to demonstrate how to programmatically implement the three described estimator methods using R.

Practice using R

In this practice, you will be guided step by step through the application of various estimation methods to estimate the ATE from observational data. We will then apply the various methods to the use case described in the previous section. You will be guided and given a demonstration of three estimation methods: standardization, inverse probability of treatment weighting, and propensity score matching.

The dataset for our use case is included in the file `cattaneo2.dta`, which can be downloaded from the STATA website (Stata 2014). It is also included in the code folder for the chapter on the book website.

The following code uses the `haven` package to load the dataset referenced as data in the later code:

```
# Load necessary library
library(haven)  # for read_dta function
# Load the dataset
data <- haven::read_dta("cattaneo2.dta")
```

The following section describes the first estimation method, standardization, for estimating the average treatment effect.

Standardization

The standardization method uses a standard generalized linear model (glm) to perform parametric standardization by adjusting bias by including all confounders as covariates. The following is a demonstration of standardization on our use case dataset. The demonstration includes three steps: initializing the `CausalModels` package, fitting the standardization model, and visualizing the coefficient plot.

1. Initialize `CausalModels` package:

   The function `CausalModels::init_params` must be run first before any other function can run. The function requires four input parameters: `outcome = bweight`; `treatment = mbsmoke`; `covariates = confounders`, and `data`. consisting of the list of variables (`married`, `prenatal1`, `fbaby`, `medu`) and the data. The `covariates` is the list of `confounders`: "`mmarried`", "`prenatal1`", "`fbaby`", "`medu`."

```
# Initialize the parameters for the CausalModels package
data$mbsmoke <- as.factor(data$mbsmoke)
confounders <- c("mmarried", "prenatal1", "fbaby","medu")
CausalModels::init_params(
    outcome = bweight, treatment = mbsmoke,
    covariates = confounders, data
)
```

2. Fit standardization model:

   We use the standardization function `CausalModels::standardization` to fit a standardization model `std_model` with the following arguments: the dataset, `data`, and number of bootstrapping samples to calculate standard errors, `n.boot = 100`, and `simple = T` to fit a simplified version of the model. Next, we print the standardization model for viewing.

```
# Fit standardization model
std_model <- standardization(data,n.boot=100,simple = T)
std_model
 # Output:
Call: glm(formula = bweight ~ mbsmoke + mmarried +
```

```
prenatal1 + fbaby + medu, family = family, data = data)

Coefficients:
(Intercept) mbsmoke1 mmarried prenatal1 fbaby  medu
3195.673    -222.610 148.512  52.567   -56.873 6.802

Average treatment effect of mbsmoke:
Estimate -   -222.6099
SE       -    24.52151
95% CI   -   ( -270.6712 ,  -174.5487 )
```

The results show the coefficients of the effect of the treatment and the covariates on the output of the data-fitted regression model. The intercept, 3195.673, is the baseline birth weight in grams. The coefficient for the treatment 'mbsmoke' indicates that maternal smoking is associated with a decrease of 222.610 grams in birth weight, holding other variables constant. That coefficient estimates the average treatment effect, as shown in the results above, including the standard error and the 95% confidence interval. The influence of the confounders is as follows. Prenatal care, 'prenatal1', and the mother's education, 'medu,' positively impact the birth weight, but the influence of education is much less significant than that of prenatal care. The mother's first baby 'fbaby' has a negative influence, resulting in a weight reduction of 56.873 grams from the baseline.

3. Visualizing the coefficient plot for the standardization model:

   The next code shows how to use the `coefplot` and `broom` packages to visualize the standardization model coefficients and their 95% confidence intervals. First, we create **tidy data frame** for the model coefficients using the function `tidy(std_model$model)`. We use `coefplot` for visualization using the function `coefplot` with the following parameters: the standardization model, `std_model$model`; `intercept=TRUE` to include the intercept term in the plot; `outerCI=1.96` to define 95% confidence intervals.

```
# Install and load necessary packages
library(coefplot)
library(broom)

# Create a tidy data frame for the model coefficients
coef_df <- tidy(std_model$model)

# Use coefplot for visualization
coefplot(std_model$model, intercept=TRUE, outerCI=1.96) +
  ggtitle("Standardization Model Coefficients with 95% Confidence
Intervals") +
  theme_minimal() +
  theme(
```

```
    text = element_text(size = 12),
    plot.title = element_text(hjust = 0.5)
  )
```

Figure 10.10 shows the output coefficient plot for the standardization model coefficients with 95% confidence.

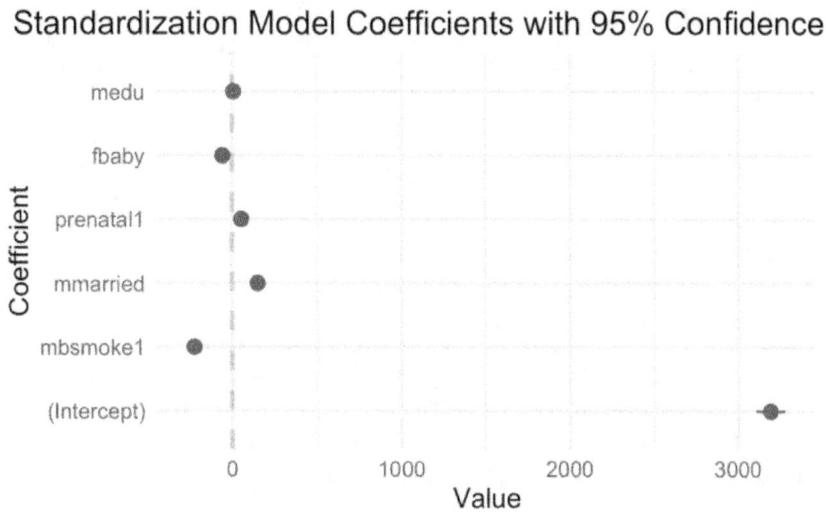

Figure 10.10—Standardization model coefficients with 96% confidence fitting the Cattaneo2 dataset

At this point, we conclude the demonstration of the standardization method. Let us proceed to the next practice of applying the inverse probability of treatment weighting to the same dataset as our use case.

Inverse probability of treatment weighting

The IPTW estimation method involves estimating the propensity score by logistic regression and then using the estimated PS to inversely weigh the data for the treated and untreated groups to balance the probability distribution of the covariates for the two groups. The following demonstration walks you through the steps and provides ample explanations. The code for the demonstration implements six steps:

1. Loading libraries and initializing the causal model:

   The following code chunk loads the `CausalModels` and `survey` libraries, and next initializes the causal model to assess the causal effect of maternal smoking ('`mbsmoke`') on birth weight ('`bweight`'), accounting for various confounders ('`mmarried`', '`prenatal1`', '`mage`', '`fbaby`'). The results show the default formulas for the output and propensity models.

   ```
   library(CausalModels)
   library(survey)
   # Convert relevant columns to factors
   ```

```
data$mbsmoke <- as.factor(data$mbsmoke)
data$mage <- as.factor(data$mage)
confounders <- c("mmarried", "prenatal1", "mage", "fbaby")

# Initialize the parameters for CausalModels package
CausalModels::init_params(
  outcome = "bweight",
  treatment = "mbsmoke",
  covariates = confounders,
  data = data
)

# Output:
Successfully initialized!

Summary:

Outcome - bweight
Treatment - mbsmoke
Covariates - [ mmarried, prenatal1, mage, fbaby ]

Size - 4642 x 23

Default formula for outcome models:
bweight ~ mbsmoke + mage + mmarried + (mbsmoke * mmarried) + I(mmarried *
mmarried) + prenatal1 + (mbsmoke * prenatal1) + I(prenatal1 * prenatal1)
+ fbaby + (mbsmoke * fbaby) + I(fbaby * fbaby)

Default formula for propensity models:
mbsmoke ~ mage + mmarried + I(mmarried * mmarried) + prenatal1 +
I(prenatal1 * prenatal1) + fbaby + I(fbaby * fbaby)
```

The initialized default formulas for `CausalModels::init_params` primarily assume continuous covariates for their default formula generation. The `CausalModel` package documentation shows that `init_params` constructs formulas including polynomial terms for covariates, which are intended for continuous variables rather than categorical ones.

The structure of these formulas depends on the `simple` argument in the initialization function `CausalModels::init_params`. When `simple = FALSE` (default): The function builds complex formulas that include interactions and squared terms for continuous variables to capture non-linear relationships. When `simple = TRUE`: It builds basic formulas that exclude interactions, using only linear terms for the covariates. The initialized default formulas typically have the following patterns.

Outcome model formula:

$$Outcome \sim Treatment + Covariates + (Treatment * Covariates) + I(Covariates * Covariates)$$

The outcome formula is used for standardization or outcome regression to estimate counterfactual outcomes.

Propensity Model Formula:

$$Treatment \sim Covariates + I(Covariates * Covariates)$$

The propensity model formula is used for propensity score matching or inverse probability weighting (IPW) to balance confounding between treatment groups.

To use categorical covariates, you must ensure they are coded as factors in your data frame, which we have done, though the formula builder might still apply quadratic terms unless customized, as demonstrated in the code output above.

The default formulas will cause **aliasing** (singularity) in regression models (including logistic regression via `glm`) when estimating coefficients for a set of predictors assembled into a design (model) matrix X.

Consequence for our model:

- `I(mmarried * mmarried)` is a duplicate of `mmarried` if `mmarried` is 0/1.
- `I(prenatal1 * prenatal1)` is a duplicate of `prenatal1` if `prenatal1` is 0/1.
- `I(fbaby * fbaby)` is a duplicate of `fbaby` if `fbaby` is 0/1.

Removing those squared terms in the default formulas will not change the fitted model or predictions—it only removes redundancy and cleans up warnings.

Let's explain why removing the squared 0/1 terms does NOT change the fitted model. The key reason is that the fit depends on the column space of matrix X, not on a particular set of columns. The linear predictor is $\eta = X\beta$. If a column of X is a perfect duplicate of another, then adding or removing that duplicate does not change the space spanned by the columns (the column space). Maximum likelihood estimates (and all derived quantities: fitted values, log-likelihood, AIC, etc.) depend only on the column space. When R detects redundancy, it internally drops aliased columns, effectively fitting the model on a full-rank subset X^* that spans the same column space. Therefore, explicitly removing the redundant squared terms from the formula yields the same X^* that R already used implicitly. As a result:

- Fitted probabilities (propensity scores) are identical.
- Coefficients for the non-aliased terms are identical (up to trivial numerical noise from reordering), because they were estimated from the same effective design.
- Standard errors, z-values, log-likelihood, and AIC are unchanged.

The only difference is cosmetic/clarity: there are no NA coefficients and no "*not defined because of singularities*" message.

In a later section, we will demonstrate empirically that the default model is generated by the `init_params` function when both x and I(x^2) are included in the formulas on one hand and when only x is included for the three 0/1 categorical variables, `mmarried`, `prenatal1`, and `fbaby`.

2. Fitting the propensity score model:

   This fits a logistic regression model to estimate the propensity scores, which predict the probability of smoking during pregnancy (`mbsmoke`) given the confounders.

```
# Fit the propensity score model
propensity_model <- glm(
  formula = mbsmoke ~ mage + mmarried + I(mmarried * mmarried) +
            prenatal1 + I(prenatal1 * prenatal1) + fbaby +
            I(fbaby * fbaby),
  family = binomial(link = "logit"),
  data = data
)
```

3. Calculating propensity scores:

   We use the propensity model to predict the propensity scores using the function predict. Next, we compute the inverse probability weights to balance the treated and untreated groups to mimic a randomized controlled trial.

```
# Calculate propensity scores
propensity_scores <- predict(propensity_model, type = "response")
# Compute inverse probability weights
data$weights <- ifelse(
    data$mbsmoke == 1, 1 / propensity_scores,
     1 / (1 - propensity_scores)
)
```

4. Fitting the weighted outcome model using the *survey* package

   The `svyglm` and `svydesign` functions are part of the *survey* package in R, and they are used together to fit generalized linear models to data collected using complex survey sampling techniques. These include stratification, clustering, and weighting, common in survey data. The `svyglm` function fits a generalized linear model (GLM) to survey data using the design object created by `svydesign`. It properly accounts for the survey's complex design using weights and other design attributes. We use `svyglm` to compute a weighted outcome model, `outcome_model`, which takes the default outcome formula from the initialization step and the design object specified by the `svydesign` function with the parameters `ids = ~1`, `weights = ~weights`, `data = data`. The parameter `ids = ~1` indicates no clusters in the data, meaning each row is an independent observation. `weights = ~weights`: uses the previously calculated inverse probability weights; `data = data`: refers to the analyzed dataset.

```
# Fit the weighted outcome model using survey package
```

```
outcome_model <- svyglm(
  formula = bweight ~ mbsmoke + mage + mmarried + (mbsmoke * mmarried) +
            I(mmarried * mmarried) + prenatal1 +
            (mbsmoke * prenatal1) + I(prenatal1 * prenatal1) + fbaby +
            (mbsmoke * fbaby) + I(fbaby * fbaby),
  design = svydesign(ids = ~1, weights = ~weights, data = data)
)
```

5. Summary of the weighted outcome model

   The following code prints a summary of the fitted weighted outcome model, `outcome_model`, using the *survey* package.

```
# Summary of the weighted outcome model
summary(outcome_model)
                  Estimate Std. Error t value Pr(>|t|)
(Intercept)        2343.18      20.82 112.527  < 2e-16 ***
mbsmoke1           -217.74      49.55  -4.394 1.14e-05 ***
mage14              904.29     152.08   5.946 2.95e-09 ***
mage15              686.56     165.01   4.161 3.23e-05 ***
mage16              833.07      69.07  12.062  < 2e-16 ***
mage17              948.25      79.92  11.866  < 2e-16 ***
mage18              996.84      53.67  18.573  < 2e-16 ***
mage19              925.36      52.74  17.544  < 2e-16 ***
mage20              919.00      54.71  16.798  < 2e-16 ***
mage21              930.39      46.25  20.117  < 2e-16 ***
mage22              938.58      54.98  17.072  < 2e-16 ***
mage23              973.81      52.14  18.679  < 2e-16 ***
mage24              908.89      48.78  18.632  < 2e-16 ***
mage25              895.77      53.54  16.731  < 2e-16 ***
mage26              914.96      56.88  16.085  < 2e-16 ***
mage27              915.17      61.94  14.776  < 2e-16 ***
mage28              899.12      58.26  15.434  < 2e-16 ***
mage29              949.66      50.47  18.815  < 2e-16 ***
mage30              833.20      84.52   9.858  < 2e-16 ***
mage31              871.82      80.15  10.877  < 2e-16 ***
mage32              916.23      54.21  16.901  < 2e-16 ***
mage33              881.23      72.79  12.106  < 2e-16 ***
mage34              964.51      69.56  13.865  < 2e-16 ***
mage35              922.58      79.75  11.569  < 2e-16 ***
mage36              717.04     159.46   4.497 7.07e-06 ***
mage37             1144.69     153.04   7.479 8.89e-14 ***
mage38              672.19     204.33   3.290 0.001011 **
```

```
mage39               992.37      150.82    6.580 5.23e-11 ***
mage40              1118.68      159.76    7.002 2.88e-12 ***
mage41              1018.41       86.48   11.776  < 2e-16 ***
mage42               626.44      167.67    3.736 0.000189 ***
mage43              1092.13      149.02    7.329 2.73e-13 ***
mage44               889.83      121.88    7.301 3.35e-13 ***
mage45              1043.81       40.79   25.590  < 2e-16 ***
mmarried             178.69       28.80    6.204 6.00e-10 ***
prenatal1             69.64       28.52    2.442 0.014650 *
fbaby                -75.18       20.82   -3.610 0.000309 ***
mbsmoke1:mmarried    -66.70       48.21   -1.383 0.166630
mbsmoke1:prenatal1   -34.38       48.36   -0.711 0.477194
mbsmoke1:fbaby       129.47       50.33    2.572 0.010135 *
```

The summary gives statistics on the estimated model coefficients. The intercept (2343.18 grams) represents the average birth weight for the baseline group (reference categories of all factors). The estimate of the `'mbsmoke1'` coefficient, the ATE, indicates that, on average, maternal smoking is associated with a decrease of approximately 217.74 grams in birth weight, holding other variables constant. The effect of the mother's age (`'mage'`) shows positive associations with birth weight and is highly significant. For `mage14` (age category 14)- the coefficient estimate indicates that compared to the baseline maternal age category, age category 14 is associated with an increase of approximately 904.29 grams in birth weight. Similar interpretations apply to other age categories. The coefficient for the marital status (`'mmarried'`) indicates that compared to if the mother is married, the effect is an increase of approximately 178.69 grams in birth weight compared to the baseline. The coefficient for a mother having a first-trimester prenatal exam (`'prenatal1'`) shows a positive association, estimate = 69.64, which means that compared to the baseline, a mother having a first-trimester prenatal exam is associated with an increase of approximately 69.64 grams in birth weight. The coefficient for first-baby status (`'fbaby'`) indicates a negative association with birth weight. The estimate indicates that compared to the baseline, it is associated with a decrease of approximately 75.18 grams in birth weight.

The coefficients for the interaction terms in the outcome regression model like `mbsmoke1:mmarried` represent how the effect of one predictor (e.g., maternal smoking, `mbsmoke1`) on the outcome variable (birth weight, `bweight`) changes depending on the level of another predictor (e.g., marital status, `mmarried`). Their p-values and stars indicate the significance of these interaction terms. The following interprets the results for the coefficients of the interaction terms. The coefficient for `` `mbsmoke1:mmarried` = -66.70 `` indicates the difference in the effect of `mbsmoke1` on `bweight` depending on `mmarried`—the negative coefficient suggests that the negative impact of smoking on birth weight is less (i.e., less negative) when the mother is married. P-value: 0.166630 (Not significant, indicated by lack of stars); `` `mbsmoke1:prenatal1` = -34.38 `` suggests that the effect of `mbsmoke1` on `bweight` changes depending on `prenatal1`. The negative coefficient implies that the adverse effect of smoking on birth weight is reduced for those receiving prenatal care; `` `mbsmoke1:fbaby` `` = `129.47` indicates the interaction effect between `` `mbsmoke1` `` and `` `fbaby` ``. The positive coefficient suggests that the negative impact of smoking on birth weight is reduced (or even reversed) for the first-born baby.

As noted earlier, we will demonstrate empirically that removing the squared terms for the 0/1 covariates, `mmarried`, `prenatal1`, and `fbaby`, from the default formulas does not change the fitted model or its predictions.

The following code defines `fitfull` and `fitclean`: `fitfull` fits the default propensity model formula with the squared terms included, whereas `fitclean` fits the same model with those squared terms removed.

```
> fitfull <- glm( mbsmoke ~ mage + mmarried + I(mmarried^2) + prenatal1 +
I(prenatal1^2) + fbaby + I(fbaby^2), family = binomial(link = "logit"),
data = data )
> fitclean <- glm( mbsmoke ~ mage + mmarried + prenatal1 + fbaby, family
= binomial(link = "logit"), data = data )
```

When you run `fitfull` and `fitclean` in R you will get exactly the same results. For lack of space we do not show the outputs here. In the following, we demonstrate additional empirical evidence.

The following code compares the linear predictors in both models, and the result is TRUE.

```
> all.equal(fitted(fitfull), fitted(fitclean))
# [1] TRUE
```

The following code checks the maximum absolute difference, and the result is zero.

```
> max(abs(fitfull$linear.predictors - fitclean$linear.predictors))
# [1] 0
```

The following code compares the overall model fit statistics which proves they are identical.

```
> AIC(fitfull)
# [1] 4177.71
> AIC(fitclean)
# [1] 4177.71
> logLik(fitfull)
# 'Log Lik.' -2052.855 (df=36)
> logLik(fitclean)
# 'Log Lik.' -2052.855 (df=36)
```

6. Creating balance plots:

   The code for creating balance plots to compare the distribution of covariates between treated and control groups before and after matching or weighting is essential to check the quality of propensity score matching or weighting. The code for creating the balance plots is included in the code folder for this chapter but is not displayed here due to the lack of space. The code creates the balance plots in three steps: fitting the propensity score model, matching or weighting, and computing histograms representing probability mass functions for each covariate. *Figure 10.11* shows the code output consisting of four plots for the covariates 'prenatal1', 'mage,' 'mmarried,' and 'fbaby.'

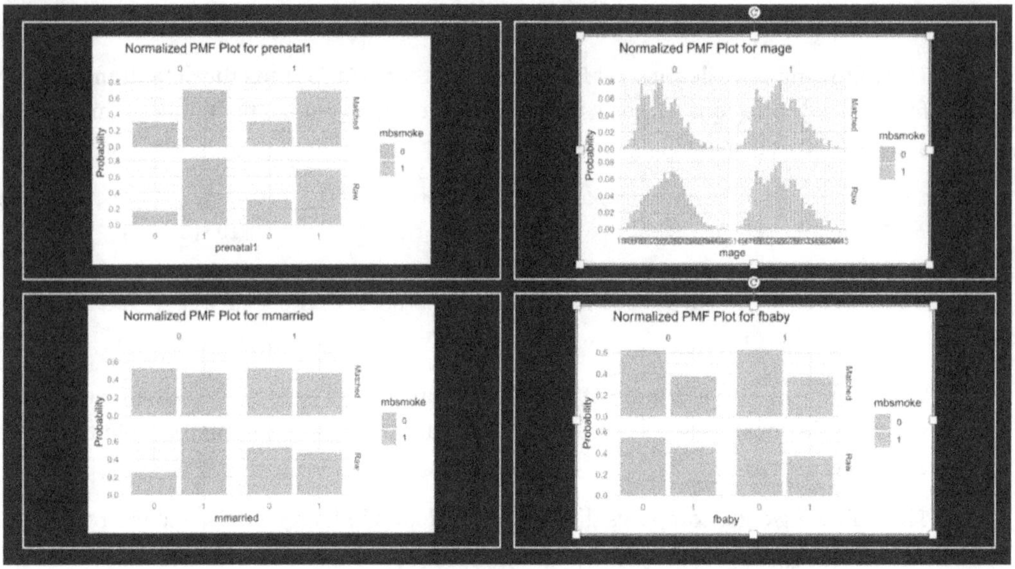

Figure 10.11—Balance plots for the confounders in the Cattaneo2 dataset- the plots show the normalized probability mass function for each covariate for the treated (the 1 column) and control group (the 0 column) in the raw data and the matched data

Each balance plot shows the probability mass function for the treatment and control groups for the raw and matched data. The plots show similar distributions for the treated and control groups for the matched data, while they are dissimilar for the raw data. The effect of matched distributions over the covariates for the control and treatment groups produces an unbiased estimate of the treatment effect.

At this point, we conclude the demonstration of the IPTW method. Let us proceed to the next practice of applying the inverse propensity score matching to the same dataset as our use case.

Propensity score matching (PSM)

PSM is creating matched sets of treated and untreated subjects similar in the covariates (e.g., `mmarried`, `prenatal1`, `mage`, `fbaby`). The creation of matched sets is to control for confounding variables and reduce selection bias when estimating the treatment effect (in this case, the effect of `mbsmoke` on some outcome variable such as `bweight`).

The following demonstration walks you through the steps and provides ample explanations. The code for the demonstration implements four steps:

1. Creating matched data:

We use the function `matchit()` from the `MatchIt` package to perform the matching. The parameters of the `matchit()` function are the formula: `mbsmoke ~ mmarried + prenatal1 + mage + fbaby`; the `data`, the matching method specified as the nearest neighbor; the distance, which is the metric for calculating propensity scores; "`logit`" indicates logistic regression to estimate the propensity scores. Next, we extract the matched data, naming the weights '`matching_weights`'.

```
# Load necessary libraries
library(MatchIt)
library(dplyr)
library(boot)

# Assuming 'data' is your dataset
data$mbsmoke <- as.factor(data$mbsmoke)
# Perform propensity score matching
psm <- matchit(mbsmoke ~ mmarried + prenatal1 + mage + fbaby,
               data = data,
               method = "nearest",
               distance = "logit")
# Extract the matched data, specifying a different name for weights
matched_data <- match.data(psm, weights = "matching_weights")
```

2. Computing the ATE:

The code calculates the mean birth weights for smoking and non-smoking mothers and computes the ATE by subtracting the mean birth weight of non-smokers from that of smokers in the matched data. The following is the code and results for computing the ATE.

```
# Calculate ATE
mean_bw_smoke <- matched_data %>%
  filter(mbsmoke == 1) %>%
  summarize(mean_bweight = mean(bweight, na.rm = TRUE)) %>%
  pull(mean_bweight)
mean_bw_nosmoke <- matched_data %>%
  filter(mbsmoke == 0) %>%
  summarize(mean_bweight = mean(bweight, na.rm = TRUE)) %>%
  pull(mean_bweight)
ATE <- mean_bw_smoke - mean_bw_nosmoke
print(paste("Mean birth weight for smoking mothers:", mean_bw_smoke))
print(paste("Mean birth weight for non-smoking mothers:", mean_bw_
nosmoke))
print(paste("Average Treatment Effect = ", ATE))

# Output:
```

```
[1] "Mean birth weight for smoking mothers: 3137.65972222222"
[1] "Mean birth weight for non-smoking mothers: 3349.72337962963"
[1] "Average Treatment Effect =  -212.063657407407"
```

3. Computing standard error and confidence interval:

   We use a non-parametric bootstrap method to estimate the standard error (SE) and confidence Interval (CI). The bootstrap involves repeatedly resampling the matched dataset with replacement, recalculating the ATE for each resample, and then using the distribution of these ATE estimates to compute the SE and CI. The following is the code and the results.

```
# Define a function for bootstrapping the ATE
bootstrap_ate <- function(data, indices) {
  sampled_data <- data[indices, ]
  mean_bw_smoke_boot <- sampled_data %>%
    filter(mbsmoke == 1) %>%
    summarize(mean_bweight = mean(bweight, na.rm = TRUE)) %>%
    pull(mean_bweight)
  mean_bw_nosmoke_boot <- sampled_data %>%
    filter(mbsmoke == 0) %>%
    summarize(mean_bweight = mean(bweight, na.rm = TRUE)) %>%
    pull(mean_bweight)
  return(mean_bw_smoke_boot - mean_bw_nosmoke_boot)
}
# Perform bootstrapping with 1000 replications
set.seed(123)  # For reproducibility
boot_results <- boot(data = matched_data, statistic = bootstrap_ate, R =
1000)
# Calculate 95% CI from bootstrapped results
boot_ci <- boot.ci(boot_results, type = "perc")
# Print results
print(paste("Bootstrap Standard Error = ", sd(boot_results$t)))
print(paste("95% Confidence Interval = (", boot_ci$percent[4], ", ",
boot_ci$percent[5], ")", sep=""))

# Output:
"Bootstrap Standard Error =  29.5673094759983"
"95% Confidence Interval = (-267.989098351548, -153.375395391343)"
```

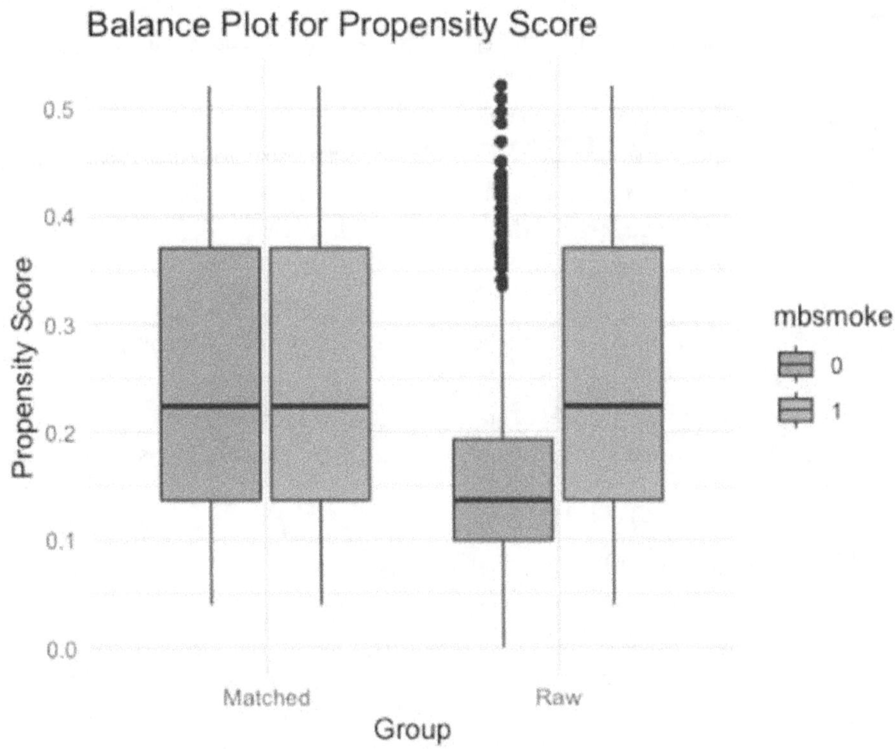

Figure 10.12—Balance plots for the propensity score in the Cattaneo2 dataset- the plots show the box plots for the propensity score for the treated (the 1 column) and control group (the 0 column) in the raw data and the matched data

4. Balance plot for propensity scores:

   The following code produces balance plots for the propensity scores comparing the raw and matched data. The purpose of the balance plot is to show that the matched data have similar propensity score distributions for the treated and control groups, unlike the raw data in the initial dataset. First, we subset `data` to common columns using `intersect` to retain only the common columns in data and `matched_data`. Next, we add the propensity score column and the computed propensity score to both raw and matched datasets. Following this, we add a group column to differentiate between raw and matched data. Finally, we combine raw and matched propensity scores using the function `rbind` after ensuring that both datasets of the two groups have the same structure. *Figure 10.12* shows the boxplot for the propensity score of the raw data and the matched data, which demonstrates that after matching, the propensity score is balanced for the treated and the control groups, while the initial data shows unbalanced scores, which would produce ATE estimation bias.

```
# Load necessary libraries
library(ggplot2)
library(tidyverse)
```

```
# Extract common columns
common_columns <- intersect(names(data), names(matched_data))
# Subset data to have the same columns
data_common <- data[common_columns]
matched_data_common <- matched_data[common_columns]
# Add Group column for comparison
data_common$Group <- "Raw"
matched_data_common$Group <- "Matched"
# Combine raw and matched propensity scores
combined_data <- rbind(data_common, matched_data_common)
# Function to create box plot for propensity scores
create_propensity_score_plot <- function(combined_data) {
  ggplot(combined_data, aes(x = Group, y = propensity_score, fill = 
mbsmoke)) +
    geom_boxplot() +
    theme_minimal() +
    labs(title = "Balance Plot for Propensity Score",
         x = "Group",
         y = "Propensity Score")
}
# Create and print the propensity score balance plot
print(create_propensity_score_plot(combined_data))
```

As we conclude this section on estimating causal effects, it's important to recap the key points we've covered.

Two estimators can be derived for estimating the ATE under the POF assumptions. The first estimator is regression adjustment, which estimates two regression functions of the observed outcome for the treated and untreated groups and then estimates the average of the differences between the two functions evaluated for the units in the respective groups. The inverse propensity score estimator estimates the propensity score, uses the inverse estimate to weigh the observed outcome for the treated and untreated groups, and then takes the arithmetic average of the difference. The doubly robust estimator, a comprehensive hybrid, combines the strengths of both estimator methods, providing a robust approach to estimating causal effects. Algorithmic statistical methods for implementing causal effect estimators include standardization, IPTW, and PSM. They also use statistical methods that include g-computation, glm for regression models, and bootstrapping. The R practice demonstrates how to programmatically employ the three statistical methods of standardization, IPTW, and PSM on a real-life observational dataset use case.

Summary

The potential outcomes framework is a formal approach for determining causality in observational studies and experimental trials using statistical methods. The framework relies on potential outcomes in counterfactual experiments performed on all individuals in a population when given treatment ($T = 1$) and when not given treatment ($T = 0$).

Under this counterfactual experiment, the causal effect measurement answers the question of what outcome can be anticipated if treatment is given or not given to everyone. In reality, we have a missing data problem: the counterfactual outcomes, the unrealized outcomes, because we can observe only the outcome for one of the assignments: treatment or nontreatment for each unit. Hence, in actual experiments, the causal effect measurement is an association measurement that looks into the question, "In what way do the treated and untreated groups vary?"

The causality effect is determined by observing how treatment impacts a single population. In contrast, the effect of association is determined by observing how treatment impacts two different subpopulations.

We can only rely on statistical methods to estimate the potential outcomes from the observed data. The accuracy of estimating the causal effect is greatly affected by the presence of confounding, the covariates representing population units that influence the probability distributions of the assignment to treatment or nontreatment and the outcome. One of the main objectives of observational studies and randomized trials is to reduce the impact of confounding variables.

Causal inference involves three phases: modeling, identification, and estimation. Modeling involves specifying model variables and the causal assumptions regarding the relationships between the variables. Causal diagrams, directed acyclic graphs, are instrumental in making the causal assumptions explicit, specifying the endogenous and exogenous model variables, and providing a nonparametric representation of the functional, structural equations governing the model behavior. For instance, causal diagrams can depict the presence of template mechanisms like causation, mediation, confounder, collider, chain, and instrument variables. Causal diagrams with the machinery of the do-calculus can decide whether causal questions are identifiable. Causal diagrams are central in identifying good and bad control or adjustment variables, and the covariates are used in regression models to predict the outcome variable in response to the treatment variable.

Causal effects describe various estimands relevant to quantifying the effect of treatment, including the ITE, the ATE, the ATT, the ATU, and the CATE.

The potential outcomes framework makes three assumptions for causal effect identifiability. The assumptions are exchangeability, positivity, and SUTVA. Exchangeability sets the condition that the potential outcomes are independent of the treatment, conditioned on knowing the confounders, the covariates X. Positivity requires that the propensity score be positive and less than 1. SUTVA is composed of two assumptions: consistency and non-interference. Consistency requires that if $T = t$ is applied to a unit, the potential outcomes $Y(t)$ for the unit be well-defined, e.g., no multiple treatment versions. Non-interference means that the potential outcomes of a unit are influenced only by the treatment of the unit and not impacted by the treatment of other units.

Two estimators can be derived to estimate the ATE under the POF assumptions. The first estimator is regression adjustment, which estimates two regression functions of the observed outcome for the treated and untreated groups and then estimates the average of the differences between the two functions evaluated for the units in the respective groups. The inverse propensity score estimator estimates the propensity score, uses the inverse estimate to weigh the observed outcome for the treated and untreated groups, and then takes the arithmetic average of the difference. The doubly robust, comprehensive hybrid estimator combines the strengths of both estimator methods, providing a robust approach to estimating causal effects.

Algorithmic statistical methods for implementing causal effect estimators include standardization, IPTW, and PSM. They also use statistical methods that include G-computation for causal inference, GLM for regression models, and bootstrap for confidence intervals. As demonstrated in the R practice, all three estimators can be employed and programmatically implemented using R libraries and the `CausalModels` R package.

References

- Almond, Douglas , Kenneth Y. Chay, and David S. Lee. 2005. *The Costs of Low Birth Weight*. The Quarterly Journal of Economics. 1031-1083.
- Athey, Susan , and Guido Imbens. 2016. *Recursive partitioning for heterogeneous causal effects*. Proceedings of the National Academy of Sciences, 7353–7360.
- Austin, P. C. 2011. *An Introduction to Propensity Score Methods for Reducing the Effects of Confounding in Observational Studies.* Multivariate Behav Res. 399-424.
- Cattaneo, M. D. 2010. *Efficient semiparametric estimation of multi-valued treatment effects under ignorability*. Journal of Econometrics 138–154.
- Cleophas, Ton J., Aeilko H. Zwinderman, Toine F. Cleophas, and Eugene P. Cleophas. 2009. *Statistics Applied to Clinical Trials*- Fourth edition. Springer.
- n.d. Clinical Research - randomized-control-trial. `https://research.mrcpsych.uk/study-designs/randomized-control-trial`.
- Cole, S. R., and C. E. Frangakis . 2009. *The consistency assumption in causal inference: a definition or an assumption?* Epidemiology 3-5.
- Drukker, David M. . 2016. *Estimating treatment effects from observational data using teffects, steffects, and eteffects*. UK Stata Users Group meeting. London: `http://fmwww.bc.edu/repec/usug2016/drukker_uksug16.pdf`, September 8-9. `http://fmwww.bc.edu/repec/usug2016/drukker_uksug16.pdf`.
- Garrido, Melissa M., Bryan Dowd, Paul L. Hebert, and Matthew L. Maciejewski. 2016. *Understanding Treatment Effect Terminology in Pain and Symptom Management Research*. Journal of Pain and Symptom Management 446–452.
- Gerber, Alan S., and Donald P. Green . 2012. *Field experiments : design, analysis, and interpretation*. New York: W. W. Norton.
- Heckman, J. J., and R. Pinto. 2022. National Bureau of Economic Research. National Bureau of Economic Research Working Paper Series. February. `https://www.nber.org/system/files/working_papers/w29787/w29787.pdf`.

- Hernán , M. A., and J. M. Robins. 2006. *Estimating causal effects from epidemiological data*. J. Epidemiol Community Health 578-586.
- Hernán, M. A., and J. M. Robins. 2020. *Causal Inference: What If*. Boca Raton, FL: CRC Press.
- Holland, Paul W. 1986. *Statistics and Causal Inference*. Journal of the American Statistical Association 945-960.
- Horvitz , D. G., and D. J. Thompson . 1952. *A generalization of sampling without replacement from a finite universe*. J Am Stat Assoc. 663–685.
- Hosmer, David W., and Stanley Lemeshow. 2000. *Applied Logistic Regression* (Second Edition). John Wiley.
- Igelström, Erik, Peter Craig, Jim Lewsey, John Lynch, Anna Pearce, and Srinivasa Vittal Katikireddi. 2022. *Causal inference and effect estimation using observational data*. Journal of Epidemiology & Community Health 1-7.
- Imbens, Guido W. 2004. *Nonparametric Estimation of Average Treatment Effects Under Exogeneity: A Review*. The Review of Economics and Statistics 4–29.
- Kunzel, S. R., J. S. Sekhon, B. J. Bickel, and B. Yu. 2019. *Metalearners for estimating heterogeneous treatment effects using machine learning*. Proceedings of the National Academy of Sciences, 4156–4165.
- LaLonde, Robert J. 1986. *Evaluating the Econometric Evaluations of Training Programs with Experimental Data*. The American Economic Review 604-620.
- Lamm, Michael, and Yiu Fai Yung. 2017. *Estimating Causal Effects from Observational Data with the CAUSALTRT Procedure*. https://api.semanticscholar.org/CorpusID:3867864.
- Lee, S., and W. Lee. 2022. *Application of Standardization for Causal Inference in Observational Studies: A Step-by-step Tutorial for Analysis Using R Software*. Journal of Preventive Medicine and Public Health 116-124.
- McCullagh, P., and J. A. Nelder. 1998. *Generalized Linear Models*. Boca Raton, Florida: Chapman and Hall/CRC.
- Moisset, X, M Perié, B Pereira, E Dumont, C Lebrun-Frenay, FX Lesage, F Dutheil, F Taithe, and P Clavelou. 2017. *Decreased prevalence of cancer in patients with multiple sclerosis: A case-control study*. PLoS One.
- Neyman, J. 1923. *On the application of probability theory to agricultural experiments*. Statistical Science 465–480.
- Nie, X., and S. Wager. 2021. *Quasi-oracle estimation of heterogeneous treatment effects*. Biometrika 299–319.
- Pearl, J. 2010. *An Introduction to Causal Inference*. Int J Biostat. 7.
- Pearl, J. 2010. *On the consistency rule in causal inference: axiom, definition, assumption, or theorem?* Epidemiology 872–5.
- Pearl, J.. 2009. *Causality: Models, Reasoning, and Inference*. Cambridge University Press.
- Rosenbaum, Paul R., and Donald B. Rubin. 1983. *The central role of the propensity score in observational studies for causal effects*. Biometrics 41-55.

- Rosenblum, M., and M. J. van der Laan. 2009. *Using regression models to analyze randomized trials: asymptotically valid hypothesis tests despite incorrectly specified models.* Biometrics 937-945.
- Rubin, D. B. 1974. *Estimating causal effects of treatments in randomized and nonrandomized studies.* Journal of Educational Psychology 688–701.
- SAS Institute Inc. 2016. SAS/STAT® 14.2 User's Guide. The CAUSALTRT Procedure. Cary, NC: SAS Institute Inc.
- Shi, Amy X. , Paul N. Zivich, and Haitao Chu. 2024. *A Comprehensive Review and Tutorial on Confounding Adjustment Methods for Estimating Treatment Effects Using Observational Data.* Applied Sciences.
- Stata. 2014. *Datasets for Stata Treatment-Effects Reference Manual,* Release 14. `https://www.stata-press.com/data/r14/te.html`.
- StataCorp. 2023. *Stata 18 Causal Inference and Treatment-Effects Estimation Reference Manual.* College Station, TX: Stata Press. `https://www.stata.com/manuals/causal.pdf`.
- Stuart, Elizabeth A. 2010. *Matching methods for causal inference: A review and a look forward.* Statistical Science 1–21.
- Valojerdi, Ameneh Ebrahim, and Leila Janan. 2018. *A brief guide to propensity score analysis.* Med J Islam Repub Iran. 122.
- VanderWeele, T. J., and M. A. Hernán. 2013. *Causal Inference Under Multiple Versions of Treatment.* Journal of Causal Inference 1-20.

Join our community on Discord

Join our community's Discord space for discussions with the author and other readers: `https://packt.link/hYCWN`

11

Causal Inference with Machine Learning

This chapter presents four meta-learning algorithms that are most prominently employed for estimating the **Conditional Average Treatment Effect** (CATE). Meta-learners are machine learning algorithms that derive information from the results of other machine learning algorithms, known as base learners. The goal of estimating the CATE is to gain insights into **heterogeneous treatment effects** (HTEs), where treatment effects vary across population individuals and subgroups. For example, some individuals may respond to treatment with more or less positive or negative effects, or no effect at all. Evaluating HTEs provides valuable information to decision-makers and policymakers in real-world applications, including medicine, epidemiology, economics, and social research. We demonstrate the implementation of the meta-learners in Python. We describe a causal model for generating synthetic data, which is used throughout to evaluate the performance of the implemented meta-learner algorithms. Illustrations and explanations of the meta-learners and their performance analyses are provided in respective interactive practice sessions using Python. We will be covering the following topics:

- Elucidating heterogeneous treatment effects
- Explaining meta-learning for causal inference
- Specifying generative data model
- Understanding S-Learner
- Understanding T-Learner
- Understanding X-Learner
- Understanding R-Learner

Technical requirements

The technical requirements for this chapter are included in the *Python* project files, along with instructions on setting up the software environment and installing the required libraries. The project files are in the GitHub folder named "`Chapter_11`" on the book's website at the URL: `https://github.com/PacktPublishing/Causal-Inference-with-Bayesian-Networks/tree/main/Chapter_11`

Elucidating heterogeneous treatment effects

The focus in *Chapter 10* has previously been on estimating the **average treatment effects** (ATE), which indicate the expected mean of the treatment effect across the entire population. The underlying assumption is that the treatment effects are homogeneous in the entire population, implying that the distribution over the population is fairly symmetrical around the mean, and the treatment effects differ quantitatively but not qualitatively, being either all positive or all negative.

The focus in this chapter is on the causal inference of **heterogeneous treatment effects** (HTE), where the treatment effects differ qualitatively across the population as a function of the covariates (features) of individuals in the population, with individuals in subgroups having order of magnitude effect differences, or some may have positive, negative, or no effect.

The **Conditional Average Treatment Effect** (CATE) is a fundamental concept in HTE, quantifying the average treatment effect for specific subgroups characterized by individual covariates.

We begin by highlighting the distinction between associative questions, the subject of statistical analysis, and causal questions, the subject of causal inference, our main topic. Next, we formally state the CATE estimation problem and review the potential outcomes causal model.

Causal vs. associative questions

Causal questions center on the notion of **intervention**, investigating the effect of altering a variable within our control on the outcome of an individual in a population. Examples of causal questions:

- Does smoking reduce fetal growth for a pregnant mother?
- Does raising the minimum wage decrease the unemployment rate of workers in a state?

Associative questions center on the notion of **observation** by investigating changes in data patterns where a variable takes different values in different data regimes. Examples of associative questions:

- Is there a correlation between the birth weight of infants and the pregnant mother's smoking?
- Does the unemployment rate in a state correlate with increases in that state's minimum wage?

Having delineated the differences between associative and causal questions, we will state the CATE estimation problem in the following section, which addresses causal questions related to assessing heterogeneous treatment effects.

The CATE estimation problem

To state the problem of estimating the CATE, we consider the **potential outcomes framework (POF)**, also known as the **Neyman-Rubin model**, which is based on statistical methods (Neyman 1923, Rubin 1974). As illustrated in *Chapter 10*, the POF formulates causal inference as a statistical analysis problem with missing data or counterfactual outcomes.

Figure 11.1 depicts the potential outcome model in which an individual (or unit) i has covariates $X_i \in \mathbb{R}^d$ and is assigned a binary treatment indicator $T_i \in \{0,1\}$, which takes the value one if the individual i receives the treatment and zero if i receives the control (indicating not treated or receiving a placebo). Let $Y_i(1) \in \mathbb{R}$ be the "potential outcome" if we give the treatment to i. Let $Y_i(0) \in \mathbb{R}$ be the "potential outcome" if we give the control to i.

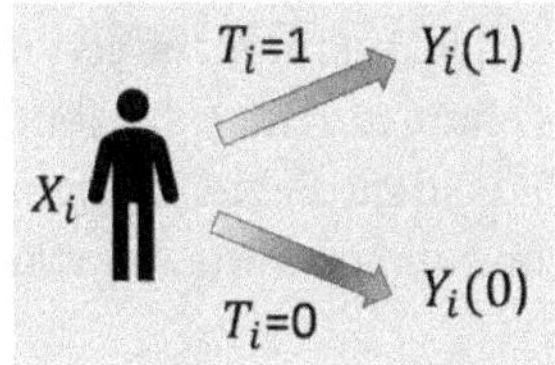

Figure 11.1—Potential outcome model

The **conditional average treatment effect** (CATE) is the treatment effect estimand focusing on subpopulations in which all units have the same value x assigned to a covariate vector X. The CATE is a function $\tau(x)$ defined as the expected difference between the two POs conditioned on the covariates $X = x$.

$$\tau(x) = E[Y(1) - Y(0)|X = x] \quad (11.1)$$

The CATE provides insight into the effect of treatment cases when the effect differs between subgroups, also known as the **heterogeneous treatment effect (HTE)** (Nie and Wager 2021, Kunzel, et al. 2019, Athey and Imbens 2016).

The CATE estimation problem is to compute an estimate $\hat{\tau}(x)$ of the true $\tau(x)$ of *Equation 11.1* using as input the observational data consisting of 3-tuples for the treatment T_i, the covariates X_i, the observed outcome Y_i, for every unit i indexed from one to n, where n is the total sum of all units in both the treatment and control group.

$$(T_i, X_i, Y_i): i = 1, \cdots n \quad (11.2)$$

The **core issue in causal inference** is that real data can possess the potential outcome for either the treatment or the control condition, but not simultaneously for any given unit. The missing data problem is that for any unit i receiving the treatment ($T_i = 1$), we only observe the potential outcome $Y_i(1)$ and lack information on the potential outcome had the unit received the control $Y_i(0)$. For any unit i administered the control ($T_i = 0$), we only know the potential outcome $Y_i(0)$ and lack knowledge of the potential outcome had the unit received the treatment, $Y_i(1)$.

Estimating the CATE is identifiable under the three *Neyman-Rubin* assumptions: the **Exchangeability** assumption, the **Positivity** assumption, and the **Stable Unit Treatment Value Assumption (SUTVA).** See the *Identifying Causal Effects* section in *Chapter 10*.

Estimating the CATE quantifies the causal effect of a treatment on specific subgroups of a population, providing more specific and actionable insights than what is obtained from estimating the overall **average treatment effect** (ATE). Application areas of CATE estimation include:

- **Personalized medicine** to customize therapy to the individual patient;
- **Targeted marketing** and **recommendation systems** to identify customers most likely to respond to marketing offers and to product suggestions;

- **Social science research** by identifying subgroups who will benefit most from targeted social initiatives;
- **Economic policy setting** by assessing workforce development initiatives effect on various demographics employment and labor markets.

Wrapping up this section, let us summarize the main points. Estimating the conditional average treatment effects involves evaluating the average treatment effects for individuals or subgroups of the population as a function of the individual's characteristics. Under the assumptions of the Neyman-Rubin model, the task of estimating the CATE is identifiable, implying it can be computed using statistical machine learning methods, which is the subject of the following section.

Explaining meta-learning for causal inference

We begin this section by describing the concept of using meta-learners for estimation of the conditional average treatment effects. We follow this by reviewing machine learning approaches commonly employed in meta-learning algorithms for CATE estimation.

What is in a meta-learner?

A meta-learner algorithm is a machine learning algorithm that derives knowledge from the outputs of other machine learning algorithms. A meta-learner for estimating conditional average treatment effects takes as input: a) the observational data, consisting of the triple, treatment T, the covariates X, the observed outcome Y, and b) base learners which are any arbitrary machine learning methods and computes as output an estimation of the conditional average treatment effect, $\hat{\tau}(x)$, as a non-parameterized function of the covariates $X = x$. See *Figure 11.2*.

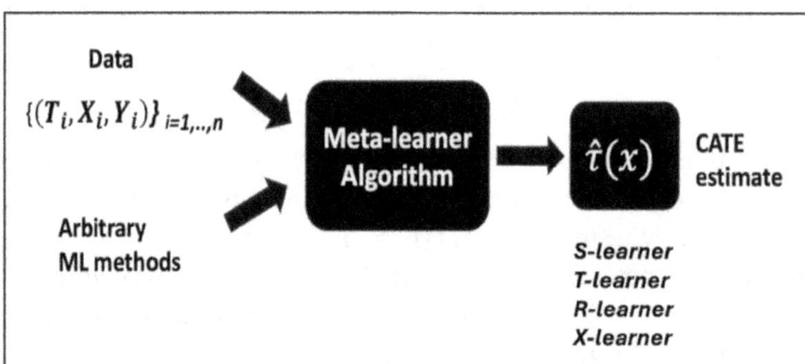

Figure 11.2—Function block diagrams of the meta-learner algorithms

The meta-learner algorithms for estimating the CATE do not commit to which base learners to use. A base learner is any supervised machine learning algorithm or regression analysis method.

In the following, we provide an overview of machine learning and regression methods that are viable approaches for use as base learners in various meta-learner algorithms. We also highlight their pros and cons for estimating the CATE on heterogeneous treatment effects.

Machine learning and regression methods

In this section, we provide an overview of various machine learning and regression methods that yield good models for adoption by meta-learner algorithms for estimating conditional average treatment effects. The software for this chapter, available on the book's website, implements several meta-learners in Python with the capability to use the methods overviewed in this section as base learners. The methods are **Random Forests (RF)**, **Light Gradient Boosting Machine (LightGBM)**, **Extreme Gradient Boosting (XGBoost)**, and **Linear Regression (LR)**. We discuss those methods next.

Random forest (RF)

Random Forest is an ensemble learning technique introduced by Breiman (Breiman 2001). Random Forests belong to the ensemble methods category (Re and Valentin 2012), which are statistical and computational learning techniques that mimic social learning and involve integrating diverse expert opinions to derive a consolidated ensemble decision. An illustrative example is the majority vote ensemble, where the decisions from various learning machines are aggregated. The class that obtains the majority of "votes"—specifically, the class predicted by most of the learning machines—becomes the final prediction of the ensemble.

The RF algorithm employs the Random **Subspace Method**, also referred to as **Feature Bagging** or **Feature Randomization**, which constructs N distinct decision trees by limiting each to a random subset of features through **Bootstrap Aggregating**. Each tree is constructed using a distinct random subset of features from the original dataset, employing a selection with replacement methodology. The ultimate prediction employs averaging by calculating the mean of forecasts from all trees. See *Figure 11.3*. For an excellent primer on decision trees and ensemble methods, see (Chadha 2025).

T	X1	X2	X3	Y
0	0.54	0.70	0.60	4.00
0	0.57	0.30	0.90	2.00
1	0.40	0.80	0.24	3.00
0	0.11	0.93	0.70	9.00
1	0.60	0.30	0.70	7.00
1	0.80	0.50	0.34	1.00

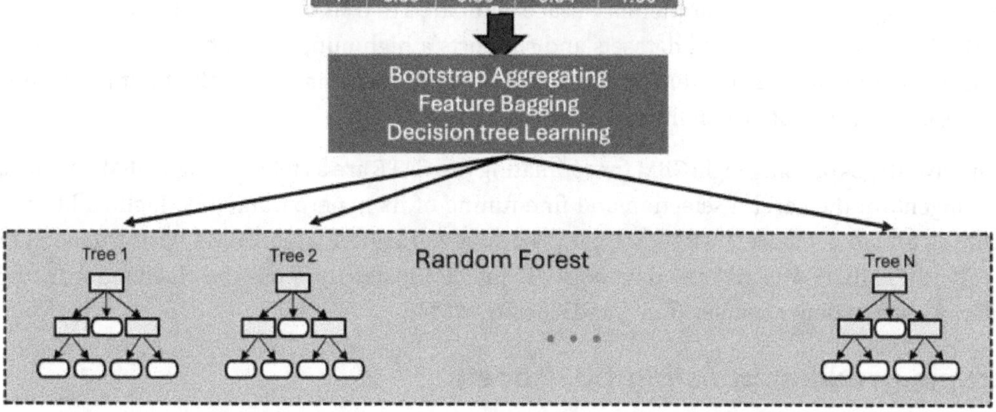

Figure 11.3—Random Forest uses bootstrap aggregating and feature bagging for predicting the output Y from the treatment T and the features given by the covariates X

The advantages of utilizing Random Forest for CATE estimation include the following. Random Forests can model intricate, non-linear treatment effects without explicit specification and do not require prior assumptions on the data's statistical distribution (e.g., normality or homoscedasticity). Random Forests can automatically identify relationship interactions between treatment and covariates. Furthermore, Random Forests are robust, having reduced sensitivity to the effects of outliers in the data. Studies have shown that Random Forests are capable of producing insights into the influence of covariates on treatment effect heterogeneity.

The disadvantages of using Random Forest for estimating the CATE are as follows:

- Random Forests represent a black box whose internals are less interpretable than other approaches that adopt linear models.
- Random Forests have higher computational complexity due to the training and management of multiple trees.
- Random Forests are sensitive to the setting of hyperparameters, and their efficacy depends on the precise tuning of these parameters.
- Random Forests' results from training data may not generalize well to new data, which can occur in cases with limited data.

Light gradient boosting machine (LightGBM)

LightGBM is an open-source framework developed by Microsoft (Microsoft 2022). LightGBM is an ensemble learning framework that builds trees sequentially, with each tree correcting errors of previous trees. LightGBM uses leaf-wise growth, where trees grow leaf-wise (best-first) rather than level-wise (depth-first). LightGBM employs histogram-based learning, which categorizes continuous features into discrete bins to accelerate training. LightGBM adopts **Gradient-based One-Side Sampling (GOSS)**, which focuses on instances with larger gradients during training. For an overview of LGBM, see (Avanwyk 2018).

The advantages of utilizing LightGBM for CATE estimation include the following. LightGBM often achieves superior prediction accuracy to all other boosting algorithms. LightGBM is efficient, with higher training speed and lower memory usage compared to traditional gradient-boosting methods. LightGBM scales well with large datasets and supports a high number of observations. LightGBM's built-in L1 (Lasso) and L2 (Ridge) regularization prevent overfitting. LightGBM naturally supports categorical features—native handling of categorical covariates.

The disadvantages of using LightGBM for estimating the CATE are as follows. LightGBM performance is contingent on the careful selection and fine-tuning of its hyperparameters. LightGBM may risk overfitting on small datasets without appropriate regularization. LightGBM's performance may be less robust due to its sensitivity to data noise. LightGBM installation may be challenging, requiring additional system dependencies (Chen and Guestrin 2016).

Extreme gradient boosting (XGBoost)

Extreme Gradient Boosting (XGBoost), developed by Chen and Guestrin, is a supervised learning approach that utilizes gradient descent for an enhanced implementation of **gradient-boosted decision**

trees. XGBoost is known for its speed, efficiency, and ability to scale well with large datasets. XGBoost performs regularized learning, incorporating L1 and L2 regularization terms inside the objective function to prevent overfitting and improve generalization performance. XGBoost employs an approximate greedy algorithm based on a quantile sketch algorithm to find approximate best splits. XGBoost efficiently handles missing values and sparse data via sparsity-aware split finding. For a good overview and introduction to XGBoost, see (Kavlakoglu and Russi 2024).

XGBoost consistently demonstrates strong performance across various prediction tasks, thanks to its built-in regularization learning mechanisms that prevent overfitting. XGBoost handles missing values through its native support of sparsity-aware split finding. XGBoost is computationally efficient, supporting multi-threading and parallel processing. XGBoost has built-in cross-validation capabilities, enabling robust model evaluation and hyperparameter tuning.

The disadvantages of using XGBoost for estimating the CATE are as follows. XGBoost requires fine-tuning of many hyperparameters. XGBoost yields black-box prediction models, obscuring the interpretability of the rationale behind their predictions. XGBoost can be memory-intensive for large datasets. XGBoost dependencies require the installation of additional libraries, which can be challenging.

Linear regression (LR)

Linear regression utilizes linear models, assuming that the effects of features contribute independently to the outcome, which is modeled as a linear function of (a weighted sum of) input features—the weights being the regression coefficients. Linear regression utilizes the **Ordinary Least Squares (OLS)** algorithm to compute the regression coefficients by minimizing the sum of the squared errors between the observed values and the values predicted by the model. For a technical text on regression, see (Fahrmeir, et al. 2013).

The advantages of using linear regression for estimating the CATE include the following. Linear regression yields interpretable prediction models where regression coefficients quantify how the covariates and the treatment influence the outcome, up to an unknown random error representing external factors. A linear regression model can explicitly model interactions between treatment and covariates (Tian, et al. 2014). Linear regression is fast to train on training data and make predictions on new data. Linear regression yields results that are stable and reproducible. Linear regression enables the making of statistical inferences, providing p-values and confidence intervals for predictions of treatment effects.

The disadvantages of linear regression for estimating the CATE are as follows. Linear regression cannot capture non-linear treatment effects without manual feature engineering. The linear regression tacit homogeneity assumption, which assumes constant error variance across all observations, may not be generally justified. Linear regression is sensitive to outliers, which can significantly impact model coefficient estimates. Linear regression has limited flexibility, which can cause it to underfit complex relationships in the data. Linear regression performance degrades with highly correlated features.

To summarize this section, here are the key takeaways. The choice of base learner has a significant impact on the CATE estimation by meta-learners. Four distinct supervised machine learning methods to adopt as base learners are: Random Forest, LightGBM, XGBoost, and Linear Regression. Each base learner method has its advantages and disadvantages. The optimal base learner depends on your specific use case and the observational dataset. Random Forests are a good default choice that strikes a balance between flexibility and robustness. LightGBM/XGBoost are the best for large datasets where prediction accuracy is paramount. Linear Regression is ideal when interpretability is critical or when the treatment effect is known to be linear. Tree-based methodologies (Random Forest, LightGBM, XGBoost) generally surpass linear models in handling complicated, heterogeneous treatment effects. Nonetheless, linear models provide superior interpretability and a clear representation of treatment interactions.

With this, we conclude this section and proceed to the next, where we will describe a causal model that will be used as a running use case throughout the chapter to demonstrate and evaluate the performance of various meta-learner algorithms for estimating heterogeneous treatment effects.

Specifying generative data models

In this section, we introduce a causal model that serves as a generative data model for generating a synthetic dataset, which we use throughout the remainder of this chapter. The causal model is designed to simulate use case causal effect relations with heterogeneous effects. Following this, we will explain the steps for implementing the causal model using Python. Afterward, we will engage you in a practice session using Python, where we provide a walkthrough of the steps to generate a synthetic dataset with heterogeneous treatment effects and to analyze the dataset's statistical properties to illustrate the heterogeneity of the simulated causal effects represented by the dataset.

Causal model with heterogeneous treatment effects

We previously mentioned the missing data problem, which is the core issue in causal inference with observational data. The gist of the problem is that the actual individual treatment effects are never directly observable because we can only observe one potential outcome, the outcome observed as a result of the treatment assigned to the individual. Still, the other potential outcomes, the counterfactual outcomes from other treatment assignment decisions, are never observed on the same individual.

Due to the missing data problem, we lack a gold standard to evaluate the performance of meta-learning algorithms on observational datasets unless we possess knowledge of or make assumptions about the data generative model, the underlying causal model.

It has been recognized that evaluating meta-learning algorithms for estimating the CATE on observational data represents a challenge, even when the identifying assumptions of the Neyman-Rubin model are satisfied. That is why researchers have suggested that to provide proof of concept for meta-learning algorithms in CATE estimators, we may rely on synthetic or semi-synthetic datasets (with real covariates) that possess a causal model, in which both potential outcomes are simulated and hence known, to showcase their properties (Curth, et al. 2021).

In the software for this chapter, two causal models for heterogeneous treatment effects, named `model1` and `model2`, have been implemented along with a wrapper class as a front end to interface with the causal models. Both models have the same graph, but their functional equations differ.

In the next section, we formally describe one of the two causal models, model2, which we utilize for the empirical analyses of the metalearners presented throughout this chapter.

Describing causal models

In this section, we describe a causal model for generating synthetic datasets to demonstrate and evaluate the meta-learning algorithms described in this chapter. The causal model is designed to exhibit several key properties that make it suitable for evaluating and comparing different CATE estimation methods. The aim is to create a synthetic dataset with highly heterogeneous treatment effects that vary based on covariates. The causal model is designed to simulate heterogeneous treatment effects with complex confounding where covariates nonlinearly influence both the treatment assignments and the outcomes of the treatments.

Figure 11.4 depicts the causal graph of the synthetic data model. The graph illustrates the causal structure used to generate our synthetic data. The graph has three nodes: X: Covariates $(X_1, \cdots, X_5)$ represented as a box; T: Treatment variable - represented as a circle; Y: Outcome variable - represented as a circle. The directed graph has three edges: $X \rightarrow T$: Covariates influence treatment assignment (confounding); $X \rightarrow Y$: Covariates directly affect the outcome; $T \rightarrow Y$: Treatment affects the outcome.

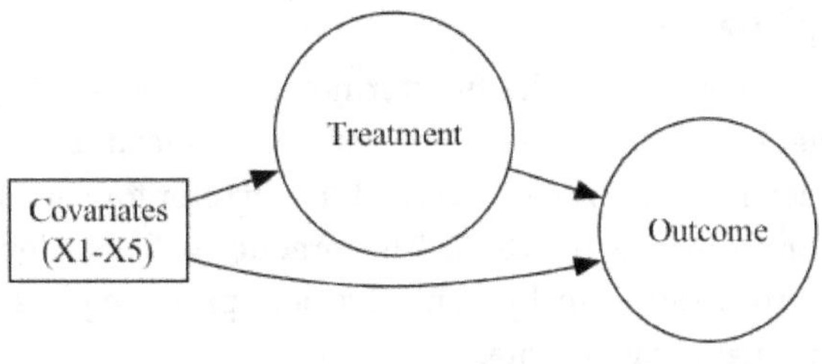

Figure 11.4—Causal graph for the synthetic data model

The following specifies the functional relationships of the causal model. Covariates $X = (X_1, X_2, \cdots, X_5)$, consists of five independent features drawn from a uniform distribution over [0,1]. Treatment is a binary variable (0 or 1) assigned with nonlinear confounding using the following expression for the propensity score:

$$p(X) = \frac{1}{1+\exp\left(-(-2+4X_1+2X_2^2-3X_3X_4)\right)} \quad (11.3)$$

The true CATE for an individual with covariates X is given by the function $\tau(X)$, which serves as the ground truth when evaluating estimation methods. We prescribe $\tau(X)$ for highly heterogeneous treatment effects given by the following formula using a threshold function.

$$\tau(X) = 4.0 \cdot \mathbb{1}(X_1 > 0.5) - 3.0 \cdot \mathbb{1}(X_2 > 0.7) +$$

$$5.0 \cdot \mathbb{1}(X_3X_4 > 0.5) - 2.0 \cdot \mathbb{1}(X_5 < 0.3) \quad (11.4)$$

The formula of $\tau(X)$ creates **heterogeneous treatment effects** with the following components: positive effect of 4.0 if X1 > 0.5; negative effect of 3.0 if X2 > 0.7; positive effect of 5.0 if X3*X4 > 0.5; and negative effect of 2.0 if X5 < 0.3.

To model **nonlinear baseline effects**, we use the following formula for the baseline outcome, the expected output for the control group conditioned on the covariates:

$$\mu_0(X) = 1.0\sin(2X_1) + 0.5X_2^2 + 0.8X_3X_4 + 0.3\exp(X_5) \quad (11.5)$$

To model the noise term, we consider a formula for *heteroskedastic noise* (noise level varies with X_1):

$$\varepsilon \sim \mathcal{N}(0{,}0.5 + 0.5X_1) \ (11.6)$$

The model's final equation for the outcome Y as a function of the treatment and the covariates is given by the equation:

$$Y = \mu_0(X) + T \cdot \tau(X) + \varepsilon \ (11.7)$$

Ending this section, let us summarize the main points. The data generation process involves a causal model that includes covariates X drawn from a uniform distribution, treatment assignment T with nonlinear confounding, a nonlinear functional form for heterogeneous treatment effects and baseline effects, and heteroskedastic noise (where the noise level varies with one covariate).

In the following section, we describe the Python code that simulates the causal model and generates the synthetic datasets to be used for evaluating the metalearners' performance in the coming sections.

Practice using Python

The code for this practice is included in the **Jupyter** notebook '`synthetic_data_for_cate2.ipynb`'. In this practice, we will generate the synthetic dataset from the causal model described in the preceding section. The dataset will be utilized throughout the chapter for evaluating the performance of metalearner algorithms for estimating the CATE. This practice will also demonstrate how to analyze the dataset, compute descriptive statistical parameters, and produce plots to visualize the heterogeneous treatment effects simulated by the data.

We will begin by generating the synthetic dataset.

Generating synthetic datasets

A class `synthetic_data_for_cate` is defined in the Python file '`synthetic_data_for_cate_class.py`'. The class provides a unified interface for different synthetic data generation models, allowing users to experiment with different data models without the need to write additional Python modules of the same structure. The class has two synthetic data models, *'model1'* and *'model2'*, with *'model2'* being the default data model used, whose equations are listed in the previous section.

The parameters to initialize the class are: `model_type: str, default='model2'`; `n: int, default=1000` (number of samples to generate); `d: int, default=5` (number of covariates/features to generate). The class has the method `get_synthetic_data()` that generates the synthetic data and returns the triple: `X: ndarray of shape (n, d)` (matrix of covariates/features); `treatment: ndarray of shape (n,)` (binary treatment assignment (*1=treated, 0=control*)); `y: ndarray of shape (n,)` (outcome variable) The class also has the method `get_true_cate()`, which takes the matrix of covariates/features X, an array of shape (n, d), calculates the true CATE for the given features, and returns `true_cate`, an array of shape (n,).

The code for generating the synthetic data and creating the dataframe, which contains the treatment, covariates, and outcome using 'model2' with the default values for the number of samples (n = 1000) and the number of covariates (d = 5), is as follows:

```
# Create an instance of synthetic_data_for_cate with model2
data_generator = synthetic_data_for_cate(model_type='model2')
# Generate synthetic data
X, treatment, y = data_generator.get_synthetic_data()
# Create DataFrame with all variables
data = pd.DataFrame({
    'treatment': treatment,
    'y': y,
    'X1': X[:, 0],
    'X2': X[:, 1],
    'X3': X[:, 2],
    'X4': X[:, 3],
    'X5': X[:, 4]
})
```

The following list shows the first five records of the generated data.

```
treatment y         X1       X2       X3       X4       X5
1   2.011870   0.548814 0.715189 0.602763 0.544883 0.423655
0   0.589514   0.645894 0.437587 0.891773 0.963663 0.383442
1   9.690471   0.791725 0.528895 0.568045 0.925597 0.071036
0   2.296981   0.087129 0.020218 0.832620 0.778157 0.870012
1   1.394680   0.978618 0.799159 0.461479 0.780529 0.118274
```

Having generated a synthetic dataset for this task, we are now ready to proceed with the statistical analysis of the dataset to gain insight into the heterogeneous treatment effects of our generative model.

The next task will focus on the correlation analysis between the model variables.

Performing correlation analysis

This practice section analyzes the correlations between all variables in our synthetic dataset. We calculate the **Pearson correlation coefficients** between all pairs of variables and visualize these correlations using a heatmap where the intensity of color represents correlation strength; numerical values show the exact correlation coefficients. Red indicates a positive correlation, and blue indicates a negative one.

Note that correlation does not imply causation. The causal graph shown in *Figure 11.4* illustrates the presumed causal relationships, whereas this correlation analysis reveals only the statistical associations. The code for generating the correlation heatmap is shown below.

```
import seaborn as sns
import matplotlib.pyplot as plt
```

```
# Analyze correlations between variables
plt.figure(figsize=(10, 6))
correlation_matrix = data[['treatment', 'y', 'X1', 'X2', 'X3', 'X4', 'X5']].
corr()
sns.heatmap(correlation_matrix, annot=True, cmap='coolwarm', center=0)
plt.title('Correlation Heatmap')
plt.tight_layout()
plt.show()
```

Executing the code computes and plots the heatmap for the generated synthetic data for CATE, as shown in *Figure 11.5*.

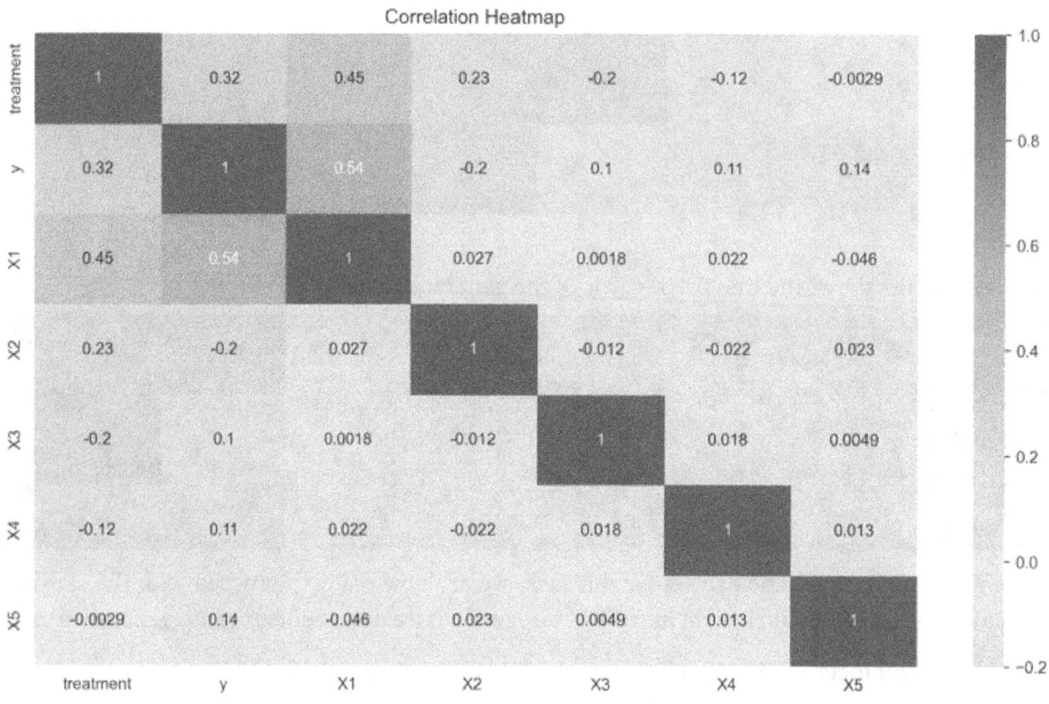

Figure 11.5—Heat map for the generated synthetic data for CATE

The correlation heatmap shows that the treatment has a positive correlation with the outcome, with a correlation coefficient of 0.32. This raw correlation is confounded primarily by X1 and, to a lesser extent, by the other four covariates, as reflected by X1 having a correlation coefficient of 0.45 with treatment and 0.54 with the outcome. The correlation heatmap shows a weak correlation between the covariates, indicating limited multicollinearity between the covariates. Note that, despite the low correlation between covariates, there may still be non-linear relationships between covariates that aren't captured by the Pearson correlation coefficient.

Having completed the correlation analysis task, let us proceed to the next task of computing basic statistics about the true CATE.

Computing descriptive statistics on the True CATE

The true CATE represents the actual causal effect of the treatment on the outcome for each individual based on their covariates. For our synthetic data, the true CATE is calculated using the `get_true_cate` method of the `synthetic_data_for_cate` class. The following code computes the true CATE and prints basic statistics, including the mean, standard deviation, minimum, maximum, and proportions of positive, negative, and neutral effects.

```
# Compute true CATE values using the data generator
true_cate = data_generator.get_true_cate(X)

# Add true CATE to the DataFrame
data['true_cate'] = true_cate

# Print basic statistics about the true CATE
print(f"True CATE Statistics:")
print(f"  Mean: {true_cate.mean():.4f}")
print(f"  Std Dev: {true_cate.std():.4f}")
print(f"  Min: {true_cate.min():.4f}")
print(f"  Max: {true_cate.max():.4f}")

# Count the number of positive and negative treatment effects
positive_effects = (true_cate > 0).sum()
negative_effects = (true_cate < 0).sum()
neutral_effects = (true_cate == 0).sum()
print(f"\nTreatment Effect Distribution:")
print(f"  Positive Effects: {positive_effects} ({positive_effects/len(true_
cate)*100:.1f}%)")
print(f"  Negative Effects: {negative_effects} ({negative_effects/len(true_
cate)*100:.1f}%)")
print(f"  Neutral Effects: {neutral_effects} ({neutral_effects/len(true_
cate)*100:.1f}%)")
```

Running the above code produces the following results:

```
True CATE Statistics:
  Mean: 1.2560
  Std Dev: 3.1344
  Min: -5.0000
  Max: 9.0000

Treatment Effect Distribution:
  Positive Effects: 518 (51.8%)
```

```
Negative Effects: 260 (26.0%)
Neutral Effects: 222 (22.2%)
```

The results show that the synthetic data simulate a use case of heterogeneous treatment effects, where the treatment effect varies qualitatively across individuals and subgroups in the population. Nearly half of the population experiences positive effects, while the other half is split between negative effects and neutral effects, having no effect.

Having computed basic statistics on the true CATE, we will proceed in the next section to plotting the true CATE distribution, which will provide a fine-grained picture of the heterogeneous nature of the treatment effects in our synthetic dataset.

Plotting the True CATE distribution

This task involves plotting the distribution of the true **Conditional Average Treatment Effect** (CATE). The code for this task is contained in the notebook but not shown here due to space limitations. The execution of the code produces the plot in *Figure 11.6*.

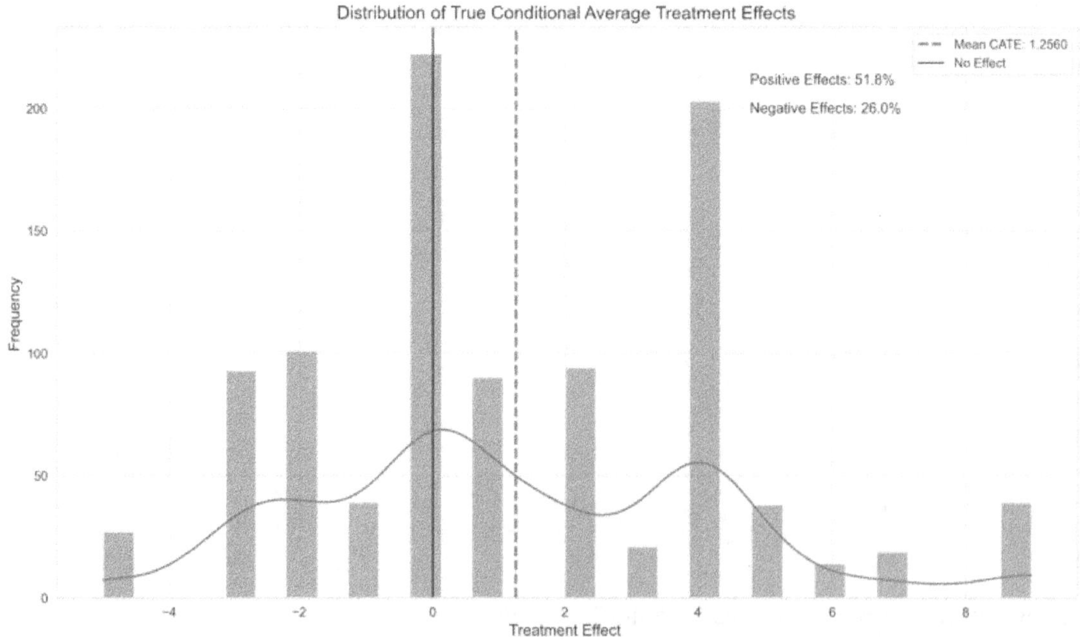

Figure 11.6—The true CATE distribution for the synthetic data

The distribution appears to have multiple peaks, suggesting distinct subgroups with different treatment responses. The heterogeneity of the treatment effects is apparent from the partitioning of population subgroups into those unaffected by the treatment and others who are positively or negatively affected.

Having visualized the true CATE distribution, the next task is to visualize the relationship between the true CATE and the covariates.

Visualizing the relationship between the True CATE and the covariates

In this section, we visualize the relationship between each covariate and the true CATE values. Due to space restrictions, the code for this task is not shown. Executing the code in the notebook will output the five plots shown in *Figure 11.7*. Each plot shows one covariate (X-axis) against the true CATE (Y-axis), with a horizontal line at y = 0 separating positive from negative effects. The plots reveal key patterns in the relationship between covariates and the true CATE, including threshold effects for X1 and X2 (values jump at X1 ≈ 0.5 and X2 ≈ 0.7), interaction effects for X3 and X4 (where X3*X4 > 0.5), and step functions for X5 (lower CATE values when X5 < 0.3).

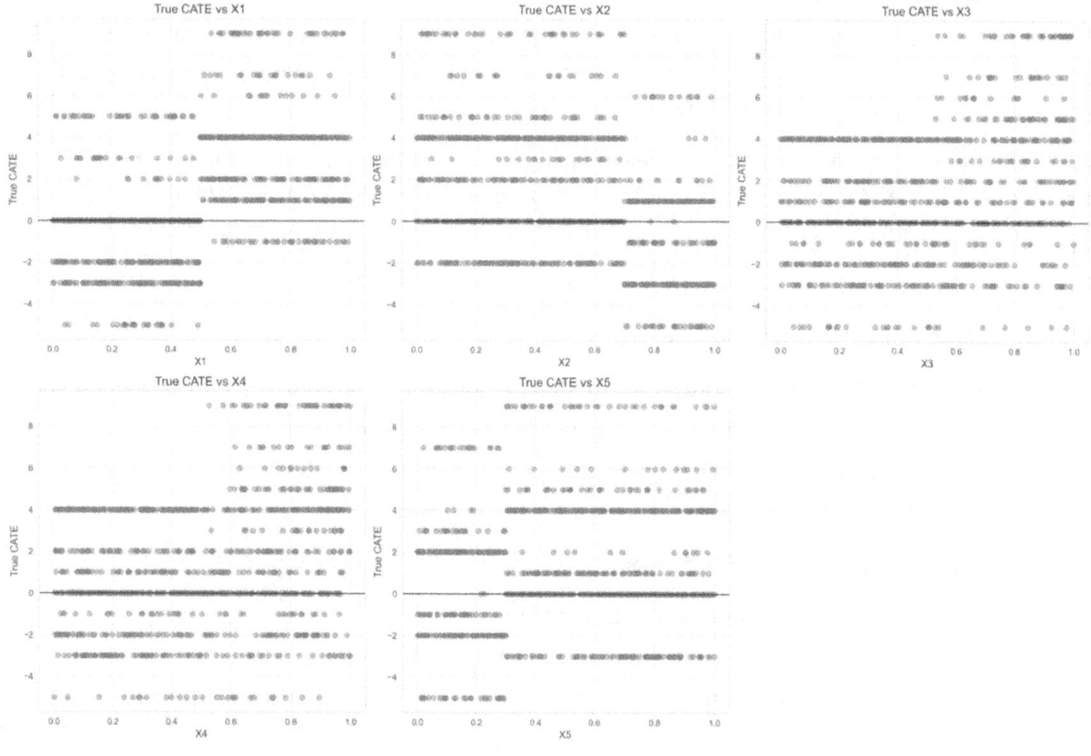

Figure 11.7—Plots for the True CATE versus each of the five covariates in the synthetic data (This image is for illustrative purposes only)

Due to the missing data problem in causal inference from observational data, a causal model is necessary to generate synthetic datasets for evaluating the performance of meta-learners in estimating conditional average treatment effects. The causal model enables the simulation of heterogeneous treatment effects and the generation of datasets where the potential outcomes of treatments are available. In contrast, real-life scenarios typically provide only the actual outcome of the treatment, not the other potential outcomes, known as counterfactuals. The causal model described in this section, along with the datasets generated from the Python code demonstrated in the practice section, will be used throughout the chapter.

With this, we conclude this section and proceed next to present the first meta-learner, namely the S-Learner.

Understanding S-Learner

In this section, we begin with a description of S-Learner (Single-Learner). We then follow this with a description of implementing S-Learner using Python. Subsequently, we provide practice using Python for the empirical evaluation of the implemented S-Learner on the synthetic dataset generated by the causal model described in the previous section.

Describing S-Learner

S-Learner is a meta-learner approach for estimating CATE. The S-learner employs a single model that incorporates the treatment indicator as a feature, i.e., it treats the treatment assignment as a single feature (*Foster, Taylor and Ruberg 2011*).

The S-learner uses any suitable machine learning algorithm to fit the outcome model $M(Y{\sim}(X,T))$ for features X,T. Here M is any suitable machine learning algorithm. The CATE estimator provided by S-learner is then given as: $\hat{\tau}(x) = \hat{\mu}(x,1) - \hat{\mu}(x,0)$. and estimate the combined outcome as:

$$\hat{\mu}(x,t) = E[Y|X = x.T = t] \quad (11.8)$$

The key characteristics of S-Learner are:

- Single Model Approach: Uses a single model that includes the treatment indicator as a feature;
- Flexible Base Learners: Can use any regression machine learning model;
- Simple Implementation: Straightforward to implement and understand;
- Efficient Computation: Only requires training one model.

The S-Learner approach follows these steps:

1. Combine the treatment indicator with the feature matrix;
2. Train a single model to predict outcomes using the combined features;
3. Estimate the CATE by taking the difference between predicted outcomes with *treatment = 1* and *treatment = 0*.

The main advantage of the S-Learner is its simplicity. Still, it may struggle with treatment effect heterogeneity if the treatment indicator is not given enough importance by the base learner. In the S-learner, the treatment indicator is included as a feature, like all the other features, without being given any special role.

Formally, the S-learner first estimates the combined response function, *Equation 11.8* using any base learner on the entire dataset. The base learner is any supervised ML algorithm or regression algorithm. The S-learner then estimates the CATE by evaluating the $\hat{\mu}(x,t)$ for $t = 0$ and $t = 1$, then takes the difference to estimate the CATE function as follows:

$$\hat{\tau}_S(x) = \hat{\mu}(x,1) - \hat{\mu}(x,0) \quad (11.9)$$

Having described S-Learner, let us proceed to describe how to implement it using Python next.

Implementing S-Learner using Python

The project associated with this chapter on the book's website implements the S-Learner as the *SLearner class* defined in the Python file 's_learner.py' contained in the **meta-learners** folder. We provide a highlight of the code in the 's_learner.py' file, explaining the class definition with code snippets step by step.

Importing libraries and defining the class

First, we load the required libraries.

```
from sklearn.base import BaseEstimator, RegressorMixin
from sklearn.ensemble import RandomForestRegressor
from typing import Optional, Union
import numpy as np
```

The following statement defines the class:

```
class SLearner(BaseEstimator, RegressorMixin):
```

The `BaseEstimator` parameter specifies the base learner that will be used for prediction. A base learner must implement `fit()` and `predict()` methods. Common choices for a base learner include: `RandomForestRegressor()`, `LGBMRegressor()`, `XGBRegressor()`, `LinearRegression()`. A base learner in the Python implementation should be from scikit-learn or follow scikit-learn's API.

The `RegressorMixin` parameter is a `mixin` class for regressors in *scikit-learn*. It provides default implementations for methods common to all regressors.

The code for the class constructor is listed below.

```
def __init__(self, model: Optional[BaseEstimator] = None):
    if model is None:
        self.model = RandomForestRegressor()
    else:
        self.model = model
```

The *Slearner* class has one attribute `model_instance_ : BaseEstimator` which represents a fitted model instance.

Defining methods of the SLearner class

The *SLearner* class defines three methods: `fit()`, `predict()`, and `effect()` as described next.

Defining the fit() method

The *SLearner* class defines the `fit()` method for fitting the S-learner model. The method takes three parameters: 1) the feature matrix X: array-like of shape (`n_samples, n_features`), 2) the treatment assignments (0 or 1) t : array-like of shape (`n_samples,`), and 3) the observed outcomes y: array-like of shape (`n_samples,`). The `fit()` method returns the instance itself.

```
def fit(self, X: np.ndarray, t: np.ndarray, y: np.ndarray) -> 'SLearner':
    # Concatenate treatment with features
        X_with_t = np.column_stack([X, t])
    # Fit the model
    self.model_instance_ = self.model.fit(X_with_t, y)
    return self
```

We describe the `predict()` method code next.

Defining the predict() method

The *SLearner* defines the method `predict()` to predict outcomes for given features and treatment assignments. The method takes two parameters: a feature matrix. `X` : array-like of shape (`n_samples, n_features`), and treatment assignments (0 or 1) `t` : array-like of shape (`n_samples`). The method returns the predicted outcomes, `y_pred`, as an array-like object of shape (n_samples).

```
def predict(self, X: np.ndarray, t: np.ndarray) -> np.ndarray:
            X_with_t = np.column_stack([X, t])
            return self.model_instance_.predict(X_with_t)
```

The code defining the `effect()` method is described next.

Defining the effect() method

The *SLearner* defines the method `effect()` for estimating the CATE. It takes the feature matrix as input, `X` : array-like of shape (`n_samples, n_features`), and returns as output, cate, the estimate of the CATE. The method returns the estimated conditional average treatment effects, `cate` : array-like of shape (`n_samples,`), which represents the expected impact of the treatment on the outcome for an individual with specific characteristics. Computing this involves 1) creating two copies of each individual's features, one with $treatment = 1$ and one with $treatment = 0$; 2) predicting outcomes for both scenarios; 3) computing the difference between the predicted outcomes (treatment minus control), which gives us an estimate of the treatment effect for each individual, conditional on their covariates.

```
def effect(self, X: np.ndarray) -> np.ndarray:
    # Create copies with treatment 0 and 1
    X_control = np.column_stack([X, np.zeros(X.shape[0])])
    X_treated = np.column_stack([X, np.ones(X.shape[0])])
    # Compute CATE as difference in predictions
    cate = self.model_instance_.predict(
        X_treated
```

```
        ) - self.model_instance_.predict(X_control)
        return cate
```

With this, we have completed describing the implementation of the S-Learner.

This approach allows the S-Learner framework to flexibly model individual-level effects by leveraging a single predictive model that incorporates both treatment and covariate information. By concatenating the treatment indicator to the feature set, the model learns to distinguish differences in outcomes attributable to the presence or absence of treatment, conditional on observed characteristics. The method is versatile and can be applied with any supervised learning estimator compatible with scikit-learn's predict interface.

With the core mechanism behind the S-Learner explained, we now turn to a practical example to illustrate its application.

In the following section, we provide a walkthrough on how to use the *SLearner* class to estimate the CATE using the synthetic dataset of our heterogeneous treatment effects causal model, which we computed in the previous section.

S-Learner practice using Python

This practice demonstrates how to use the S-Learner implementation using Python. The code for this practice is included in the file `s_learner.ipynb` found in the `notebooks` folder. Ensure that you have installed all required dependencies before running this Notebook, which include the necessary libraries: `NumPy`, `Matplotlib`, `Seaborn`, and `scikit-learn`. The Notebook uses the local packages: `synthetic_data_for_cate_class` for data generation and `s_learner` for the SLearner implementation.

In the following section, we give a step-by-step demonstration on how to use the *SLearner* class to perform tasks related to estimating the CATE. We will begin with the generation of synthetic data next.

Generating synthetic data

The following code generates synthetic data with heterogeneous treatment effects and prints basic information about the generated data.

```
# Create an instance of synthetic_data_for_cate with model2
data_generator = synthetic_data_for_cate(model_type='model2')
# Generate synthetic data with heterogeneous treatment effects
features, treatment_vector, outcomes = data_generator.get_synthetic_data()
# Print basic information about the generated data
print(f"Generated data with {features.shape[0]} samples and {features.shape[1]} features")
print(f"Treatment assignment rate: {treatment_vector.mean():.2f}")
print(f"Outcome mean: {outcomes.mean():.2f}")
```

The code produces the following output:

```
Generated data with 1000 samples and 5 features
Treatment assignment rate: 0.49
Outcome mean: 2.26
```

Next, we proceed with the main task of creating an *SLearner* object to carry out follow-up tasks.

Creating an S-Learner object and fitting it to the data

The code below illustrates how to create an *SLearner* object using the **Random Forest** model as the base learner and fit the model with the synthetic data.

```
# Initialize S-Learner with Random Forest as base learner
sl = SLearner(RandomForestRegressor(n_estimators=100, random_state=42))
# Fit the model
print("Fitting S-Learner with Random Forest...")
sl.fit(
    X=features,             # Feature matrix
    t=treatment_vector,     # Treatment assignments (0/1)
    y=outcomes              # Observed outcomes
)
print("Model fitting complete.")
```

The code should output the following after it runs successfully.

```
Fitting S-Learner with Random Forest...
Model fitting complete.
```

Upon creating the *SLearner* object, we are prepared to execute numerous jobs, the primary one being the estimation of the CATE, which we will proceed with next.

The S-Learner CATE estimation

The following code will estimate the CATE for each individual in our dataset.

```
# Estimate treatment effects (CATE)
cate = sl.effect(X=features)
# Print basic statistics about the estimated CATE
print(f"CATE Statistics:")
print(f"  Mean: {cate.mean():.4f}")
print(f"  Std Dev: {cate.std():.4f}")
print(f"  Min: {cate.min():.4f}")
print(f"  Max: {cate.max():.4f}")
```

The code, when run, will output the following summary statistics:

```
CATE Statistics:
  Mean: 1.1137
  Std Dev: 2.8711
  Min: -5.6518
  Max: 10.6275
```

The next task is to plot the CATE distribution to depict how the estimated CATE varies across the population visually.

The S-Learner CATE distribution visualization

We do not list the visualization code here; you can find it in the Jupyter Notebook for this section. The code will output the CATE distribution depicted in *Figure 11.8*.

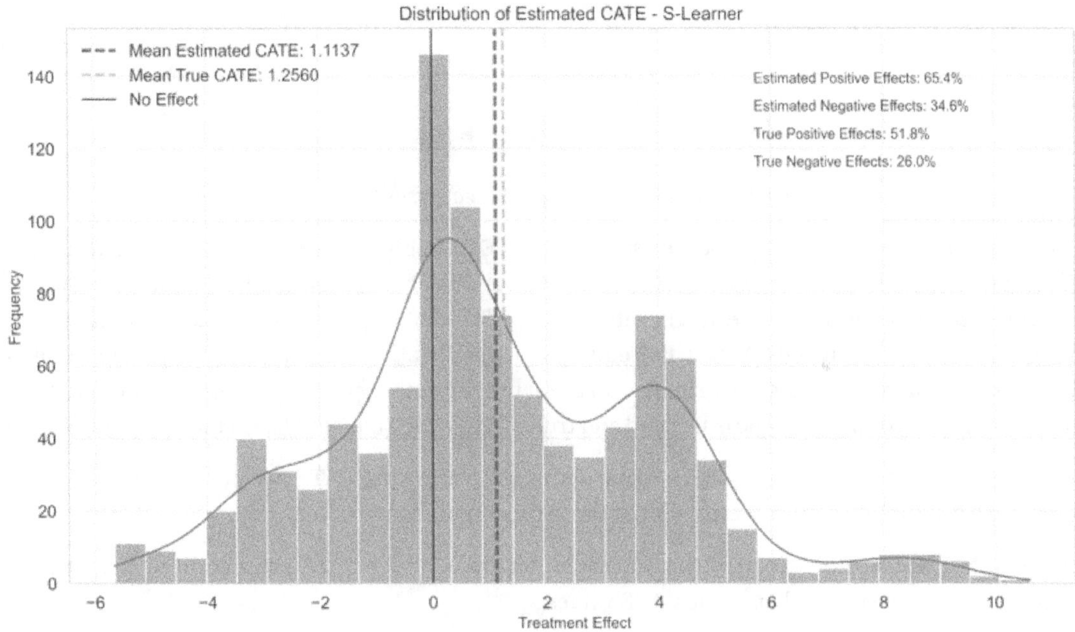

Figure 11.8—S-learner Estimated CATE distribution

The next task will be to compare the estimated CATE with the true CATE distribution by having them overlaid on the same plot, which we do next.

The S-Learner overlaid CATE distributions: estimated vs. True

The notebook for this practice contains the code for this task, which compares the degree of proximity between the real and estimated distributions. The code will create the plot shown in *Figure 11.9*, which displays the estimated and true CATE distributions on the same graph, using different colors.

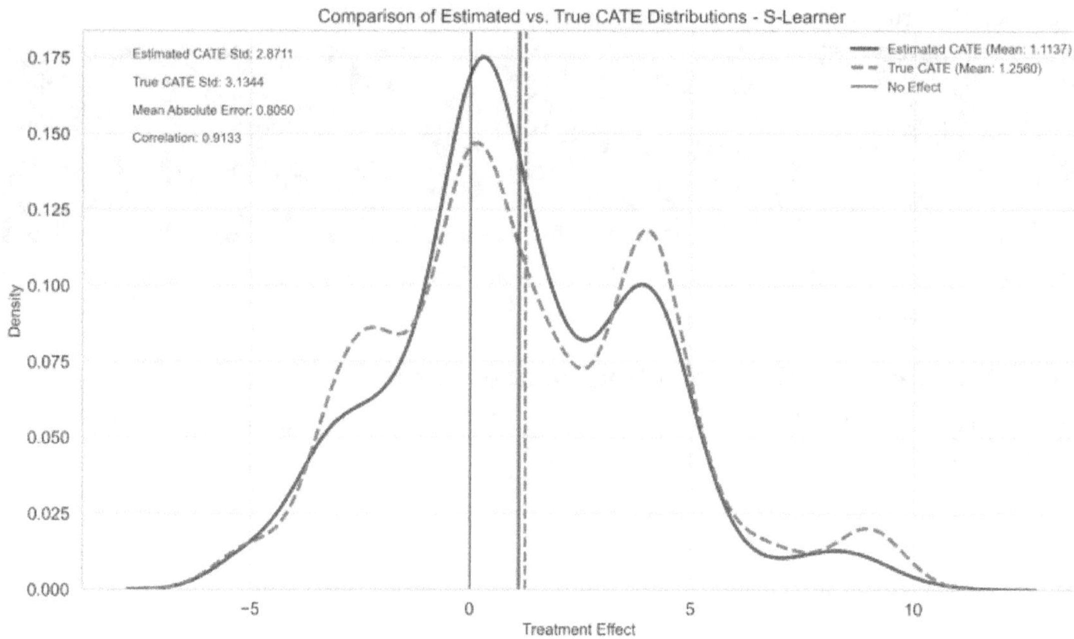

Figure 11.9—Comparison of the S-Learner estimated versus the true CATE distribution

The plot indicates that the shape of the estimated CATE distribution generally aligns with the shape of the true CATE distribution. The plot clearly shows that the estimated CATE tracks the true CATE on average over the range of positive and negative effects. The plot displays the computed correlation between the true and estimated CATE, which equals 0.9133, indicating a very positive empirical result on the evaluation of the learner's performance on the synthetic data. The plot also displays the computed mean absolute error between the estimated and true CATE, which equals 0.805, a relatively small value.

To further analyze the accuracy of the S-Learner in estimating the CATE, we perform one final task in the next section: plotting a scatter plot of the estimated and true CATE.

Our final task will further analyze the S-Learner accuracy of the CATE estimation by producing a scatter plot of the estimated and true CATE values.

The S-Learner CATE accuracy evaluation: predicted vs. actual treatment effects

The notebook contains code, not listed here due to space restrictions, for computing the scatter plot shown in *Figure 11.10*. The scatter plot in the figure illustrates the projected CATE values on the y-axis and the actual treatment effects on the x-axis. The figure utilizes separate symbols to denote the treated and control groups. The diagonal line (y = x) in the graph signifies flawless prediction. Proximity to the diagonal line signifies precise predictions, while deviations imply estimation inaccuracies.

Figure 11.10 visually illustrates how well the S-Learner can predict the true treatment effects, whether systematic biases exist in the predictions, and whether the predictions differ in accuracy between the treated and control groups. The plot also reveals whether there are outliers for whom the s-learner predictions are particularly inaccurate (points far off from the diagonal line).

The scatter plot reveals an important characteristic of our data and model. The True CATE Values (x-axis) appear as discrete values, forming vertical clusters, while the Estimated CATE Values (y-axis) appear as continuous values, spread across a range from approximately -6 to +10. This pattern occurs because in `'model2'`, true CATE values are generated using threshold functions, as shown in *Equation 11.4*. Each term in the equation is a binary condition (0 or 1) multiplied by a coefficient. With four binary conditions, there are $2^4 = 16$ possible combinations, creating a discrete set of possible CATE values. The S-Learner utilizes Random Forest Regressor, which generates continuous predictions by averaging the outputs of multiple decision trees, resulting in a continuous range of estimated values.

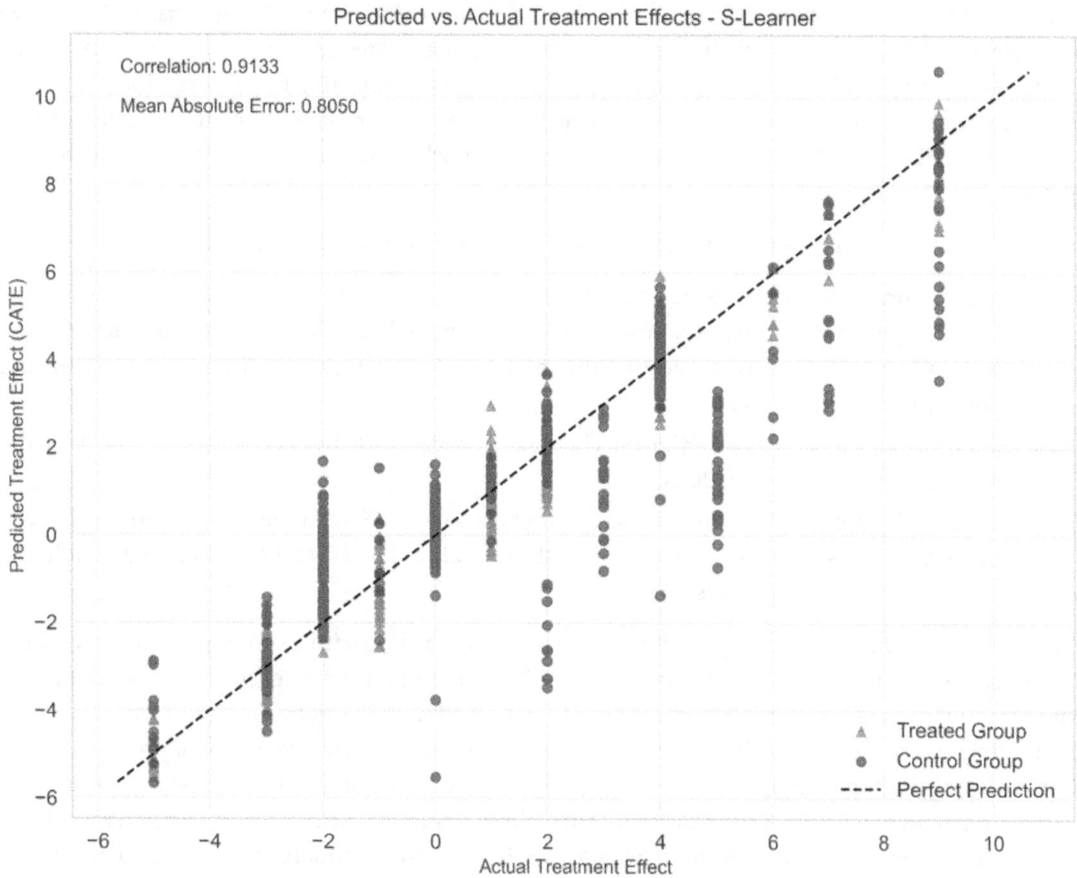

Figure 11.10— S-Learner Scatter Plot: Estimated Versus Actual Treatment Effect for Treatment and Control Groups

The visual assessment of the scatter plot shows points clustering around the diagonal line but with considerable spread, suggesting moderate accuracy. Given the difficulty of recovering discrete threshold-based effects using a continuous prediction model, the S-Learner implementation with a Random Forest base learner detects the broad direction of treatment effects (positive vs. negative) but struggles to quantify their magnitude accurately.

The scatter plot reveals the presence of several types of outliers. Some points deviate significantly from the perfect prediction line. A few points have estimated CATE values below -5 or above +8, which are more extreme than most predictions. Some points with a positive true CATE have a negative estimated CATE, and vice versa, representing qualitative errors in the effect direction. These outliers likely occur when multiple feature interactions create complex patterns that the model fails to capture, rare feature combinations appear in the data that weren't well represented in the training process, or the inherent noise in the data generation process leads to challenging cases.

Comparing the red triangles (treated group) and blue circles (control group), we can make the following remarks. One, both groups appear similarly distributed across the true CATE spectrum. Two, there is no obvious difference in prediction accuracy between the treated and control groups. Three, the treated group may have slightly more points in certain regions, reflecting the non-random treatment assignment in the data generation process. Fourth, the similar prediction accuracy across groups suggests the S-Learner is not biased toward either group, which is a positive finding for fairness in treatment effect estimation.

In sum, the scatter plot reveals several systematic patterns in errors that are visible:

- **Regression to the Mean:** Extreme true CATE values tend to have less extreme predictions, a common phenomenon in predictive modeling; **Horizontal Banding:** Predictions cluster around certain horizontal bands, suggesting the model is identifying some discrete effect levels but not capturing the full complexity.
- **Overestimation of Negative Effects:** The model tends to predict less negative values for the most negative true CATE values.
- **Underestimation of Positive Effects:** The highest true CATE values are often underestimated. These patterns suggest the model is smoothing out the sharp threshold effects present in the true data generation process.

With this, we conclude this practice and bring this section to a close with a summary of the key points. The S-Learner is a meta-learner technique that utilizes any suitable base learner to estimate the outcome as a function of the combination of the features X and the treatment indicator T. The S-learner then estimates the CATE by taking the difference between the values of the outcome prediction function where the treatment assignment indication is switched from control to treatment while keeping all other attributes constant. Empirical results of an implemented S-Learner using Python reveal that utilizing a Random Forest base learner yields plausible yet flawed estimates of the actual conditional average treatment effects (CATE) values. The discrete characteristics of the actual causal effects pose a significant difficulty for any continuous prediction model.

In the next section, we introduce another category of meta-learning algorithms for estimating the CATE, namely the T-Learner.

Understanding T-Learner

This section commences with an exposition of T-Learner. Subsequently, we provide a description of the implementation of T-Learner utilizing Python. We subsequently offer a practice session using Python to empirically assess the implemented T-Learner on the synthetic dataset generated by the causal model outlined in the preceding section.

Describing T-Learner

T-learner (Two-Learner) is a meta-learner approach for estimating the CATE by following these steps:

1. Split the data into treatment and control groups;
2. Train separate models on each group;
3. Estimate CATE by taking the difference between predictions from both models.

Let $Y(1)$ denote the outcomes for the treatment group for which $T = 1$; $Y(0)$ denote the outcomes for the control group for which $T = 0$; and M denote any suitable machine learning algorithm. The T-learner uses M to a) fit the outcome model $M(Y(1){\sim}X)$ for the treatment group, using features X, and b) fit the outcome model $M(Y(0){\sim}X)$ for the control group, also using features X.

The fitted models for the treatment and control groups enable computing the conditional expectation of the outcome in the treatment and control groups as functions of the covariates X, defined respectively as follows:

$$\hat{\mu}_1(x) = E(Y(1)|X = x] \text{ (11.10)}$$

$$\hat{\mu}_0(x) = E[Y(0)|X = x] \text{ (11.11)}$$

The CATE estimator provided by T-learner is then given by:

$$\hat{\tau}(x) = \hat{\mu}_1(x) - \hat{\mu}_0(x) \text{ (11.12)}$$

Several contributions to the literature have utilized variations of T-learners. Examples using decision trees, random forests, and gradient-boosted trees as base learners are provided, respectively, by (*Athey and Imbens 2016*), (*Lu , et al. 2018*), and (*Powers, et al. 2018*).

The primary advantage of T-Learner is its ability to capture heterogeneous treatment effects when the treatment and control response functions differ significantly. However, it may be less efficient than other meta-learners when data is limited.

T-learners, as opposed to S-learners, model response variable Y with two treatment groups separated, assuming that group-specific conditional average potential outcomes are derived from independent regression models with distinct conditional means given by the function $\mu_0(x)$ for the control group and $\mu_1(x)$ for the treatment group. The disadvantage of T-Learners is that they divide the sample into two groups, which keeps the groups from sharing common underlying knowledge when estimating the two surfaces. When the distributional characteristics of the individuals in the two groups are identical, as in a randomized trial, this is particularly troublesome.

Having explained T-Learner, we shall now proceed to delineate its implementation in Python.

Implementing T-learner using Python

This chapter's project on the book's website implements the T-Learner through the `TLearner` class specified in the Python file `t_learner.py` located in the `meta-learners` directory. We use a series of code snippets to explain the `TLearner` class definition in the parts that follow.

Adding libraries and defining the class

First, we add the necessary libraries.

```
import numpy as np
from sklearn.base import clone, BaseEstimator, RegressorMixin
from sklearn.linear_model import LinearRegression
from typing import Optional, Union
```

The following declares the class:

```
class TLearner(BaseEstimator, RegressorMixin):
```

BaseEstimator specifies the base learner that will be used for prediction, which must implement `fit()` and `predict()` methods. RegressorMixin is a mixin class for regressors in `scikit-learn`. It provides default implementations for methods common to all regressors.

The code for the class constructor listed below initializes the class. It accepts an optional parameter named 'model' that, when unspecified, defaults to `LinearRegression`. Otherwise, it uses the provided model. The code initializes three attributes of a Tlearner object. `model_t_ : BaseEstimator` is the fitted model for the treatment group. `model_c_ : BaseEstimator` is the fitted model for the control group. `is_fitted_ : bool` indicates whether the model has been fitted.

```
def __init__(self, model: Optional[BaseEstimator] = None):
    self.model = model if model is not None else LinearRegression()
    self.model_t_ = None
    self.model_c_ = None
    self.is_fitted_ = False
```

Defining methods of the TLearner class

The *TLearner* class defines three methods: `fit()`, `predict()`, and `effect()` as described next.

Defining the fit() method

The *TLearner* class defines the `fit()` method for fitting the T-learner model. The fit method fits separate models for the treatment and control groups. It creates clones of the base model to ensure separate instances are used. The method takes three parameters: the feature matrix `X` : array-like of shape `(n_samples, n_features)` ; the treatment assignments (0 or 1); `t` : array-like of shape (`n_samples,`); and the observed outcomes `y` : array-like of shape (`n_samples,`). The method returns the fitted estimator `self : TLearner`

```
def fit(self, X: np.ndarray, t: np.ndarray, y: np.ndarray) -> 'TLearner':
     # Validate input
    X = np.asarray(X)
    t = np.asarray(t)
    y = np.asarray(y)
    if X.shape[0] != t.shape[0] or X.shape[0] != y.shape[0]:
```

```
            raise ValueError("X, t, and y must have the same number of samples")
        # Split data into treatment and control groups
        treatment_mask = t == 1
        control_mask = t == 0
        X_t = X[treatment_mask]
        y_t = y[treatment_mask]
        X_c = X[control_mask]
        y_c = y[control_mask]
        # Create and fit models for treatment and control groups
        self.model_t_ = clone(self.model)
        self.model_c_ = clone(self.model)
        self.model_t_.fit(X_t, y_t)
        self.model_c_.fit(X_c, y_c)
        self.is_fitted_ = True
        return self
```

We describe the `predict()` method code next.

Defining the predict() method

The `TLearner` defines the method `predict()` to predict outcomes for given features and treatment assignments. The method takes two parameters: a feature matrix. `X` : array-like of shape (`n_samples`, `n_features`), and treatment assignments (0 or 1) `t` : array-like of shape (`n_samples`). The method returns the predicted outcomes `y_pred` : array-like of shape (`n_samples`). This method employs the treatment model for samples with `t = 1` and the control model for samples with `t = 0`.

```
    def predict(self, X: np.ndarray, t: np.ndarray) -> np.ndarray:
        if not self.is_fitted_:
            raise ValueError("This TLearner instance is not fitted yet. "
                             "Call 'fit' before using this method.")
        X = np.asarray(X)
        t = np.asarray(t)
        if X.shape[0] != t.shape[0]:
            raise ValueError("X and t must have the same number of samples")
        # Initialize predictions array
        y_pred = np.zeros(X.shape[0])
        # Predict using treatment model for treated units
        treatment_mask = t == 1
        if np.any(treatment_mask):
            y_pred[treatment_mask] = self.model_t_.predict(X[treatment_mask])
        # Predict using control model for control units
        control_mask = t == 0
        if np.any(control_mask):
```

```
        y_pred[control_mask] = self.model_c_.predict(X[control_mask])
    return y_pred
```

The code defining the `effect()` method is described next.

Defining the effect() method

The `TLearner` defines the method `effect()` for estimating the CATE. It takes the feature matrix as input, `X`: array-like of shape (`n_samples, n_features`), and returns as output the estimated conditional average treatment effects, `cate`: `ndarray` of shape (`n_samples`,). The CATE is estimated as the difference between the predicted outcomes under treatment and control conditions.

```
def effect(self, X: np.ndarray) -> np.ndarray:
    if not self.is_fitted_:
      raise ValueError("This TLearner instance is not fitted yet."
                       "Call 'fit' before using this method.")
    X = np.asarray(X)
    # Predict outcomes under treatment and control conditions
    y_t_pred = self.model_t_.predict(X)
    y_c_pred = self.model_c_.predict(X)
    # Calculate treatment effect as the difference
    cate = y_t_pred - y_c_pred
    return cate
```

Thus, we have finalized our exposition of the T-Learner implementation.

The following section provides a guide on utilizing the *TLearner* class to estimate the CATE using the synthetic dataset derived from our heterogeneous treatment effects causal model, which was computed in the preceding section.

T-learner practice using Python

This practice demonstrates how to use the T-Learner implementation using Python. The code for this practice is included in the file `t_learner.ipynb` found in the `notebooks` folder. Ensure that you have installed all required dependencies before running this Notebook, including the local packages: `synthetic_data_for_cate_class` for data generation and `t_learner` for the *TLearner* implementation.

In the following, we give a step-by-step demonstration of how to use the *TLearner* class to perform tasks related to estimating the CATE.

The first step is to create an instance of the `synthetic_data_for_cate` class and run the `synthetic_data_for_cate()` method with `model2`, as done in the SLearner practice. The details of the data generation step are not duplicated here. The subsequent step is to create a *Tlearner* object and fit it with the data from the first step.

Creating a T-Learner object and fitting it to the data

For this demonstration, we use the Random Forest-based T-Learner as our primary model. The fitting process involves:

Splitting the data into treatment and control groups

Training separate models on each group

```
# Initialize T-Learner with different base models
t_learner_rf = TLearner(RandomForestRegressor(n_estimators=100, random_
state=42))
# Select Random Forest as our primary model
tl = t_learner_rf
# Fit the model
print("Fitting T-Learner with Random Forest...")
tl.fit(
    X=features,             # Feature matrix
    t=treatment_vector,     # Treatment assignments (0/1)
    y=outcomes              # Observed outcomes
)
print("Model fitting complete.")
```

The code should output the following after it runs successfully.

```
Fitting T-Learner with Random Forest...
Model fitting complete.
```

Having created the *TLearner* object, we are now ready to perform various tasks, the first of which is to estimate the CATE, which we do next.

The T-Learner CATE estimation

The code block for the CATE estimation is identical to the one in the S-learner practice except for the first line of code, which is replaced by the following.

```
# Estimate treatment effects (CATE)
cate = tl.effect(X=features)
```

The code when run will output the following summary statistics.

```
CATE Statistics:
  Mean: 1.1326
  Std Dev: 2.9263
  Min: -6.0083
  Max: 10.3394
```

The next task is to plot the CATE distribution to depict how the estimated CATE varies across the population visually.

The T-Learner CATE distribution visualization

We do not list the visualization code here, which you can find in the Jupyter Notebook for this section. The code will output the CATE distribution depicted in *Figure 11.11*.

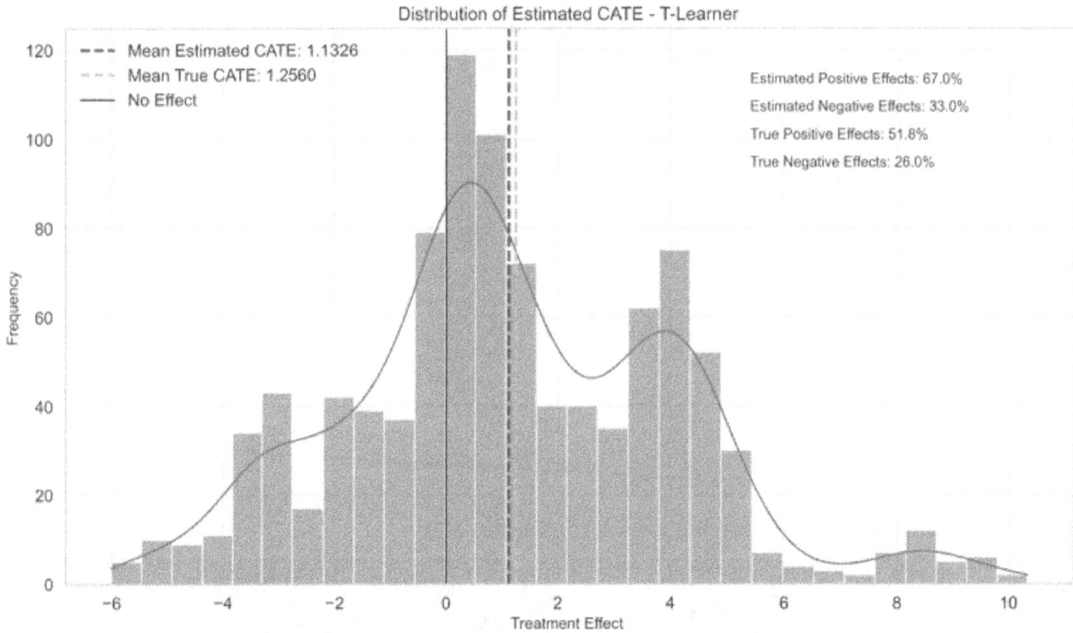

Figure 11.11—The T-Learner Estimated CATE distribution on the Synthetic Data

The subsequent job involves comparing the estimated CATE with the true CATE distribution by overlaying them on the same plot, which we will proceed to perform next.

The T-Learner overlaid CATE distributions: estimated vs. True

The notebook for this practice also includes code for plotting the t-learner estimated CATE distribution overlaid on the real CATE distribution to compare the degree of proximity between the real and the estimated distributions. The code is not shown but can be viewed in the `t-learner.ipynb` file of this section's notebook. See *Figure 11.12*. The plots in the figure are nearly identical to those for the S-Learner, and all comments made in the S-Learner practice section apply here; therefore, they are not repeated.

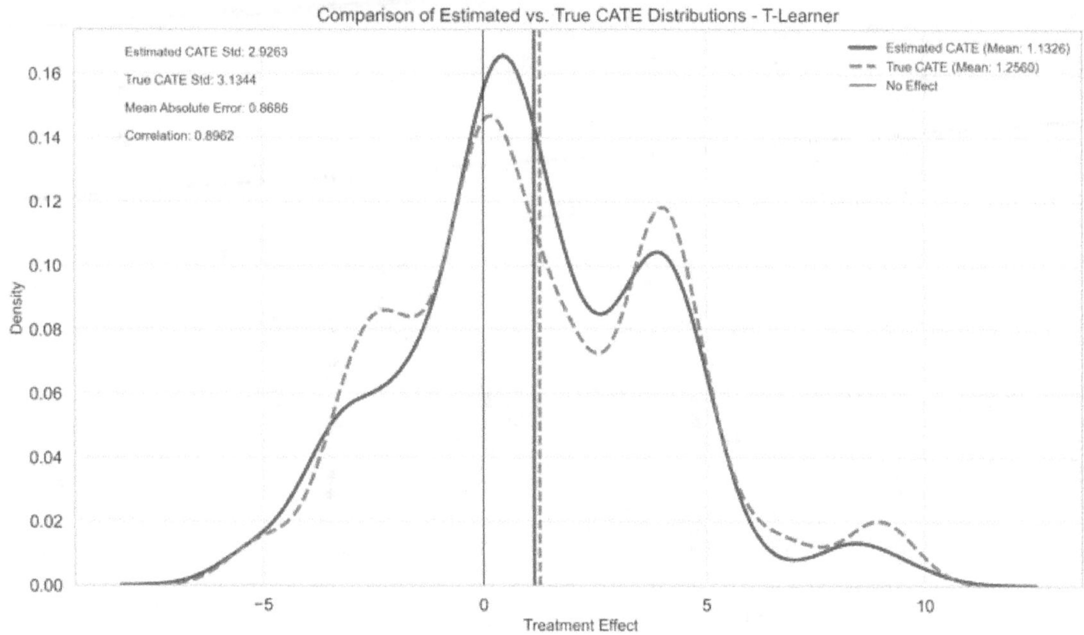

Figure 11.12— Comparison of the s-learner estimated versus the true CATE

Our final task will further analyze the T-Learner accuracy of the CATE estimation by producing a scatter plot of the estimated and true CATE values.

The T-Learner CATE accuracy evaluation: predicted vs. actual treatment effects

The Notebook code for this task, not listed due to space limitations, is used to compute the scatter plot in *Figure 11.13*. The scatter plot for the T-Learner, as well as its estimated CATE distribution, shares striking similarities with those obtained for the S-Learner. The scatter plots for the S-Learner and T-Learner have comparable clustering around the diagonal line. The following explains why the empirical performance of the T-Learner resembles that of the S-Learner on the synthetic data used.

1. The synthetic data contains sufficient samples in both the treatment and control groups for the S-Learner to learn the treatment effect accurately.
2. The Random Forest Regressor used in both approaches can capture complex relationships, even when the treatment is just another feature.
3. The synthetic data true CATE is based on threshold functions that create distinct subgroups with different effects, which both models can capture equally well.
4. The underlying functional forms for treatment and control outcomes may not be sufficiently different to warrant the separate modeling approach of T-Learner.

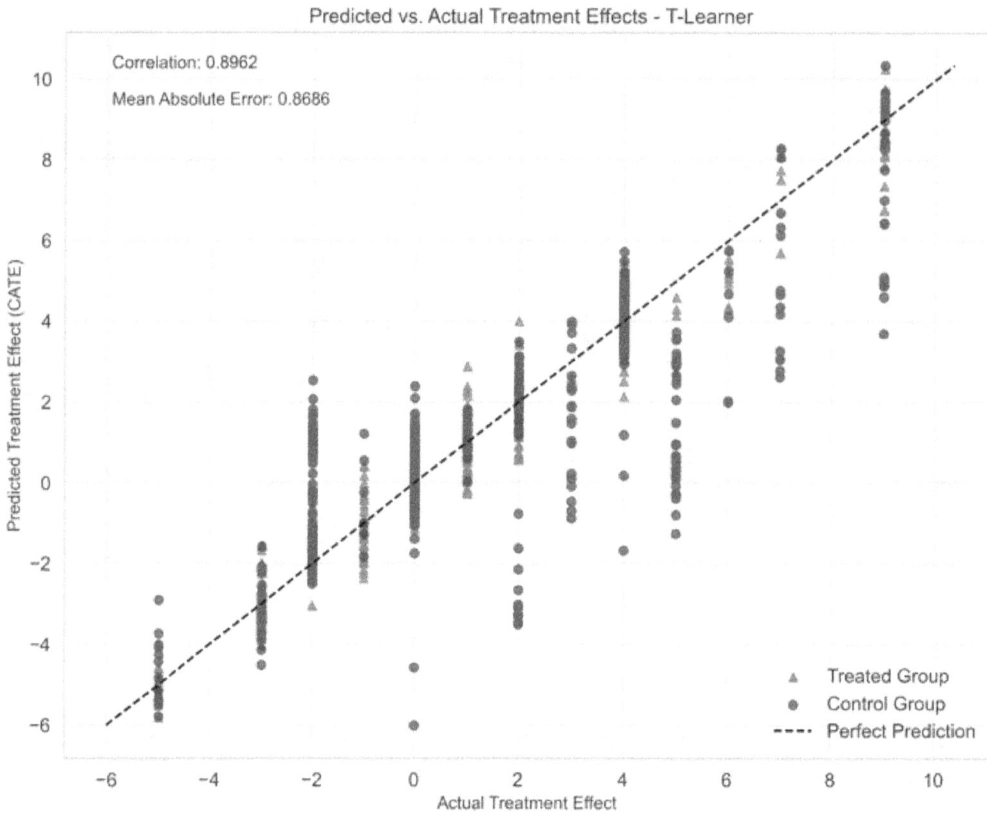

Figure 11.13— T-Learner Scatter Plot: Estimated Versus Actual Treatment Effect for Treatment and Control Groups

To conclude this section, let us summarize the key takeaways. The T-learner employs base learners to train two models to predict outcomes for the control and treatment groups, unlike the S-learner, which uses a single model. The T-learner can adjust to different response surfaces within the groups and is theoretically more advantageous. The empirical results for our generated synthetic data model suggest that the S-learner may be preferred due to its simplicity and computational efficiency.

With this, we conclude this section, which covers the second meta-learner, the T-Learner, and proceed to the next section to present the third meta-learner, namely the X-Learner.

Understanding X-Learner

This part commences with an exposition of X-Learner. We subsequently delineate the implementation of X-Learner utilizing Python. We follow up with a practice session employing Python to empirically assess the implemented X-Learner on the synthetic dataset generated in the preceding section.

Describing X-Learner

The X-Learner is a meta-learner approach for estimating heterogeneous treatment effects. It was introduced by Künzel et al. (2019) in their paper "*Metalearners for estimating heterogeneous treatment effects using machine learning.*"

The X-learner algorithm has the following key characteristics. The algorithm employs a three-stage approach. The algorithm utilizes flexible base learners, enabling the use of any suitable supervised learning algorithm or regression methods. It employs separate models, one for the control group and another for the treatment group. The algorithm uses crossover estimation to estimate treatment effects by crossing over between groups. The algorithm is robust to imbalance, as it can effectively handle different treatment and control group sizes. The X-Learner often outperforms other meta-learners when treatment groups are imbalanced. The X-learner is robust to confounding when good predictive models are used and provides more stable estimates in many practical scenarios.

The following describes the X-Learner algorithm steps:

X-learner three-stage approach

The X-learner algorithm estimates the CATE in three stages. The computation in each stage is described in the following sections.

First stage

In the first stage, the algorithm adopts two separate outcome models, one for the control subgroup $M_1(Y^0 \sim (X^0, T = 0)$ and another $M_2\left(Y^1 \sim (X^1, T = 1)\right)$ for the treatment subgroup. The algorithm computes estimate $\hat{\mu}_1(x)$ and $\hat{\mu}_0(x)$ for the response functions of the treatment and the control subgroups,

$$\mu_0(x) = E[Y(0)|X = x] \ (11.13)$$

$$\mu_1(x) = E(Y(1)|X = x] \ (11.14)$$

by fitting the two respective models M_1 for control and M_2 for treatment using a *first stage base learner*, any supervised learning or regression algorithm. The algorithm uses the estimated response functions for both fitted models to predict counterfactual outcomes for everyone in the subgroups. The counterfactual response of an i-th individual in the control subgroup, had the individual been assigned the treatment, is computed using the estimated treatment response function.

$$Y_i^1 = \hat{\mu}_1(X_i^0) \ (11.15)$$

The counterfactual response of an i-th individual in the treatment subgroup had the individual been assigned the no-treatment is computed using the estimated control response function.

$$Y_i^0 = \hat{\mu}_0(X_i^1) \ (11.16)$$

Second stage

The algorithm computes "imputed" treatment effects for everyone using the counterfactual response functions. For treated individuals, the imputed treatment effect is the actual outcome minus the predicted control outcome.

$$\widetilde{D}_i^0 = \hat{\mu}_1(X_i^1) - Y_i^0 \ (11.17)$$

For control individuals, the imputed treatment effect is the predicted treatment outcome minus the actual outcome.

$$\widetilde{D}_i^1 = Y_i^1 - \hat{\mu}_0(X_i^0) \text{ (11.18)}$$

The imputed treatment effects for the control and treatment groups provide two ways to estimate the conditional average treatment effect (CATE) function, $\tau(x)$, by using two separate models, $M_3(\widetilde{D}^0 \sim X^0)$ and $M_4(\widetilde{D}^1 \sim X^1)$, and employing a *second stage base learner*—any supervised learning or regression algorithm—to predict the imputed treatment effects for the control subgroup to obtain $\hat{\tau}_0(x)$, and to predict the imputed treatment effects for the treatment subgroup to obtain $\hat{\tau}_1(x)$.

Third stage

In the third stage, the algorithm combines predictions $\hat{\tau}_0(x)$ and $\hat{\tau}_1(x)$ obtained by fitting two separate models for the imputed treatment effects by the second-stage base learners to get the final CATE estimate by taking the simple average.

$$\tau(x) = (\hat{\tau}_0(x) + \hat{\tau}_1(x))/2 \text{ (11.19)}$$

The final CATE estimate can be represented using a general formula that incorporates a weighted average: $\tau(x) = g(x)\hat{\tau}_0(x) + (1 - g(x))\hat{\tau}_1(x)$. In this equation, g(x) is restricted to the range [0, 1] for all values of x within the domain of X. A notable case occurs when $g(x)$ equals $1/2$, resulting in the weighted average simplifying to a straightforward (unweighted) average.

The above completes the exposition of the X-Learner algorithm, setting the stage for illustrating its implementation using Python, which is the topic of the next section.

Implementing X-Learner using Python

This chapter's project on the book's website implements the X-Learner via the `XLearner` class defined in the Python file `x_learner.py` situated in the `metalearners` directory.

In the subsequent sections, we provide an in-depth explanation of the `XLearner` class using a series of code snippets.

Adding libraries and defining the class

Initially, we import the requisite libraries.

```
import numpy as np
from sklearn.base import clone
from sklearn.linear_model import LinearRegression
from sklearn.ensemble import GradientBoostingRegressor
```

The `XLearner` class takes one parameter, `model`: an estimator object, optional (`default=None`). If None, the class uses `LinearRegression` for outcome models and `GradientBoostingRegressor` for treatment effect models. The class parameter must implement `fit()` and `predict()` methods. Here `model` is the **base learner** used for the class *four attributes*.

- `model_c` : estimator object is used to predict outcomes for the control group;
- `model_t` : estimator object used to predict outcomes for the treatment group;

- `te_model_c` : estimator object is used to predict treatment effects for the control group;
- `te_model_t` : estimator object is used to predict treatment effects for the treatment group.

```
def __init__(self, model=None):
    if model is None:
        self.model_c = LinearRegression()
        self.model_t = LinearRegression()
        self.te_model_c = GradientBoostingRegressor()
        self.te_model_t = GradientBoostingRegressor()
    else:
        self.model_c = clone(model)
        self.model_t = clone(model)
        self.te_model_c = clone(model)
        self.te_model_t = clone(model)
```

Defining methods of the XLearner class

The *XLearner* class defines two methods: `fit()` and `predict()`, as described next.

Defining the fit() method

The `fit()` method takes as input three parameters: the **feature matrix** of covariates, X : array-like, shape `(n_samples, n_features)`; the **outcome variable**, y: array-like, shape `(n_samples)`; and the **treatment indicator** (0 for control, 1 for treatment) t: array-like of shape `(n_samples)`. The `fit()` method returns self: object.

The fitting process follows these steps:

1. Fit separate models for the control and treatment groups;
2. Predict counterfactual outcomes for each group;
3. Compute imputed treatment effects;
4. Train models to predict the treatment effects as a function of covariates.

```
def fit(self, X, y, treatment):
    # Fit control and treatment models
    self.model_c.fit(X[treatment == 0], y[treatment == 0])
    self.model_t.fit(X[treatment == 1], y[treatment == 1])
    # Predict counterfactual outcomes
    control_outcome_t = self.model_c.predict(X[treatment == 1])
    treatment_outcome_c = self.model_t.predict(X[treatment == 0])
    # Treatment effect estimations
    tau_t = y[treatment == 1] - control_outcome_t
    tau_c = treatment_outcome_c - y[treatment == 0]
    # Train treatment effect models
    self.te_model_t.fit(X[treatment == 1], tau_t)
```

```
        self.te_model_c.fit(X[treatment == 0], tau_c)

        return self
```

Defining the predict() method

The `predict()` method takes as input the **feature matrix,** X: array-like, shape (`nsamples, nfeatures`), and returns the predicted Conditional Average Treatment Effects (**CATE**) as output. The prediction process is in three steps.

1. Predict treatment effects using the treatment group model.
2. Predict treatment effects using the control group model.
3. Take a simple average of the two predictions for the treatment and control groups to obtain the final CATE estimate. Simple averaging can be replaced with weighted averaging based on propensity scores for improved performance.

```
    def predict(self, X):
        # Predict treatment effects
        tau_hat_t = self.te_model_t.predict(X)
        tau_hat_c = self.te_model_c.predict(X)
        # Average treatment effect
        return (tau_hat_t + tau_hat_c) / 2
```

As such, we have concluded our presentation of the X-Learner implementation.

In the next section, we demonstrate how to employ the *XLearner* class, which implements the X-Learner, to estimate the CATE for the synthetic dataset generated from our heterogeneous treatment effects causal model, as calculated in the prior section.

X-Learner practice using Python

This practice demonstrates how to use the X-Learner implementation using Python. The code for this practice is included in the file `x_learner.ipynb` found in the *notebooks* folder. Ensure that you have installed all required dependencies before running this Notebook, including the local packages: `synthetic_data_for_cate_class` for data generation and `x_learner` for the *XLearner* implementation.

In the following, we give a step-by-step demonstration on how to use the *XLearner* class to perform tasks related to estimating the CATE.

The first step is to create the synthetic data triple (`features`, `outcomes`, `treatment_vector`), as demonstrated in the code listed for the *SLearner practice*. The details of this step are not duplicated here. This synthetic data triple is used as input to carry out the computations in the following sections of this practice. The next step is to instantiate an *XLearner* object and fit it with the data.

Fitting the X-Learner with the data

For this demonstration, we create an `Xlearner` object, `xl`, using the `RandomForestRegressor` as the base learner. Next, we fit `xl` with the synthetic data from the previous step using the `fit()` method of the `Xlearner` object.

```
# Select Random Forest as our primary model
x_learner_rf = XLearner(RandomForestRegressor(n_estimators=100, random_
state=42))
xl = x_learner_rf
# Fit the model
print("Fitting X-Learner with Random Forest...")
xl.fit(
    X=features,                  # Feature matrix
    y=outcomes,                  # Observed outcomes
    treatment=treatment_vector        # Treatment assignments (0/1)
)
print("Model fitting complete.")
```

Running the above code cell in the workbook produces the following output.

```
Fitting X-Learner with Random Forest...
Model fitting complete.
```

Having fitted the *Xlearner* with the data, we are now ready to estimate the CATE in the subsequent step.

The X-Learner estimation of the CATE

After fitting the model, we can estimate the Conditional Average Treatment Effect (CATE) using the `predict()` method of the `Xlearner` object, `xl`, with the input parameter `X` set to the `features`.

```
# Estimate treatment effects (CATE)
cate = xl.predict(X=features)
# Print basic statistics about the estimated CATE
print(f"CATE Statistics:")
print(f"  Mean: {cate.mean():.4f}")
print(f"  Std Dev: {cate.std():.4f}")
print(f"  Min: {cate.min():.4f}")
print(f"  Max: {cate.max():.4f}")
```

Running the above code produces the following output:

```
CATE Statistics:
  Mean: 1.1327
  Std Dev: 2.8346
  Min: -5.5628
  Max: 9.5982
```

The result of this step provides a summary statistic of the estimated treatment effect across the entire population. In the next step, we will create a histogram plot to visualize the distribution of the estimated treatment effect.

The X-Learner CATE distribution visualization

We do not list the visualization code here, which you can find in the **Jupyter** Notebook for this section. The code will output the CATE distribution depicted in *Figure 11.14*.

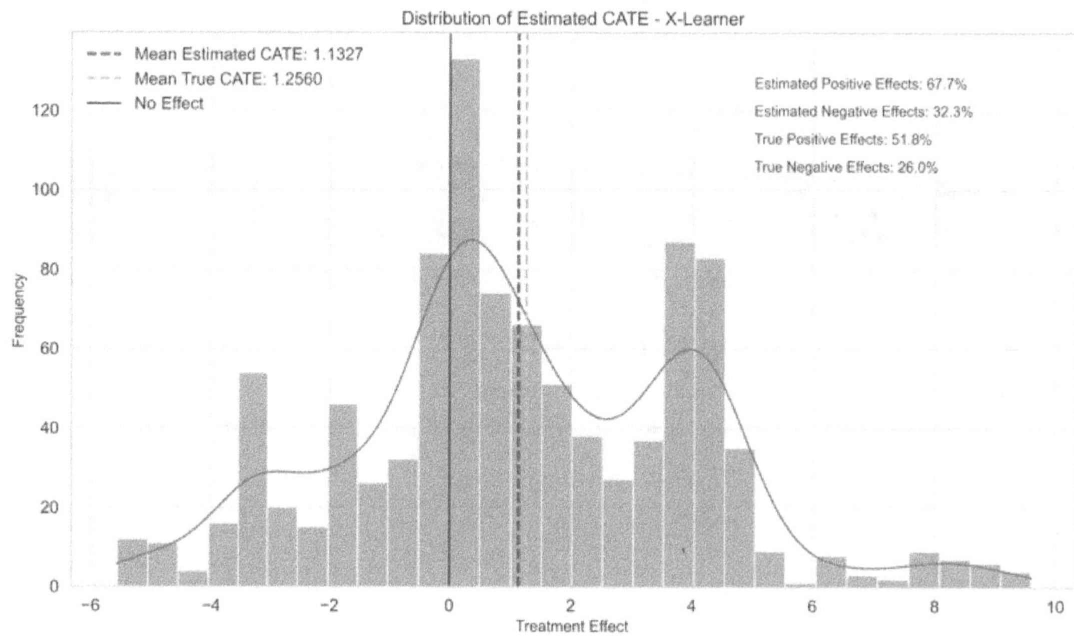

Figure 11.14—The X-Learner Estimated CATE distribution on the Synthetic Data

The next step will be to assess the proximity of the estimated CATE distribution to the true CATE distribution by superimposing both distributions on the same plot.

The X-Learner overlaid CATE distributions: estimated vs. True

The code for this section can be viewed in the *'x-learner.ipynb'* file, which is not listed due to space limitations. Executing the code will output the plot of *Figure 11.15,* which shows the X-Learner estimated CATE distribution overlaid on the real CATE distribution, allowing for a comparison of the degree of proximity between the real and estimated distributions.

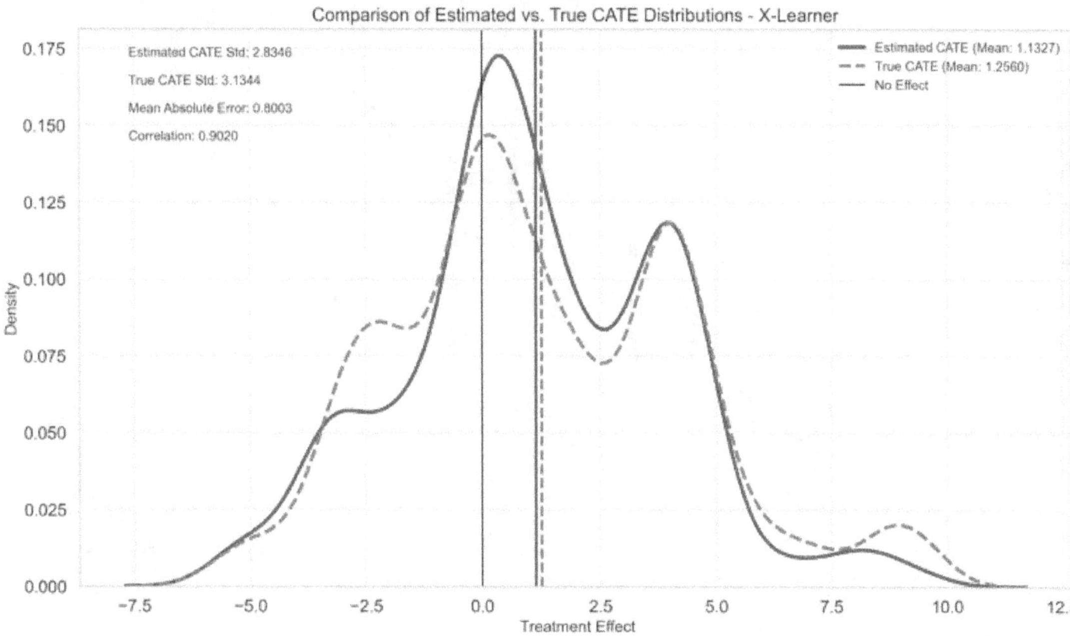

Figure 11.15— Comparison of the X-learner estimated versus the true CATE

In the subsequent phase, we conduct a thorough comparison of the estimated and actual treatment effects for the complete population entities using a scatter diagram.

The X-Learner CATE accuracy evaluation: predicted vs. actual treatment effects

The Notebook code for this task, not listed due to space limitations, is used to compute the scatter plot in Figure *11.16*. The scatter plot for the X-Learner, as well as its estimated CATE distribution, shares striking similarities with those obtained for the S-Learner and the T-Learner, as displayed in the previous sections.

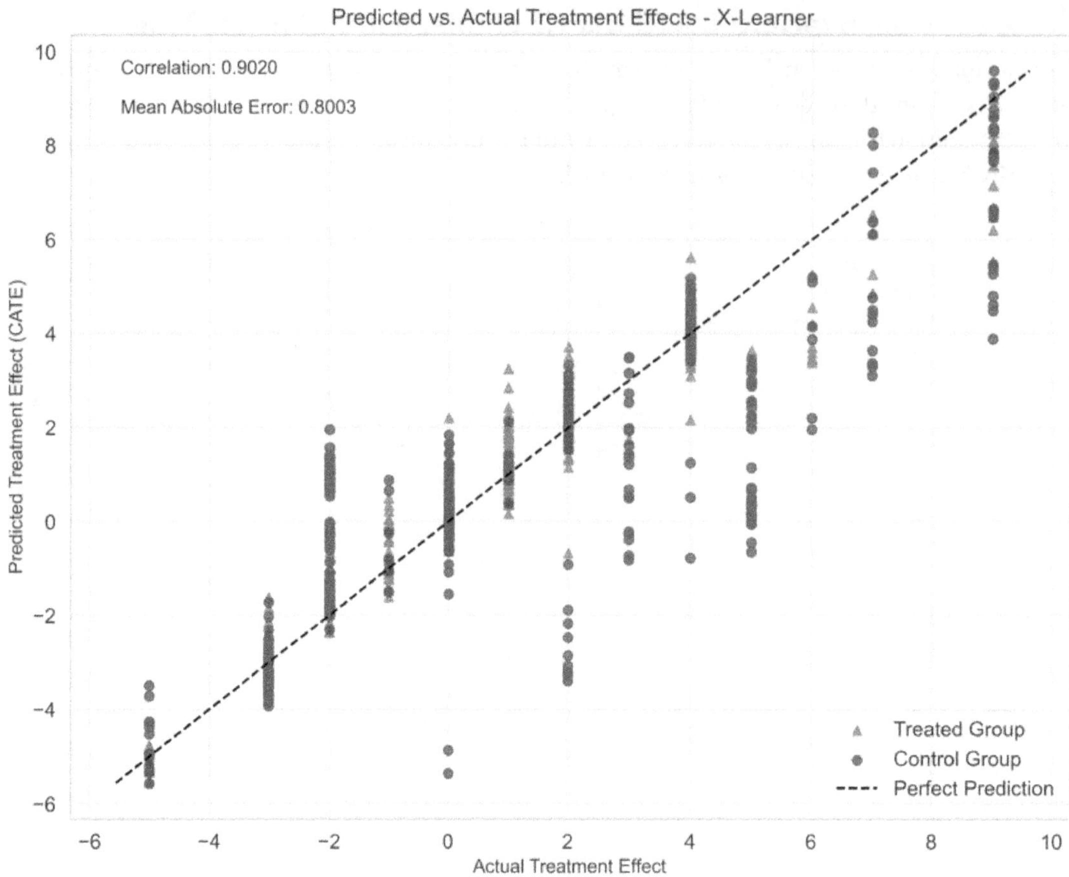

Figure 11.16– X-Learner Scatter plot: estimated vs. actual for the treatment and control group

In the following section, we elaborate on the analysis and comparison of the empirical results for the performance of our implementations of the X-learner, T-learner, and S-learner, using random forest regression as the base learner.

Comparison of X-Learner results with S-Learner and T-Learner

A look at the estimated CATE distributions for the X-Learner, S-Learner, and T-Learner (*Figures 11.5, 11.8, 11.11*) shows similarities in the distribution shape (all three share the same bell-shaped multi-modal pattern), the mean CATE values (all three have mean CATE values close to the mean of the true CATE), the proportions of positive and negative effects (all three find the same groups of people who do or do not benefit from treatment), as well as the spread of distribution (all three have about the same standard deviation of the estimated CATE values).

A look at the CATE Accuracy Evaluation for all three meta-learners (*Figures 11.6, 11.9, 11.12*) also shows close similarities in the correlation between the estimated and the true CATE; all three meta-learners have nearly identical mean absolute error for the estimated CATE and have similar prediction patterns for the spread of the estimated CATE around the ideal prediction line, with similar patterns for both the control and treatment groups.

To wrap up this section, the following highlights the key takeaway points. The X-Learner estimates treatment effects using a multi-stage approach:

1. Fit separate models for control and treatment groups;
2. Estimate counterfactual outcomes and compute imputed treatment effects;
3. Train models to predict these imputed treatment effects on the treatment and control groups;
4. Combine predictions to get the final CATE estimate. The empirical results on our synthetic dataset model yield close similarities between the estimated CATE distribution and the accuracy evaluation plots for all three meta-learners, X-Learner, S-Learner, and T-Learner, implying that choosing between them may not have a significant effect on the quality of the CATE estimate for this dataset.

We now end the discussion on the third meta-learner, the X-Learner, and go to the subsequent section, which presents the final meta-learner, the R-Learner.

Understanding R-Learner

We begin this section with a formal description of the meta-learning approach of the R-Learner algorithm. We subsequently delineate the implementation of R-Learner utilizing Python. We follow up with a practice session employing Python to empirically assess the implemented R-Learner on the synthetic dataset created in the preceding section.

Describing R-Learner

The R-Learner is a meta-learner approach for estimating heterogeneous treatment effects, introduced by Nie and Wager (*Nie and Wager 2021*) in their paper, "*Quasi-Oracle Estimation of Heterogeneous Treatment Effects*." It employs the **Robinson transformation** (*Robinson 1988*), a **residualization approach,** to separate the estimation of treatment effects from that of baseline outcomes and propensity scores. In the following section, we define the R-Learner algorithm.

R-Learner algorithm

The R-Learner algorithm comprises three stages. In the first stage, the algorithm fits an *outcome model* to predict outcomes and a *propensity model* to predict treatment assignments. In the second stage, the algorithm computes residuals from both models. For every unit in the dataset, the outcome residual is the actual outcome minus the predicted outcome, and the treatment residual is the actual treatment minus the predicted treatment probability. In the third stage, the algorithm employs any regression method to regress the outcome residuals on the treatment residuals. The regression coefficient then represents the treatment effect.

First stage: fit outcome and propensity models

In the first stage, the algorithm fits two models: the *outcome model* and the *propensity model*. We fit the outcome model to estimate the expected output conditioned on the covariates, $m(x) = E[Y|X = x]$, using Ordinary Least Squares (OLS) regression:

$$\widehat{m}(x) = argmin_m \frac{1}{n} \sum_{i=1}^{n} (Y_i - m(X_i))^2 \quad (11.20)$$

Additionally, we fit the propensity model to estimate the propensity score, $e(x) = P(T = 1|X = x)$, which is the probability of treatment assignment given the covariate, using OLS regression:

$$\hat{e}(x) = argmin_e \frac{1}{n}\sum_{i=1}^{n}\left(T_i - e(X_i)\right)^2 \text{ (11.21)}$$

Second stage: compute residuals

In the second stage, the algorithm computes the outcome residuals, given by,

$$R_i^Y = Y_i - \hat{m}(X_i) \text{ (11.22)}$$

Additionally, the algorithm computes the treatment residuals, given by,

$$R_i^T = T_i - \hat{e}(X_i) \text{ (11.23)}$$

The outcome and treatment residuals represent modeling errors, which are the components of the observed outcomes and residuals that depend on external factors other than the covariates.

Third stage: regress outcome residuals on treatment residuals

In this third and final stage, the algorithm estimates the treatment effect by regressing the outcome residuals on the treatment residuals:

$$\hat{\tau}(x) = argmin_\tau \frac{1}{n}\sum_{i=1}^{n}(R_i^Y - \tau(X_i) \cdot R_i^T)^2 \text{ (11.24)}$$

In the next section, we lay out the mathematical justification of the R-Learner estimation $\hat{\tau}(x)$ of the conditional average treatment effect.

The R-Learner mathematical justification

The R-Learner is based on rewriting the outcome Y in the following decomposed form:

$$Y = m(X) + \tau(X) \cdot \left(T - e(X)\right) + \varepsilon \text{ (11.25)}$$

where ε is a noise term with $E[\varepsilon|X, T] = 0$.

This decomposition leads to the R-Learner objective:

Equation 11.25 implies that $\tau(X)$ can be estimated by regressing the outcome residual $Y - m(X)$ on the treatment residual $T - e(X)$.

That is, $\hat{\tau}(x)$ Is given by:

$$\hat{\tau}(x) = argmin_\tau E\left[\left(Y - m(X) - \tau(X) \cdot \left(T - e(X)\right)\right)^2 \middle| X = x\right] \text{ (11.26)}$$

Here, $m(X)$ and $e(X)$ are estimated from observational data. *Equation 11.26* mathematically expresses the computation of the third stage of the R-Learner algorithm.

Informally, the main intuition behind the algorithm is that if the treatment has an effect, then the unexplained variation in the outcome (outcome residuals) should be related to the unexplained variation in the treatment (treatment residuals).

In the following section, we outline the R-Learner's advantages and limitations.

R-Learner advantages and limitations

Among the advantages of the R-Learner is that it is both flexible and robust. The R-Learner is doubly robust; it can reliably predict the treatment effect when either the outcome model or the propensity model has accurate predictions. The algorithm is flexible; it can employ any supervised learning regression method as the base learner at each stage, enabling the analysis of complex, non-linear causal relationships.

The R-Learner has several limitations. Its algorithm requires more complex implementation than other meta-learners, e.g., the S-Learner or T-Learner. It requires careful handling of division by zero in residual calculations. The algorithm is sensitive to extreme propensity scores (values close to 0 or 1).

At this point, we conclude the exposition of the R-Learner algorithm, paving the way for demonstrating its implementation in Python, which will be addressed in the subsequent section.

Implementing R-Learner using Python

The project for this chapter has multiple R-Learner implementations located in the `metalearners` folder. For an overview of the roles and rationale of the various R-Learner implementation modules, refer to the Markdown file `explanation.md` included in the project repository. The various implementations aim to address specific challenges related to propensity scores and non-linear effects, striking a balance between theoretical correctness and practical performance.

The `r_learner.py` file defines the `RLearner` class, which gives the base implementation of the R-Learner algorithm. We do not list the Python code for the class definition due to space limitations. The following provides a high-level description of the logic of the `RLearner` class.

The class takes optional parameters consisting of estimator objects: `outcome_model`, `treatment_model`, and `effect_model`, which are the base learners used to predict outcomes, treatment assignments, and treatment effects, respectively. Estimator objects must implement `fit()` and `predict()` methods. The class uses **Random Forest Regressor** if the estimator object value is `None`. The class has a parameter, a Boolean flag on whether to use propensity scores in the prediction step. If its value is `False`, the class uses a constant treatment residual of 1 for all samples, and if its value is `True`, the class uses `1 - propensity_scores` as the treatment residual. The class has another Boolean flag parameter indicating whether to clip propensity scores to avoid extreme values, which is relevant only if the flag for using propensity scores for prediction is set to `True`. The last parameter is an optional tuple, with a default value of (0.1, 0.9), which assigns lower and upper bounds for the clipping propensity scores, relevant only if the Boolean flag to clip propensity is set to `True`.

The `RLearner` class defines two methods, `fit()` and `predict()`.The `fit()` method takes three parameters: the feature matrix `X`, an array-like, shape (`n_samples`, `n_features`), the outcome variable `y`, array-like, shape (`n_samples,`), and the treatment indicator (0 for control, 1 for treatment), `treatment`, array-like, shape (`n_samples,`). The `fit()` method performs the following process steps. First, it splits the data into training and validation sets and fits the outcome model on the training set. Next, it computes propensity scores and predicts outcomes for the validation set. Following this, it computes residuals for both outcomes and treatments and fits the effect model on the ratio of residuals. The rationale for data splitting is necessary to avoid overfitting and ensure that the residuals are computed on data not used for fitting the outcome and treatment models. The `predict()` method takes one parameter, the feature *matrix X* for which to predict treatment effects and returns the predicted CATE.

In the following practice section, we demonstrate how to run the R-Learner on a synthetic dataset.

R-learner practice using Python

This practice demonstrates how to use the R-Learner implementation using Python. The code for this practice is included in the file `r_learner.ipynb` found in the `notebooks` folder. Ensure that you have installed all required dependencies before running this Notebook, including the local packages: `synthetic_data_for_cate_class` for data generation and `r_learner_imporoved` for the `RLearner` implementation.

In the following, we give a step-by-step demonstration on how to use the `RLearner` class to perform tasks related to estimating the CATE.

The first step is to create the synthetic data triple (`features`, `outcomes`, `treatment_vector`), as demonstrated in the code listed for the *SLearner practice*. The details of this step are not duplicated here. This synthetic data triple is used as input to carry out the computations in the following sections of this practice. The next step is to instantiate an `Rlearner` object and fit it with the data.

Fitting R-learner

The following code first creates an `RLearner` instance with the default primary model, **Random Forest**. Next, it converts the **covariates matrix features** to a dense array if it's sparse. Afterwards, it fits the `RLearner` object with the data using the dense feature matrix.

```
r_learner_rf = RLearner(
    use_propensity_for_prediction=False
)
rl = r_learner_rf
try:
    print("Fitting R-Learner with Random Forest...")
    if hasattr(features, 'toarray'):
        features_dense = features.toarray()
    else:
        features_dense = features
    rl.fit(
        X=features_dense,  # Dense feature matrix
        y=outcomes,  # Observed outcomes
        treatment=treatment_vector  # Treatment assignments
    )
    print("Model fitting complete.")
except AttributeError as e:
    print(f"Error fitting model: {str(e)}")
```

Executing the above code should give us the following output:

```
Fitting R-Learner with Random Forest...
Model fitting complete.
```

Having the *Rlearner* fitted with the data, we are ready to estimate the CATE in the subsequent step.

Estimating the CATE

After fitting the model, the following code snippet estimates the CATE, providing an estimate of the treatment effect for each individual, conditional on their covariates. The code prints basic statistics about the estimated CATE.

```
cate = rl.predict(X=features_dense)
print(f"CATE Statistics:")
print(f"  Mean: {cate.mean():.4f}")
print(f"  Std Dev: {cate.std():.4f}")
print(f"  Min: {cate.min():.4f}")
print(f"  Max: {cate.max():.4f}")
```

When executed, the code produces the following output:

```
CATE Statistics:
  Mean: 0.9329
  Std Dev: 1.7444
  Min: -2.4474
  Max: 5.4238
```

The result of this step provides a summary statistic of the estimated treatment effect across the entire population. In the next step, we will create a histogram plot to visualize the distribution of the estimated treatment effect.

CATE distribution visualization

Figure 11.17 displays the R-Learner estimated CATE distribution plot. Space constraints preclude listing the `Notebook` code for this step. The plot in *Figure 11.17* provides valuable visual information on how the estimated CATE captures the treatment effect heterogeneity of the synthetic data, reflected by the estimated treatment effect variability and spread. *Figure 11.17* displays quantitative parameters comparing the mean estimated CATE (0.9329) versus the mean true CATE (1.2560), and the estimated positive effects (74.2%) versus the true positive effects (51.8%), as well as the estimated negative effects (25.8%) versus the true negative effects (26%).

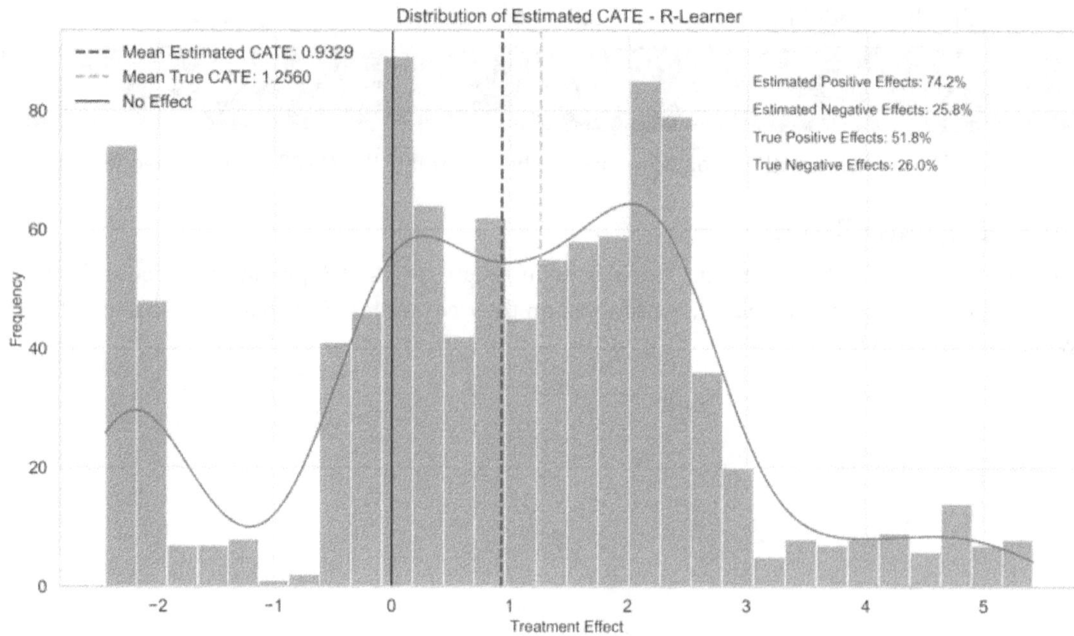

Figure 11.17—R-Learner distribution of estimated CATE

The next step will be to assess the proximity of the estimated CATE distribution to the true CATE distribution by superimposing both distributions on the same plot.

Overlaid CATE distributions: estimated vs. True

Figure 11.18 displays the R-Learner estimated CATE distribution overlaid on the true CATE distributions. The code listing for this step is suppressed due to a lack of space. Comparing the two distributions, we note that the estimated CATE distribution has similar peaks and modes to the true distribution. The estimated distribution is markedly narrower than the true distribution, indicating an under-estimated bias for larger positive treatment effects and an over-estimated bias for highly negative treatment effects. *Figure 11.18* shows the mean absolute error between the estimated and true CATE is 1.495, and the correlation between the estimated and the true CATE is 0.8462. Comparing this plot with those obtained for the other three meta-learners, we note that the R-Learner yields inferior empirical results on the used synthetic data relative to the other meta-learners, despite its complexity.

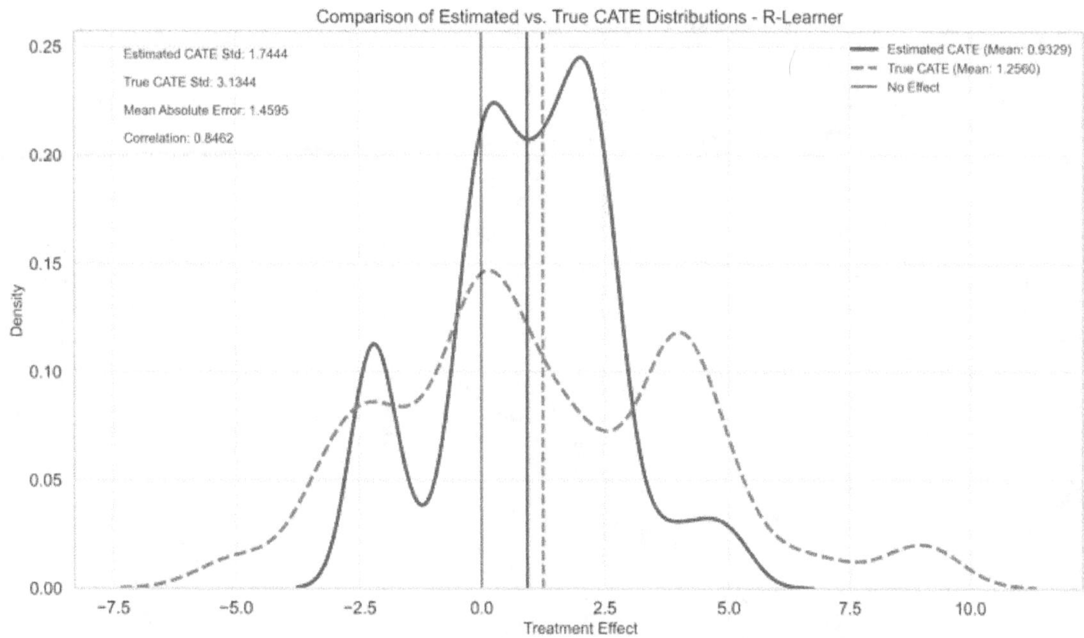

Figure 11.18—R-Learner estimated CATE distribution overlaid on the true CATE distribution

In the next step, we delve deeper into comparing the estimated and true treatment effects for the entire population entities using a scatter diagram.

Evaluating R-Learner estimated CATE accuracy

Figure 11.19 displays a scatter plot of the estimated and the true treatment effects for all population entities in both control and treatment subgroups. The code is located in the Notebook and is not shown due to space limitations. The scatter plot displays the estimated treatment effect on the y-axis and the corresponding true treatment effects on the x-axis, using different symbols to distinguish between the treated and control groups. A diagonal line represents perfect prediction (y = x). The scatter plot shows underestimated values for larger positive treatment effects and overestimated values for highly negative effects. The estimated CATE accuracy is higher for CATE values closer to the estimated **average treatment effect** (ATE), which equals 0.9329 (see *Figure 11.18*), than for extreme CATE values toward the minimum and maximum on the actual treatment effect of the x-axis. The scatter plot for the R-Learner shows inferior empirical results on the synthetic data compared to those of the other three meta-learners displayed in the previous sections.

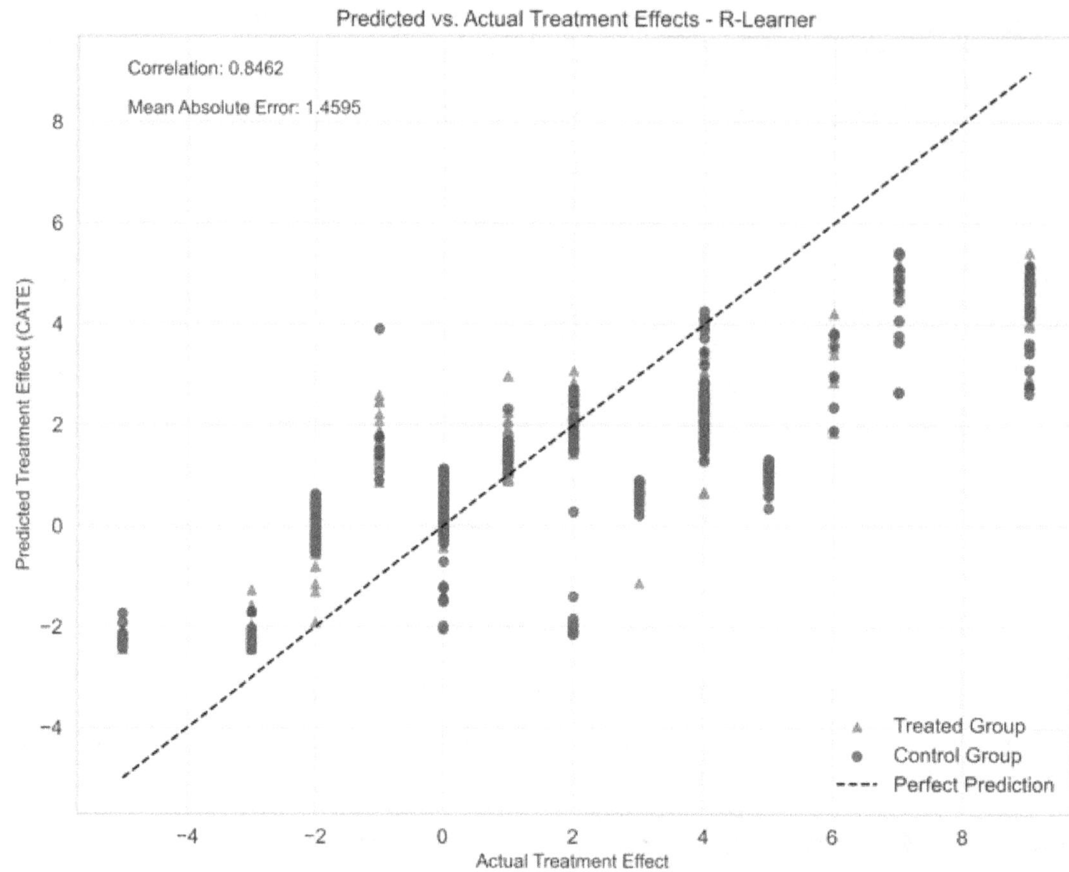

Figure 11.19— R-Learner Scatter plot: treatment effects estimated on the y-axis versus actual on the x-axis- blue color for the control group and red color for the treatment group

Concluding this section on the R-Learner, let us summarize the takeaway points. The R-Learner approach has three stages: in the first stage, fitting two models to predict outcomes and another to predict treatment assignment; in the second stage, computing residuals from both models; and in the third stage, regressing outcome residuals on treatment residuals to estimate treatment effects. The R-Learner performance on our synthetic dataset model, manifested by the scatter plot comparing the estimated and actual treatment effects and the statistical parameters, the mean square error, and the correlation coefficient between the estimated and true treatment effects, all show a lesser quality of the CATE estimate than all three other meta-learners, X-Learner, T-Learner, and S-Learner. The empirical results on our synthetic data for the R-Learner indicate that, despite its theoretical properties and potential to leverage complexity, its performance may be inferior to that of other meta-learners in certain contexts. The potential for it not to outperform other meta-learners in specific situations is an important consideration.

Summary

In this chapter, we explored the fundamentals and theoretical foundations of four widely used meta-learning algorithms for making causal inferences from observational data. These algorithms include the S-Learner, T-Learner, X-Learner, and R-Learner. Additionally, we demonstrated how to implement these four meta-learners using Python, and during the practice sessions, we conducted a step-by-step walkthrough on applying the implemented meta-learners and evaluating their performance with a causal generative model designed to create synthetic observational data.

Understanding how meta-learning algorithms operate is essential for estimating the **Conditional Average Treatment Effect** (CATE) from observational data, particularly within the framework of the Neyman-Rubin model. Estimating the CATE offers helpful advice for policy setting and decision-making by assessing how a treatment or intervention impacts individuals and subgroups within a population based on their characteristics (covariates). Analyzing heterogeneous treatment effects has widespread applications in areas such as personalized medicine, targeted marketing, customized product recommendations, targeted advertising, tailored loan offers, and personalized educational interventions.

The upcoming chapters in the book will provide real-world examples of estimating causal effects using observational data, with a focus on use cases in economics, epidemiology, and social studies.

References

- Athey, Susan , and Guido Imbens. 2016. *Recursive partitioning for heterogeneous causal effects.* Proceedings of the National Academy of Sciences, 7353–7360.
- Avanwyk. 2018. An Overview of LightGBM. May 18. `https://www.avanwyk.com/an-overview-of-lightgbm/`.
- Breiman, Leo. 2001. *Random Forests.* Machine Learning 5–32.
- Caron, Alberto, Gianluca Baio, and Ioanna Manolopoulou. 2022. *Estimating individual treatment effects using non-parametric regression models: A review.* J R Stat Soc Series A, 1115-1149.
- Chadha, Aman. 2025. Primers. *Decision Trees and Ensemble Method.* Accessed 2025. `https://aman.ai/primers/ai/decision-trees-and-ensemble-methods/`.
- Chen, Tianqi , and Carlos Guestrin. 2016. "XGBoost: A Scalable Tree Boosting System." CoRR.
- Curth, Alicia , David Svensson, James Weatherall, and Mihaela van der Schaar. 2021. *Really Doing Great at Estimating CATE? A Critical Look at ML Benchmarking Practices in Treatment Effect Estimation.* 35th Conference on Neural Information Processing Systems (NeurIPS).
- Fahrmeir, Ludwig , Thomas Kneib, Stefan Lang, and Brian Marx. 2013. *Regression: Models, Methods and Applications.* Berlin Heidelberg: Springer-Verlag.
- Foster, J. C., J.M.G. Taylor, and S.J. Ruberg. 2011. *Subgroup identification from randomized clinical trial data.* Statistics in Medicine 2867–2880.
- Jacob, Daniel . 2921. *CATE meets ML.* Digital Finance 99–148.
- Kavlakoglu, Eda, and Erika Russi. 2024. *What is XGBoost?* May 9. `https://www.ibm.com/think/topics/xgboost`.

- Kunzel, S. R., J. S. Sekhon, B. J. Bickel, and B. Yu. 2019. *Metalearners for estimating heterogeneous treatment effects using machine learning*. Proceedings of the National Academy of Sciences, 4156–4165.
- Lu, M., S. Sadiq, D.J. Feaster, and H. Ishwaran. 2018. *Estimating individual treatment effect in observational data using random forest methods*. Journal of Computational and Graphical Statistics 209–219.
- Microsoft. 2022. "microsoft/LightGBM". July 7. `https://github.com/microsoft/LightGBM`.
- Morton, V, and D Jorgerson. 2005. *Regression to the mean: treatment effect without the intervention*. J Eval Clin Pract. 59-65.
- Neyman, J. 1923. *On the application of probability theory to agricultural experiments*. Statistical Science 465–480.
- Nie, X., and S. Wager. 2021. *Quasi-oracle estimation of heterogeneous treatment effects*. Biometrika 299–319.
- Powers, S., J. Qian, K. Jung, A. Schuler, N.H. Shah, T. Hastie, and R. Tibshirani. 2018. *Some methods for heterogeneous treatment effect estimation in high dimensions*. Statistics in Medicine 1767–1787.
- Re, Matteo, and Giorgio Valentin. 2012. *Ensemble methods: A review*. In Advances in Machine Learning and Data Mining for Astronomy, 563-594. Chapman & Hall.
- Robinson, P. M. 1988. *Root-N-Consistent Semiparametric Regression*. Econometrica 931-954.
- Rubin, D. B. 1974. *Estimating causal effects of treatments in randomized and nonrandomized studies*. Journal of Educational Psychology 688–701.
- Tian, L., .A. A. Alizadeh, A. J. Gentles AJ, and R. Tibshirani. 2014. *Estimating Interactions between a Treatment and a Large Number of Covariates*. J Am Stat Assoc. 1517-1532.

12

Causal Inference in Economic Research

This chapter examines the **National Supported Work** (NSW) **Demonstration**, an experimental program that was designed to assess the impact of subsidized and transitional work experience on the lives of disadvantaged individuals. The data of this experiment was first analyzed by Lalonde, so it is often referred to as the LaLonde dataset. We look at the LaLonde dataset in this chapter and discuss different approaches to estimate the causal effect. First, we try standardization using a linear regression model to control the confounders and estimate the causal effect. Next, we calculate the propensity scores to assess the extent of overlap between the distribution of the propensity scores in the treatment and control groups. The causal effect will then be calculated using the method of **inverse probability of treatment weighting** (IPTW). Finally, we use the method of **propensity score matching** to calculate the causal effect. All these methods are implemented in Python, and we will demonstrate how the `causallib` package in Python can be used to estimate the causal effect.

In this chapter we're going to cover the following main topics:

- Causal inference in Economics
- National Supported Work (NSW) Demonstration
- Lalonde dataset
- Causal model
- Estimation using standardization
- Estimation using inverse probability of treatment weighting
- Estimation using propensity score matching

Technical requirements

To run the Python code, you must: (1) install Python 3 following the installation and setup guide at the URL: `https://realpython.com/installing-python/`. (2) Install the Jupyter ecosystem using the installation and setup guide at the URL: `https://docs.jupyter.org/en/stable/install.html`

The Python code in this chapter is in a Jupyter Notebook that can be downloaded from the GitHub URL: `https://github.com/PacktPublishing/Causal-Inference-with-Bayesian-Networks/tree/main/Chapter_12`

You need to install the following Python packages to run the Jupyter Notebook for this chapter:

- `pandas`: `https://pandas.pydata.org/docs/getting_started/install.html`
- `statsmodels`: `https://pypi.org/project/statsmodels/`
- `causallib`: `https://pypi.org/project/causallib/`
- `scikit-learn`: `https://scikit-learn.org/stable/install.html`

We will use the `statsmodels` package for standardization and the `causallib` package (Shimoni et al. 2019) for **inverse probability of treatment weighting** (IPTW) and propensity score matching.

Causal inference in economics

In this section, we talk about the importance of causal inference in economics. In the field of economics, causal inference allows economists to move beyond correlations and study cause-and-effect relationships. Using this approach, researchers can evaluate the effectiveness of economic policies, programs, and interventions in complex systems. Understanding true causal relationships allows the researchers and policymakers to make informed economic decisions.

One of the main applications of causal inference in economics is policy evaluation. An economic policy or program is a set of actions that governments or organizations take to affect economic activity. For example, causal inference methods can be used to determine the effect of trade agreements on the economic growth of a country, or the effect of a minimum wage increase policy on employment levels.

Causal inference can also be used in labor economics. Labor economics focuses on how the labor force interacts with production processes. For example, researchers can use causal inference to study the effect of job training programs on labor markets. In the next section, we will use causal inference to evaluate an economic program called the **National Supported Work** (NSW) **Demonstration.**

National supported work (NSW) demonstration

In labor economics, the general belief is that training and work experience programs can improve participants' chances in the labor market. The **National Supported Work** (NSW) **Demonstration** was a federally and privately funded program initiated to test this idea. It was operated from 1974 to 1979 in 15 locations across the US, and provided work experience for a period of 6-18 months to individuals who were suffering from social and economic problems like ex-offenders, ex-addicts, etc. In this program, the focus was more on gaining practical work experience rather than formal training sessions.

The NSW Demonstration program was a **randomized controlled trial** (RCT). The participants eligible for this program were randomly assigned into a treatment group and a control group. The treatment group received the work experience, while the control group did not participate in the program. This design was used to ensure that differences in the results between the treatment and control groups could be attributed to the program rather than other factors.

The participants' preintervention income and demographic information (such as age, education, and marital status) were gathered through initial surveys and Social Security Administration data. To evaluate the outcome of the program, the treatment and control groups underwent follow-up interviews at specific intervals.

The data from the evaluation of the NSW demonstration was analyzed by the economist Robert Lalonde (*Lalonde 1986*). The analysis of the experimental dataset from the NSW program showed that it was an effective program and increased earnings. In his paper, Lalonde estimated the causal effect of the NSW demonstration on participants' earnings using an observational dataset. Since the NSW program was an RCT, the causal effect was easily calculated and served as the benchmark. Lalonde replaced the experimental control group with individuals from survey datasets. These people did not participate in the experiment, so the data became observational. With this new dataset, Lalonde tried to estimate the causal effect of the program on earnings using standard econometric methods to see if these methods could recover the experimental benchmark.

The results of his study showed that the econometric methods failed and the estimates that they produced were far from the experimental benchmark. The dataset used in Lalonde's paper serves as a benchmark in causal inference and is known as the *Lalonde dataset*. In the next section, we will take a deep look at the Lalonde dataset and then we will see how we can use the principles of causal inference to evaluate the outcomes of the NSW demonstration using this dataset.

Lalonde dataset

We learned about the Lalonde dataset in the previous section. In this section, we will perform an **exploratory data analysis (EDA)** on this dataset using Python. The EDA analysis allows us to investigate this dataset and understand its structure before building the causal model and estimating the causal effect.

EDA analysis

The entire Python code for this chapter is contained in a Jupyter notebook which can be found in the GitHub repository for the book: `https://github.com/PacktPublishing/Causal-Inference-with-Bayesian-Networks/tree/main/Chapter_12`

Loading the libraries

First, we need to load all the required Python libraries for this case study:

```
import numpy as np
import pandas as pd
import matplotlib.pyplot as plt
import statsmodels.api as sm
from causallib.estimation import IPW
from causallib.evaluation import evaluate
from sklearn.linear_model import LogisticRegression
from sklearn.utils import resample
from sklearn.neighbors import NearestNeighbors
%matplotlib inline
```

Loading the datasets

The Lalonde dataset can be downloaded from *Rajiv Dehejia 's* webpage: `https://users.nber.org/~rdehejia/nswdata2.html`

The data was drawn from the papers by Dehejia and Wabha (Dehejia and Wabha 1999) (Dehejia and Wabha 2002), and Lalonde (Lalonde 1986). There are two versions of the Lalonde dataset commonly found in the literature and on the web. The first version has the original RCT data from the NSW demonstration program, hence, it is an experimental dataset. The treatment and control groups have 185 and 260 participants respectively (totally 445 participants). This dataset is purely experimental, which means that the participants were randomly assigned to treatment and control groups. In this chapter, we will call this dataset the NSW dataset.

The second version is an extended dataset created by Lalonde for his study, and we simply call it the Lalonde dataset. It contains the same 185 treated participants from the original NSW dataset. However, the control group contains 429 individuals, drawn from survey datasets (*Population Survey of Income Dynamics and Current Population Survey*). This change turns the original experimental NSW dataset into the observational Lalonde dataset. This dataset was created to study the challenges of using non-experimental control groups. In this chapter we will focus on the Lalonde dataset and only use the experimental NSW dataset as a benchmark to evaluate the causal inference results.

We can use the URLs of data files on Dehejia's webpage to read the files for the treatment and control groups and concatenate them to construct both datasets. First, we create the experimental NSW dataset and store it in a dataframe called `nsw`:

```
columns = ["treat", "age", "education", "black",
           "hispanic","married", "nodegree", "re74",
           "re75", "re78"]
urls = ["http://www.nber.org/~rdehejia/data/nswre74_treated.txt",
        "http://www.nber.org/~rdehejia/data/cps3_controls.txt"]
data = [pd.read_csv(url, sep='\s+', header=None,
        names=columns) for url in urls]
nsw = pd.concat(data, ignore_index=True)
nsw
```

The output is shown in *Figure 12.1*.

	treat	age	education	black	hispanic	married	nodegree	re74	re75	re78
0	1.0	37.0	11.0	1.0	0.0	1.0	1.0	0.000	0.00	9930.0460
1	1.0	22.0	9.0	0.0	1.0	0.0	1.0	0.000	0.00	3595.8940
2	1.0	30.0	12.0	1.0	0.0	0.0	0.0	0.000	0.00	24909.4500
3	1.0	27.0	11.0	1.0	0.0	0.0	1.0	0.000	0.00	7506.1460
4	1.0	33.0	8.0	1.0	0.0	0.0	1.0	0.000	0.00	289.7899
...	...	...	...	...	...	...	...	...	...	...
440	0.0	21.0	9.0	1.0	0.0	0.0	1.0	31886.430	12357.22	0.0000
441	0.0	28.0	11.0	1.0	0.0	0.0	1.0	17491.450	13371.25	0.0000
442	0.0	29.0	9.0	0.0	1.0	0.0	1.0	9594.308	16341.16	16900.3000
443	0.0	25.0	9.0	1.0	0.0	1.0	1.0	24731.620	16946.63	7343.9640
444	0.0	22.0	10.0	0.0	0.0	1.0	1.0	25720.920	23031.98	5448.8010

445 rows × 10 columns

Figure 12.1—NSW dataset

This dataframe has 335 rows and 10 columns. Next, we create the Lalonde dataset and store it in a dataframe called `lalonde`:

```
urls = ["http://www.nber.org/~rdehejia/data/nswre74_treated.txt",
        "http://www.nber.org/~rdehejia/data/cps3_controls.txt"]
data = [pd.read_csv(url, sep='\s+', header=None, names=columns)
        for url in urls]
lalonde = pd.concat(data, ignore_index=True)
lalonde
```

The output is shown in *Figure 12.2*.

	treat	age	education	black	hispanic	married	nodegree	re74	re75	re78
0	1.0	37.0	11.0	1.0	0.0	1.0	1.0	0.0	0.0	9930.0460
1	1.0	22.0	9.0	0.0	1.0	0.0	1.0	0.0	0.0	3595.8940
2	1.0	30.0	12.0	1.0	0.0	0.0	0.0	0.0	0.0	24909.4500
3	1.0	27.0	11.0	1.0	0.0	0.0	1.0	0.0	0.0	7506.1460
4	1.0	33.0	8.0	1.0	0.0	0.0	1.0	0.0	0.0	289.7899
...	...	...	...	...	...	...	...	...	...	...
609	0.0	18.0	11.0	0.0	0.0	0.0	1.0	0.0	0.0	10150.5000
610	0.0	24.0	1.0	0.0	1.0	1.0	1.0	0.0	0.0	19464.6100
611	0.0	21.0	18.0	0.0	0.0	0.0	0.0	0.0	0.0	0.0000
612	0.0	32.0	5.0	1.0	0.0	1.0	1.0	0.0	0.0	187.6713
613	0.0	16.0	9.0	0.0	0.0	0.0	1.0	0.0	0.0	1495.4590

614 rows × 10 columns

Figure 12.2—Lalonde dataset

This dataset has 614 rows and 10 columns. Both the NSW and Lalonde datasets share the same features.

Feature description

Let's see the description of each column in these datasets:

- **treat:** 1 for the treatment group and 0 for the control group
- **age:** Age of the participant in years
- **education:** Number of years of education of the participant
- **black:** 1 if the participant was black and 0 otherwise
- **hispanic:** 1 if the participant was hispanic and 0 otherwise
- **married:** 1 if the participant was married and 0 otherwise
- **nodegree:** 1 if the participant had no high school diploma and 0 otherwise
- **re74:** Earnings of the participant in 1974 before starting the program (USD)
- **re75:** Earnings of the participant in 1975 before starting the program (USD)
- **re78:** Earnings of the participant in 1978 after completing the program (USD)

Remember that the treatment group (rows with treat=1) in both NSW and Lalonde datasets is the same. But the control groups (rows with treat=0) are different.

Next, we perform an EDA analysis for the Lalonde dataset. We get a statistical summary of this dataset using the `describe()` function:

```
lalonde.describe()
```

The output is shown in *Figure 12.3*.

	treat	age	education	black	hispanic	married	nodegree	re74	re75	re78
count	614.000000	614.000000	614.000000	614.000000	614.000000	614.000000	614.000000	614.000000	614.000000	614.000000
mean	0.301303	27.363192	10.268730	0.395765	0.117264	0.415309	0.630293	4557.546569	2184.938207	6792.834483
std	0.459198	9.881187	2.628325	0.489413	0.321997	0.493177	0.483119	6477.964479	3295.679043	7470.730792
min	0.000000	16.000000	0.000000	0.000000	0.000000	0.000000	0.000000	0.000000	0.000000	0.000000
25%	0.000000	20.000000	9.000000	0.000000	0.000000	0.000000	0.000000	0.000000	0.000000	238.283425
50%	0.000000	25.000000	11.000000	0.000000	0.000000	0.000000	1.000000	1042.330000	601.548400	4759.018500
75%	1.000000	32.000000	12.000000	1.000000	0.000000	1.000000	1.000000	7888.498250	3248.987500	10893.592500
max	1.000000	55.000000	18.000000	1.000000	1.000000	1.000000	1.000000	35040.070000	25142.240000	60307.930000

Figure 12.3—The statistical summary of the Lalonde dataset

We see that almost 30% of the rows belong to the treatment group. The average age of the participants was 27 years, and their average years of education were 10 years. This is consistent with the fact that many of the participants had no high school diploma (63%). In addition, most of them were not married (59%).

Comparing the treatment and control groups

Let's compare the mean of the earnings of the treatment and control groups after completing the program. First, we calculate it for the NSW dataset:

```
nsw.groupby('treat')['re78'].mean().round(2)

## Output
treat
0.0    4554.80
1.0    6349.14
```

We see that the treatment group has higher earnings on average, suggesting that the NSW demonstration was a successful program. Next, we compare the earnings for the Lalonde dataset:

```
lalonde.groupby('treat')['re78'].mean().round(2)

## Output
treat
0.0    6984.17
1.0    6349.14
```

We see that the control group has a higher earning on average which may suggest that the NSW demonstration program was not a successful program. However, it is important to note that in this observational dataset, we have an experimental treatment group versus a non-experimental control group coming from survey data. Hence, the treatment data is very different from the control data. The next code compares the histograms of **re74**, **age** and **education** in treatment and control. *Figure 12.4* shows the output plots.

```
treated = lalonde[lalonde['treat'] == 1]
control = lalonde[lalonde['treat'] == 0]
for col in ['re74', 'education', 'age']:
    plt.figure(figsize=(8, 4))
    min_val, max_val = np.min(lalonde[col]), np.max(lalonde[col])
    bins = np.linspace(min_val, max_val, 21)
    plt.hist(control[col], bins=bins, alpha=0.8, label='Control')
    plt.hist(treated[col], bins=bins, alpha=0.6, label='Treatment')
    plt.xlabel(col, fontsize=14)
    plt.ylabel('Frequency', fontsize=14)
    plt.legend(fontsize=11)
    plt.show()
```

We can also compare the binary features **black, hispanic**, married and **nodegree** in treatment and control groups in this dataset.

```
lalonde.groupby('treat')[['black', 'hispanic', 'married',
                          'nodegree']].mean().round(2)
```

```
## Output
+-------+------+---------+--------+----------+
| treat | black | hispanic | married | nodegree |
+=======+======+=========+========+==========+
|   0   | 0.20 |   0.14   |  0.51  |   0.60   |
+-------+------+---------+--------+----------+
|   1   | 0.84 |   0.06   |  0.19  |   0.71   |
+-------+------+---------+--------+----------+
```

The output shows that the treatment and control groups are very different in terms of ethnicity and marital status.

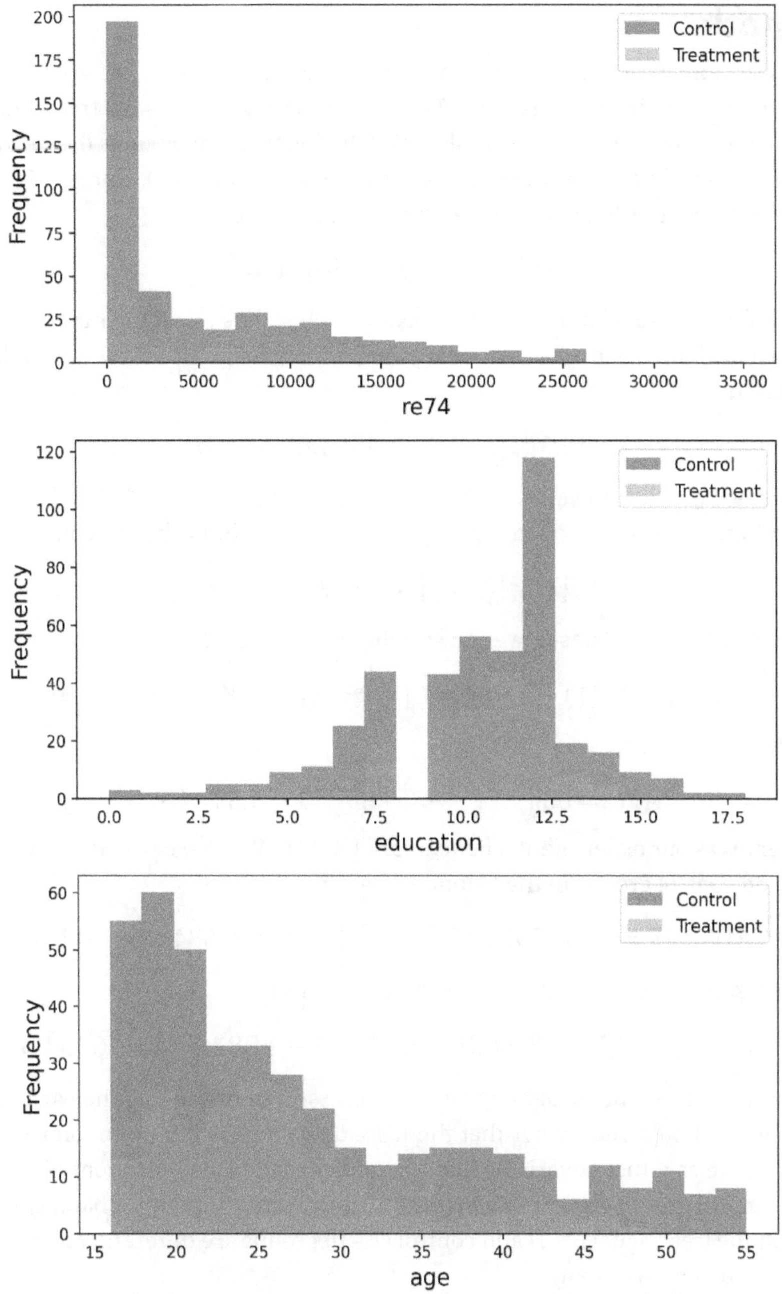

Figure 12.4—Comparing the histograms of treatment and control groups

In this section, we loaded the Lalonde dataset and performed an EDA analysis on it. In the next section, we will discuss the causal estimation problem for the Lalonde dataset.

Causal model

In the previous section, we looked at the Lalonde dataset. In this section, we are going to define our estimand for the causal estimation problem. Let's assume that T denotes the treatment (the treat column) and Y denotes the outcome (the re78 column). Additionally, X represents the vector of covariates (the remaining columns like age, education, etc. combined into a vector). Our goal is to calculate the **Average Treatment Effect (ATE)**, which was defined in *Chapter 10*:

$$ATE = E[Y(1) - Y(0)] \quad (12.1)$$

We first do that for the experimental NSW dataset which serves as a benchmark. In the previous section, we calculated the mean of re78 in treatment and control groups for the NSW dataset. We saw that in this dataset:

$$E[Y^{obs}|T = 0] = 4554.80 \quad (12.2)$$

where Y^{obs} denotes the observed values of Y that form the dataset. So, $E[Y^{obs} |T=0]$ is the mean of the outcome (the column re78) of the control group in the dataset. Similarly, we had:

$$E[Y^{obs}|T = 1] = 6349.14 \quad (12.3)$$

Since this is an experimental dataset, we can simplify *Equation 12.1* as:

$$ATE = E[Y(1) - Y(0)] = E[Y|T = 1] - E[Y|T = 0] \quad (12.4)$$

Hence, we have:

$$ATE = 6349.14 - 4554.80 = 1794.34 \quad (12.5)$$

This number serves as our experimental benchmark for ATE. We also calculated the mean of re78 in the treatment and control groups in the Lalonde dataset:

$$E[Y^{obs}|T = 0] = 6984.17, \; E[Y^{obs}|T = 1] = 6349.14 \quad (12.6)$$

Based on these results, we can see that in the Lalonde dataset:

$$E[Y^{obs}|T = 1] - E[Y^{obs}|T = 0] = 6349.14 - 6984.17 = -635.03 \quad (12.7)$$

The result may suggest that the program was not successful in improving the earnings of the treatment group. But we should remember that the Lalonde dataset is not a randomized experimental dataset anymore. We saw that covariates like age and education had different distributions in the treatment and control groups (refer to *Figure 12.4*). This means that the probability distributions of the covariates in the treatment ($T = 1$) and control ($T = 0$) groups are different, and the covariates (X) are confounders. So, it follows that:

$$ATE = E[Y(1) - Y(0)] \neq E[Y|T = 1] - E[Y|T = 0] \quad (12.8)$$

This equation indicates that the result of *Equation 12.7* is not a correct estimation for ATE since it ignores the confounding present in the Lalonde dataset.

In the following sections, we use the three different methods mentioned in *Chapter 10* to estimate the ATE for the Lalonde dataset. These methods are standardization, IPTW, and propensity score matching.

Estimation using standardization

In this section we see how we can use standardization to estimate the causal effect. We will use standardization with a linear regression model to estimate the ATE of the Lalonde dataset.

Linear regression

As mentioned before, the presence of confounders doesn't allow us to estimate the ATE of the Lalonde dataset since the treatment and control groups are systematically different. This is where standardization can help us. We can statistically control the confounders by including them in a linear regression model. Linear regression allows us to isolate the relationship between the treatment and the outcome groups by keeping all the confounders constant. First, we need to create a multiple linear regression model in which the dependent variable is the outcome (*Y*) and the regressor variables are the treatment variable (*T*) and the covariates (*X*):

$$Y = \beta_0 + \tau T + \beta_1 X_1 + \beta_2 X_2 + \cdots + \epsilon \quad (129)$$

Here, τ *and* β_i are the model's coefficients, and ε is the error term. Additionally, X_i denotes each of the covariate variables that form the covariate vector *X* (in our dataset X_i can be age, education, etc.).

After training the regression model, τ (the coefficient of *T*) gives the average change in *Y* for a one-unit increase in *T* while keeping all the other variables (X_i) constant. We know that *T* is a binary variable (0 for the control group and 1 for the treatment group). Therefore, a one-unit increase in *T* means switching from the control group to the treatment group. This means that τ estimates the average difference in the outcome (*Y*) between the treatment and control group by keeping all the covariates constant. Mathematically, τ estimates:

$$E[Y|T=1] - E[Y|T=0],\ X_i = const \quad (12.10)$$

If we assume confounding can be removed by keeping all the covariates constant, then τ is an estimate of ATE.

Estimating the causal effect

Now, we use standardization with a linear regression model to estimate the ATE of our dataset. The next code uses the `OLS` class in the `statsmodels` package to fit a linear model to the dataset. We use the **Ordinary Least Squares (OLS)** method to estimate the parameters of a linear regression model. The covariate columns are stored in the dataframe `X`, and the series `T` and `Y` store the treatment and outcome columns (treat and re78). In OLS model, `Y` is the target array, and `X` and `T` form the regressors array. By passing `cov_type='HC0'` to OLS, we use the heteroscedasticity robust covariance that results in more reliable confidence intervals. The estimated ATE is the coefficient of `T` (treat). We can also get its standard error, p-value and 95% confidence interval.

```
covariates = ['age', 'education', 'black', 'hispanic',
              'married', 'nodegree', 're74', 're75']
```

```
X = lalonde[covariates]
X1 = sm.add_constant(X)
T = lalonde['treat']
Y = lalonde['re78']

model = sm.OLS(Y, pd.concat([X1, T], axis=1))
results = model.fit(cov_type='HC0')
cols = {"Estimation": [results.params['treat']],
        "Standard error": [results.bse['treat']],
        "p-value": [results.pvalues['treat']],
         "95% conf interval": [results.conf_int().loc["treat"].round(3).
tolist()]}
output = pd.DataFrame(cols, index=["ATE"])
output.round(3)
```

The output is shown in *Figure 12.5*.

	Estimation	Standard error	p-value	95% conf interval
ATE	1548.244	734.521	0.035	[108.61, 2987.878]

Figure 12.5—Output of the linear model

We see that the linear model predicts an ATE of $1548.24, which is close to our benchmark in *Equation 12.5*.

Sensitivity analysis using E-value

After calculating the causal effect, we can do a sensitivity analysis to assess how robust our estimation is. We can use **E-value** for this purpose. Remember that in causal inference, we try to isolate the relationship between the treatment and the outcome groups by keeping all the measured confounders constant. That is why we kept all the covariates (X_i) constant in *Equation 12.10*. Here the main assumption is that there were no unmeasured confounders. However, this assumption is not testable. The E-value can tell us how sensitive the result is to violations of this assumption. In fact, it is a measure of how strong an unmeasured confounder would have to be to explain away the observed causal effect.

To calculate the E-value when we have a continuous outcome, first we need to standardize the causal effect estimate (here it is ATE) by dividing it by the standard deviation of the outcome:

$$d = \frac{ATE}{sd(Y)} \quad (12.11)$$

Next, we calculate the approximate risk ratio:

$$RR \approx e^{0.91d} \quad (12.12)$$

Finally, the E-value is calculated using *the* following formula:

$$E - value = \begin{cases} RR + \sqrt{RR(RR-1)} & if\ RR \geq 1 \\ RR^{-1} + \sqrt{RR^{-1}(RR^{-1}-1)} & if\ RR < 1 \end{cases} \quad (12.13)$$

A small E-value (1 to 1.5) is an indication of low robustness to additional unmeasured confounding. An E-value between 1.5 to 3 means moderate robustness and an E-value of 3 and above is an indication of strong robustness. The next code calculates the E-value for our ATE estimate:

```
def calculate_evalue(estimate, sd_outcome):
    d = estimate / sd_outcome
    rr = np.exp(0.91 * d)

    # E-value for point estimate
    if rr >= 1:
        e_value = rr + np.sqrt(rr * (rr - 1))
    else:
        rr_inv = 1 / rr
        e_value = rr_inv + np.sqrt(rr_inv * (rr_inv - 1))

    return e_value

# Calculate E-value
sd_outcome = Y.std()
e_value = calculate_evalue(
    estimate=results.params['treat'],
    sd_outcome=sd_outcome
)

print("E-Value: ", np.round(e_value, 2))

## Output
E-Value:  1.71
```

The result shows that we have moderate robustness, and the causal estimation is vulnerable to moderately strong confounding.

So far, we have discussed causal estimation using the standardization method. In the next section, we will look at another estimation method.

Estimation using inverse probability of treatment weighting

In the following sections, we discuss an approach which was first proposed by Dehejia and Wahba (Dehejia and Wahba 1999). In this approach, we first use a machine learning model to estimate the propensity scores of each participant, which is the probability that they receive the NSW work experience. We can use the propensity scores to estimate ATE and check the overlap of the propensity score distribution between the treatment and control groups. In this section, we use the IPTW to estimate the ATE from propensity scores.

Calculating propensity scores

As mentioned in *Chapter 10*, propensity score is the conditional probability of a unit being assigned to the treatment group given a set of observed covariates.

$$\pi(x) = Pr(T = 1|X = x) = E[T|X = x] \quad (12.14)$$

In observational studies, we need to estimate the propensity scores using a model. Here, we use a logistic regression model which is defined by the following equation:

$$\pi(x) = Pr(T = 1|X = x) = \frac{1}{1+e^{-(\beta_0+\beta_1 x_1+\cdots+\beta_k x_k)}} \quad (12.15)$$

Here X is a vector of the covariate variables (X_1, X_2, ...X_k), and β_i are the model's coefficients which are determined after training the logistic regression model using the observed data. After training, the logistic regression model can estimate the propensity score for a given set of observed covariates x=(x_1, x_2, ...x_n).

We can create a logistic regression model using the `Logit` class in the `statsmodels` package:

```
class StatsModelsLogit:
    def fit(self, X, y):
        X_const = sm.add_constant(X)
        self.model = sm.Logit(y, X_const).fit(disp=0)
        return self

    def predict_proba(self, X):
        X_const = sm.add_constant(X)
        p = self.model.predict(X_const)
        return np.column_stack([1 - p, p])

    def predict(self, X):
        proba = self.predict_proba(X)[:, 1]
        return (proba >= 0.5).astype(int)

learner = StatsModelsLogit()
```

Then we create an `IPW` object in the `causallib` package that takes the logistic regression model and fits it to the dataset using the `fit()` method. IPW expects the model to follow the scikit-learn estimator API. So, the model must implement methods like `fit()`, `predict()` and `predict_proba()`. That is the reason that we wrapped the `statsmodels` model in `StatsModelsLogit` class that mimics the scikit-learn interface. In the `fit()` method of IPW, the vector of the covariates (`X`) is the training vector, and the treatment column (`T`) is the target.

```
ipw = IPW(learner)
X = lalonde[covariates]
ipw.fit(X, T)
```

Finally, we calculate the propensity scores using the `compute_propensity()` method. Then we create a new column in the Lalonde dataset to store the propensity scores:

```
propensity_scores = ipw.compute_propensity(X, T, treatment_values=1)
lalonde['propensity_score'] = propensity_scores
lalonde
```

The result is shown in *Figure 12.6.*

	treat	age	education	black	hispanic	married	nodegree	re74	re75	re78	propensity_score
0	1.0	37.0	11.0	1.0	0.0	1.0	1.0	0.0	0.0	9930.0460	0.638770
1	1.0	22.0	9.0	0.0	1.0	0.0	1.0	0.0	0.0	3595.8940	0.224634
2	1.0	30.0	12.0	1.0	0.0	0.0	0.0	0.0	0.0	24909.4500	0.678244
3	1.0	27.0	11.0	1.0	0.0	0.0	1.0	0.0	0.0	7506.1460	0.776324
4	1.0	33.0	8.0	1.0	0.0	0.0	1.0	0.0	0.0	289.7899	0.701639
...	...	...	...	...	...	...	...	...	...	...	...
609	0.0	18.0	11.0	0.0	0.0	0.0	1.0	0.0	0.0	10150.5000	0.123144
610	0.0	24.0	1.0	0.0	1.0	1.0	1.0	0.0	0.0	19464.6100	0.034560
611	0.0	21.0	18.0	0.0	0.0	0.0	0.0	0.0	0.0	0.0000	0.183351
612	0.0	32.0	5.0	1.0	0.0	1.0	1.0	0.0	0.0	187.6713	0.383032
613	0.0	16.0	9.0	0.0	0.0	0.0	1.0	0.0	0.0	1495.4590	0.089712

614 rows × 11 columns

Figure 12.6—Lalonde dataset with the propensity scores

Now, we have the estimated propensity score of each participant in this dataset, and we can use them to estimate the ATE.

ATE estimation

The IPTW method estimates the ATE using the propensity scores based on the following equation:

$$\widehat{ATE} = E\left[\frac{Y^{obs}T}{\pi(x)}\right] - E\left[\frac{Y^{obs}(1-T)}{1-\pi(x)}\right] \quad (12.16)$$

This equation defines the weight of each unit in the treatment and control groups. The weights of the units in the treatment group (w_T) and control group (w_C) are defined as follows:

$$w_T = \frac{1}{\pi(x)}, \; w_C = \frac{1}{1-\pi(x)} \quad (12.17)$$

In fact, in *Equation 12.16*, these weights are used to calculate the weighted mean of the outcomes in the treatment and control groups. We can implement this equation in Python as follows to estimate the ATE:

```
outcome_1 = ((T*Y / propensity_scores) / (T / propensity_scores).sum()).sum()
outcome_0 = ((1-T)*Y / (1-propensity_scores) / ((1-T) / (1-propensity_scores)).
sum()).sum()
effect = outcome_1 - outcome_0
print("Outcomes:\n 0: {} \n 1: {}".format(outcome_0, outcome_1))
print("Effect: ", effect)

## Output
Outcomes:
 0: 6422.849
 1: 6647.515
Effect:  224.676
```

We could also use the `causallib` to get the same results. The function `estimate_population_outcome()` estimates the outcomes, and the function `estimate_effect()` gives their difference as the ATE estimation.

```
outcomes = ipw.estimate_population_outcome(X, T, Y)
effect = ipw.estimate_effect(outcomes[1], outcomes[0])
print("Outcomes:")
print(outcomes)
print("Effect: ", effect)

## Output
Outcomes:
0.0    6422.84
1.0    6647.51
```

```
dtype: float64
Effect:  diff    224.67
dtype: float64
```

Hence, IPTW estimates the value of ATE to be $224.67. We can also calculate the 95% confidence interval and standard error of this estimation using bootstrapping. Here we resample the initial dataset with replacement to create a large number of bootstrap samples, and for each bootstrap sample, the ATE is estimated:

```
np.random.seed(10)
n_iterations = 500
ate_samples = []
learner1 = StatsModelsLogit()

for _ in range(n_iterations):
    # Create a bootstrap sample
    X_resampled, T_resampled, Y_resampled = resample(X, T, Y)

    # Fit IPW model and estimate ATE
    ipw_s = IPW(learner1)
    ipw_s.fit(X_resampled, T_resampled)
    outcomes = ipw_s.estimate_population_outcome(
        X_resampled, T_resampled, Y_resampled)
    ate_resampled = ipw_s.estimate_effect(outcomes[1], outcomes[0])
    ate_samples.append(ate_resampled)

# Calculate the confidence intervals
ate_samples = np.array(ate_samples)
lower_bound = np.percentile(ate_samples, 2.5)
upper_bound = np.percentile(ate_samples, 97.5)
# Calculate the standard error as the standard deviation of bootstrap ATE
estimates
standard_error = np.std(ate_samples, ddof=1)

print("95% Confidence Interval:",
      [np.round(lower_bound, 3), np.round(upper_bound, 3)])
print("Standard Error:", np.round(standard_error, 3))
```

```
## Output
95% Confidence Interval: [-1521.248, 2275.346]
Standard Error: 951.961
```

The collection of the estimated ATEs from all bootstrap samples forms the bootstrap distribution, and this distribution is used to estimate the confidence interval and the standard error. Finally, we calculate the E-value:

```
e_value = calculate_evalue(
    estimate=effect.values[0],
    sd_outcome=sd_outcome
)
print("E-Value: ", np.round(e_value, 2))

## Output
E-Value:  1.2
```

The E-value shows a weak robustness. We also see that the estimation ($224.67) is far from the benchmark ($1794.34). In the next section, we will see why the IPTW does not provide a good causal estimate.

Balance plots

A balance plot is generally used to compare the distribution of the propensity scores between the treatment and control groups of the Lalonde dataset. We can plot the histograms of the propensity scores of the treatment and control groups together to see if they have a significant overlap or not. In the `causallib` package, we can easily create this plot using the `plot_weight_distribution()`:

```
plt.figure(figsize=(7, 4))
evaluation_results = evaluate(ipw, X, T, Y)
evaluation_results.plot_weight_distribution()
plt.show()
```

The next code creates such a plot, and *Figure 12.7* shows the result.

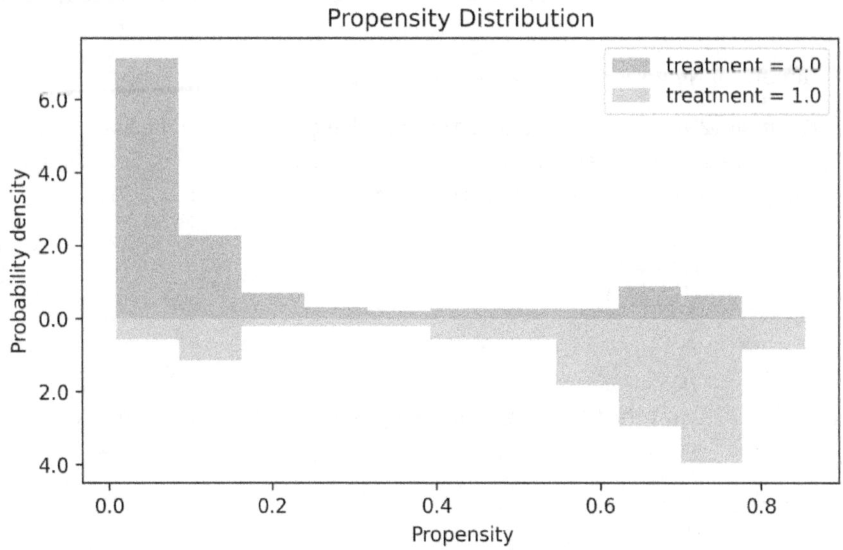

Figure 12.7—Balance plot for the propensity scores of the Lalonde dataset

This plot shows that the distribution of propensity scores for the treatment and control groups has a significant *overlap* (also called a *common support*). This means that for any given propensity score, there are some participants in both treatment and control groups. In *Figure 12.7*, we see that for any value of the propensity score, the probability density of both histograms is greater than zero. When there is no overlap, it indicates that there are certain ranges of propensity score for which only one group (treatment or control) exists. In that case, we cannot make a valid comparison between the treatment and control groups.

In propensity score analysis, the assumption of overlap (or common support) is very important since it guarantees having comparable units in both treatment and control groups across the range of propensity scores.

We can also check if the distributions are *balanced* or not. Here, balanced means that the distributions of propensity scores for the treatment and control groups are sufficiently similar. The balanced distribution of treatment and control groups means that these groups are comparable in terms of observed covariates. Hence, we can have a more accurate estimation of the causal effect. *Figure 12.7* shows that distributions are not balanced since they don't look similar. The balance between the distribution of propensity scores results from the balance between the covariates in the treatment and control groups. We can also check this balance numerically. We can use a metric called **standardized mean difference (SMD)** defined as follows:

$$SMD = \frac{\bar{X}_{i,T} - \bar{X}_{i,C}}{\sqrt{s_{i,T}^2 + s_{i,C}^2}} \quad (12.18)$$

where $\bar{X}_{i,T}$ and $\bar{X}_{i,C}$ and are the mean of the covariate X_i in the treatment and control groups respectively, and $s_{i,T}$ and $s_{i,C}$ denote the standard deviation of X_i in the treatment and control groups respectively.

Important Note

Equation 12.18 is used in the `causallib` package to calculate the SMD values. However, there are alternative equations that can be used to calculate it. For example, in some packages the following equation is used for SMD calculation:

$$SMD = \frac{\bar{X}_{i,T} - \bar{X}_{i,C}}{\sqrt{\frac{(s_{i,T}^2 + s_{i,C}^2)}{2}}}$$

The following code calculates the SMD of age in the Lalonde dataset as an example:

```
covariate = 'age'
mean_xi_t = treated[covariate].mean()
mean_xi_c = control[covariate].mean()
std_xi_t = treated[covariate].std(ddof=0)
std_xi_c = control[covariate].std(ddof=0)
smd = (mean_xi_t - mean_xi_c) / np.sqrt(std_xi_t**2 + std_xi_c**2)
print("SMD for age:", np.round(smd, 4))

## Output
SMD for age: -0.1713
```

This is the so-called unweighted absolute SMD value for age. We can also calculate the weighted SMD of a covariate. Here, similar to *Equation 12.16,* we use the propensity scores to create a weight for each value of the covariate in the dataset. We use $1/\pi(x)$ and $1/(1-\pi(x))$ as the weight of each unit in $X_{i,T}$ and $X_{i,C}$, respectively. These weights are then used to calculate the weighted mean of X_i for the treatment and control groups in *Equation 12.18*. The following code calculates the weighted SMD of age in our dataset:

```
treated = lalonde[lalonde['treat'] == 1]
control = lalonde[lalonde['treat'] == 0]
w_t = 1/treated['propensity_score']
w_c = 1/(1-control['propensity_score'])

numerator = np.average(treated[covariate], weights=w_t) -
np.average(control[covariate], weights=w_c)
denominator = np.sqrt(np.var(treated[covariate]) + np.var(control[covariate]))
weightes_smd = numerator / denominator
```

```
print("Weighted SMD for age:", np.round(weightes_smd, 4))

## Output
Weighted SMD for age: -0.1187
```

A **Love plot** is a graphical way to show the balance between the covariates in the treatment and control groups. It plots the unweighted absolute SMD of each covariate beside its weighted absolute SMD. Using this plot, we can check the balance of the covariates in the raw data. Additionally, we can see how the weights defined by the propensity scores improve the balance. The following code creates the love plot for the Lalonde dataset:

```
plt.figure(figsize=(7, 4))
evaluation_results.plot_covariate_balance(kind="love", thresh=0.1)
plt.show()
```

The resulting plot is shown in *Figure 12.8*.

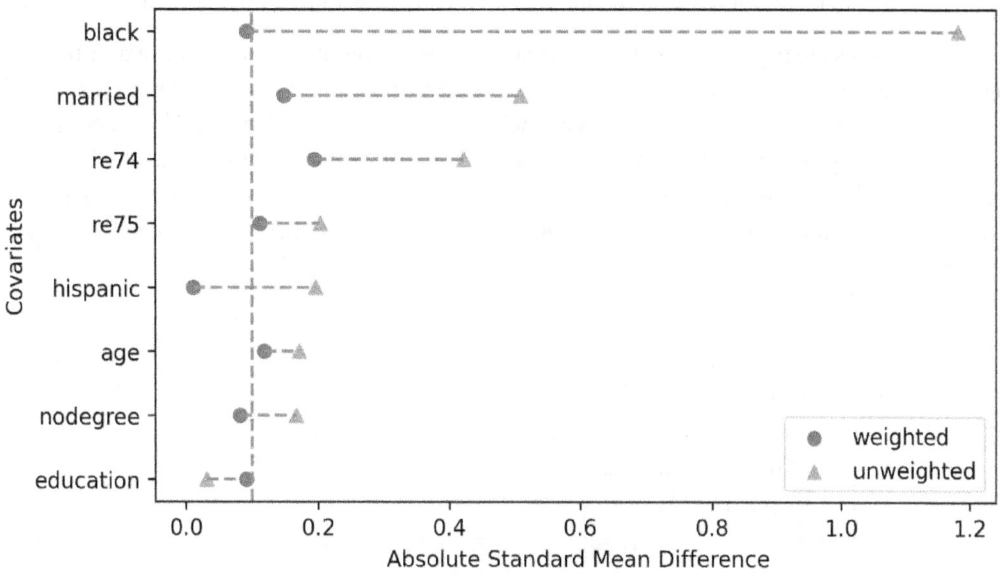

Figure 12.8—Love plot of the Lalonde dataset

For each covariate, a lower absolute SMD means more balance between the treatment and control groups. A threshold of 0.1 is usually used as a cutoff for imbalanced covariates. If the weights calculated using the propensity scores (*Equation (12.18)*) are effective, the imbalance for the covariates should be lower after weighting. Therefore, the weighted absolute SMD for each covariate should be less than its unweighted counterpart, and ideally less than 0.1. Based on *Figure 12.8*, the weights could improve the balance of the covariates (except for education). However, the weighted SMD of some of them is still much above the threshold. In this section, we showed how propensity scores can be used to estimate the causal effect in the IPTW method. However, using the propensity scores for causal estimation is not limited to IPTW. In the next section, we will introduce another method that uses propensity scores for causal estimation.

Estimation using propensity score matching

In the previous section, we calculated the propensity scores of the Lalonde dataset. We also saw that the distribution of propensity scores for the treatment and control groups has a significant overlap (*Figure 12.7*). In this section, we will use the method of propensity score matching to estimate the ATE. This method aims to balance the distribution of treatment and control groups. It creates a matched sample from the original dataset in which the distribution of treatment and control groups is similar. This matched sample mimics the conditions of a **randomized controlled trial** (RCT).

Creating matched data

In propensity score matching, we first find matches for each treated unit from the control group. Each unit in the treatment group is matched with one or more units from the control group based on their similarity, and the similarity is measured by the propensity score. Here, the main assumption is that two units become more similar as their propensity scores approach each other. There are different matching methods, and here we use the method of nearest neighbor matching in which each unit in the treatment group is matched with one unit in the control group that has the closest propensity score.

The next code takes the trimmed dataset and creates a matched control group for the treatment group using the nearest neighbor method. It uses the `NearestNeighbors` class in the `scikit-learn` package to find the nearest neighbor of each treated participant from the control group. Then it combines the original treatment group with these nearest neighbors to create the matched dataset. Since we have 180 participants in the treatment group (after trimming), the code finds 180 matched participants for them from the control group. Hence, the final matched dataset (`lalonde_matched`) will have 360 participants:

```
treated = lalonde[lalonde['treat'] == 1]
control = lalonde[lalonde['treat'] == 0]
knn = NearestNeighbors(n_neighbors=1, algorithm='ball_tree')
knn.fit(control[['propensity_score']])
_, indices = knn.kneighbors(treated[['propensity_score']])
control_matched = control.iloc[indices.flatten()]
lalonde_matched = pd.concat([treated, control_matched])
print("Shape of the matched dataset:", lalonde_matched.shape)

## Output
Shape of the matched dataset: (370, 11)
```

Now that we've constructed the matched dataset, we can see if its balance has improved or not.

Checking the balance

Now, let's check the balance of the distribution of the propensity scores in the treatment and control groups of the matched dataset. The following code creates the balance plot of the matched dataset:

```
X_matched = lalonde_matched[covariates]
Y_matched = lalonde_matched['re78']
T_matched = lalonde_matched['treat']

plt.figure(figsize=(7, 4))
evaluation_results = evaluate(ipw, X_matched, T_matched, Y_matched)
evaluation_results.plot_weight_distribution()
plt.show()
```

You can see the output in *Figure 12.9*.

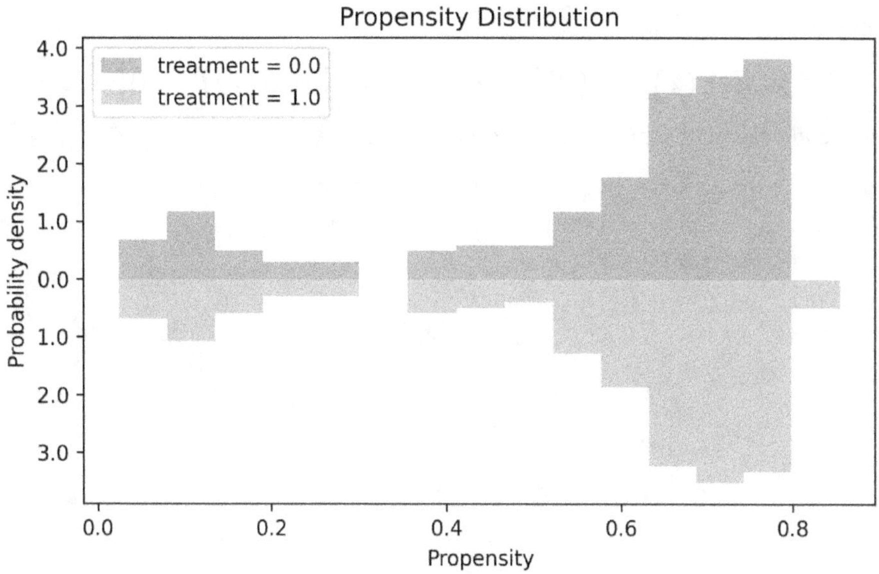

Figure 12.9— Balance plot for the propensity scores of the matched Lalonde dataset

As *Figure 12.9* shows, the distribution of treatment and control groups in the matched dataset has become more similar to each other. We can also check the SMD of the covariates. The next code calculates the absolute SMD of all the covariates in the matched dataset. Please note that these are the unweighted values.

```
evaluation_results = evaluate(ipw, X_matched, T_matched, Y_matched)
evaluation_results.evaluated_metrics.covariate_balance['unweighted']

## Output
Covariate
```

```
age          0.144343
education    0.009836
black        0.010440
hispanic     0.015831
married      0.115223
nodegree     0.008385
re74         0.035404
re75         0.007245
Name: unweighted, dtype: float64
```

We see that age and marital status are only marginally higher than 0.1, while the remaining SMDs are less than that. Overall, the dataset is balanced, and now we are ready to calculate the ATE.

ATE estimation

Since the matched dataset is balanced now, we can assume that confounding is not present anymore, and it follows that:

$$ATE = E[Y(1) - Y(0)] = E[Y|T = 1] - E[Y|T = 0] \quad (12.19)$$

The following code calculates the ATE based on this equation:

```
treated_matched = lalonde_matched[lalonde_matched['treat']==1]
control_matched= lalonde_matched[lalonde_matched['treat']==0]
EY_1 = treated_matched['re78'].mean()
EY_0 =  control_matched['re78'].mean()
effect = EY_1 - EY_0

print("Outcomes:\n 0: {} \n 1: {}".format(EY_0, EY_1))
print("Effect: ", effect)

## Output
Outcomes:
 0: 4367.369
 1: 6349.143
Effect:  1981.774
```

In the end, an ATE of $1981.774 is predicted by the propensity score matching. Like the IPTW approach, we can calculate the 95% confidence interval and standard error of this estimation using bootstrapping.

```
np.random.seed(10)
n_iterations = 500
ate_samples = []

for _ in range(n_iterations):
```

```
    data_resampled = resample(lalonde_matched)
    data_resampled_treated = data_resampled[data_resampled['treat'] == 1]
    data_resampled_control = data_resampled[data_resampled['treat'] == 0]
    ate_resampled = data_resampled_treated['re78'].mean() - data_resampled_
control['re78'].mean()
    ate_samples.append(ate_resampled)
ate_samples = np.array(ate_samples)
lower_bound = np.percentile(ate_samples, 2.5)
upper_bound = np.percentile(ate_samples, 97.5)
standard_error = np.std(ate_samples, ddof=1)
e_value = calculate_evalue(effect, sd_outcome)
print(f"95% Confidence Interval: [{np.round(lower_bound, 3)}, {np.round(upper_
bound, 3)}]")
print(f"Standard Error: {np.round(standard_error, 3)}")
print("E-Value: ", np.round(e_value, 2))

## Output
95% Confidence Interval: [650.823, 3476.383]
Standard Error: 709.25
E-Value:  1.86
```

We can see that the standard error of propensity score matching is less than IPTW and standardization, and the E-value indicates moderate robustness.

In this section, we looked at propensity score matching and how we can utilize the propensity scores to create a matched dataset. We also learned how to check the balance of a dataset. Finally, we used the matched dataset to calculate the ATE.

Comparing the results

Table 12.1 compares the results of the different causal estimation methods that we used with the benchmark.

	ATE Estimate	Standard error	95% confidence interval	E-value
Benchmark	$1794.34	-	-	-
Standardization	1548.24	734.52	[108.61, 2987.88]	1,71
IPTW	224.68	951.96	[-1521.25, 2275.35]	1.2
Propensity score matching	1981.77	709.25	[650.823, 3476.383]	1.86

Table 12.1—Comparing the causal inference methods

The IPTW has the worst estimation and largest standard error. This is also consistent with a low E-value. The **Love plot** in *Figure 12.8* showed that the balance between the covariates in the treatment and control groups was not very good; the weighted SMD of some of the covariates was still much above the threshold. Standardization can provide a better estimation, since covariates like age, education, and pre-treatment earnings (re74, re75) are strong predictors for the outcome. In addition, for this dataset, the outcome behaves approximately linearly with respect to the covariates included. The propensity score matching gives the closest estimation to the benchmark since it matches each treated unit to the most similar control.

Summary

In this chapter, we analyzed the Lalonde dataset, which is a benchmark in the field of economics. This dataset comes from the NSW demonstration, which was an experimental program designed to assess the impact of subsidized work experience on the lives of disadvantaged individuals. The Lalonde dataset was used to estimate the causal effect of the NSW demonstration on participants' earnings. The goal was to determine whether or not the earnings of the participants improved after completing the program.

We first performed an EDA analysis on this dataset and observed that the control group on average had higher earnings after completing the program compared to the treatment group. We also noted that the treatment data is very different from the control data since we have an experimental treatment group versus a non-experimental control group coming from survey data. Hence, we had to apply causal inference methods to estimate the true causal effect of this program on earnings.

We used three different methods to estimate the ATE of earnings after the end of the NSW program: **standardization**, IPTW and **propensity score matching**. Standardization was performed using a linear regression model. We used the `causallib` package to calculate the propensity scores and then used them to calculate the ATE using the IPTW method. We checked the overlap and balance of the distribution of the propensity scores in the treatment and control groups using the love plot and SMD values. The method of nearest neighbor was used to perform the propensity score matching, and the ATE of the matched dataset was calculated after that. Finally, we used bootstrapping to calculate the confidence interval and standard error of the ATE estimations resulting from IPTW and propensity score matching. Comparing the results showed that propensity score matching gives the best estimation for the ATE.

References

- Dehejia, Rajeev H., and Sadek Wahba. 1999. *Causal Effects in Non-Experimental Studies: Reevaluating the Evaluation of Training Programs. Journal of the American Statistical Association* 94 (448): 1053-1062.
- Dehejia, Rajeev H., and Sadek Wahba. 2002. *Propensity Score Matching Methods for Non-Experimental Causal Studies. Review of Economics and Statistics* 84 (1): 151-161.
- Lalonde, Robert J. 1986. *Evaluating the Econometric Evaluations of Training Programs with Experimental Data. The American Economic Review* 76 (4): 604-620.
- Shimoni, Y., Karavani, E., Ravid, S., et al. 2019. *An Evaluation Toolkit to Guide Model Selection and Cohort Definition in Causal Inference*. arXiv preprint arXiv:1906.00442.

Join our community on Discord

Join our community's Discord space for discussions with the author and other readers: `https://packt.link/hYCWN`

Join our community on Discord

[illegible]

13

Causal Inference in Epidemiology

Epidemiology is a branch of the medical sciences that investigates the spread and causes of diseases within groups of people. The goal of many epidemiological studies is to estimate the causal effect of a specific exposure on a particular outcome. Survival analysis is concerned with estimating treatment effects on the expected time until the occurrence of an event. In this chapter, we present a detailed analysis of the **NHEFS (National Health Epidemiologic Follow-up Study**) dataset and use causal survival analysis methods to estimate the effect of smoking cessation on mortality. We will demonstrate how the `lifelines` and `causallib` packages in Python can be used to perform the causal survival analysis.

In this chapter, we're going to cover the following main topics:

- Causal inference in epidemiology
- Survival analysis
- NHEFS dataset
- Survival analysis of NHEFS dataset
- Estimation using inverse probability of treatment weighting
- Estimation using standardization

Technical requirements

To run the Python code, you must:

- Install Python 3 following the installation & setup guide at the URL: `https://realpython.com/installing-python/`
- Install Jupyter ecosystem using the installation and setup guide at the URL: `https://docs.jupyter.org/en/stable/install.html`

The Python code in this chapter is in a Jupyter Notebook that can be downloaded from the GitHub URL: `https://github.com/PacktPublishing/Causal-Inference-with-Bayesian-Networks/tree/main/Chapter_13`

You need to install the following Python packages to run the Jupyter Notebook for this chapter:

- pandas: https://pandas.pydata.org/docs/getting_started/install.html
- statsmodels: https://pypi.org/project/statsmodels/
- causallib: https://pypi.org/project/causallib/
- scikit-learn: https://scikit-learn.org/stable/install.html
- lifelines: https://pypi.org/project/lifelines/

We will use the `statsmodels` package for standardization and the `causallib` package (Shimoni et al. 2019) for **inverse probability of treatment weighting (IPTW)** and **propensity score matching**. The `lifelines` package is used for survival analysis in Python.

Causal inference in epidemiology

In this section, we discuss the importance of causal inference in the field of epidemiology. Epidemiology is a branch of the medical sciences that studies the distribution and determinants of diseases or disorders within groups of people. Here, distribution describes the frequency and patterns of a disease in a population, and determinants are the causes and factors that influence its occurrence. In other words, epidemiology investigates how, when, and why diseases occur and spread, and who is most at risk of contracting them.

In epidemiology, causal inference is mainly used to study the determinants of diseases. It can be used to determine whether an observed correlation between an exposure and a disease is a true cause-and-effect relationship. This is important since it allows epidemiologists to identify the risk factors that cause a disease and create effective health interventions. Causal inference can help with determining the different factors that contribute to disease development, such as the effect of smoking on lung cancer. It can also be used to evaluate health policies and interventions, for example, the effect of vaccination on preventing a certain disease.

In the previous chapter (causal inference in economic research), we saw an application in which we wanted to estimate the causal effect on an outcome occurring at a particular time (earnings of the participants in 1978). However, in this chapter, we will see an example of causal survival analysis which combines causal inference with survival analysis to determine the causal effect of an exposure, treatment or intervention on the time until an outcome happens. Causal survival analysis is crucial in epidemiology because it helps us understand the probability of an event (like the start of a disease) over time and the factors that cause it to evaluate the effectiveness of treatments and interventions.

Survival analysis

In this section, we will talk about survival analysis. We first review the statistical details of this method and then talk about causal survival analysis which combines the power of causal inference with traditional survival analysis.

Background and important concepts

Survival analysis is a branch of statistics that deals with analyzing the time until an event occurs. It studies a variable that has a start time and an end time when a specific event occurs. Here, the focus is on the time between the start time and the end time (the corresponding event for that). One example is the time to death after a disease. In this case, the start time can be the diagnosis of the disease, and the end time is death. Here, the researchers want to know the effect of a specific drug on the survival time.

Let's review some important concepts and tools of survival analysis using an example. Suppose that we have studied the survival time of some subjects after diagnosing a fatal disease. The data is shown in *Table 13.1*. Each subject was studied for 10 years. The binary column death indicates if the subject died during the study or not (1: yes, 0: no). For each subject, we know the time until death in years (survival time after diagnosis), except subjects 9 and 10 who were still alive when the study ended after 10 years. In survival analysis, **censoring** occurs when the exact time of an event (like death in this example) is not observed for all the subjects in a study. When a study with a fixed duration ends, any subjects who have not experienced the event are considered **administratively censored.** Hence, in this example, the survival time of subjects 9 and 10 is administratively censored. Administrative censoring is not the only type of censoring that can happen in survival analysis. Sometimes, we need to deal with non-administrative censoring. For example, a subject may drop out of the study, which doesn't allow us to determine his time of death.

Subject	Survival time after diagnosis (years)	Death
1	2	1
2	3	1
3	3	1
4	4	1
5	5	1
6	5	1
7	5	1
8	7	1
9	10	0
10	10	0

Table 13.1—Survival time of subjects after disease diagnosis

In the following section, we will discuss three important statistical measures that are very useful in survival analysis.

Survival probability, risk and hazard

In this section, we will define some statistical measures that can help with survival analysis. These measures are functions of the survival time *T*. Suppose that the random variable *T* denotes the time to the occurrence of the event that we want to study in the survival analysis (in our example *T* denotes the survival time after diagnosis or simply the time to death). We define the **survival probability** at time *t* as:

$$P(T > t) \quad (13.1)$$

This is the proportion of the subjects whose survival time is greater than *t*. We have calculated the survival probability of the subjects listed in *Table 13.1* for each year, and the result is shown in *Table 13.2*. The survival probability at time zero (*t*=0) is equal to 1, and it decreases monotonically with *t*.

We also define the **risk** at time *t* as:

$$1 - P(T > t) = P(T \leq t) \quad (13.2)$$

Which is simply the **cumulative distribution function** (CDF) of *T* at *t*. The risk at time zero is equal to zero and monotonically increases with *t*.

t	Number of deaths at t	Number of subjects surviving at t	Number of subjects surviving just before t	Survival probability	Risk	Hazard
0	0	10	10	10/10 =1	0	0/10 = 0
2	1	9	10	9/10 = 0.9	0.1	1/10 = 0.1
3	2	7	9	7/10 = 0.7	0.3	2/9 = 0.22
4	1	6	7	6/10 = 0.6	0.4	1/7 = 0.14
5	3	3	6	3/10 = 0.3	0.7	3/6 = 0.5
7	1	2	3	2/10 = 0.2	0.8	1/3 = 0.34
10	0	2	2	2/10 = 0.2	0.8	0/2 = 0

Table 13.2—Survival probability and risk of subjects

Finally, the discrete time **hazard** at time *t* is a conditional probability defined as:

$$P(T = k | T > k - 1) \quad (13.3)$$

This is the proportion of subjects who die at time *k* among those who had survived just before *k*, which is denoted by *k*-1 here. The risk and hazard values are also calculated and shown in *Table 13.2*. Unlike risk, the hazard may increase or decrease over time.

The hazard function is generally defined as:

$$h(t) = \lim_{\Delta t \to 0} \frac{P(t \leq T \leq t + \Delta t | T \geq t)}{\Delta t}$$

It measures the instantaneous event rate at time *t* among subjects who have survived up to time *t*. However, this formula is useful when *t* is a continuous variable. The discrete time hazard function *Equation 13.3* is often used when *t* is a discrete variable (such as days or months).

As we see later, sometimes we need to add a weight for each participant when we calculate the survival probability. These weights are added to adjust for confounding variables. For example, suppose that in *Table 13.1*, the weight of subject 4 is 1.2, and the weight of all the remaining subjects is 1. *Table 13.3* shows how including these weights changes the calculations. Let's see how the first weighted survival probabilities are calculated in this table. Here, subject 4 represents more than one participant. In fact, when we count the number of subjects in this table, subject 4 adds 1.2 to the sum, not 1. Hence at the beginning, we can assume that we have 10.2 participants, not 10, and at *t*=1, when one of the other participants dies, we assume that 9.2 participants survived at this time. So, the weighted survival probability is simply: 9.2/10.2=0.90.

At t=4, subject 4 dies, so the number of subjects that survived is 6 (all have a weight of 1). Now the weighted survival probability is 6/10.2=0.59.

t	Number of deaths at t	Number of subjects surviving at t	Number of deaths at t (weighted)	Number of subjects surviving at t (weighted)	Weighted Survival probability
0	0	10	0	9 + 1.2	10.2 / 10.2 = 1
2	1	9	1	8 + 1.2	9.2/10.2=0.90
3	2	7	2	6 + 1.2	8.2/10.2=0.71
4	1	6	1.2	6	6/10.2=0.59
5	3	3	3	3	3/10.2=0.29
7	1	2	1	2	2/10.2=0.20
10	0	2	0	2	2/10.2=0.20

Table 13.3—Survival probability using weights

In the following section, we will introduce a statistical plot used in survival analysis to show how the probability of survival changes over time.

Kaplan–Meier curve

If we plot the survival probability versus time, the result is called a **Kaplan-Meier curve**. *Figure 13.1* shows the Kaplan-Meier curve of the data in *Table 13.2*. The curve starts at 1 and steps down at each event (death). A flat section in this curve means that no events occurred during that period, and a steeper drop means that more events (deaths) occurred in a short time.

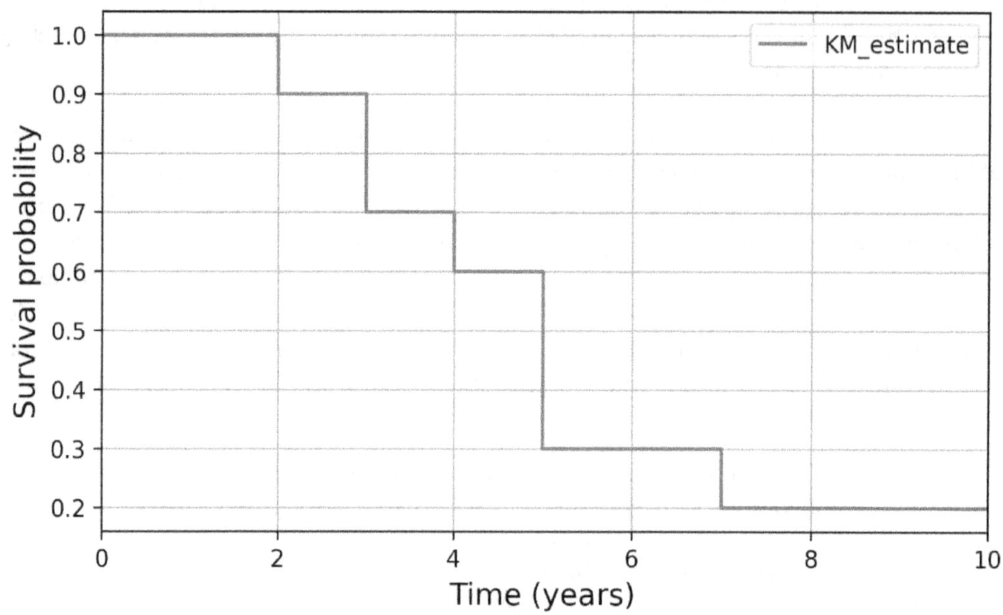

Figure 13.1— Kaplan-Meier curve

Remember that the subjects were studied for 10 years. These years are covered in the time axis on the plot. Initially, the survival probability is 1 and decreases at each event. The maximum drop is in the fifth year, in which 3 deaths occurred. No deaths occurred after the seventh year; therefore, the curve remained flat after that.

In this section, we discussed the survival analysis concepts. In the next section we will talk about the application of causal inference in survival analysis.

Causal survival analysis

Causal survival analysis combines conventional survival with causal inference. It allows us to estimate how a treatment or intervention affects the time until an event occurs, like time to death, failure, recovery, etc. For example, using causal survival analysis, we can study the effect of a new drug on the survival time of the subjects in *Table 13.1*. Here, causal survival analysis allows us to evaluate the effectiveness of this new drug. We can use different methods like standardization, propensity score matching, and IPTW. In the next sections, we will see how we can apply some of these methods to perform a causal survival analysis. Now that we are familiar with causal survival analysis, we take a look at the NHEFS dataset: a public dataset which is a benchmark in causal inference.

NHEFS dataset

In this section, we will take a detailed look at the NHEFS dataset. We will demonstrate how to load this dataset and perform an EDA analysis on it.

NHEFS study

The **first National Health and Nutrition Examination Survey** (**NHANES I**) was conducted by the **National Center for Health Statistics** (**NCHS**) between 1971 and 1974. This study was designed to measure the health and nutritional status of the U.S. population. The **National Health Epidemiologic Follow-up Study** (**NHEFS**) was a study that followed up on the participants of NHANES I.

Follow-up data collection spanned from 1982 to 1992. The main objective of the NHEFS was to study the connections between the clinical, nutritional, and behavioral data collected in NHANES I and subsequent health outcomes such as sickness, death, and hospitalization. It was also planned to monitor the changes in participants' health characteristics over time and better understand the natural course of chronic diseases. In summary, the NHEFS dataset allowed researchers to investigate how different factors from earlier stages of life may influence long-term health and mortality. For a more detailed description of the NHEFS study and the publicly available datasets and documentation, you can refer to: `wwwn.cdc.gov/nchs/nhanes/nhefs/`

In this chapter, we will use this dataset to estimate the effect of smoking cessation on mortality using causal survival analysis methods. We first load the dataset and describe its important features. Next, we perform an EDA analysis. Finally, we try to estimate the average causal effect of smoking cessation on the time to death.

Loading the dataset

First, we need to load all the required Python libraries for this case study:

```
import pandas as pd
import numpy as np
import matplotlib.pyplot as plt
from causallib.datasets.data_loader import load_nhefs_survival
import lifelines
from sklearn.linear_model import LogisticRegression
from causallib.estimation import IPW
from causallib.evaluation import evaluate
from causallib.survival.weighted_survival import WeightedSurvival
from causallib.survival import StandardizedSurvival
from lifelines import CoxPHFitter
from lifelines.statistics import proportional_hazard_test
import warnings
warnings.filterwarnings('ignore')
```

The NHEFs dataset is included in the causallib library, and we can use the following code to load it:

```
# Load and pre-process NHEFS data
data = load_nhefs_survival()
nhefs = pd.concat([data.a, data.X, data.t, data.y], axis=1)
nhefs.shape

## Output
(1629, 21)
```

As the output shows, this dataset has 1629 rows and 21 columns. Here is a list of all columns:

```
nhefs.columns

## Output
Index(['qsmk', 'age', 'race', 'sex', 'smokeintensity', 'smokeyrs', 'wt71',
'active_1', 'active_2', 'education_2', 'education_3', 'education_4',
'education_5', 'exercise_1', 'exercise_2', 'age^2', 'wt71^2',
'smokeintensity^2', 'smokeyrs^2', 'longevity', 'death'], dtype='object')
```

Feature description

The description of all the columns is included in the dataset and can be obtained using this code:

```
data['descriptors']
```

Here, we only review the description of the columns:

- `qsmk`: Did the participant quit smoking between 1st questionnaire and 1982? 1: yes, 0: no.
- `death`: Did the participant die by 1992? 1: yes, 0: no.
- `longevity`: Time of death (in months) measured from the beginning of the follow-up study. It can take values subsequent from 1 (January 1983) to 119 (November 1992) for participants with death=1. It is 120 for participant who survived the study (death=1).
- `age`: Age of participant in 1971.
- `smokeyrs`: Years of smoking.
- `race`: 0: white 1: black or other.
- `sex`: 0: male 1: female.
- `education`: Amount of education by 1971: 1: 8th grade or less, 2: high school dropout, 3: high school, 4: college dropout, 5: college or more.
- `exercise`: In recreation, how much exercise? in 1971, 0: much exercise, 1: moderate exercise, 2: little or no exercise.
- `active`: In your usual day, how active are you? in 1971, 0: very active, 1: moderately active, 2: inactive.
- `smokeintensity`: Number of cigarettes smoked per day in 1971.
- `wt71`: Weight in kilograms in 1971.

Please note that dummy encoding has been used for the categorical columns active, education, and exercise. For example, exercise has 3 possible values: 0, 1 and 2. So, it is converted to two binary columns `exercise_1` and `exercise_2`. Hence, `exercise_1=True` indicates that `exercise=1`, while `exercise_2=True` indicates that `exercise=2`. When both are False, it implies that `exercise=0`. In addition, the dataset includes the squared values of the columns `age`, `wt71`, `smokeintensity`, and `smokeyrs`.

The start of the follow-up study was January 1, 1983, and the administrative end of the follow-up study was December 31, 1992, so the follow-up study took 120 months.

Exploratory data analysis (EDA)

Let's assume that *A* denotes the treatment (the qsmk column) and *Y* denotes the outcome (the death column). Please note that *T* denoted the treatment variable so far; however, in this chapter, *T* denotes the time to death in months measured from the beginning of the follow-up study (the longevity column). We also denote the remaining covariate features by *X* as seen in the following code block. We saw that the dataset has 1629 participants in it (number of rows).

```
X = data.X
A = nhefs['qsmk']
Y = nhefs['death']
T = nhefs['longevity']
```

First, we can see how many of them died during the study:

```
(Y==1).sum()

## Output
318
```

Hence 318 participants died between 1983 and 1992, and 1311 participants were still alive when the follow-up study terminated in 1992. This means that the survival time of these remaining 1311 participants is administratively censored.

Effect of smoking on death rate

In this case study, we want to estimate the average causal effect of smoking cessation on the time to death based on the follow-up study data. Based on the qsmk column we can divide the participants into two groups: those who quit smoking between 1st questionnaire and 1982 (`qsmk=1`), and those who did not (`qsmk=0`). We can calculate the percentage of each group:

```
A.value_counts(normalize=True)

## Output
qsmk
0    0.737262
1    0.262738
```

So, 26% of the participants quit smoking between the 1st questionnaire and 1982. The next code shows how many participants survived in each group:

```
np.round(pd.crosstab(A, Y, normalize='index')*100, 2)

## Output
Death     0        1
--------------------
0    82.01       17.99

1    76.17       23.83
```

We can see that 76% of the participants who quit smoking survived after 10 years, while 82% of non-quitters survived. Given the well-established health risks associated with smoking, this is an unexpected result.

The next code compares the average value of some of the covariates between these groups:

```
nhefs.groupby('qsmk')[['age', 'sex', 'race', 'exercise_2', 'active_2']].mean().
round(2)

## Output
         age       sex  race  exercise_2  active_2
qsmk
-------------------------------------------------
0     42.92  0.53  0.15    0.38        0.09
1     46.70  0.45  0.09    0.41        0.11
```

As we see, the quitters' group were older, less active, and did less exercise on average, and this might explain the higher death rate in this group. In this section, we loaded the NHEFS dataset and performed an EDA analysis on it. In the next section, we perform a survival analysis on this dataset.

Survival analysis of NHEFS dataset

First, in the following code, we plot the Kaplan-Meier curves of the smoke quitters and non-quitters using the `lifelines` package. We first create the `KaplanMeierFitter()` object. This object has a `fit()` method that takes the time to event (*T*) and outcome columns (*Y*) and a label for the plot. Finally, the method `plot_survival_function()` plots the curve. It also shows the confidence interval of the curve by default. We need to call these methods for each group (A=0 and A=1) separately.

```
kmf = lifelines.KaplanMeierFitter()
kmf.fit(durations=T[A == 0], event_observed=Y[A == 0], label="A=0")
kmf.plot_survival_function()
kmf.fit(durations=T[A == 1], event_observed=Y[A == 1], label="A=1")
kmf.plot_survival_function()
```

```
plt.xlabel("T (months)", fontsize=12)
plt.ylabel("Survival probability", fontsize=12)
plt.show()
```

The result is shown in *Figure 13.2*.

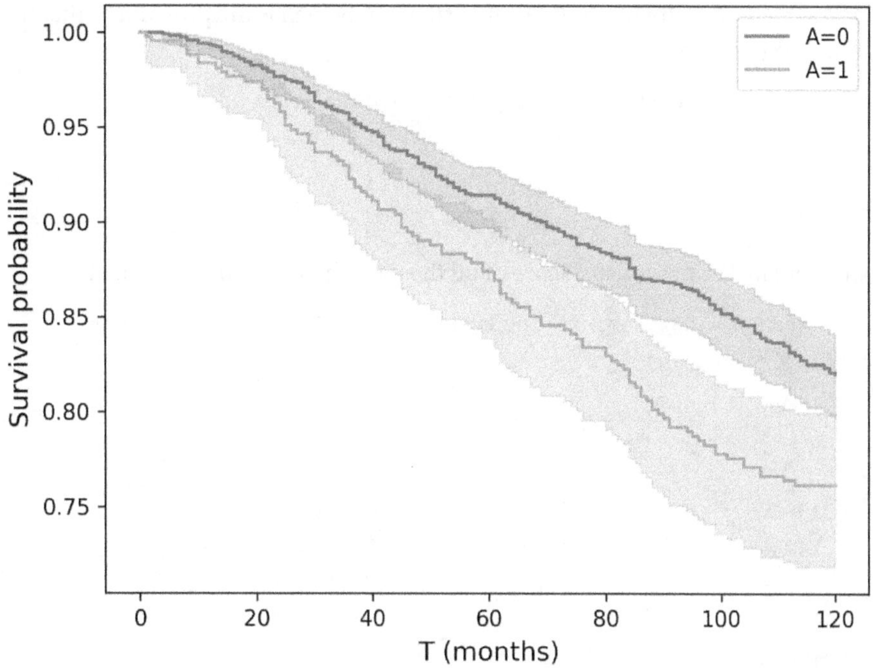

Figure 13.2— Kaplan-Meier curve of smoke quitters (A=1) vs non-quitters (A=0)

Remember that the Kaplan-Meier curve is a plot of survival probability versus time. Hence, we plotted *P[T>t|A=0]* versus *P[T>t|A=1]* for *t=0* to 120. So, based on the curves shown in *Figure 13.2*, the survival probability of non-quitters seems to be higher than that of quitters for all values of *T*. This is again a surprising result. But it is important to note that the quitters and non-quitters groups are not exchangeable. When two groups are exchangeable, it means that any observed difference in outcomes between the groups can be confidently attributed to the treatment, rather than to pre-existing differences between the participants in each group. In other words, if they hadn't been treated, their outcomes would have looked the same on average as the untreated group. However, in this dataset, the quitters and non-quitters are systematically different due to the presence of confounding.

We saw previously that 76% of the participants who quit smoking survived after 10 years, while 82% of non-quitters survived. However, this result does not necessarily imply that smoking cessation increases mortality since the quitters and non-quitters groups are not exchangeable. We also observed that the quitters group was older on average. Older people are more likely to quit smoking, but they are also more likely to die. Hence, age is a confounding variable that makes smoking cessation look bad.

In the next sections, we will follow the analyses suggested by Hernán and Robins in their book on causal inference (Hernán and Robins 2020) to estimate the causal effect of smoking cessation on mortality.

Estimation using inverse probability of treatment weighting

In *Chapter 12*, we used the inverse probability of treatment weighting (IPTW) method to estimate the ATE using the propensity scores. Here we follow the same steps. First, a machine learning model is used to estimate the propensity scores of each participant, which is the probability that they will quit smoking given the covariates:

```
learner = LogisticRegression(penalty=None, solver='lbfgs',
                             max_iter=5000)
ipw = IPW(learner)
ipw.fit(X, A)
```

Next, we can calculate the propensity score of all the participants using the model:

```
propensity_scores = ipw.compute_propensity(X, A, treatment_values=1)
print(propensity_scores)

## Output
id
0        0.109695
1        0.158895
2        0.138457
           ...
1628     0.179839
Name: 1, Length: 1629, dtype: float64
```

Then the weight of each participant is calculated using the following equation for the participants of each group (quitters and non-quitters):

$$w_{A=1} = \frac{1}{\pi(x)}, w_{A=0} = \frac{1}{1-\pi(x)} \quad (13.4)$$

Where $\pi(x)$ denotes the propensity score. The weights of participants are calculated using the following code:

```
weights = ipw.compute_weights(X, A)
print(weights)

## Output
id
0        1.123210
1        1.188912
2        1.160709
```

```
        ...
1628    5.560518
Length: 1629, dtype: float64
```

Now we can easily use these weights to plot the IPTW-adjusted Kaplan-Meier curves using the next code (We showed how the weights change the survival probabilities in *Table 13.3*). We use the same `fit()` method, but this time pass the `weights` to that:

```
kmf = lifelines.KaplanMeierFitter()
kmf.fit(durations=T[A==0], event_observed=Y[A==0],
        weights=weights[A==0], label="A=0")
kmf.plot_survival_function()
kmf.fit(durations=T[A==1], event_observed=Y[A==1],
        weights=weights[A==1], label="A=1")
kmf.plot_survival_function()

plt.xlabel("T (months)", fontsize=12)
plt.ylabel("Survival probability", fontsize=12)
plt.show()
```

The plot is shown in *Figure 13.3*.

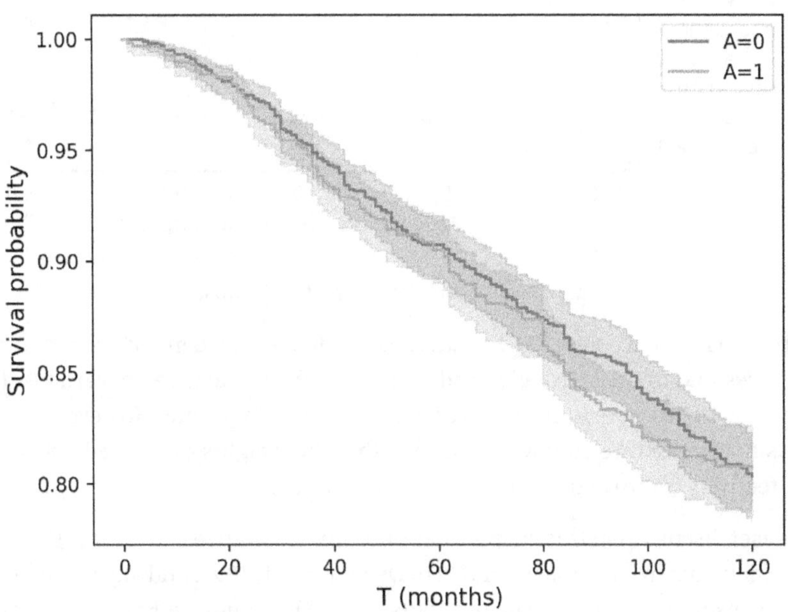

Figure 13.3— IPTW-adjusted Kaplan-Meier curve of smoke quitters (A=1) vs non-quitters (A=0)

Now we see that the difference in survival probability after 10 years diminishes after the IPTW adjustments. We see that the confidence interval also overlaps. Hence, the new plot does not suggest that smoking cessation increases mortality anymore.

Remember that in the previous chapter we used a love plot to show the balance between the covariates in the treatment and control groups. Using this plot, we could check the balance of the covariates in the raw data. Additionally, we could see how the weights defined by the propensity scores improve the balance. The next code creates the love plot of the raw and weighted data:

```
evaluation_results = evaluate(ipw, X, A, Y)
evaluation_results.plot_covariate_balance(kind="love")
plt.show()
```

The result is shown in *Figure 13.4*.

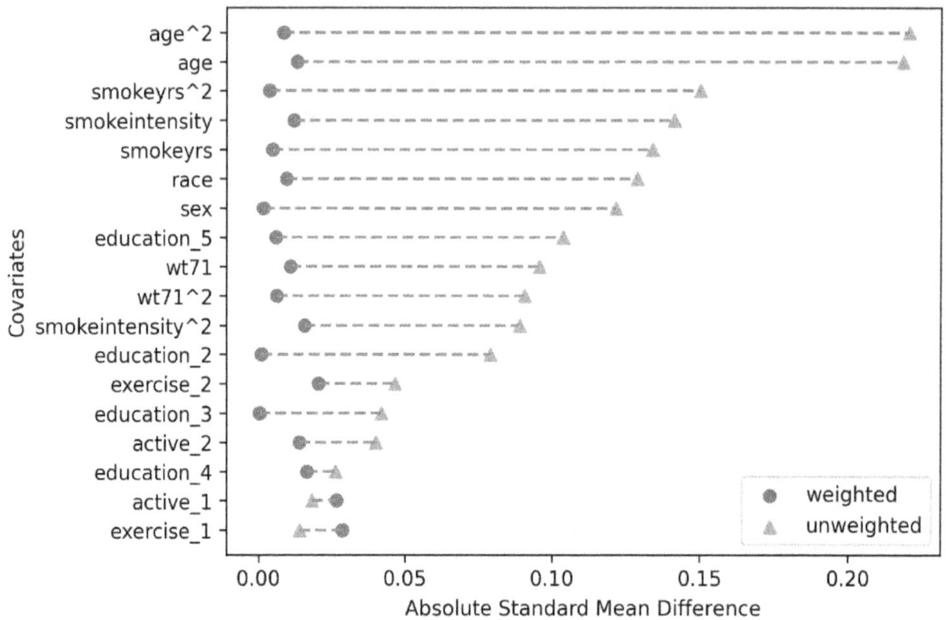

Figure 13.4— Love plot of the NHEFS dataset

For each covariate, a lower absolute SMD means more balance between the quitters and non-quitters. We saw that a threshold of 0.1 is usually used as a cutoff for imbalanced covariates. Hence, we see that some of the covariates were imbalanced in the raw unweighted data. However, the weighted SMD of all of them is below the threshold which means that the weights calculated using the propensity scores could effectively improve the balance of the covariates.

The NHEFS dataset discussed in this chapter has only administrative censoring. The follow-up study followed the participants from January 1, 1983, to December 31, 1992, and any participants who didn't die by 1922 was censored. There were no dropouts. The IPTW-adjusted Kaplan-Meier can handle administrative censoring. However, if we have non-administrative censoring, then we need to use the **Inverse Probability of Censoring Weighting (IPCW)** method that adjusts for bias caused by non-administrative (informative) censoring. IPCW re-weights the observed participants to represent those who were censored. Participants who remain uncensored but have a high probability of censoring get larger weights. Now that we covered the estimation using IPTW, we will look at another estimation method: standardization.

Estimation using standardization

In this method, first we fit a survival regression model. This model can predict hazard as a function of covariate features (*X*) and the treatment variable (*A*). The **Cox proportional hazards model** is one of the most widely used models in survival analysis. This model is defined as follows:

$$h(t|X_i, A) = h_0(t)\exp\left(\beta_1 X_1 + \beta_2 X_2 + \cdots \beta_p X_p + \beta_{p+1} A\right) \quad (13.5)$$

Here $h(t|X_i, A)$ is the hazard of a participant at time t given all the covariates X_i and treatment A, and $h_0(t)$ is the baseline hazard function (hazard when the treatment and all covariates are zero). Finally, $\beta_1, \dots \beta_{p+1}$ are the model parameters. Once this model is trained on the dataset, it can predict the hazard at time t for each participant. The following code fits a Cox model to our dataset using the `CoxPHFitter()` class in `lifelines` package. After creating an object from this class, we can call its `fit()` method to train the model. It takes the dataset and the name of the time-to-event and outcome columns in that.

```
cph = CoxPHFitter()
cph.fit(nhefs, duration_col='longevity', event_col='death')
```

Now we can print the parameters of the model:

```
print(cph.params_)

## Output
covariate
qsmk                  0.028795
age                   0.081665
race                  0.047359
sex                  -0.436450
smokeintensity       -0.000086
smokeyrs              0.019047
wt71                 -0.061347
active_1              0.160250
active_2              0.234973
education_2          -0.132926
education_3          -0.433116
education_4          -0.228105
education_5          -0.381206
exercise_1           -0.140340
exercise_2            0.155111
age^2                -0.000027
wt71^2                0.000399
smokeintensity^2      0.000085
smokeyrs^2            0.000015
```

Next, we print the baseline hazard function. Since it is a function of *t*, it has a separate value for each value of *t*.

```
print(cph.baseline_hazard_)

## Output
baseline hazard
1.0             0.000260
2.0             0.000261
4.0             0.000262
...                  ...
120.0           0.000549
[109 rows x 1 columns]
```

Please note that it has 109 values since we have 109 unique values in the longevity column in the NHEFS dataset.

```
print(len(np.unique(nhefs['longevity'])))
## Output
109
```

The Cox proportional hazards model assumes that the exponential term in *Equation 13.5* is constant over time. This is called the **Proportional Hazards (PH) assumption**. Hence, before using this model, we need to check this assumption. We can use the method `proportional_hazard_test()` in `lifelines.statistics` to test this assumption. This method first calculates the Schoenfeld residuals for the treatment and covariates at each event time. The Schoenfeld residual is the observed variable minus its expected value. Then it applies a transformation to event times. Finally, it tests if residuals are correlated with transformed time using a chi-squared test. Here the null hypothesis means no correlation and the alternative hypothesis means that a correlation exists.

If a covariate such as *Xi* (or the treatment variable *A*) has a p-value less than 0.05, the null hypothesis can be rejected, indicating that there is a correlation between the Schoenfeld residuals and transformed time. Hence, its coefficient (β_i) in *Equation 13.5* is a function of time, and the exponential term of this equation is not constant over time. This indicates a violation of the PH assumption. To ensure that the PH assumption holds, the p-value of the treatment and all the covariates must be greater than or equal to 0.05. The following code tests the PH assumption and shows the p-value of all the covariates:

```
ph_test = proportional_hazard_test(cph, nhefs, time_transform='rank')
ph_test.summary[[ph_test.summary[['p']], 'p']]

## Output
                       p
active_1        0.485864
active_2        0.189770
age             0.795786
```

```
age^2              0.857662
education_2        0.118560
education_3        0.578225
education_4        0.449149
education_5        0.187891
exercise_1         0.746152
exercise_2         0.318405
qsmk               0.082535
race               0.305018
sex                0.181851
smokeintensity     0.142566
smokeintensity^2   0.088912
smokeyrs           0.573526
smokeyrs^2         0.447550
wt71               0.593023
wt71^2             0.775919
```

We see that all p-values are greater than 0.05, so the PH assumption holds and the Cox model is valid for this dataset.

But how can we calculate the survival probability from the hazard function estimate by the Cox model? First, we need to calculate the baseline cumulative hazard at time *t*. It is the sum of all $h_0(t_j)$ for $t_j \le t$:

$$H_0(t) = \sum_{t_j \le t} h_0(t_j) \quad (13.6)$$

Then, we use the following equation to estimate the survival probability at time *t*:

$$P(T > t) = \exp\left(-H_0(t)\exp\left(\beta_1 X_1 + \beta_2 X_2 + \cdots \beta_p X_p + \beta_{p+1} A\right)\right) \quad (13.7)$$

Using this equation, we can calculate the survival probability of each participant at time *t* for either *A*=0 or *A*=1. The next code shows how the survival probabilities of the first participant (the first row) in the NHEFs data can be calculated:

```
survival_curve = cph.predict_survival_function(nhefs[cph.params_.index].
iloc[[0]])
print(survival_curve)

## Output
             0
--------------
1.0     0.999601
2.0     0.999201
4.0     0.998799
```

```
5.0      0.998397
...      ...
119.0    0.831251
120.0    0.830552

109 rows × 1 columns
```

The output shows the survival curve of the first participant. This is the survival probability of the first participant for all the times of event in the study that were estimated (these times are the unique values of the longevity column).

Next, we calculate the **population-averaged survival curve** for each value of A (0 or 1). We assume that all the participants in the dataset had the same treatment (same value of A=a) but kept their own covariates (X). For each participant i, we use the Cox model to compute its survival curve at all times with A=a.

$$S_i^{(a)}(t) = \exp\left(-H_0(t)\exp\left(\beta_1 X_{i1} + \beta_2 X_{i2} + \cdots \beta_p X_{ip} + \beta_{p+1} a\right)\right) \quad (13.8)$$

where $S_i^{(a)}(t)$ is the survival probability of participant i at time t under A=a. Then at each time (t), we average across the population of all the participants in the dataset to get the population-averaged survival curve under A=a:

$$\bar{S}^{(a)}(t) = \frac{1}{n}\sum_{i=1}^{n} S_i^{(a)}(t) \quad (13.9)$$

where n is the total number of participants. We can use this equation to calculate the population-averaged survival curve for A=0 and A=1. In summary, first we fit the Cox model and predict each participant's counterfactual survival under A=0 and A=1, and then average those predictions to get the population-level survival curves. The next code implements these calculations in Python:

```
cph = CoxPHFitter()
cph.fit(nhefs, duration_col="longevity", event_col="death")
pop_avg_survival_curves = {}
for a in [0, 1]:
    temp = nhefs.copy()
    temp["qsmk"] = a
    ind_surv_curves = cph.predict_survival_function(temp)
    pop_avg_survival_curves[a] = ind_surv_curves.mean(axis=1)
```

Now we have the population-averaged survival curve for A=0 and A=1. For example, for A=0, we get:

```
pop_avg_survival_curves[0]
```

```
## Output
1.0       0.999393
```

```
2.0       0.998786
4.0       0.998176
            ...
120.0     0.807241
Length: 109, dtype: float64
```

And for A=1, we have:

```
pop_avg_survival_curves[1]

## Output
1.0       0.999375
2.0       0.998750
4.0       0.998123
            ...
120.0     0.803252

Length: 109, dtype: float64
```

We could also use the `causallib` package to directly calculate the population-averaged survival curve. This is shown in the next code. Here we use the `StandardizedSurvival` class in the `causallib` package and initialize it with a `CoxPHFitter` object of the `lifelines` package. Then we use its `fit()` method to train the model on the dataset. Finally, the population-averaged survival curves are retrieved using the `estimate_population_outcome()` method.

```
standardized_survival_cox = StandardizedSurvival(survival_model=lifelines.
CoxPHFitter())
standardized_survival_cox.fit(X, A, T, Y)
pop_avg_survival_curves = standardized_survival_cox.estimate_population_
outcome(X, A, T)
```

Next, we need to plot the survival curves. The following code plots them:

```
plt.step(pop_avg_survival_curves.index,
         pop_avg_survival_curves[0].values,
         where="post", label="A=0")
plt.step(pop_avg_survival_curves.index,
         pop_avg_survival_curves[1].values,
         where="post", label="A=1")
plt.legend()
plt.xlabel("T (months)", fontsize=12)
plt.ylabel("Survival probability", fontsize=12)
plt.show()
```

The output is shown in *Figure 13.5*.

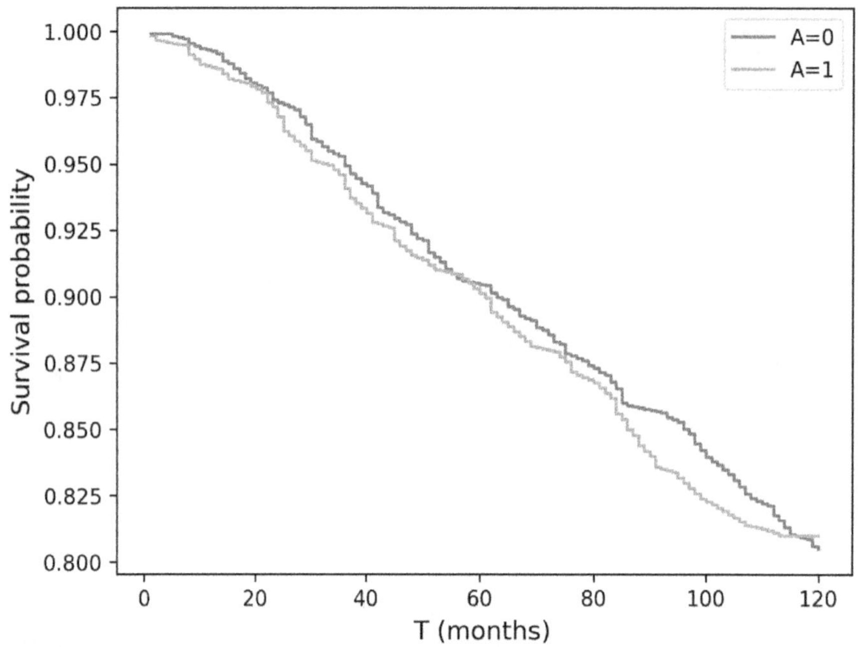

Figure 13.5— Standardized Kaplan-Meier curve of smoke quitters (A=1) vs non-quitters (A=0)

As this figure shows, the results are very similar to what we had with the IPTW method. The survival curves get much closer to each other, and the difference in survival probability after 10 years diminishes after standardization. Therefore, the new plot does not suggest that smoking cessation increases mortality anymore.

So far, we have tried different causal estimation methods (IPTW and standardization), and both of them suggest that the mortality remains the same for both quitters and non-quitters. Of course, the results provide no evidence of the benefits of quitting smoking.

The fact is that many people quit smoking because they already got sick. So, in this dataset, quitters may be sicker than non-quitters at baseline. Benefits of quitting take many years to appear. If you only observe survival for 10 years, the immediate effect might be dominated by people who quit because they were already ill. A longer follow-up study (20+ years) possibly could show more evidence for the benefits of quitting smoking and its true effect on mortality. However, as the results showed, causal inference is still a powerful tool to analyze such datasets. Without causal inference one could conclude that quitting smoking increases mortality!

Summary

In this chapter, causal survival analysis in epidemiology was introduced. We first learned about survival analysis and its core concepts like survival probability, risk, and hazard. We saw how to use a Kaplan-Meier curve to visualize the survival probabilities over time. Then we analyzed the NHEFS dataset, a benchmark dataset in survival causal inference. The NHEFS dataset can be used to estimate the causal effect of quitting smoking on mortality.

An EDA analysis was performed on this dataset, and the surprising observation was that 76% of the participants who quit smoking survived after 10 years, while 82% of non-quitters survived. Hence, we had to apply causal inference methods to estimate the true causal effect of quitting smoking on mortality.

We used two different methods to estimate the survival curve of quitters and non-quitters: **IPTW** and **standardization**. We used the `causallib` package to calculate the propensity scores using a logistic regression model and then used them to calculate the survival curves using the IPTW method. We checked the balance of the distribution of the propensity scores in the quitters and non-quitters using a love plot. Finally, we used the standardization method with a **Cox proportional hazards model** to estimate the survival curves. The results from both methods showed that the mortality remains the same for both quitters and non-quitters.

References

- Hernán, M.A., and Robins, J.M. 2020. *Causal Inference: What If*. Boca Raton: Chapman & Hall/CRC.
- Shimoni, Y., Karavani, E., Ravid, S., et al. 2019. *An Evaluation Toolkit to Guide Model Selection and Cohort Definition in Causal Inference*. arXiv preprint arXiv:1906.00442.

Get this book's PDF version and more

14

Causal Inference in Social Science Research

This chapter examines the famous **Card and Krueger minimum wage study** from 1994, which investigated the impact of a minimum wage increase on fast-food employment in New Jersey and Pennsylvania. Using surveys of fast-food outlets, the study contrasted employment in New Jersey, which raised its minimum wage, to neighboring Pennsylvania, where it remained unchanged. The main idea behind this causal analysis is to use the trend of the control group (those in Pennsylvania) as a counterfactual for the unobserved trend in the treatment group (those in New Jersey). The causal analysis employs linear regression and the **Inverse Probability of Treatment Weighting (IPTW)** to estimate the causal effect. All these methods are implemented in Python, and the `causallib` package will be used to estimate the causal effect.

In this chapter, we're going to cover the following main topics:

- Causal inference in social science research
- Card and Krueger dataset
- Causal model
- Estimation using linear regression
- Estimation using inverse probability of treatment weighting

Technical requirements

To run the Python code, you must:

- Install Python 3 following the installation & setup guide at the URL: `https://realpython.com/installing-python/`
- Install Jupyter ecosystem using the installation and setup guide at the URL: `https://docs.jupyter.org/en/stable/install.html`

The Python code in this chapter is in a Jupyter Notebook that can be downloaded from the GitHub URL: `https://github.com/PacktPublishing/Causal-Inference-with-Bayesian-Networks/tree/main/Chapter_14`

You need to install the following Python packages to run the Jupyter Notebook for this chapter:

- `pandas`: `https://pandas.pydata.org/docs/getting_started/install.html`
- `statsmodels`: `https://pypi.org/project/statsmodels/`
- `causallib`: `https://pypi.org/project/causallib/`
- `scikit-learn`: `https://scikit-learn.org/stable/install.html`
- `seaborn`: `https://pypi.org/project/seaborn/`
- `wildboottest`: `https://pypi.org/project/wildboottest/`

We will use the `statsmodels` package for **linear regression** and the `causallib` package (Shimoni et al. 2019) for **inverse probability of treatment weighting** (IPTW). The `wildboottest` package will be used for wild cluster bootstrap.

Causal inference in social science research

Social science is the scientific study of societies and human relationships within those societies. It attempts to understand how humans interact, how societies work, and how social structures influence individual and collective behavior. In this field, scientific methods are used to investigate how individuals and societies function and make decisions. Social science researchers try to understand why and how people and societies act the way they do.

Causal inference is central to social science research since researchers want to go beyond correlations and make reliable claims about cause and effect. Researchers use different causal inference methods to estimate the impact of social factors or interventions on outcomes using concepts like counterfactuals, confounding variables, or average treatment effects. Causal inference can help with evaluating the effectiveness of policies and programs by providing estimates of their causal impacts. In this section, we will see an example for the application of causal inference in social science research: the famous *Card and Krueger minimum wage study*.

Minimum wages and fast-food employment

The Card and Krueger study on the effect of minimum wage on fast-food employment (Card and Krueger 1994) is one of the most influential natural experiment papers in social science research. Prior to the early 90s, most economists believed that minimum wage laws reduced employment among low-skilled workers. This view was based on the **competitive labor market model** which states that if you raise the price of labor, demand for it goes down.

In April 1992, the state of New Jersey raised its minimum wage from \$4.25 to \$5.05/hour. This increase gave New Jersey the highest state minimum wage at that time and was heavily opposed by business leaders and industry groups, particularly those representing restaurants and fast-food chains. Many economists expected employment losses, especially for low-wage and low-skill workers in New Jersey.

However, the neighboring Pennsylvania did not raise its minimum wage. This created the possibility for a scientific study by Card and Krueger. They had two otherwise similar states, but only one had changed the policy; hence, one could be used as a treatment group and the other one as a control group. Card and Krueger surveyed 410 fast-food restaurants (Burger King, KFC, Wendy's, and Roy Rogers) in New Jersey (treatment group) and Pennsylvania (control group). They collected data on employment, wages, and menu prices before and after the policy change. If the classical belief was right (higher minimum wage results in fewer jobs), then after April 1992, employment in New Jersey had to drop in comparison to Pennsylvania.

Card and Krueger used causal inference in their 1994 study. The main objective was to determine the causal effect of a minimum wage increase on fast-food employment, and their methodology which was based on using a natural experiment and **difference-in-differences (DiD)** was a pioneering example of this approach in empirical economics.

Important note

In their 1994 paper, Card and Krueger didn't call their method causal inference since the phrase was not common in economics literature at the time. However, their approach is part of the causal inference toolbox today. The explicit use of "causal inference" as a methodology emerged later, built on the pioneering works of researchers like Card and Krueger. David Card later won the 2021 Nobel Prize in Economic Sciences for his empirical contributions to labor economics, including the famous minimum wage study.

In the next section, we will take a deep look at the dataset used for the study of Card and Krueger. Later, we will follow the analyses suggested by Card and Krueger to estimate the causal effect of a minimum wage increase on fast-food employment.

Card and Krueger dataset

The data of Card and Krueger's paper are available on David Card's UC Berkeley faculty page: `https://davidcard.berkeley.edu/data_sets/njmin.zip`

We first load the data and discuss the important features. After performing an EDA analysis, we focus on the causal estimation problem. Our goal is to study the effect of a minimum wage increase on fast-food employment in New Jersey using different causal inference methods.

First, we need to load all the required Python libraries for this case study:

```
import numpy as np
import pandas as pd
import matplotlib.pyplot as plt
import seaborn as sns
import statsmodels.api as sm
import requests
import zipfile
```

```
from io import BytesIO, TextIOWrapper
from causallib.estimation import IPW
from causallib.evaluation import evaluate
from sklearn.linear_model import LogisticRegression
from wildboottest.wildboottest import wildboottest
%matplotlib inline
```

The following code downloads this data, unzips it, and stores it in the Pandas dataframe `df`. The column descriptions are also stored in the dictionary `column_descriptions`.

```
url = "https://davidcard.berkeley.edu/data_sets/njmin.zip"
response = requests.get(url)
z = zipfile.ZipFile(BytesIO(response.content))

df = (
    pd.read_csv(
        z.open("public.dat"),
        sep=r"\s+",
        header=None,
        engine="python"
    )
    .applymap(lambda x: pd.to_numeric(x, errors="coerce"))
)

with TextIOWrapper(z.open("codebook"), encoding="cp437") as f:
    codebook = f.readlines()

# Extract relevant lines (skip formatting/headers)
relevant_lines = (
    codebook[7:11] + codebook[13:19] + codebook[21:38] + codebook[40:59]
)
cols = [line.split()[0] for line in relevant_lines]
descriptions = [" ".join(line.split()[4:]).strip() for line in relevant_lines]
df.columns = cols
# Create dictionary of column -> description
column_descriptions = dict(zip(cols, descriptions))
```

After loading the dataset, we get the shape of `df`:

```
print(df.shape)
## Output
(410, 46)
```

So, we have 410 rows and 46 columns in this dataset. In the next section, we look at the description of the features in this dataset. Each row represents one of the surveyed fast-food restaurants, and each column describes a feature of these restaurants.

Feature description

Let's see the description of some of the more important columns in this dataset:

- `SHEET`: Sheet number (unique store id for the fast-food restaurant)
- `CHAIN`: chain 1= Burger King, 2=KFC, 3= Roy Rogers, 4= Wendy's
- `CO_OWNED`: 1 if company owned
- `STATE`: 1 if New Jersey (NJ); 0 if Pennsylvania (PA)
- `WAGE_ST`, `WAGE_ST 2`: Starting wage in $/hr (pre-treatment, post-treatment)
- `EMPFT`, `EMPFT2`: Number of full-time employees (pre-treatment, post-treatment)
- `NMGRS`, `NMGRS2`: Number of managers or asset managers (pre-treatment, post-treatment)
- `EMPPT`, `EMPPT2`: Number of part-time employees (pre-treatment, post-treatment)

> The description of all the columns can be found in the dictionary `column_descriptions`.

Remember that Card and Krueger surveyed 410 fast-food restaurants (Burger King, KFC, Wendy's, and Roy Rogers) in New Jersey (treatment group) and Pennsylvania (control group). Hence, each row of this dataset represents one of those fast-food restaurants. Each restaurant has a unique id (`SHEET` column). The `STATE` column gives the state in which the restaurant is located.

Card and Krueger's study used two waves of surveys of fast-food restaurants to investigate the effects of New Jersey's April 1992 minimum wage increase (from $4.25 to $5.05). The first wave was conducted in February–March 1992, just before New Jersey's minimum wage increase went into effect on April 1, 1992. In this wave, they collected baseline data on employment, wages, and some other features from 410 fast-food restaurants. The first wave created the pre-treatment data.

The second wave was conducted in November–December 1992, about 7–8 months after the minimum wage increase in New Jersey. They re-surveyed the same restaurants to measure how employment, wages, and prices had changed, and the result was the post-treatment data.

> The dataset's column names are structured to distinguish between the first and second waves. If the column name contains no suffix, the data in the column is related to the first wave (pre-treatment), whereas a 2 suffix denotes the second wave of data (post-treatment). For example, `EMPFT` represents the number of full-time employees before the treatment (minimum wage increase), while `EMPFT2` gives the same number after the treatment.

EDA analysis

In this section, we will perform an **exploratory data analysis** (**EDA**) on this dataset using Python. First, we define some new features. The Card and Krueger study defined the **Full-Time Equivalent** (**FTE**) employment hours (FTE_EMP column) as the sum of hours worked by managers, regular employees, and part-time employees. The latter was given a weight of 0.5 to account for their often shorter working hours. The percentage of full-time employees is also calculated based on that (`fulltime_employee_perc` column). The price of the full meal is defined as the sum of the price of medium soda, small fries, and an entree (including all taxes). The next step adds these new features to the dataset:

```
df['FTE_EMP'] = df['EMPFT'] + df['NMGRS'] + 0.5 * df['EMPPT']
df['FTE_EMP2'] = df['EMPFT2'] + df['NMGRS2'] + 0.5 * df['EMPPT2']
df['price_full_meal'] = df['PENTREE'] + df['PFRY'] + df['PSODA']
df['price_full_meal2'] = df['PENTREE2'] + df['PFRY2'] + df['PSODA2']
df['fulltime_employee_perc'] = df['EMPFT'] / df['FTE_EMP']
df['fulltime_employee_perc2'] = df['EMPFT2'] / df['FTE_EMP2']

df['salary1'] = df['WAGE_ST'] == 4.25
df['salary2'] = df['WAGE_ST2'] == 4.25
df['salary3'] = df['WAGE_ST2'] == 5.05

state_map = {0: "PA", 1: "NJ"}
df["state"] = df["STATE"].map(state_map)
```

The following code calculates the average store types in both states:

```
df = pd.get_dummies(df, prefix="C", dtype=int, columns=["CHAIN"]).rename(
        columns={'C_1': "Burger King", 'C_2': "KFC",
                 'C_3': "Roy Rogers", 'C_4': "Wendy's"})

chains = ['Burger King', 'KFC', 'Roy Rogers', 'Wendy\'s']
result_chains = df.groupby(['state'])[chains].mean()
result_chains.index.name = None
result_chains = result_chains.T*100
result_chains.round(2)
```

The output is shown in *Table 14.1*:

	NJ	PA
Burger King	41.09	44.30
KFC	20.54	15.19
Roy Rogers	24.77	21.52
Wendy's	13.60	18.99

Table 14.1—Distribution of store types in New Jersey and Pennsylvania

We see that most stores belonged to Burger King. We can also calculate the distribution of the ownership status (company-owned versus franchisee-owned) in New Jersey and Pennsylvania:

```
result_coowned = df.groupby(['STATE'])['CO_OWNED'].mean().to_frame()
result_coowned.index.name = None
result_coowned = result_coowned.rename(index={0: "PA", 1: "NJ"}, columns={'CO_
OWNED': 'Company-owned (%)'}).T*100
result_coowned.round(2)

## Output
                       PA      NJ
-----------------------------------
Company-owned (%)    35.44   34.14
```

As the output shows, the results were close to each other.

Next, we calculate the means of some key features in the pre-treatment period (February–March 1992).

```
pre_treat_cols = ['FTE_EMP', 'fulltime_employee_perc', 'WAGE_ST',
                  'salary1', 'price_full_meal', 'HRSOPEN']
result_w1 = df.groupby(['state'])[pre_treat_cols].mean()
result_w1.index.name = None
result_w1 = result_w1.rename(
            columns={'FTE_EMP': 'FTE Employment',
                 'fulltime_employee_perc': 'Full-time employees (%)',
                 'WAGE_ST': 'Starting Wage',
                 'salary1': 'Wage = $4.25 (percentage)',
                 'price_full_meal': 'Price of full meal',
                 'HRSOPEN': 'Hours open on weekdays'}).T
result_w1.loc['Full-time employees (%)'] *= 100
result_w1.loc['Wage = $4.25 (percentage)'] *= 100
result_w1.round(2)
```

Table 14.2 shows the output table.

	NJ	PA
FTE Employment	20.44	23.33
Full-time employees (%)	32.85	35.04
Starting Wage	4.61	4.63
Wage = $4.25 (percent)	30.51	32.91
Price of full meal	3.35	3.04
Hours open on weekdays	14.42	14.53

Table 14.2—Mean of features in the pre-treatment period

We can also calculate the mean of the same features in the post-treatment period (November–December 1992), after the minimum wage increase in New Jersey.

```
post_treat_cols = ['FTE_EMP2','fulltime_employee_perc2','WAGE_ST2',
                   'salary2', 'salary3', 'price_full_meal2', 'HRSOPEN2']
result_w2 = df.groupby(['state'])[post_treat_cols].mean()
result_w2.index.name = None
result_w2 = result_w2.rename(
       columns={'FTE_EMP2': 'FTE Employment',
               'fulltime_employee_perc2': 'Full-time employees (%)',
               'WAGE_ST2': 'Starting Wage',
               'salary2': 'Wage = $4.25 (percentage)',
               'salary3': 'Wage = $5.05 (percentage)',
               'price_full_meal2': 'Price of full meal',
               'HRSOPEN2': 'Hours open on weekdays'}).T
result_w2.loc['Full-time employees (%)'] *= 100
result_w2.loc['Wage = $4.25 (percentage)'] *= 100
result_w2.loc['Wage = $5.05 (percentage)'] *= 100
result_w2.round(2)
```

The results are given in *Table 14.3*.

	NJ	PA
FTE Employment	21.03	21.17
Full-time employees (%)	35.87	30.38
Starting Wage	5.08	4.62
Wage = $4.25 (percent)	0.00	25.32
Wage = $5.05 (percent)	85.50	1.27
Price of full meal	3.41	3.03
Hours open on weekdays	14.42	14.65

Table 14.3—Mean on features in the post-treatment period

Based on the data in *Table 14.2*, in the pre-treatment period, the average employment was 23.33 full-time equivalent (FTE) employment hours per store in Pennsylvania, while it was 20.44 in New Jersey. Average starting wages and average hours of operation were very close in the two states before the wage increase in New Jersey. There were no significant cross-state differences in the percentage of full-time workers; however, the average price of a full meal was higher in New Jersey.

As the data in *Table 14.3* shows, after the wage increase, the average starting wage at fast-food restaurants in New Jersey increased by 10 percent. The next code compares the distribution of starting wages in both states before and after the wage increase in New Jersey. The results are shown in *Figure 14.1*.

```
cols = ["WAGE_ST", "WAGE_ST2"]
titles = [
    "Pre-treatment (February-March 1992)",
    "Post-treatment (November-December 1992)"
]
plt.figure(figsize=(8, 10))
for i, (col, title) in enumerate(zip(cols, titles), start=1):
    ax = plt.subplot(2, 1, i)
    sns.histplot(
        data=df,
        x=col,
        hue="state",
        binwidth=0.1,
        stat="percent",
        common_norm=False,
        multiple="dodge",
        ax=ax
    )
    ax.set_title(title, fontsize=13)
    ax.set_xlabel("Wage Range", fontsize=13)
    ax.set_ylabel("Percentage of Stores", fontsize=13)
    ax.set_xticks(np.arange(4.25, df[col].max(), 0.2))

plt.subplots_adjust(hspace=0.35)
plt.show()
```

This figure provides more insight about the starting wage distribution in New Jersey and Pennsylvania. Before the wage increase in New Jersey, the distributions were very similar (*Figure 14.1* top). However, after that, all the restaurants in New Jersey that had previously paid less than $5.05 per hour raised their starting wage to match the new rate.

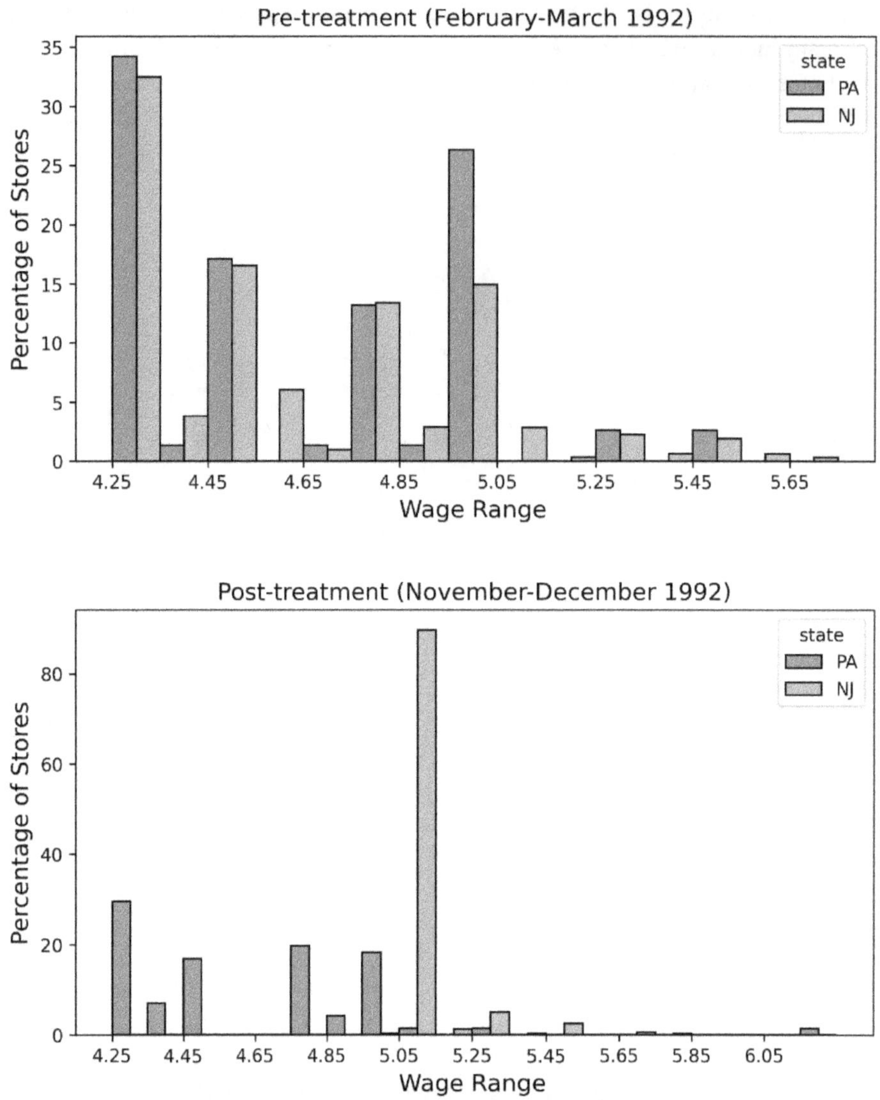

Figure 14.1— Distribution of starting wage rates

In this section, we loaded the Card and Krueger dataset and performed an EDA analysis on it. In the next section, we will discuss the causal estimation problem for this dataset.

Causal model

In this section, we focus on the causal estimation problem. First, we need to find the treatment and outcome columns in this dataset. Here the treatment is the minimum wage increase that only happened in New Jersey. Hence, the column `STATE` represents the treatment variable, as `STATE=0` indicates that the store is in the control group (Pennsylvania) and `STATE=1` implies that the store is in the treatment group (New Jersey).

The outcome is represented by the column FTE_EMP2 (FTE employee hours) that was added during the EDA analysis. This quantity represents the employment level in a store, and its average value in a state can represent the fast-food employment level in that state. Let's assume that T denotes the treatment and Y denotes the outcome. Additionally, X represents the vector of covariates (the remaining columns). As usual, we want to calculate the **Average Treatment Effect** (**ATE**):

$$ATE = E[Y(1) - Y(0)] \quad (14.1)$$

If we could assume that no confounding is present, then ATE could be estimated as

$$E[Y^{obs}|T = 1] - E[Y^{obs}|T = 0] \quad (14.2)$$

The following code calculates the value of *Equation 14.2*.

```
T = df['STATE']
Y = df['FTE_EMP2']
Y[T==1].mean()-Y[T==0].mean()

## Output
-0.14
```

Based on this result, we would conclude that the increase in minimum wage decreased employment levels at fast-food restaurants.

A confounder variable usually has an imbalance probability distribution, which means that its probability distribution differs between the treatment and control groups. Remember that we used the **standardized mean difference** (**SMD**) in *Chapter 12* to check the balance of covariates. In the EDA section, we saw that the average `FTE_EMP` and the average number of KFC and Wendy's stores were significantly different in New Jersey and Pennsylvania. Let's calculate their SMD using the following code to check their balance.

```
covariates = ['FTE_EMP', 'KFC', 'Wendy\'s']
for covariate in covariates:
    mean_xi_t = df[df['STATE']==1][covariate].mean()
    mean_xi_c = df[df['STATE']==0][covariate].mean()
    std_xi_t = df[df['STATE']==1][covariate].std(ddof=0)
    std_xi_c = df[df['STATE']==0][covariate].std(ddof=0)
    smd = (mean_xi_t - mean_xi_c) / np.sqrt(std_xi_t**2 + std_xi_c**2)
    print(f"SMD for {covariate}:", np.round(smd, 3))

## Output
SMD for FTE_EMP: -0.194
SMD for KFC: 0.099
SMD for Wendy's: -0.104
```

Assuming a threshold of 0.1 as a cutoff for an imbalanced covariate, we see that `FTE_EMP` is certainly imbalanced, and the other ones are very close to this threshold, so confounding is likely present in this dataset, and we cannot rely on our result from *Equation 14.2*. We need to apply the causal inference methods to have a reliable estimate of ATE. In the next section, we will use linear regression to estimate the causal effect.

Estimation using linear regression

Card and Krueger used linear regression to estimate the causal effect of minimum wage increase in their paper (Card and Krueger 1994). In this section, we follow their method and implement their regression models in Python. The regression models that they used had two general forms:

$$\Delta E_i = a + bX_i + cNJ_i + \varepsilon_i \quad (14.3)$$

And

$$\Delta E_i = a' + b'X_i + c'GAP_i + \varepsilon'_i \quad (14.4)$$

where ΔE_i is the change in FTE employment for restaurant i before and after treatment. NJ_i is an indicator variable equal to 1 if observation i is in New Jersey and 0 otherwise (it is equivalent to column `STATE` in our dataset), and GAP_i is the gap between the initial observed wage for observation i (denoted by W_{1i}) and the minimum wage ($5.05):

$$GAP_i = \begin{cases} \dfrac{5.05 - W_{1i}}{W_{1i}} & \mathit{if\ store\ i\ is\ in\ NJ\ and}\ W_{1i} \geq \$5.05 \\ 0 & \mathit{Otherwise} \end{cases} \quad (14.5)$$

In these equations, X_i represents other covariates like store type or location. The next code creates a regression model based on *Equation 14.3*. We first create ΔE_i as the difference between `FTE_EMP2` and `FTE_EMP` and then create a regression model (`model1`) in which the independent variables are represented by the binary columns `Roy Rogers`, `KFC`, `Wendy's`, `STATE`, and `CO_OWNED`. We use the `fit()` method with the parameters `cov_type='cluster'` and `cov_kwds={'groups': final_df['STATE']}`. Hence, it fits the regression model using clustered standard errors that are robust to within-cluster correlation. The data is partitioned into clusters based on the `STATE` column. Observations with the same `STATE` will be treated as part of the same cluster. After fitting the model, the coefficient of `STATE`, its standard error, p-value, and 95% confidence interval are shown in the output.

```
df['delta_emp'] = df['FTE_EMP2'] - df['FTE_EMP']
final_df = df.dropna(subset=['FTE_EMP2', 'FTE_EMP',
                              'WAGE_ST', 'WAGE_ST2'])

model1 = sm.OLS(final_df['delta_emp'],
                sm.add_constant(final_df[['Roy Rogers', 'KFC',
                'Wendy\'s', 'STATE', 'CO_OWNED']]))

results = model1.fit(cov_type='cluster', cov_kwds={'groups': final_
df['STATE']})
```

```
cols = {"Estimation": results.params['STATE'],
        "Standard error": results.bse['STATE'],
        "p-value": results.pvalues['STATE'],
        "95% conf interval": results.conf_int().loc["STATE"].round(3).tolist()}
output = pd.DataFrame(cols, index=["STATE"])
output.round(3)
```

The output is shown in *Table 14.4*:

Coefficient	Estimation	Standard error	p-value	95% conf interval
STATE	2.282	0.15	0.0	[1.988, 2.575]

Table 14.4—The output of model1

Please note that we restricted our analysis to the set of stores with available FTE employment and wage data both before and after the treatment. Here the results suggest that changing the STATE from Pennsylvania to New Jersey increases the change in FTE employment by $2.28. One problem with the clustered standard errors method is that it requires many clusters to work properly, making it unreliable when there are just two clusters (two states). We can also use the wild cluster bootstrap method that works better than clustered standard errors when you have few clusters. The next code implements it using the `wildboottest` package. This package can only report the p-value of the estimation.

```
boot = wildboottest(
    model1,
    param="STATE",
    cluster=final_df["STATE"],
    B=9999
)

## Output
| param    |    statistic |    p-value |
|:---------|------------:|-----------:|
| STATE    |      15.222 |      0.000 |
```

Here the p-value is less than 0.05, indicating a statistically significant result; however, inference should be interpreted cautiously given only two clusters. Next, we create a second model based on *Equation 14.4*. Here, we want to see what happens with the FTE employment of the stores in New Jersey that were forced to increase their wages to meet the new minimum rate. We first create the GAP column using *Equation 14.5* and then create a regression model (`model2`) in which the independent variables are represented by the binary columns `Roy Rogers`, `KFC`, `Wendy's`, `GAP`, and `CO_OWNED`. After fitting the model, the coefficient of GAP, its standard error, p-value, and 95% confidence interval are shown in the output.

```
df['GAP'] = np.where(
    (df['STATE'] == 1) & (df['WAGE_ST'] < 5.05),
    (5.05 - df['WAGE_ST']) / df['WAGE_ST'],
```

```
    0
)
final_df = df.dropna(subset=['FTE_EMP2', 'FTE_EMP',
                              'WAGE_ST', 'WAGE_ST2'])

model2 = sm.OLS(final_df['delta_emp'],
                sm.add_constant(final_df[['Roy Rogers',
               'KFC', 'Wendy\'s', 'CO_OWNED', 'GAP',
               ]]))

results = model2.fit(cov_type='cluster', cov_kwds={'groups': final_
df['STATE']})
cols = {"Estimation": results.params['GAP'],
        "Standard error": results.bse['GAP'],
        "p-value": results.pvalues['GAP'],
        "95% conf interval":
       results.conf_int().loc["GAP"].round(3).tolist()}
output = pd.DataFrame([cols], index=["GAP"])
output.round(3)
```

The result is shown in *Table 14.5*.

Coefficient	Estimation	Standard error	p-value	95% conf interval
GAP	16.363	1.927	0.0	[12.592, 20.134]

Table 14.5—The output of model2

This model gives a greater standard error than the previous one; however, its implications for the New Jersey-Pennsylvania comparison are similar. We see that the coefficient of GAP is positive, and it looks statistically significant, suggesting that in the stores affected by the minimum wage increase, the FTE employment increases. But as mentioned before, the clustered standard errors method may be unreliable with only two clusters. Again, we can use the wild cluster bootstrap to get the p-value.

```
boot = wildboottest(
    model2,
    param="GAP",
    cluster=final_df["STATE"],
    B=9999
)

## Output
| param    |    statistic |    p-value |
|:--------|------------:|----------:|
| GAP      |        8.504 |      0.000 |
```

The p-value is still less than 0.05 (statistically significant) but must be interpreted cautiously with only two clusters.

The following code can estimate this average increase based on the mean value of GAP among New Jersey stores:

```
final_df[final_df['STATE']==1]['GAP'].mean() * 16.363

## Output
1.71
```

And the output implies a 1.71 increase in FTE employment in New Jersey compared to Pennsylvania. The results of this section showed that the minimum wage increase had slightly increased employment in New Jersey. In the next section, we use a different causal method to confirm this result.

Estimation using inverse probability of treatment weighting

In this section, we will use the **inverse probability of treatment weighting** (**IPTW**) method to estimate the ATE using the propensity scores. Like the previous section, we restrict our analysis to the set of stores with available FTE employment and wage data both before and after the treatment. In the next code, a logistic regression model is used to estimate the propensity scores of each store in the dataset. The columns `FTE_EMP2` and `STATE` represent the outcome and treatment variables respectively, and the covariates are `Roy Rogers`, `KFC`, `Wendy's`, `FTE_EMP`, `WAGE_ST`, and `CO_OWNED`.

```
final_df = df.dropna(subset=['FTE_EMP2', 'FTE_EMP',
                             'WAGE_ST', 'WAGE_ST2'])

cov_list = ['FTE_EMP', 'KFC', 'Roy Rogers',
            'Wendy\'s', 'WAGE_ST', 'CO_OWNED']

X = final_df[cov_list]
Y = final_df['FTE_EMP2']
T = final_df['STATE']

learner = LogisticRegression(penalty=None, solver='lbfgs',
                             max_iter=500)

ipw = IPW(learner)
ipw.fit(X, T)
```

After fitting the model, we can calculate the causal effect:

```
outcomes = ipw.estimate_population_outcome(X, T, Y)
effect = ipw.estimate_effect(outcomes[1], outcomes[0])
print("Outcomes:")
print(outcomes)
print("Effect: ", effect['diff'])

## Output
Outcomes:
0    20.714
1    21.398
dtype: float64
Effect:  0.684
```

We see that the FTE employment after the minimum wage increase is slightly higher in New Jersey compared to Pennsylvania. We can also check the balance of covariates using a love plot in the next code:

```
evaluation_results = evaluate(ipw, X, T, Y)
plt.figure(figsize=(7, 4))
evaluation_results.plot_covariate_balance(kind="love",
                                          thresh=0.1)
plt.show()
```

This plot is shown in *Figure 14.2*.

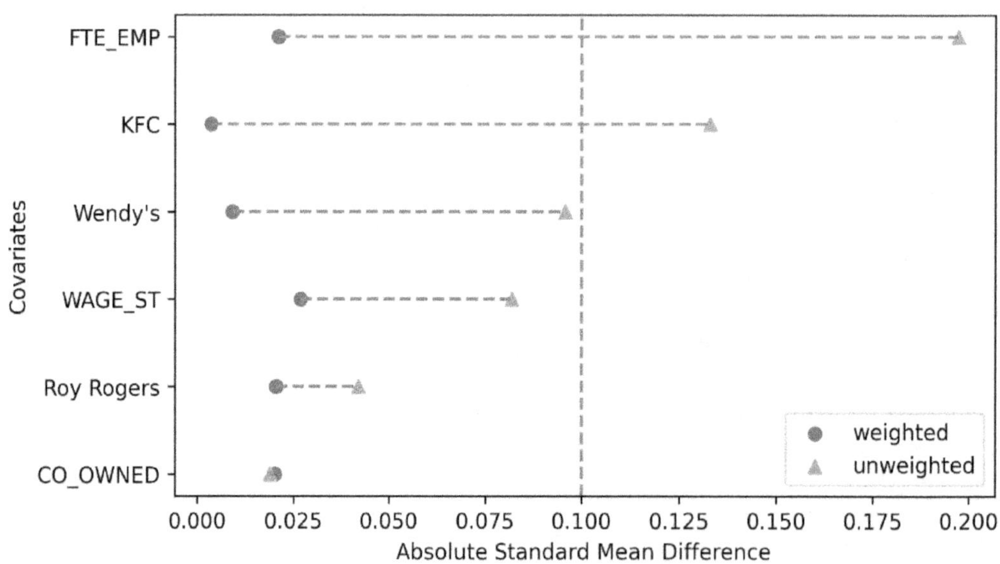

Figure 14.2— Love plot of the covariates

As this figure shows, the SMD of some covariates was initially above the threshold (0.1); however, the weights calculated from the propensity scores could improve the balance of the covariates significantly, and the weighted SMD of all of them is below the threshold.

As mentioned before, the standard competitive labor market model predicts that if you raise the price of labor, demand for it goes down. However, the results of the Card and Krueger study were inconsistent with the predictions of the standard competitive model. The results showed that the increase did not lead to a reduction in fast-food employment in New Jersey; in fact, they suggested a slight increase. This study fundamentally questioned the long-held assumption that minimum wage increases lead to job losses and changed the academic and political debate on the issue.

Summary

In this chapter, we discussed the famous Card and Krueger minimum wage study. In 1992, New Jersey raised its state minimum wage from $4.25 to $5.05 per hour, while the minimum wage in neighboring Pennsylvania remained at $4.25. Card and Krueger surveyed over 400 fast-food restaurants in both states before and after New Jersey's wage increase and compared the employment trends in New Jersey (as the treatment group) to those in Pennsylvania (as the control group). At that time, most economists believed that an increase in the minimum wage would reduce fast-food employment in New Jersey. However, the results of this study showed that a minimum wage increase in New Jersey did not lead to a decrease in employment and may have even slightly increased it.

We first loaded the Card and Krueger dataset in Python and performed an EDA analysis on that. We used two different methods to estimate the effect of minimum wage increase on fast-food employment in New Jersey compared to Pennsylvania: linear regression and IPTW. The results showed a slight increase in employment in New Jersey, which was consistent with Card and Krueger's study.

References

- Card, D., and Alan B. Krueger. 1994. *Minimum Wages and Employment: A Case Study of the Fast-Food Industry in New Jersey and Pennsylvania. The American Economic Review* 84 (4): 772-793.
- Shimoni, Y., E. Karavani, S. Ravid, S., et al. 2019. *An Evaluation Toolkit to Guide Model Selection and Cohort Definition in Causal Inference.* arXiv preprint arXiv:1906.00442.

Join our community on Discord

Join our community's Discord space for discussions with the author and other readers: `https://packt.link/hYCWN`

15

Unlock Access to the Code Bundle and the PDF Version

Your copy of this book includes the following exclusive benefits:

Complete Code Bundle

Download the full source code for this book's examples and projects.

DRM-Free PDF Version

Download DRM-free PDF and ePub copies of this book.

7-Day Packt Library Access

Get 7-day unlimited access to 8,000+ books and videos. No credit card required.

Available for first-time Packt+ trial users only.

Next-Gen Reader Access

Read this book on Packt Reader with progress sync, dark mode and note-taking.

Follow this guide to unlock them. The process takes only a few minutes and only needs to be completed once.

Unlock this book's free benefits in three easy steps

Step 1

Have your purchase invoice ready for *step 3*. If you have a physical copy, scan it using your phone and save it as a PDF, JPG, or PNG.

For more help on finding your invoice, visit `https://www.packtpub.com/unlock-benefits/help`.

Note: If you bought this book directly from the Packt website, no invoice is required. After *step 2*, you can access your exclusive content right away.

Step 2

Scan the QR code or go to `packtpub.com/unlock`.

On the page that opens (similar to *Figure 15.1* on desktop), search for this book by name and select the correct edition.

Figure 15.1 – Packt unlock landing page on desktop

Step 3

After selecting the book, sign in to your Packt account or create one for free. Then, upload your invoice (PDF, PNG, or JPG, up to 10 MB). Follow the on-screen instructions to finish the process.

Need help?

If you get stuck and need help, visit `https://www.packtpub.com/unlock-benefits/help` for a detailed FAQ on how to find your invoice and more. This QR code will take you to the help page.

Note: If you are still facing issues, reach out to `customercare@packt.com`.

Other Books You May Enjoy

If you enjoyed this book, you may be interested in these other books by Packt:

The Causal Mindset Handbook

Dr. Quentin Gallea

ISBN: 9781806117857

- Understand causality and why correlation can mislead decisions
- Distinguish predictive models from causal inference
- Use causal graphs to reason about cause and effect
- Evaluate impact with experiments and quasi-experiments
- Spot flawed causal claims in business and everyday life
- Apply causal thinking confidently to real decisions

Causal Inference in R

Subhajit Das

ISBN: 9781837639021

- Get a solid understanding of the fundamental concepts and applications of causal inference
- Utilize R to construct and interpret causal models
- Apply techniques for robust causal analysis in real-world data
- Implement advanced causal inference methods, such as instrumental variables and propensity score matching
- Develop the ability to apply graphical models for causal analysis
- Identify and address common challenges and pitfalls in controlled experiments for effective causal analysis
- Become proficient in the practical application of doubly robust estimation using R

Packt is searching for authors like you

If you're interested in becoming an author for Packt, please visit `authors.packt.com` and apply today. We have worked with thousands of developers and tech professionals, just like you, to help them share their insight with the global tech community. You can make a general application, apply for a specific hot topic that we are recruiting an author for, or submit your own idea.

Share your thoughts

Now you've finished *Causal Inference with Bayesian Networks*, we'd love to hear your thoughts! Scan the QR code below to go straight to the Amazon review page for this book and share your feedback or leave a review on the site that you purchased it from.

`https://packt.link/r/1835084982`

Your review is important to us and the tech community and will help us make sure we're delivering excellent quality content.

Index

A

B

D

E

F

G

H

I

J

N

O

P

S

T

U

V

W

X

www.ingramcontent.com/pod-product-compliance
Lightning Source LLC
LaVergne TN
LVHW060010210926
848481LV00013B/3681
* 9 7 8 1 8 3 5 0 8 4 9 8 4 *